Metacommunity Ecology

MONOGRAPHS IN POPULATION BIOLOGY

SIMON A. LEVIN AND HENRY S. HORN, SERIES EDITORS

List of titles follows index

Metacommunity Ecology

MATHEW A. LEIBOLD

AND

JONATHAN M. CHASE

PRINCETON UNIVERSITY PRESS

Princeton and Oxford

Published by Princeton University Press, 41 William Street,
Princeton, New Jersey 08540

In the United Kingdom: Princeton University Press,
6 Oxford Street, Woodstock, Oxfordshire OX20 1TR

press.princeton.edu

ISBN 978-0-691-04916-8

Library of Congress Control Number 2017957237

British Library Cataloging-in-Publication Data is available

This book has been composed in Times LT Std

10 9 8 7 6 5 4 3 2 1

Contents

Preface   xi

1. Introduction: The Rise, Fall, and Rise Again of
     Metacommunity Ecology   1
     1.1. The indelible influence of scale   3
     1.2. The metacommunity framework allows simultaneous
         consideration of multiple processes at multiple scales   5
     1.3. Discrete versus continuous metacommunities   9
     1.4. A brief history of metacommunity ecology   13
     1.5. MacArthur's paradox: Determinism versus stochasticity
         and the null model wars   15
     1.6. Reemergence of the metacommunity perspective:
         1990s and 2000s   18
     1.7. A prospectus for metacommunity ecology: What are we trying
         to do in this book?   19
     1.8. Moving forward: Overview of the rest of the book   20
     1.9. A final note on our assumptions   21

2. The Theories of Metacommunities   23
     2.1. An overview and critical analysis of the four archetypes
         of metacommunity ecology   27
     2.2. Comparing and contrasting the archetypes   34
     2.3. Exploring metacommunity ecology at the interface
         of the archetypes   37
     2.4. Is there an "ubermodel"?   42
     2.5. The influence of local coexistence mechanisms   46
     2.6. Conclusions   47

3. Processes in Metacommunities   49
     3.1. Resource utilization and response   51
     3.2. Evidence for demographic stochasticity and drift   54
     3.3. Dispersal limitation   62

3.4. Interactions between stochasticity, dispersal, and
    interspecific effects   74
3.5. The influence of habitat heterogeneity   77
3.6. Interactions between habitat heterogeneity and dispersal   84
3.7. Implications for local versus regional controls on diversity   87
3.8. Conclusions and synthesis   88

4. Metacommunity Patterns in Space   90
4.1. Patterns of SADs and related diversity metrics   92
4.2. Null models and co-occurrence in metacommunities   94
4.3. Elements of metacommunity structure   99
4.4. Using variation partitioning to diagnose spatial,
    environmental, and random effects   106
4.5. Variation in the relative importance of metacommunity-
    structuring processes   112
4.6. Factors that influence variation in metacommunity patterns   117
4.7. Contributions of different species and different localities
    to the overall metacommunity pattern   126
4.8. How well can we hope to do in explaining metacommunity
    structure?   127
4.9. Caveats and conclusions   129

5. Interactions between Time and Space in Metacommunities   131
5.1. Temporal turnover: What does theory predict?   133
5.2. Patterns of temporal turnover   135
5.3. How time can influence deterministic community assembly   140
5.4. Priority effects and multiple stable equilibria in
    metacommunities   141
5.5. Endpoint assembly cycles   144
5.6. Frequency-dependent coexistence in spatially continuous
    metacommunities   147
5.7. Conclusions   149

6. What Can Functional Traits and Phylogenies Tell Us about
    Coexistence in Metacommunities?   151
6.1. A brief history of trait- and phylogeny-based "assembly rules"   154
6.2. The correlation between phylogenetic and trait-based
    information and "real" metacommunity processes   158
6.3. Trait and phylogenetic over- or underdispersion: What does
    coexistence theory predict?   162

6.4. Phylogenetic and functional-trait dispersions in a simple
    SS metacommunity   167
6.5. Phylogenetic and trait dispersions in non-SS metacommunity
    archetypes   173
6.6. Conclusions   175

7. Combining Taxonomic and Functional-Trait Patterns to Disentangle
    Metacommunity Assembly Processes   177
7.1. Using functional information to enhance taxonomic pattern
    analysis   179
7.2. Using functional information to enhance analyses of change
    through space or time   184
7.3. Toward a trait-based theory of metacommunity assembly   189
7.4. Closing the loop: Predicting species abundance and
    distribution from traits   197
7.5. Conclusions   200

8. Eco-evolutionary Dynamics in Metacommunities   202
8.1. Building an evolutionary ecology of metacommunities   204
8.2. Adaptive evolution in metapopulations and metacommunities 205
8.3. The community monopolization hypothesis   207
8.4. Evolution toward neutrality   219
8.5. Frequency-dependent evolution   226
8.6. The interaction of community monopolization and neutral
    evolution   228
8.7. The interaction between community monopolization and
    neutral evolution in the "real" world?   233
8.8. Conclusions   236

9. Macroevolution in Metacommunities   239
9.1. How metacommunity processes influence phylogeny
    and radiations   241
9.2. Historical effects   245
9.3. Synthesis: A research agenda for integrating evolutionary
    and ecological processes that affect biodiversity   250
9.4. Conclusions   254

10. The Macroecology of Metacommunities   255
10.1. What is macroecology?   257
10.2 Synthesizing biodiversity macroecology processes and
    patterns   262

10.3 Dissecting biodiversity macroecology patterns 267
10.4. The role of metacommunity assembly processes
    in biodiversity macroecology patterns 274
10.5. Diversity partitioning and the SAR 276
10.6. Metacommunity assembly and the nested SAR 277
10.7. Metacommunity assembly and the ISAR 279
10.8. Biodiversity in the Anthropocene 283
10.9. Metacommunities and contemporary biogeography:
    Scale-dependent patterns of species diversity along
    ecological gradients 294
10.10. Other macroecological patterns 300
10.11. Conclusions 301

11. Food Webs in Metacommunities 303
11.1. How do spatial processes and trophic interactions combine
    to influence coexistence in simple metacommunities? 304
11.2. How do trophic interactions influence metacommunity
    processes? 306
11.3. Toward a theory for trophically structured metacommunities 310
11.4. Spatial processes and May's diversity-stability theory 319
11.5. Frequency dependence and feedbacks between trophic
    interactions and spatial processes 324
11.6. Food-web metacommunity assembly processes and the
    scale-dependent productivity-diversity relationship 327
11.7. The influence of metacommunity processes on food-web
    structure and indirect interactions 332
11.8. Conclusions 333

12. Community Assembly and the Functioning of Ecosystems
    in Metacommunities 335
12.1. The role of spatial processes in mediating BEF relationships 339
12.2. A simple framework based on resource competition
    in a metacommunity context 342
12.3. Ecosystems within metacommunities as CASs? 359
12.4. Do ecosystems have regular features? 363
12.5. Conclusions 368

13. From Metacommunities to Metaecosystems 369
13.1. Why spatial dynamics are so important in ecosystems 371
13.2. Elements of metaecosystems ecology 373

13.3. An emerging set of principles? 377
13.4. Conclusions 379

14. A Coming Transition in Metacommunity Ecology 380
14.1. The accomplishments of metacommunity ecology
version 1.x 382
14.2. Synthesis through metacommunity ecology 384
14.3. The current status and limitations of metacommunity
version 1.9 385
14.4. Going from version 1.9 to version 2.0 387
14.5. From basic to applied metacommunity ecology 389
14.6. Conclusions 391

References 393
Index 465

Preface

One of the most stimulating moments in our two careers was being involved with the Metacommunity Ecology Working Group in 2001. It was sponsored by the National Center for Ecological Analysis and Synthesis (NCEAS), which tasked itself with trying to reconcile different approaches to the burgeoning idea of metacommunities. When the group first met, we thought we were going to reconcile two different sets of models. The first was based on mechanistic niche models that tried to understand what the endpoint of community assembly might be among heterogeneous localities in a metacommunity given a particular species pool; we were writing our first book on that very topic (Chase and Leibold 2003; in the current book, this approach corresponds to what we will call the species sorting approach). The second set of models was based on extensions of the metapopulation models of Hanski and colleagues (Hanski 1999), which placed a greater emphasis on dispersal based on a possible trade-off between colonization and competitive ability (in this book this corresponds to what we will call the patch dynamic approach). On the first morning of our first meeting, however, we realized that we hadn't yet thought enough about Hubbell's (2001) then new neutral theory approach, nor about the mass effects models that superficially seemed similar to patch dynamics but actually have a number of different assumptions and predictions (see Chap. 1 for details on the history of metacommunity ecology).

This diversity of approaches and ideas was fascinating to us for many reasons. For one thing, it opened our eyes to possible explanations we just hadn't considered before. For another it drove home the point that there wasn't a single way to think about community assembly, providing an early avenue for some synthesis among studies and systems. For example, it might explain why experiments and patterns related to one group of organisms (say zooplankton, for example) might not seem as useful in understanding another (e.g., tropical forest trees). The results of the NCEAS working group, especially the paper by Leibold et al. (2004) and the edited book by Holyoak et al. (2005), were very satisfying, and the former has been among the most cited conceptual papers in ecology in this century (so far).

When we decided to write this book (longer ago than we care to admit), we had a desire to evaluate the progress that had been made in metacommunity ecology since then and to consolidate this evaluation into one convenient synthesis. We found so many exciting ideas, results, findings, and conjectures in our review that we struggled to be able to write this book. Indeed, we'd finish a chapter and start to feel good about it, move on to the next chapters, and find that several months later we needed to make important edits to our earlier chapters to reflect important publications that were continuously emerging. Writing the book became a Sisyphean task!

Our other main goal with the book was to address head-on a number of important gaps in metacommunity ecology that were becoming increasingly obvious but not being comprehensively addressed. We would, for example, get asked to review countless papers that were obsessed with devised tests to determine which of the four metacommunity paradigms (as we called them then) was best able to explain the patterns in a given system but that didn't necessarily consider alternatives or more nuanced views about metacommunities.

In the end this book tries to do both: review past successes and identify shortcomings that could lead to future opportunities. We hope it reads that way. We hope that young scientists will find a useful synopsis and critique of the field that may help them to devise research programs that exploit the successes and opportunities; we hope senior scientists will find a thoughtful synthesis that helps them understand the context for the work that is being done in this and related fields; and we hope that less specialized readers find inspiration in the general sense of excitement we have experienced while doing metacommunity ecology and in its synthetic connections to other fields. In the end we hope to reenergize this area of ecology.

During the writing of this book, we quickly realized that a key issue was the link between modern metacommunities and evolutionary processes (both micro and macro). Indeed, in the works of many of the "forefathers" of metacommunity ecology, including G. Evelyn Hutchinson, Robert MacArthur, and E. O. Wilson, ecological and evolutionary processes were often discussed hand in hand, although the fields have diverged through time. In this book, we quickly realized how critical this link is and that it deserves much more careful synthesis. This perception brought to mind a Christmas card that a colleague and friend of ours, Shahid Naeem, sent in the 1990s that we reproduce here. His comments echo the feeling of intimate linkages between ecology and evolution and respect for the history of the field that has shaped our thoughts. It's also a lot of fun and we wanted to share this image with others.

At the outset, we would like to acknowledge the work of the metacommunity ecology research community at large. It is truly magical to read so many papers, often by researchers we didn't know at all or didn't previously know were working in this area, and learn novel lessons about nature and ecology. It is really true that without these contributions, this book wouldn't be what it is. We also have a

FIGURE 0.1. The Naeem Triptych. As an explanation of the images, Shahid included the following text:

"The outer panels of a triptych open to reveal an inner scene of inspirational images, usually religious in nature. The outer panels differ in quality, either having less color than the inner panels or focused on only a few subjects for whom the triptych was commissioned or inspired by. The paintings are richly iconographic in nature, using animals, plants, objects and religious figures as symbols of the characters and their virtues found in the tales the triptych wishes to portray.

In this triptych, the outer panel represents both evolution and ecology, the processes that inspired the inner panels. People serve as the symbolic representations of these processes, to make the outer panels distinct from the inner. On the left is Wallace, represented by his placement behind Darwin. He is further symbolized by the mosquito on his hand that gave him the malaria that caused the fever that gave him the dream that inspired him to come up with natural selection and evolution as the process by which species originated from ancestors. Darwin is symbolized by

(*continued*)

(Figure 0.1. continued)

the well known tale of his eagerness to collect beetles in which, having already a beetle in each hand, he placed one in his mouth to catch a third. The beetle in his mouth released a noxious chemical onto his tongue which caused him to spit it out. He is shadowed by Wallace since at least one historian believes that Darwin's synthesis on evolution was not coincidentally simultaneous, but directly inspired by Wallace's letter that Darwin had in his possession for longer than he confessed.

On the right is ecology, symbolized by Hutchinson noting that two corixids have a size ratio of 1.30, shown by holding up one finger on hand and one and three on the other hand (much the way in Christian iconography the three upheld fingers symbolize the Trinity). This simple idea, blown out of proportion to its original intent, is one of the legendary beginnings of ecology's first paradigm—competition-based community structure. As he was inspired by the corixids found in the pool named in honor of the patron saint Santa Rosalia, the saint appears behind Hutchinson. However, the bones of Santa Rosalia have been scientifically determined to be the bones of a goat, so Santa Rosalia is portrayed as a goat. The lemur on Hutchinson's shoulder with "FSU" on its sweatshirt symbolizes Florida State University's challenge to the paradigm that would arise many years later.

In each panel the animals represent creatures that are often used as examples of ecology and evolution. The Galapagos finches appear to show Hutchinsonian ratios in bill dimensions and the giant tortoise of the Galapagos Islands demonstrate how creatures can become modified through selection and other evolutionary processes.

The inner panels reveal the three special aspects of living systems that inspire biologists. The left panel portrays the remarkable shared history living things have, going back over three quarters of the earth's history, as known from the fossil record. The central panel shows development, reproduction, and life history strategies symbolized by different developmental stages and birth in a clutch of salamander eggs. Surrounding this central figure are simply other creatures, big and small, that have in common ecologies driven by development, reproduction, and life history processes. The animals were each chosen in part from their iconographic symbolism in Christian art (such as the lion, giraffe, mouse, blackbird) or for connections to adjacent panels (e.g., the dragonfly). Their somewhat fanciful appearance stems from my own mixed perceptions of living things that live justifiably both in the hearts of population dynamic equations and stories, folk tales, and children's literature.

The right panel symbolizes the commonality of death among living things. Again, the creatures were chosen from old triptychs and Christian religious art. For example, the blackbird symbolizes the temptations that draw mortals into the underworld (hence their linkage of the right and central panels). The leopard and bear, as further examples, symbolize cruelty.

(*Figure 0.1. continued*)
In the left panel, the Austin Mini Cooper heading off to the castle in the background represents my leaving for Silwood Park, England (not till 24 February 1992) in yet another attempt to learn more about the things in life that inspire me (ecology in an academic setting and little cars)."

great number of colleagues (including mentors, both official and unofficial, students and postdocs, and collaborators), far too many to list here, who have helped us (often inadvertently) to develop what we present here. While we cite a great many of them in this book, unfortunately we could not cite everything without getting into even more hot water with our editors.

Much of the work we present here and the ideas formed therein were supported by funding from the National Science Foundation of the United States and other entities. This includes numerous grants that we have both received but especially an NSF-OPUS award (DEB 0640302) and an Alexander von Humboldt Research Award to MAL. In addition, during much of the writing of this book, JMC was supported by the German Centre for Integrative Biodiversity Research (iDiv) Halle-Jena-Leipzig, which is funded by the German Research Foundation (award FZT 118).

We acknowledge several colleagues who gave us feedback and inspiration on the parts of the book at various stages—including Tiffany Knight, Tadashi Fukami, Brian McGill, Dominque Gravel, Pedro Peres-Neto, Luc De Meester, and an anonymous reviewer—and several colleagues who provided data and assistance with previously unpublished results presented here, including Pedro Peres Neto, Christine Parent, Felix May, and Bernadette Pinel-Alloul. Finally, we are incredibly indebted to Catalina Cuellar, who helped us with figures and references and also acted as an outstanding sounding board for us.

MAL would like to dedicate this book to Eve Whitaker and to Gabriel and Neil Howe. JMC would similarly like to thank Tiffany Knight and Jack and Christina Chase for their continued support and tolerance during the writing of this book.

CHAPTER ONE

Introduction

The Rise, Fall, and Rise Again

of Metacommunity Ecology

Prospectus

1. Traditional perspectives of community ecology, including species interactions, coexistence, and biodiversity, have focused on local-scale processes and have met with a great deal of controversy and disagreement.
2. The recognition of the importance of spatial (and temporal) processes has risen dramatically in recent years, although threads of ideas (importance of dispersal) and controversies (stochasticity vs. determinism) are evident throughout the history of ecology.
3. Metacommunity ecology, by explicitly incorporating scale as a critical feature of the outcomes of coexistence and biodiversity, among other variables, has the potential to unify what seems like a largely unresolved field.
4. This unification will require explicitly incorporating spatiotemporal heterogeneities, dispersal, the interactions between stochasiticity and determinism, and a number of complicating variables (e.g., food webs, evolution).

> The major weakness of traditional community ecology, and
> why it has so conspicuously failed to come up with many patterns, rules and workable contingent theory, is its overwhelming emphasis on localness.
> —Lawton (1999)

Community ecology is the study of how species interact with each other in ways that determine patterns in the distributions and abundances of different species. It represents the nexus at which individual traits, fitness, and population dynamics scale up to influence the distribution and coexistence among species on local to biogeographic scales and from months to millennia. It influences the role that species play in ecosystems and how they evolve. And it plays a critical role in understanding the destruction and conservation of biodiversity, as well as its restoration, as the human footprint becomes more pervasive. Unfortunately, community ecology has not yet fully lived up to its potential (Lawton 1999, Ricklefs 2008).

Are there any broadly applicable concepts and approaches that can help to resolve the clear limitations of community ecology as much of it continues to be practiced? Elton (1927) suggested four important ones—the niche, the food web, body size relations, and the trophic pyramid—that still serve as key concepts of community ecology (Chase and Leibold 2003). Lotka (1925) introduced the idea that energy relations and the laws of physics (thermodynamics) and chemistry (stoichiometry) could serve as a foundation for biology in general, and although these ideas fell largely silent, their core aspects have been championed in putative "unified" ecological theories of neutral coexistence (Hubbell 2001), metabolism (Brown et al. 2004), and stoichiometry (Sterner and Elser 2002), as well as mathematical principles such as body-size relations (Ritchie 2010), maximum entropy (Harte 2011), and neutrality (Hubbell 2001). Attempts have even been made to unify the unified theories (McGill 2010, 2011). Despite these efforts, there does not seem to be a strong sense that any one of these perspectives is able to adequately address the full scope of the questions at hand.

In this book we argue that we already know many of the key aspects of community ecology but that we do not have a framework that adequately links these in an appropriate context. We argue that the missing link that can provide this context is the combination of spatial and interaction processes that characterize metacommunity ecology. To us, the metacommunity approach allows one to explicitly transit from fitness and population dynamics to community- and ecosystem-level processes, as well as from smaller to larger scales, without the need to artificially designate where one community ends and another begins. Our goal for this book is to motivate others to share this vision of metacommunity ecology as a "synthetic hub" for understanding community and ecosystems ecology. We aim to contribute to a synthesis that is akin to the modern synthesis achieved many decades ago in evolutionary biology, which embraced the multiple roles of selection, drift, mutation, and gene flow.

Some elements of this synthesis have already been vetted. For example, Vellend (2010, 2016) developed an important conceptual connection between the major drivers of diversity in community ecology—niche selection, ecological drift, speciation, and dispersal—and the major drivers of diversity in population genetics—natural selection, genetic drift, mutation, and gene flow. In some ways, our goal is to more deliberately "look under the hood" of the relatively simple framework described by Vellend to identify just how niche selection, stochastic drift, speciation, and dispersal interact with eco-evolutionary processes (Hendry 2016), geometric

scaling processes (McGill 2010, 2011), and constraints of energy flow and conservation of matter (Loreau 2010) to influence pattern and process at multiple
spatial and temporal scales. We use this framework to discuss and synthesize
numerous levels of organization ranging from pairwise interactions, to guilds of
multiple competing taxa across scales of space and time, to micro- and macroevolutionary processes, to macroecological patterns, to food webs and ecosystems-
level processes and patterns.

1.1 THE INDELIBLE INFLUENCE OF SCALE

Before we begin exploring the advantages of the metacommunity approach, it is useful to first ask, what is a community? When we talk of what a community is, we
usually think of an idealized case in which multiple species have populations that
interact by affecting each other's birth and death rates at *a particular place and
time* (Fig. 1.1). A great deal of effort has been aimed at understanding the patterns of
species composition, relative abundance, and diversity within such communities, as
well as the processes leading to those patterns (e.g., the role of interspecific interactions, spatial effects, and environment). And many would argue that the field of
community ecology with this focus has gained considerable insight into the patterns
and processes by which species interact and coexist (Morin 2011, Mittelbach 2012).

Unfortunately, the definition of community is always qualified by some phrase
like "at a particular place and time" (also "in the same geographic area," "in the
same location," "coexist together," etc.). Such qualification is not easily operationalized (and perhaps it should not be) in anything more than an arbitrary way; that
is, the qualification "at a particular place and time" is ambiguous and user defined.
Recognizing this problem, some ecologists have suggested that perhaps the construct of a community is too artificial to be of use and should be abandoned (Ricklefs
2004, 2008; Fahrig 2013).

Because the delineation of the extent of a community might be user defined,
community-level patterns such as coexistence, relative abundance, composition, and
diversity could be context dependent, as are the mechanisms that create them. Thus
one community ecologist might explore the patterns of coexistence and species
interactions among species within a delimited area with a constituent subset of
species and associated movement and heterogeneity patterns, while another might
ask the same questions but define a community that encompasses more area and
thus types of species, as well as different degrees of movements and heterogeneity
patterns (Fig. 1.1).

Although any decision that community ecologists make in designating the
spatiotemporal extent of their communities may seem innocuous, the substantial

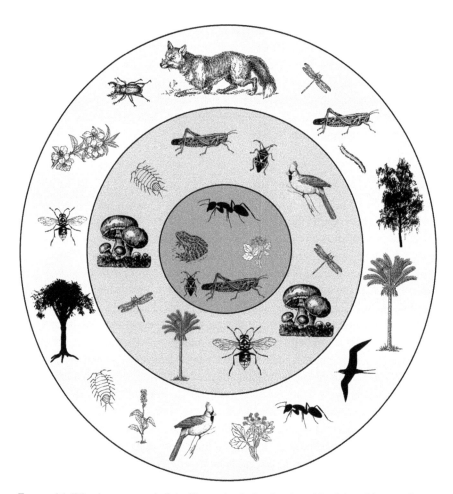

FIGURE 1.1. What is a community? An illustration indicating the subjectivity of the term "community" and which species are encompassed within that community. Lighter shaded circles indicate increasing scale of inclusion within a community, including larger areas, more species types, and associated heterogeneity and movement patterns. Species interacting at each scale are symbolized by the different illustrations of species.

"apples and oranges" of scales that occur when they are combined into a singular perspective on the community concept has led to much confusion and debate. The answers to most questions in community ecology turn out to be "it depends" (Lawton 1999, Simberloff 2004). Is competition an important structuring force for coexistence communities? It depends. Are communities dispersal limited? It depends. Are niche or neutral processes more important in driving species abundances and distributions? It depends. And so on.

One of the main reasons that the answers to community ecology's most fundamental questions have not been very well resolved is because community ecologists have not adequately embraced the pervasive influence of scale in the questions
they ask and the results they observe (Chave 2013), even though they appreci-
ate that scale is such an important problem. Indeed, recent explorations have
begun to show that almost all of the patterns and processes that they study are in-
extricably embedded within a scaling framework: patterns and processes at
smaller spatial scales are better described by smaller-scale processes (e.g., environmental filters, stochastic drift, and interspecific interactions), whereas patterns
at broader spatial scales are better described by larger-scale processes (e.g., the
regional species pool, climate, and dispersal limitation) (Rahbek and Graves 2001,
Condit et al. 2002, White and Hurlbert 2010, Belmaker and Jetz 2011, Jetz and
Fine 2012, Keil et al. 2012). Nevertheless, processes and patterns frequently affect
each other across scales as well, and it is this aspect of ecology that makes it such
a potentially important concept and the one we claim requires a metacommunity
approach.

Community ecology's history is rife with examples in which a simple recognition
of the dependence of the outcomes on scale could have resolved volumes of debate
and consternation. How could interspecific competition among similar species
produce such strong negative effects on the abundances of species in experimental
manipulations (Connell 1983) and yet not appear to influence the spatial distributions of species (Connor and Simberloff 1979)? It depends . . . *on scale* (Peres-
Neto et al. 2001). How could species partition their niches to enable local coexistence (Schoener 1974) and yet diversity not be saturated when the size of the
regional pool increases (Cornell 1985)? It depends . . . *on scale* (Loreau 2000).
How could neutral processes like ecological drift and dispersal limitation be largely
consistent with biodiversity patterns observed in tropical forest plots (Hubbell 1979,
Condit et al. 2012) and yet geographic distributions among these same forest plots
be highly niche-structured (Pitman et al. 2001, Jones et al. 2013). It depends . . .
on scale (Chase 2014).

1.2. THE METACOMMUNITY FRAMEWORK ALLOWS SIMULTANEOUS
CONSIDERATION OF MULTIPLE PROCESSES AT MULTIPLE SCALES

In its simplest form, a metacommunity represents a larger-scale "region" that is
made up of several smaller-scale "localities" (i.e., communities); these localities
are connected by dispersal and may be heterogeneous in any number of abiotic
and biotic variables (Fig. 1.2). Thus, the metacommunity framework explicitly considers more than one scale simultaneously. Although species interactions occur at

Local community

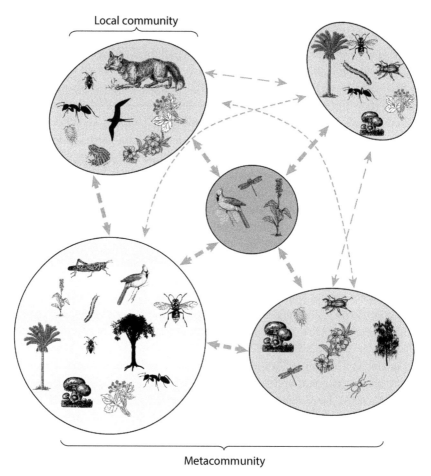

Metacommunity

FIGURE 1.2. Generalized view of a metacommunity. Each circle is a local community where populations of species grow and interact (symbolized by the different illustrations of species). Dispersal takes place among patches (symbolized by the dashed lines with arrows), and patches can be heterogeneous in environmental conditions (symbolized by the different levels of shading in each patch). Different levels of dispersal are symbolized by arrow size and line width.

relatively smaller scales, species coexistence at both smaller and larger scales results from interactions in a spatial context and can be modified by dispersal and spatial heterogeneity. Thus, to understand both the patterns and processes of coexistence and the composition and diversity of species, the interactions between scale, dispersal, and heterogeneity must be considered along with the milieu of local-scale processes (reviewed in Chesson 2000, Chase and Leibold 2003, HilleRisLambers et al. 2012).

As we will describe in more detail in Chapter 2, theoretical ideas about meta-communities have incorporated a variety of different assumptions and thus make a variety of predictions regarding the separate and combined influence of dispersal and heterogeneity. Dispersal rates are highly scale dependent; that is, the numbers of propagules that disperse from each reproductive individual in a population are typically highest nearer the parent and decline with increasing distance (though this can be highly species dependent and may, for example, be unimodal). Dispersal rates also depend on the properties of the species under consideration (i.e., some species move over much broader distances than others) and their abundances (i.e., the net number of dispersers is higher at high densities) as well as on the properties of the environment (i.e., some places are more isolated than others depending on the intervening matrix). Within this general context, however, there are three conceptually distinct ways by which dispersal rates can influence patterns in metacommunities, often interacting with the spatial heterogeneity in the system.

1. ***Dispersal limitation.*** If dispersal rates are low for at least some of the species in the metacommunity, those species will not be able to be present in all of the possible microsites where they could otherwise maintain positive growth. Dispersal limitation is the premise behind metacommunity theories such as the theory of island biogeography (MacArthur and Wilson 1967), neutral theory (Hubbell 2001), and competition-colonization-based coexistence (Hastings 1980, Tilman 1994, Chave et al. 2002). Observations and experiments that show increased diversity and spatial structuring of species composition support the view that dispersal is often a limiting process in natural communities (Cadotte 2006, Myers and Harms 2009, Condit et al. 2012). In addition to influencing patterns of diversity, dispersal limitation can alter the nature and strengths of species interactions in a local community. For example, dispersal limitation can allow similar (or even equivalent) competitors to coexist for long periods of time. Likewise, when predators are dispersal limited, prey that would otherwise not persist with a predator locally can do so as long as there is a lag period during which prey can exist in a patch without the predator present (Huffaker 1958, Holyoak and Lawler 1996), and this situation can alter the dynamics of the prey species coexistence (Shurin and Allen 2001) and diversity (Chase et al. 2010).

2. ***Dispersal sufficiency.*** Here, dispersal rates are intermediate such that most species are present in most of the habitats that they find suitable. In this case, metacommunity models of species sorting in which species partition habitats according to their preferences and tolerances to abiotic and biotic conditions provide an appropriate framework for understanding

patterns of coexistence and distribution (Chase and Leibold 2003). In comparison with (1) above, when dispersal is sufficient to allow species to persist in habitats that they find favorable, some local coexistence opportunities are eliminated (e.g., colonization-competition trade-offs, predator-prey cycling) while others are enhanced (e.g., resource partitioning).

3. ***Dispersal surplus.*** Dispersal rates for at least some species may be high enough that they can persist in a locality even when their local demographic rates would indicate they should not do so (i.e., a species can persist in a locality even if its death rate exceeds its birth rate as long as its dispersal rate from elsewhere is sufficiently high). This is known as a source-sink, or mass effect (Holt 1985, Shmida and Wilson 1985, Pulliam 1988, 2000). Here, coexistence can be either enhanced if high dispersal allows species to persist in habitats where they otherwise could not (Loreau and Mouquet 1999) or diminished if dispersal enhances species that are superior regional competitors (Mouquet and Loreau 2003) or strong predators (Huffaker 1958, Holyoak et al. 2005).

Spatial heterogeneity can also take different forms and interact with the scale and properties of the species under consideration. Heterogeneity can occur at the smallest microscale within a habitat (i.e., from the locality of a single individual to that of its neighbor) to climatic and geological gradients at large scales (e.g., from the tropics to the artic zones or from sea level to the top of the highest mountains or deepest oceans). It is not surprising, then, that heterogeneity measured on a small scale might have little influence on species that operate on a larger scale (and vice versa) (Ritchie 2010) and that different groups of species that vary in body size or dispersal mode might respond differently to the same variations in heterogeneity (Ricklefs and Lovette 1999, Kaspari et al. 2010). In addition, heterogeneity can be generated in two distinct ways.

First, heterogeneity can be *extrinsically* generated. This type of heterogeneity is imposed on a metacommunity from outside of the interactions of the species—when environmental conditions vary among localities for physical, chemical, and geological reasons. For example, soils vary in their nutrient and moisture content as a result of their underlying parent rock, and shallower parts of an aquatic ecosystem have more light and higher temperatures than deeper parts (at least until the latter freeze). This type of heterogeneity underlies the frequent distinction and divisions among broad habitat, ecosystem, and biome types. For example, many terrestrial plant species segregate among soils that vary in moisture content (e.g., hydric, mesic, or xeric species), and many aquatic species segregate between benthic and pelagic zones. Externally imposed disturbance regimes (e.g., flooding, fires) also create heterogeneities that alter patterns of metacommunity structure. It is reasonably well established that this sort of heterogeneity can strongly influence diversity (Hortal et al. 2009, Stein et al. 2014) and species compositional variation (Soininen 2014), among other factors.

Second, heterogeneity can be generated *intrinsically*. Here, heterogeneity is generated from processes among the species in the metacommunity. This can be due to simple differences in the composition of species in different localities, such as that of a competitor or predator, that might alter whether or not other species can live in that locality (Hastings 1980, Tilman 1994, Shurin and Allen 2001). A well-known example is the case in which specialist enemies can reduce the abundances of their prey, but by being dispersal limited, the enemies create a heterogeneous landscape of enemy-free and enemy-full space across different patches in the metacommunity. This situation can potentially allow many species to coexist at the metacommunity scale due to these heterogeneities, even with an underlying homogeneous landscape (Huffaker 1958, Connell 1971). Other examples emerge when the presence of species in a metacommunity alters the availability or distribution of resources, such as nutrients, light, or even mutualists, in a frequency-dependent way. For example, negative frequency dependence can create heterogeneity whereby a species alters the local resources in such a way that it favors other species over itself, allowing coexistence at the metacommunity scale (Molofsky et al. 1999, 2002). Alternatively, positive frequency dependence can create heterogeneity when a species present in a locality alters the resource conditions that favor its own existence over that of other species (e.g., priority effects), and coexistence can emerge at the metacommunity scale under certain conditions, even if it prevents its occurrence at local scales (Douglas and Wilson 2001, Molofsky et al. 2001, Yu et al. 2001, Molofsky and Bever 2002, Shurin et al. 2004, Calcagno et al. 2006). In extreme cases, species that alter the intrinsic heterogeneity of a metacommunity due to positive frequency-dependent interactions are sometimes referred to as ecosystem engineers (Jones et al. 1994, Wright and Jones 2006), or niche constructors (Odling-Smee et al. 2003). Importantly, this form of organism-induced heterogeneity will also depend on the rates of dispersal of organisms among environments and the resulting degree of dispersal limitation; if none of the species are at least somewhat dispersal limited, there would be no opportunity for some species to establish and create endogenous heterogeneity that would influence later establishing species.

1.3. DISCRETE VERSUS CONTINUOUS METACOMMUNITIES

Because the metacommunity concept largely emerged as a direct outgrowth of the traditional community-based perspective, the classic view of a metacommunity is discrete and additive (Fig. 1.2). We will term this a *metacommunity sensu stricto* (meaning "in the strict sense"). This view of metacommunity assumes that localities are discrete entities at the local scale and that they add together to create a larger, regional scale. This approach is quite amenable to theoretical investigations of species interactions and coexistence within metacommunities that emerge from two well-known analytical frameworks that treat space as a discrete series of patches connected by dispersal: (1) the metapopulation framework developed by Levins (1969), whereby discrete patches can be occupied or unoccupied by any given species but can be modified by colonization rates and extinction events (see also Hanski 1983 and 1999); and (2) the equilibrium theory of island biogeography (MacArthur and Wilson 1967), where diversity on an island patch represents the balance between extinction rates within patches of different sizes and colonization rates of species from a spatially distinct "mainland" source.

Like the theories themselves, a majority of empirical studies that explicitly test the various metacommunity theories have been conducted in habitats that can be reasonably viewed as discrete entities. Figure 1.3 shows several examples of naturally and artificially patchy systems used to examine metacommunity theory, either observationally or experimentally. The requirement for strict discrete patches in the sensu stricto version of a metacommunity likely explains why many more empirical tests of metacommunity theory have taken place in both natural and artificial systems with discrete boundaries, such as lakes, ponds, and glades, than habitats that are more appropriately viewed as continuous, such as grasslands, streams, and oceans (Cadotte 2006, Logue et al. 2011, Brown et al. 2017).

Despite the theoretical and empirical interest in treating patches as discrete habitats connected by dispersal and surrounded by an inhospitable matrix, it is clear that this is a caricature of a metacommunity that in fact is often (perhaps even always to some degree) structured in a much more continuous way. These continuities can be incorporated into the sensu stricto view of metacommunities in a variety of ways. In some cases, the adjustment can be quite simple, where the "matrix" between "core" habitat patches of interest can be examined as another discrete type of habitat even if some species can utilize both "core" and "matrix" habitats to varying extents (Fig. 1.4a). In other cases, however, although habitat heterogeneity and dispersal may be frequent, there is no clear way to delineate what is local and what is regional, and instead these are usually imposed in a more arbitrary way; for example, by sampling a given unit of area in a grassland (e.g., 1 m^2) or forest (e.g., 0.1 or 1 ha) or a volume of water in a lake or ocean (Fig. 1.4b). These types of situations are in fact much more frequently encountered and used to examine various aspects of metacommunity theory. We term this looser definition a *metacommunity sensu lato* (meaning "in the broad sense").

Though the differences may be in the minds of researchers when choosing appropriate study systems or sampling regimes, the distinction between sensu stricto

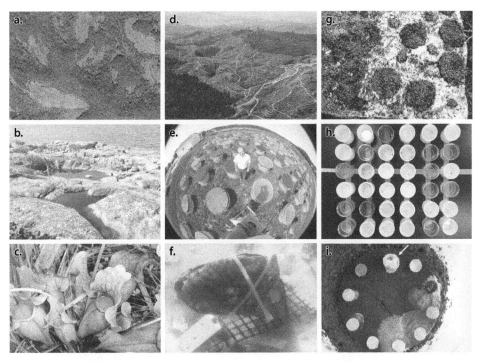

FIGURE 1.3. Some metacommunity approaches that have been explicitly examined using habitats reasonably described as discrete. On the left, three systems where observational approaches have been used: (a) an Ozark glade in central Missouri, which is a small opening in the forest where soil is shallow and dry, allowing an edaphically limited plant and animal community to persist (photo credit: Google Earth: DigitalGlobe ©); (b) a rock pool adjacent to the Baltic Sea, which contains several species of (sometimes) co-occurring *Daphnia* spp.; (c) water-filled leaves of pitcher plants (*Saraccenia* purpurea) that house a food web of bacteria, several species of protists, and some insect larvae. In the middle column, three field experiments: (d) experimentally fragmented secondary forest within a landscape cleared for oil palm production in Borneo; part of the Stability of Altered Forest Ecosystems (SAFE) project; (e) small cattle tanks and wading pools manipulating pond environments for amphibians and a variety of invertebrates; (f) a pen shell serving as a habitat for a variety of other species in the subtidal off the coast of Florida. On the right, three laboratory experimental systems: (g) experimental patches of moss used to examine the responses to a number of naturally occurring microscopic invertebrates; (h) laboratory bottles and connectors examining the responses of protists to experimental treatments; (i) experimental plugs of wood on which decaying fungi are manipulated.

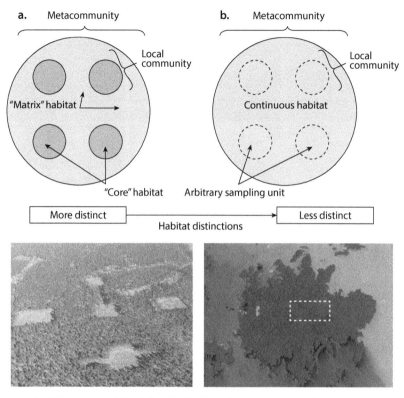

FIGURE 1.4. Metacommunities defined in habitats that are more continuous than generally conceptualized in metacommunity theory. (a) A case in which "core" habitats are defined as the community (and metacommunity) of interest, but "matrix" habitats can provide resources, propagules and other factors that might influence metacommunity processes. Illustrated by the corridor project at the Savannah River Site in South Carolina (photo credit: Jennie Haskell). (b) A case in which sampling areas are arbitrarily defined within an otherwise continuous habitat, despite the potential for considerable heterogeneity and dispersal within the defined area. Illustrated with the example of the 50-ha forest dynamics plot (approximately placed with white dashed line) on Barro Colorado Island, Panama (photo credit Google Earth: DigitalGlobe ©).

and sensu lato metacommunities is rarely made explicit. In this book, we will make use of both conceptualizations of metacommunities, depending on the context. For example, much (but not all) of the theoretical construct of metacommunities (Chap. 2) takes a sensu stricto view, in which local processes occur within a defined locality and regional processes (e.g., dispersal) occur among defined localities. However, much of the empirical work examining the processes (Chap. 3) and patterns in space (Chap. 4) and time (Chap. 5) takes place in sensu lato metacommunities. This might help to explain some of the discrepancies among studies.

The sensu lato metacommunity also emphasizes issues of scale that might be productively viewed continuously rather than discretely (i.e., local vs. regional; see Chap. 10).

1.4. A BRIEF HISTORY OF METACOMMUNITY ECOLOGY

In Lawton's (1999) search for generality in ecology, he noted that "the major weakness of traditional ecology, and why it has so conspicuously failed to come up with any patterns, rules, and workable theory, is its overwhelming emphasis on localness." (See also Simberloff 2004 and Ricklefs 2008.) While we agree that this is largely true, the legacy of spatial perspectives in community ecology is in fact quite old.

Some of the earliest explorations of competition and coexistence included a spatial perspective, even if it was not explicitly discussed as such. Gause's (1934, 1936) laboratory experiments provided one of the foundations for the competitive exclusion principle (Hardin 1960), whereby two species of *Paramecium* (*P. aurelia* and *P. caudatum*) grew just fine in the absence of competition but could not coexist when placed together (Fig. 1.5a). This result has been reproduced in countless textbooks. However, less widely recounted and rarely discussed is that in the same set of studies, Gause found that a different *Paramecium* species (*P .bursaria*) could coexist with *P. caudatum* in the same jars and that this was a likely result of spatial habitat partitioning (i.e., viewing the jar as a heterogeneous metacommunity from the perspective of the protists): *P. caudatum* primarily consume bacteria floating in the water column, whereas *P. bursaria* tends to live on the bottom of the jar in what is effectively a distinct habitat (Fig. 1.5b).

Many other classic cases of competition and coexistence actually include a spatial component to coexistence. Tansley's (1917) foundational study on the competition and (regional) coexistence of two species of *Gallium* focused on their differential performance on heterogeneously distributed soil types. Likewise, Connell's (1961a, 1961b) classic experiments showed how two species of intertidal barnacles differed in their spatial distribution with respect to tidal height; tradeoffs in performance in the face of desiccation at higher elevation in the intertidal zones and predation at lower elevations allowed these two species to coexist spatially via habitat heterogeneity.

Even G. Evelyn Hutchinson and Robert MacArthur, who are often considered founders of the local "niche-centric" view on competition and coexistence, were well aware of the importance of spatial processes for understanding patterns of species coexistence and diversity. In his "Homage to Santa Rosalia," Hutchinson (1959) described two species of ecologically and morphologically similar water

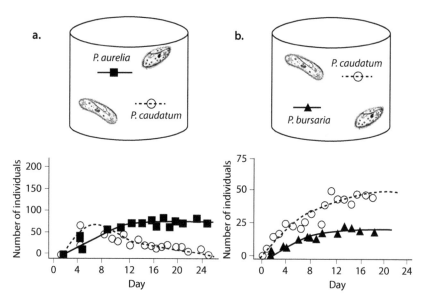

FIGURE 1.5. Data from Gause's (1934, 1936) classical experiments examining competitive interactions among *Paramecium* spp. in experimental microcosms. (a) The frequently reproduced results showing competitive exclusion when *P. caudatum* is grown with *P. aurelia*, a superior competitor. The cartoon jar illustrates that these two species both utilize resources in the same habitat (water column). (b) The rarely discussed results of competitive coexistence between *P. caudatum* and *P. bursaria*; the latter species in this case utilizes a different habitat type (the bottom of the jar), as shown in the cartoon.

boatmen (*Corixa punctata* and *C. affinis*) that he suggested could coexist in a small freshwater pool via niche partitioning involving "limiting similarity" of body-size ratios (which he tentatively suggested should be 1:1.3; see Chap. 6). However, often forgotten is that just a few paragraphs later in that paper, Hutchinson also described the distribution of two other congeneric species that co-occurred in the region but not in individual pools. One was a species (*C. macrocephala*) similar in size to the smaller *C. affinis* that was rarely found in the same pools with *C. affinis* but instead appeared to replace it and coexist with the larger *C. punctata* in deeper and less brackish pools. The other congener, *C. dentipes*, is about as large as *C. punctata*, which Hutchinson thought might exist as a "fugitive" species, being an inferior competitor locally that could persist regionally due to its higher dispersal capacity. In fact, generalizing among a variety of observations, Hutchinson concluded that not only did these species coexist locally as a result of niche (body size) partitioning, but they also coexisted regionally as a result of habitat heterogeneity and competition-colonization dynamics. Hutchinson in fact noted, "A final aspect of the limitation of possible diversity, and one that perhaps is of greatest importance, concerns what may be called the mosaic nature of the environment" (1959, 154). Clearly, Hutchinson was talking about the role of the metacommunity.

A closer read of the MacArthur's (1958) seminal study of northeastern US warblers (*Dendroica* spp.) shows that he, like Hutchinson, also had a more nuanced and spatiotemporal perspective on their coexistence than is usually depicted. Ecologists may be familiar with the role that this study played in elucidating how similar species could coexist as a result of local niche partitioning; in this case, fine-scale partitioning of foraging zones in trees. However, MacArthur was also interested in how spatial differences in habitat availability and temporal variation in food availability influenced the species' coexistence. Rather than conducting his study in a single homogenous forest type, he chose four forest plots that differed in the availability of the foraging microhabitats within the trees—specifically because of differences in forest age and tree height. Blackburnian warblers (*Dendroica fusca*), for example, feed near the tops of trees, and MacArthur found that they were proportionately more abundant in older forests with taller trees. Cape May (*D. tigrina*) and bay-breasted (*D. castanea*) warblers, on the other hand, appeared to be able to take better advantage of spatiotemporally variable outbreaks of insects (e.g., spruce budworm) by having different clutch sizes and rapid population-level responses. MacArthur argued that these spatiotemporal fluctuations likely allowed these locally inferior competitors to persist.

The volumes of research that followed focused primarily on the roles of local niche partitioning for species coexistence (Cody and Diamond 1975), which subsequently led to a considerable amount of disharmony in the field (Strong et al. 1984). Perhaps if the spatial components of these foundational studies had made it into the community ecology lexicon in addition to the local competitive factors, some of this disharmony could have been avoided.

1.5. MACARTHUR'S PARADOX: DETERMINISM VERSUS STOCHASTICITY AND THE NULL MODEL WARS

As we will discuss in more detail in Chapter 2, there are several categories of metacommunity theories that differ in their assumptions regarding the roles of species traits and the species' responses to environmental heterogeneity, stochasticity in colonization-extinction dynamics, and the importance of dispersal (Holyoak et al. 2005, Leibold and McPeek 2006). Interestingly, and perhaps somewhat paradoxically, MacArthur played a role in the development of each. MacArthur's paradox refers to the highly divergent foci he used as he helped to launch two seemingly diametrically opposed viewpoints on how communities are structured—niche assembly and neutral assembly (Schoener 1989a, Loreau and Mouquet 1999).

On the one hand, MacArthur and colleagues developed a series of theories typically involving competition for limiting resources and the ability of species to coexist at equilibrium (MacArthur and Levins 1964, 1967; MacArthur and Wilson 1967; MacArthur 1972). Much of this work laid the foundations for the metacommunity archetype known as *species sorting*, whereby species traits determine whether they can coexist either locally or regionally due to a combination of deterministic responses to variation in environmental conditions and interspecific interactions (see also Chase and Leibold 2003).

On the other hand, MacArthur, with E. O. Wilson, developed the theory of island biogeography, which in its simplest form eliminated species traits and interactions all together and instead focused on the role of dispersal (and dispersal limitation) and stochasticity in colonization and extinction events (MacArthur and Wilson 1963, 1967). This theory formed the basis for the modern *neutral theory*, which has proven useful for understanding a variety of metacommunity-level patterns only considering dispersal and stochasticity, as well as speciation, but assumes no differences in species traits or habitat heterogeneity (see also Caswell 1976 and Bell 2000 for other early neutral-model formulations).

Inspired by MacArthur and others' deterministic theories of species interactions, particularly competition, in the 1960s and 1970s many investigations examined biogeographic patterns, such as niche-overlap and the numbers and types of co-occurring species, with an eye toward species sorting as the underlying mechanism. Among the most well-known of these studies was Diamond's (1975a) description of "assembly rules," which ascribed highly deterministic mechanisms (competition, traits) to explain the patterns of bird distributions he observed among islands (Chaps. 4 and 6).

While Diamond and others focused on the deterministic side of MacArthur's paradox, Dan Simberloff and his colleagues are more strongly associated with the perspective that stochasticity and unpredictability can pervade. Inspired by his dissertation work testing the theory of island biogeography, Simberloff (Simberloff and Wilson 1969, 1970) was swayed by the potential importance of stochasticity and resulting unpredictability that pervaded many complex sciences (e.g., physics, evolution), including ecology, and argued for a clearer recognition of these stochastic forces (Simberloff 1980). Armed with observations of systems that appeared highly dynamic and a strong mathematical/statistical approach, Simberloff and his associates pressed for a more rigorous approach to discerning deterministic structure in natural systems (i.e., the use of null models and other statistical approaches) and dispatched a series of attacks on the dominant deterministic paradigm (Simberloff and Abele 1976, 1982; Connor and Simberloff 1978, 1979; Simberloff 1978; Strong et al. 1979; Simberloff and Boecklen 1981).

The backlash that emerged following this discourse set in motion a revolution that led in part to the experimental and statistical rigor that is the hallmark of community ecology today (Morin 2011, Mittelbach 2012). It also led to the perhaps undue focus on "localness" (Lawton 1999, Ricklefs 2008), because it shifted the focus away from broad patterns and toward more detailed experiments in ways that seriously constrained the field of community ecology for nearly two decades (see also Maurer 1999).

The debate about the relative structuring roles of stochastic versus deterministic forces did not, however, start in the 1970s. In fact, the debate concerning the relative importance of species interactions versus the vagaries of individual-species responses harken back to ideas that were already firmly ensconced among practitioners of ecology during the field's infancy (McIntosh 1986). For example, early controversies in community ecology pitted Clements's (1916) deterministic paradigm, which visualized communities bound together by environment and interactions as a superorganism, against Gleason's (1926, 1927) perspective, which recognized that communities were more likely the result not only of their individualistic responses to interactions and environment but also of a significant element of "chance." Palmgren (1926) also had a similar perspective on the importance of stochasticity and dispersal limitation in the distribution of species, but it has not been as widely recognized.

More than twenty years later, a symposium held by the British Ecological Society in 1944 focused on the issue of "coexistence of closely allied species" and explored the generality of Gause's (1934) then recently developed "competitive exclusion principle," which posited that two or more species could not coexist if they utilized the same niche. A report on the discussions during this symposium (British Ecological Society 1944) noted a "vigorous" debate between those who generally favored Gause's view that similar species could not coexist due to competitive exclusion, most notably Charles Elton and David Lack, and those who felt this perspective was highly oversimplified for a number of reasons. Most notable was the attack on Gause's principle by Captain Cyril Diver, a geneticist and ecologist who primarily worked on a group of land snails whose distributions appeared to be due to random processes and dispersal limitation more than to any sort of niche-differentiated interactions.

And, twenty years later still, in a symposium sponsored by the American Institute for Biology Sciences (Riley 1963), Hutchinson began one conversation saying, "I would like Dr. Riley to give us some ideas of the possibility of organisms that are obviously different being exactly the same." At this point, Riley described his idea that is remarkably similar to those typically attributed to Hubbell's (2001) neutral theory. Riley noted that particularly in tropical marine waters with exceptionally diverse communities that "there is a possibility—and

I repeat, I am setting myself up as a target—that these organisms are so nearly equal in their ability to compete that they do not eliminate each other." To this, Hutchinson invoked a metacommunity perspective by suggesting that this could work in the ocean, which "can be regarded as a large number of partially interconnected sub-areas," such that "species will disappear in one place and not in some other region."[1]

1.6 REEMERGENCE OF THE METACOMMUNITY PERSPECTIVE: 1990S AND 2000S

Spatial processes remained largely on the back burner during the 1980s, while a majority of community ecologists were busily responding to early critiques about the roles of competition and other interspecific interactions by performing careful, primarily local, experimentation (Wilbur 1997, Brown 1998, Werner 1998; but also see Hanski 1983, and Ricklefs 1987). However, space reemerged as a prominent feature of community ecology about a decade later (Ricklefs and Schluter 1993, Tilman and Kareiva 1997). Controlled experiments reminiscent of much earlier work (Huffaker 1958) reminded ecologists how spatial processes could mediate species interactions and coexistence (Holyoak and Lawler 1996), and community-level manipulations of habitat connectivity (Kruess and Tscharntke 1994, Warren 1996, Gonzalez et al. 1998) and propagule additions (Tilman 1997, Shurin 2001) showed that dispersal was often limiting for populations and communities. Simultaneously, species coexistence theory continued to incorporate spatial processes in the different forms that we now recognize as species sorting due to habitat heterogeneity (Chase and Leibold 2003), patch dynamics due to colonization-extinction dynamics (Tilman 1994, Hanski and Gyllenberg 1997), and mass effects with heterogeneity and dispersal playing a role (Amarasekare and Nisbet 2001).

All of these empirical and theoretical trends led community ecology as a whole to increase its spatiotemporal perspective. A major stimulus was the publication of Hubbell's (2001) neutral theory. Several features of the neutral theory helped to establish it as one of the dominant paradigms in community ecology today. First and foremost, it provided a rallying cry for those who were dissatisfied with the often-oversimplified niche-based view (with each species tidily occupying a distinct niche), and it appealed to mathematical elegance in its ability to explain a variety of macroecological phenomena (Bell 2001, Chave 2004). It also forced proponents of the niche perspective to recognize the limitations of its early iterations

---

[1] Chris Klausmeier has animated the notes of this conversation and uploaded it at https://www.youtube.com/watch?v=xtMjcxFnRCk.

and develop more realistic approaches (Chave et al. 2002, Wilson et al. 2003). Second, it emphasized the need for a more pluralistic approach and for the previously dominant view to recognize the sometimes important roles of dispersal limitation and ecological drift, even if not in purely neutral systems (Leibold et al. 2004, Alonso et al. 2006, Gravel et al. 2006, Leibold and McPeek 2006). Third, by incorporating an explicit speciation component, which was necessary in order for a diversity of otherwise neutral species to be maintained, the NT provided a connection with studies that examined the link between larger-scale biogeographic processes and more local-scale diversity, and a testable link to estimates of speciation rates, species phylogenies, and patterns of abundances (Ricklefs 2006, Ricklefs and Renner 2012).

Just as evolutionary biologists largely resolved the selection-versus-neutrality debate decades ago by recognizing the importance of both processes, ecologists are largely leaving behind the dogma of "all-or-nothing" tests of neutral versus niche theories (McGill et al. 2006a, Vellend 2010, 2016). Instead, we now recognize the myriad ways in which the relative importance of dispersal, stochastic processes, and deterministic processes interact with one another to form natural communities (Gravel et al. 2006, Leibold and McPeek 2006, Chase and Myers 2011). Despite some key differences during nearly a hundred years of debate, starting with Clements and Gleason and reaching fever pitch with the null model wars in the 1970s and 1980s and again with the niche-versus-neutral debate in the 2000s, the essence of the argument is the same—*Are patterns of diversity and composition predictable based on local biological features of the organisms and environment, or do chance, ecological drift, and dispersal limitation reign supreme?* The answer, clearly, is yes . . . to both sides.

1.7 A PROSPECTUS FOR METACOMMUNITY ECOLOGY: WHAT ARE WE TRYING TO DO IN THIS BOOK?

Our excitement about metacommunity ecology has come from realizing how many aspects of population, community, evolutionary, and ecosystems ecology are tied together by taking this perspective. Although syntheses of these areas harken back to Elton, Lotka, MacArthur, and others, the extra dimensionality and depth that come from metacommunity thinking has convinced us that it can be a hub for tying together these otherwise poorly integrated areas of ecology (Fig. 1.6). That this may be so has also been reinforced by the numerous colleagues who have shared their enthusiasm and developed their work in this direction. If there is anything we hope to accomplish in this book, it is to convey this sense of synthesis and to help facilitate it as ecology matures as a science capable of addressing the pressing demands placed upon it.

FIGURE 1.6. Metacommunity as a conceptual hub for synthetic ecology. We view metacommunity as both a central concept at the interface of ecological subdisciplines (circles) and as a major axis for novel insights into each subdiscipline (arrows). Shown are the various areas of ecology we address in the book, which correspond to Chapters 6–13, clockwise from the top.

1.8 MOVING FORWARD: OVERVIEW OF THE REST OF THE BOOK

The remainder of this book is divided into three sections. The first section gives an overview of the history and context of metacommunity ecology (this chapter), discusses the development, predictions, and current state of metacommunity theory (Chap. 2), and then describes evidence for the underlying processes (Chap. 3) and patterns through space (Chap. 4) and time (Chap. 5). We view this section as a bit of a primer on the current state of metacommunity thinking corresponding to the hub in Figure 1.6. We then address how traits and phylogeny can be used to enhance understanding of metacommunities (Chaps. 6 and 7). The next section then moves toward extending the metacommunity concept to consider how metacommunity ecology interacts with eco-evolutionary processes involving local adaptation and gene flow (Chap. 8), as well as extending it to processes involving biogeography and species diversification (Chap. 9). We then discuss how the role of a metacommunity perspective in providing a deeper understanding of macroecological patterns of biodiversity and biogeography (Chap. 10). We view this section as setting the role of metacommunity ecology within the context of some of the most important areas of the current development of community ecology.

Finally, the third section examines how the spatial perspective of metacommunity ecology can enhance our understanding of food webs (Chap. 11), the role of metacommunity assembly in the functioning of ecosystems (Chap. 12), and metaecosystems ecology (Chap. 13). We view this section as pushing the boundaries of how metacommunity ecology can contribute to processes and phenomena that extend beyond some of the basic ideas of community ecology (Vellend 2016). Finally, we conclude with a prospectus for future directions (Chap. 14).

1.9. A FINAL NOTE ON OUR ASSUMPTIONS

For heuristic purposes, in all of this work we will use the following propositions as given even if some ecologists might see them as still partially unresolved (see Morin 2011, Scheiner and Willig 2011, and Mittelbach 2012 for detailed discussions of many of these).

1. *Interspecific interactions influence abundance and distribution.* Species interact with each other via a wide array of direct interactions including competition, predator-prey interactions, facilitation, and mutualisms in ways that strongly influence each other's demographic rates, affecting their ability to coexist, displace, or be displaced by other species.
2. *Contingency.* The strengths of species interactions are usually strongly contingent on local conditions; that is, species interact differently in different places and times, often leading to different outcomes.
3. *Indirect effects.* Direct interactions among species can produce indirect effects via chains of direct effects or through contingent dependencies that alter how pairwise interactions occur.
4. *Feedback.* Species interactions, including indirect effects, in combination with intraspecific effects, are responsible for creating feedback in communities, which can either be stabilizing or destabilizing. Overall, there are constraints on overall feedback in the community that determine its stability and persistence, and these are influenced by the numbers of species and the nature of their interactions (May 1972, 1973).
5. *Food webs.* The regulation of abundance and factors that influence species coexistence are often the result of complex interactions among consumers

and resources through both direct and indirect effects in food webs. These
are often mediated by trade-offs involving competitive ability and
resistance/tolerance of enemies (e.g., predators, parasites, and pathogens), as
well as traits associated with other types of interactions (e.g., mutualisms).

6. *Ecosystem consequences.* Community interactions, diversity, and composition can often have strong influences on ecosystem-level structure
(biomass of trophic levels) and processes (rates of energy and nutrient
flows).

CHAPTER TWO

The Theories of Metacommunities

Prospectus

1. Metacommunity theory is characterized by four relatively distinct frameworks for representing metacommunities, each of which makes fundamentally different assumptions about which processes are operating and different predictions about a number of important patterns.

   a. The patch dynamics (PD) archetype focuses on how colonization-extinction processes lead to a dynamic equilibrium that can mediate how species coexist in a metacommunity even if the number of species within any community is low and every local population is guaranteed to go extinct eventually.

   b. The species sorting (SS) archetype focuses on how patch-to-patch differences in local environmental conditions can allow a diverse number of species to coexist in a metacommunity by the process of community assembly in response to environmental gradients.

   c. The neutral theory (NT) archetype focuses on how stochastic demography and dispersal among patches can influence local and metacommunity diversity while assuming that niche differences among component species are unimportant.

   d. The mass effects (ME) archetype focuses on how maladaptive (but perhaps unavoidable) exchanges of individuals among patches with different local conditions can enhance local diversity if metacommunity diversity is high by generating source-sink relations among local populations.

2. The core assumptions of the different archetypes differ in their relevance in a given metacommunity depending on the degree of environmental heterogeneity and the degree of connectance/dispersal in the metacommunity, as well as the scale on which it is observed.

3. Most natural situations are unlikely to correspond strongly to any single metacommunity archetype and more likely to involve aspects of each.

4. Emerging theory reveals that more realistic predictions emerge when multiple archetypes are jointly modeled and integrated. An "ubermodel" incorporating all of the core assumptions of the different archetypes might be emerging, at least for highly simplified cases.

5. Scenarios for local species interactions in most of these metacommunity archetypes are still highly simplistic, and more complex and realistic scenarios are likely to alter how metacommunities develop.

6. Progress in metacommunity ecology is hampered by continuing to think of the original four archetypes as alternative hypotheses rather than as interacting components. Future progress will require us to find ways to understand how multiple processes work jointly and interactively to affect diversity and composition of biotas across spatial scales.

Metacommunity ecology differs from traditional community ecology by an explicit recognition that three fundamental factors—scale, dispersal, and heterogeneity—separately and interactively produce a rich array of mechanisms that influence the likelihood of species occurrence and coexistence, which in turn create the patterns of species composition and biodiversity variation we observe in natural systems. As mentioned in Chapter 1, there is not a single metacommunity theory per se (at least not yet), but rather a family of theoretical models that differentially incorporate aspects of scale, dispersal, and heterogeneity, as well as trait differences, stochasticity, and nonlinearities such as frequency dependence, which alters interspecific interactions.

The first use of the word "metacommunity" that we know of, which was by Hanski and Gilpin (1991), focused on extending classic metapopulation theory to multiple interacting species by using multispecies metapopulation occupancy models such as those pioneered by Levins and Culver (1971; see also Cohen 1970, Horn and MacArthur 1972, Levin 1974, Levin and Paine 1974, Slatkin 1974, Hastings 1980, and Tilman 1994). The key process in these *patch dynamic* (PD) models is the differential use of space by organisms that have different dispersal rates in a metacommunity subject to stochastic extinctions. A second archetype of metacommunities emerged from traditional consumer-resource-based niche theory (Tilman 1982, Leibold 1995) that was extended to include habitat heterogeneity and thus allowed coexistence criteria to vary with spatial scale (Leibold 1998, Chase and Leibold 2003). The key process in these models is habitat heterogeneity and *species sorting* (SS) among habitat patches that are most favorable to the growth of different species. As the antithesis to the SS archetype, *neutral theory* (NT; Hubbell 1997, Bell 2001, Hubbell 2001) assumes that patches within a metacommunity and all the species within it are effectively identical, such that coexistence is influenced only by stochastic demography, colonization, extinction, and dispersal limitation (with speciation priming the pump). Finally, in a somewhat more complex scenario, *mass effects* (ME; also known as *source-sink*) metacommunity models include habitat heterogeneity favoring different species in different environments and model dispersal as a homogenizing spillover process that allows species, by maintaining high growth in adjacent patches (i.e., source-sink relations among patches), to persist in habitat patches that they otherwise would not occupy (Levin 1974, Loreau and Mouquet 1999, Amarasekare and Nisbet 2001).

In an early attempt to synthesize these rapidly emerging and diverging archetypes of metacommunities, a working group was convened at the National Center for Ecology Analysis and Synthesis (NCEAS) in Santa Barbara, California, from 2001 to 2003, the products of which highlighted, compared, and contrasted the differences among these four archetypes of metacommunities (Leibold et al. 2004, Holyoak et al. 2005). Although this synthesis provided a direct comparison of the metacommunity archetypes, an unfortunate consequence is that ecologists have since tended to treat these as discrete alternatives to be pitted against each other, using observational or experimental data in order to determine which archetype is "correct," at least for a given system (reviewed by Logue et al. 2011 and Brown et al. 2017).

Our original synthesis (Leibold et al. 2004, Holyoak et al. 2005) described each of the four metacommunity archetypes as "paradigms." Not surprisingly, the differences in the core assumptions of each of the approaches lead to different predictions regarding how communities are structured and how they respond to spatial and environmental variation (reviewed in Chase et al. 2005 and Leibold 2011). This likely contributed to the treatment of the various metacommunity theories as alternative hypotheses and suggested that one could conduct tests to determine which was most appropriate. However, the past decade has seen the recognition of a more nuanced view of the processes and patterns in metacommunities, and even the most seemingly diametrically opposed archetypes based in niche (SS) and neutral (NT) archetypes have been (somewhat) reasonably well integrated (Gravel et al. 2006, Leibold and McPeek 2006, Adler et al. 2007, Chase and Myers 2011). One of the main points we hope to make in this book is that treating the various metacommunity archetypes as alternative hypotheses has passed its utility, and it is time for a more fully embracing perspective that integrates into a more holistic framework each of the processes that influence species interactions and coexistence at different spatial scales. Rather than calling these metacommunity paradigms, we now term them "archetypes," to better reflect that, in reality, all of the processes of each approach may act together but that their relative importance might differ both within and among metacommunities and that there may even be unsuspected interactions among them.

Recent syntheses have aimed to devise a more unified framework for metacommunity structure. For example, with a focus on processes, Vellend (2010, 2016) suggested that the factors that influence species composition and diversity in a metacommunity could be categorized into four general processes—niche selection, dispersal, stochastic drift, and speciation. Setting aside the generation of species via speciation for a later chapter (Chap. 9), the other processes are differentially considered by each of the metacommunity archetypes. At one extreme, NT considers stochastic drift and dispersal but ignores any sort of niche selection, while at the other extreme, SS considers niche selection but largely ignores stochastic drift and dispersal. Like NT, PD includes stochasticity and dispersal, but it also includes some aspects of niche selection via differences in traits that influence spatial coexistence mechanisms (e.g., competition-colonization trade-offs, life-history trade-offs). However, PD models typically (but not always) ignore other components of selection associated with habitat heterogeneity. Finally, ME models include dispersal and niche selection but typically ignore stochastic drift.

In a slightly different vein and with more of a focus on patterns, McGill (2010) noted that even though the processes that underlie the metacommunity theories might seem diametrically opposed, they share a common set of rules inherent to stochastic geometry—(1) species are spatially clumped, (2) species vary considerably in their relative abundances, and (3) species vary geographically more or less independent from one another. We discuss this perspective in more detail in Chapter 10, but our main point here is that there are multiple permutations of processes (niche selection, dispersal, stochastic drift, and speciation) and models for how these processes interact (PD, SS, NT, ME) that can lead to these three features (clumping, variable abundance, and geographic independence), and that resolving their various roles requires an overarching conceptual framework.

In this chapter, we seek to more fully understand the roles of scale, dispersal, and heterogeneities, as well as the influence of stochastic processes and trait differences, to build on these syntheses and to refine our perspective on metacommunities beyond simply thinking of them as four alternative views to be pitted against one another. For example, one of the earliest uses of the term "metacommunity," by Wilson (1992), does not fit well into any of the four modalities; it instead focused more on the ways that local community assembly could be affected by stochasticity in the timing of species arrivals, followed by frequency dependence that allowed different species to dominate different habitat patches in multiple stable equilibria. Such landscape-scale frequency dependence emerges from the interaction of niche selection, dispersal, and stochastic drift and can complicate the distinction between the different processes (see Chap. 3). We first overview each of the separate archetypes and their predictions, as well as some complexities that are not captured by the simple archetypes outlined by Leibold et al. (2004), and we discuss the "landscape" for where and when each archetype is most likely to prevail. We then review more recent efforts at combining elements of the various archetypes and conclude that they are likely to still produce only a very crude vision of metacommunities. We finish this chapter highlighting the features of metacommunities that challenge future theory with an eye to identify and motivate future syntheses. This chapter is not meant to cover all of metacommunity theory—and ignores, for example, eco-evolutionary processes, food-web interactions, and ecosystem feedbacks (these elements will be discussed in the following chapters)—but is meant instead to cover the basic outline of the current state of metacommunity theory.

2.1 AN OVERVIEW AND CRITICAL ANALYSIS OF THE FOUR ARCHETYPES OF METACOMMUNITY ECOLOGY

Here we provide an overview of various metacommunity theories, many of which are described in detail in the original literature. Therefore, for the most part, we do not describe the mathematics involved and encourage interested readers to review the original papers. Instead, we highlight the conceptual basis and the main results of these models, focusing especially on results that will inform subsequent chapters.

2.1.1 Neutral Theory

The NT is arguably the simplest model with the fewest assumptions. The NT is sometimes referred to as a "null model" for communities from which deviations can provide useful information regarding the importance of processes other than stochasticity (e.g., Bell's 2001 "weak" version of the NT). Other times, NT is taken to be a reasonable depiction of how actual ecological systems are structured (Bell's 2001 "strong" version of the NT). In this case, although NT proponents do not necessarily believe that ecological systems are completely neutral or that species have absolutely no differences in their traits, they suggest that stochastic birth-death processes and dispersal limitation strongly override those differences (Rosindell et al. 2011).

Expanding on previous work (Hubbell 1979, 1997; Hubbell and Foster 1986), Hubbell's (2001) monograph, *The Unified Neutral Theory of Biodiversity and Biogeography*, is the most well-known NT. This work was inspired by the dynamical and stochastic nature of the equilibrium theory of island biogeography (MacArthur and Wilson 1967), which in its simplest form is neutral with respect to species traits, as well by insights and modeling from the neutral theory of evolution (Kimura 1968, 1983) and other earlier neutral models in ecology (Caswell

1978, Bell 2000). The early iterations of the NT were criticized because NT dynamics predicted that species with identical traits should eventually lead to extinction of all but one of them, even if this was predicted to take a long time (Chesson and Huntly 1997). Hubbell's (2001) NT included a speciation parameter that counteracted this effect and allowed a biogeographical equilibrium number of species to be predicted (the very slow rate of stochastic extinctions in the metacommunity were balanced by a similarly slow rate of speciation). In addition to speciation, the other main components of the expanded NT are dispersal limitation, whereby every species does not have access to every site in each time step, and *demographic stochasticity*, whereby individuals of each species that can disperse to a site has an equal (stochastic) probability of gaining access to that site and an equal probability of dying on that site. Since its original publication, and as a result of a variety of criticisms of the initial version, the NT has been fortified in several directions, including the addition of more analytical tractability (Volkov et al. 2003), the inclusion of more realistic forms of speciation (Rosindell et al. 2010), and explanations for why the primary assumption of ecological equivalence might be a reasonable outcome to expect, even among species that experience strong natural selection and environmental differences (Hubbell 2006).

   The success of the NT can at least partly be attributed to its emphasis on the sometimes important role that dispersal limitation and stochastic ecological drift can play in affecting community-level patterns, which are often just as well (if not better) predicted by NT than by other more complex models (reviewed in Rosindell et al. 2011; Fig. 2.1). It is also important by serving as a null model against more deterministic models in the face of the constant barrage of possible types of stochasticity (Condit et al. 2002, Dornelas et al. 2006, Chase and Myers 2011). In addition, the NT provides an intriguing link between local-scale ecological processes and larger-scale biogeographic processes, notably speciation/diversification dynamics and their phylogenetic signatures, which can be used to make and test the NT's predictions against more deterministic ones (Ricklefs and Renner 2012). This biogeographic perspective is mostly absent in the other conceptualizations of metacommunities, which are only now beginning to incorporate them (McPeek 2008, Harrison and Grace 2008; discussed in Chap. 9).

2.1.2 Species Sorting

Species sorting largely embodies traditional "niche-based" perspectives on species interactions and coexistence with roots that go back to some of the earliest explorations in community ecology (Elton 1927, Gleason 1927, Tansley 1935, Clements 1936). In this view, trade-offs allow species to coexist locally or to distribute

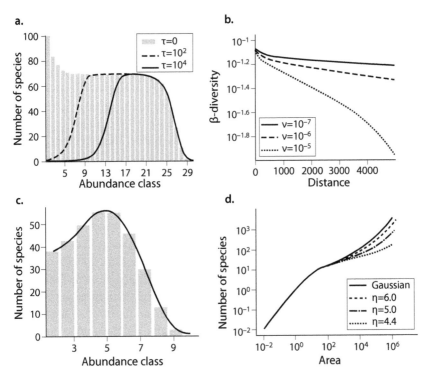

FIGURE 2.1. Predictions from neutral models. (a) Metacommunity-level species-abundance curves for neutral models with different speciation "transition times" ($\tau = 0$, 10^2, 10^4). (b) Distance-decay of community composition (expressed as β-diversity) for different values of v (the speciation rate). (c) Local species abundance curves for the same models as (a). (d) Species-area curves for different levels of η ("fatness" of the dispersal kernel). Modified from Rosindell et al. 2011.

themselves into different parts of a heterogeneous landscape because they are suitably favored in some places even though they may be at a disadvantage in others due to local conditions (Chase and Leibold 2003). Building on MacArthur's early work defining the domain of the SS archetype (MacArthur and Levins 1964, 1967; MacArthur 1972), Tilman (1982) devised a theoretical approach based on consumer-resource dynamics and coexistence at local and regional scales when species compete for limited resources. Tilman's approach defined the R^* principle, whereby a species that can reduce a limiting resource and persist on that resource at equilibrium at a lower level than that of any other species is competitively dominant. Species could coexist locally if there were two or more resources in a certain concentration and different species had the lowest R^* for each different one (i.e., a trade-off). And they could coexist regionally so long as there was heterogeneity in the spatial distributions of resources that favor each species in at least some places. This approach has since been generalized to include the differential abilities of species to respond to limiting factors other than resources—including enemies, disturbances, and spatial scales (Chase and Leibold 2003).

The SS archetype allows us to easily dispel one common misconception about niche theory and the competitive exclusion principle when placed in a spatiotemporal context; namely, that the number of resources or other limiting factors (and trade-offs in use among them) defines how many species there can be, which is often generalized to suggest that N species can coexist on N resources (Grover 1994; Holt et al. 1994; Leibold 1995, 1996, 1998; Wootton 1998; Chase and Leibold 2003, Ritchie 2010). In fact, it is frequently argued that niche theory is inadequate to explain patterns, particularly in diverse communities, because this theory often seems to imply that N species exist on fewer than N resources (Hardin 1960, MacArthur 1972, Hubbell 2001). The beginnings of the misconception can be traced back at least to Hutchinson's (1961) "paradox of the plankton," which sought to explain why more algae could coexist in lakes than the number of limiting resources. However, this is an inherently local perspective on coexistence (and also does not include temporal variation) and has long been superseded in the development of SS theory through space and time, where the ratios (rather than numbers) of resources can vary heterogeneously in space and time, allowing many more species to coexist than limiting factors. Tilman (1982) developed this prediction from simple graphical models (Fig. 2.2; see also Tilman and Pacala 1993, Leibold 1996, Chase and Leibold 2003). Chesson (2000) has shown that spatial coexistence can be substantially more complex but that it can still allow many more species to coexist than the number of limiting resources.

2.1.3 Patch Dynamics

Patch dynamics generally assumes that the environment is distributed into patches that have more or less the same (homogeneous) environmental conditions. Here, coexistence and niche selection occurs when there are differences in species dispersal and life-history traits but not among heterogeneous habitats. Stochastic change and dispersal limitation occur because species are assumed to go locally extinct at some frequency, due either to purely stochastic reasons associated with demographic processes or to some externally imposed mortality factor (e.g., disturbance) that itself is stochastic. The PD archetype can be traced back to at least Skellam (1951) and Hutchinson (1951), who independently suggested that stochastic extinctions of a competitively dominant species from otherwise suitable patches could allow members of a competitively inferior species to persist as a

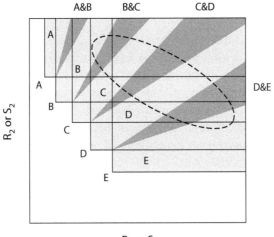

FIGURE 2.2. Representation of coexistence among five species (A-E) that compete for two essential resources (1 and 2). The elbowed lines represent the zero net growth isoclines of each species and describe their resource requirements (R_1 and R_2). The model assumes a trade-off between these two values so that species that have low requirements for resource 1 (e.g., species A) have high requirements for resource 2. The diagonal shaded areas represent the relative uptake ratios by species for the two resources, and the model assumes that each species is more efficient at taking up its more limiting resource. A given ecosystem can be defined by the input concentrations of the two resources (S_1 and S_2), and the graph shows the predicted endpoint of community assembly (either a single species or pair of coexisting species) for any given input concentration. If the supply point is in a lightly shaded area, only one species is predicted as labeled; if the supply point is in a dark-shaded area, two species are predicted to coexist as labeled. The oval shows the range of spatial variation in S_1 and S_2 for a given metacommunity. Under highly simplistic assumptions (especially lack of dispersal limitation in community assembly), the model predicts that all species that form any of the combinations predicted within the oval will coexist in the metacommunity (in this case, species B–E). See Tilman 1982.

spatiotemporal "fugitives," so long as they were more effective colonizers of empty patches. Such competition-colonization trade-offs were theoretically formalized by Cohen (1970) and Levins and Culver (1971), the latter of whom built upon Levins's (1969) development of metapopulation theory, which used patch colonization and extinction rates to derive the proportion of patches occupied by a single species. The PD framework was further refined by Slatkin (1974), Hastings (1980), and Hanski (1982, 1983), among others, who explored conditions for species coexistence at regional spatial scales and when the timescales of local processes (competitive exclusion) and regional processes (dispersal) varied. In total, these investigations studied the conditions under which species could coexist in otherwise spatially homogenous environments, even in the absence of any externally imposed disturbances, so long as they used space differentially due to differences in dispersal.

After these initial theoretical developments, variations on PD models have added disturbances to local patches (increasing local extinction rates), allowing species to coexist regionally (and locally) (Levin and Paine 1974, Hastings 1980, Tilman 1994). These variations provide a theoretical underpinning for the intermediate disturbance hypothesis (Connell 1978), which posits that diversity can be highest at intermediate levels of disturbance because they can preclude regional dominance by the superior competitor but still prevent good colonizers from excluding the good competitors. In a similar vein, which was anticipated by the empirical work by Huffaker (1958), differential colonization-extinction dynamics of predator and prey can lead to their regional coexistence even if predators would overexploit prey in the absence of spatial processes (Vandermeer 1973, Zeigler 1977).

To devise a general framework for understanding the distribution and richness of species in communities, Hanski and Gyllenberg (1997) used the PD archetype, particularly the "core-satellite" hypothesis of metapopulation dynamics (Hanski 1982), to show that both the variation in species local abundance and regional occupancy, as well as the species-area relationship could be predicted. Likewise, Chave et al. (2002) devised a simulation approach to compare the predictions of PD with those of NT, showing that in many cases, the predictions of both approaches were fairly convergent (Fig. 2.3) and that the PD approach could capture many of the patterns in community structure observed in nature.

2.1.4 Mass Effects

Mass effects models typically include niche selection via both habitat heterogeneity and life-history differences and dispersal; one of the first ME model frameworks also included aspects of stochastic drift, whereby heterogeneity among patches could be generated either abiotically or biotically due to initial colonization or extinction. The "mass effects" (sensu Shmida and Wilson 1985; Pulliam 1988, Loreau and Mouquet 1999) in ME models refers to the case in which local competitive effects can be outweighed by dispersal effects. ME models relax the assumption made in PD models that inferior competitors are displaced instantaneously when a superior competitor colonizes a patch; this allows a locally inferior species to persist in a patch with a superior competitor so long as there is spatial variance in competitive abilities that results from abiotic heterogeneities or biotic variation (e.g., due to priority effects) (Levin 1974, Amarasekare and Nisbet 2001). For example, dispersal from more favorable areas can allow a species to coexist

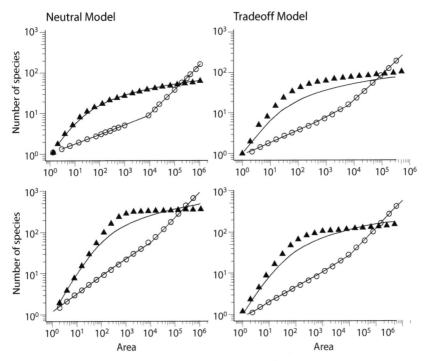

FIGURE 2.3. Models comparing species-area relations in neutral models (left panels) with models with a competition-fecundity trade-off (essentially the same as a colonization-competition trade-off, right panels) either without strong intraspecific density dependence (top panels) or with density dependence (lower panels). In each case, models with global dispersal (solid symbols) differ greatly from those with localized (nearest neighbor) dispersal (open symbols), but otherwise neutral models differ very little from patch dynamics models. Modified from Chave et al. 2002.

with a superior competitor in a locality where its local population growth is negative; if dispersal is very high, the outcome of competition could even be reversed when the inferior competitor with high dispersal from the metacommunity can swamp out a locally superior competitor (Pacala and Roughgarden 1982, Chesson 1985). Such swamping can lead to an overall reduction in local and regional diversity when dispersal rates are high because species that are more efficient dispersers can displace locally more competitive species through sheer volume (Loreau and Mouquet 1999, Amarasekare 2000b, Amarasekare and Nisbet 2001, Mouquet and Loreau 2003).

Many ME models focus on two-patch scenarios or treat space implicitly, but some aspects of ME models are sensitive to the spatial structure of habitats. This situation is illustrated by the model of Snyder and Chesson (2003), who modeled

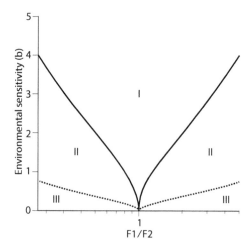

FIGURE 2.4. A mass effects model. Coexistence regions for environmental sensitivity (b) and the logarithm of adult fecundity ratio (F1/F2) for two species along a one-dimensional environmental gradient that varies as a sine function with a fixed amount of dispersal. In region I, coexistence is possible even if dispersal is global because average fitness differences are small (the ratio is close to 1), while in region II, coexistence is only possible if there is limited dispersal. Coexistence is not possible in region III. Dispersal along gradients facilitates coexistence at the metacommunity scale, but local dispersal does so more than global dispersal. Modified from Snyder and Chesson 2003.

how the coexistence of two species in a region could vary with the magnitude of environmental change that affects the relative fitness of the two species across a landscape as well as the ratio of average fitness and the nature of dispersal across the landscape for the two species (Fig. 2.4). If the average fitness is equal, any differences between the species across the landscape can allow coexistence. However, if the average fitness is not the same for the two species, then coexistence depends on the nature and degree of dispersal as well as on the magnitude of the fitness differences across habitats.

2.2 COMPARING AND CONTRASTING THE ARCHETYPES

Each of the archetypes above includes some aspects of each of the major drivers of community ecology (e.g., niche selection, dispersal, stochastic demography, and colonization-extinction processes) and have been used to generate alternative hypotheses that are often used to test the relative importance of each. Not surprisingly, different assumptions lead to different predictions (Table 2.1; Leibold 2011, Logue et al. 2011).

TABLE 2.1. Summary of Key Propositions and Predictions for Each
of the Four Archetypal Theories of Metacommunities

	Patch Dynamics	Mass Effects	Species Sorting	Neutral Theory
Propositions				
1. Dispersal affects colonists	X		X	X
2. Dispersal allows source-sink relations		X		X
3. Heterogeneity in dispersal is important	X	X		
4. Dispersal allows spatial feedbacks	X	X		
5. Interactions are direct and indirect	X	X	X	
6. Interactions depend on local environment	?	X	X	
7. Coexistence requires stabilizing effects in local communities	X		X	
8. Stochastic demography is important in allowing coexistence				X
9. Local extinctions are important	X			
Predictions				
1. Invasible local communities should be common	Y	N	N	Y
2. Fugitive species should exist in the metacommunity	Y	N	N	N
3. Community composition depends on spatial structure	Y	Y	N	Y
4. Community composition depends on environmental factors.	N/Y[a]	Y	Y	N
5. The ratio of local to regional diversity can be high.	N	Y	N	Y

Source: Modified from Table 2 in Chase et al. 2005 and from Leibold 2011.

[a]N/Y indicates that there are models without heterogeneity (i.e., competition-colonization models) and with heterogeneity (Horn and MacArthur 1972, Leibold and Loeuille 2015).

Although it is possible to compare and contrast the assumptions of these four archetypes, maybe this is not the best use of time since they likely never exist in "pure" form in natural metacommunities (Leibold et al. 2004, Chase et al. 2005, Leibold 2011, Logue et al. 2011). Instead, we point toward a conceptual synthesis that transcends the different archetypes and recognizes that important aspects of each archetype play different roles depending on the systems being considered, the organisms involved, and the scale of investigation. In the following sections, we move beyond the four archetypes by discussing approaches that (1) integrate aspects of more than one archetype into a more general approach (Sec. 2.3) and (2) examine complexities not included in the simple archetype, using the SS framework to show how dynamic community assembly in a metacommunity (Sec. 2.4) and frequency dependence (Sec. 2.5) can fundamentally alter the predicted patterns, making it more difficult to differentiate among the predominant mechanisms.

We can start by trying to understand the major factors that might affect the relative importance of each of the archetypes (Fig. 2.5). These seem to fall along two different axes (modified from Leibold 2011). First, the degree of heterogeneity among localities within the metacommunity will influence the strength of niche selection among those localities; this axis plays a key role in SS and ME models but is not considered in NT and most PD models. Second, the degree of connectivity among habitats and rates of dispersal of organisms among those habitats define the influence of dispersal for a given metacommunity. Because niche selection is fundamental in both SS and ME archetypes but ME also includes excess dispersal, rates of dispersal can modulate the relative importance of SS versus ME processes. If niche selection is strong and the exchange rates of individuals among patches are of the same order of magnitude or higher than the death rates of individuals in those patches, species can persist in a patch where they otherwise could not maintain positive growth due to ME, and this archetype's predictions will be appropriate; when that exchange rate is lower, SS will play an overriding role. If species are able to direct their dispersal, however, ME may be less likely, and patterns indistinguishable from SS can emerge even if dispersal is high because species would immigrate only into patches where they are not at a strong disadvantage relative to resident dominants (thus, the oval representing SS overlaps with that of ME in Fig. 2.5). Likewise, PDs are more likely to be important when the arrival rate of colonists into patches is on the same order of magnitude as or less than the local extinction rate of the competitively dominant species; if extinction rates of competitively dominant species are rare, the proportion of unoccupied patches will become too low for the competitively inferior but more adept colonizers to persist, and SS will predominate. Finally, the degree of stochastic drift in extinction and colonization rates regardless of species identity can vary between metacommunities, and this variation mediates the importance of NT relative to the other archetypes. For example, stochastic ecological drift is more likely to play a stronger role in smaller communities (Orrock and Fletcher 2005, Orrock and Watling 2010), where demographic stochasticity can have a proportionately larger effect on the outcome.

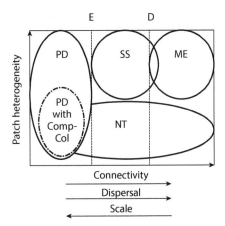

FIGURE 2.5. The four archetypes of metacommunity ecology in relation to the amount of inter-patch habitat heterogeneity and to connectivity. Patch heterogeneity refers to the degree of inter-patch differences in environmental conditions that differentially favor different species in different patches. To the left of line E, populations are dispersal limited and extinctions are high compared to colonizations. To the right of line D there is dispersal excess and immigration is high compared with local death rates. In between there is dispersal sufficiency as described in Chapter 1. A separate oval has been drawn for PD models that focus on competition-colonization (Comp-Col) models in the absence of patch heterogeneity to highlight the fact that most PD models are of this type. The oval for SS has purposefully been drawn so as to extend to the right of connectivity equal to D and overlap with ME to highlight that high dispersal can still produce patterns and processes similar to SS if dispersal is adaptive. PD = patch dynamic, SS = species sorting, ME = mass effects, NM = neutral model. See Table 2.1 for a description of the assumptions and critical processes that apply to each archetype. Modified from Leibold 2011.

2.3 EXPLORING METACOMMUNITY ECOLOGY
AT THE INTERFACE OF THE ARCHETYPES

Although the development of an "uber"-metacommunity model that seamlessly integrates all of the important processes has not yet been accomplished, several theoretical formulations have been developed that explore the interface between two (or more) of the archetypes and the relative importance of niche selection in response to habitat heterogeneity, dispersal, and stochastic drift, as presented in Figure 2.5. We specifically focus on three fairly well-developed "interface" models, each of which pairs SS with one of the other archetypes: SS and NT, SS and ME, and SS and PD.

2.3.1 The SS-NT Interface

It has been increasingly recognized that rather than being polar opposites, as they might initially seem, the essence of NT and SS in fact represent two parts of a more general continuum for metacommunities (Gravel et al. 2006, Leibold and McPeek 2006, Adler et al. 2007, Chase and Myers 2011). We highlight Adler et al.'s (2007) approach, which is based on Chesson's (2000) formalization of a generalized co-existence theory. Chesson (2000) argued that coexistence would result from the balance of "stabilizing" and "equalizing" factors that work in opposition; Adler et al. (2007) associated stabilizing and equalizing processes with SS and NT, re-spectively. Stabilizing factors produce frequency dependence in the relative fitness of species; the magnitude of such frequency dependence depends on niche dif-ferentiation, and the sign of the feedback (stabilizing or destabilizing) depends on the relative magnitudes of net intra- versus interspecific feedback. The stabilizing effect is greatest when species intraspecific limitation effects are large and nega-tive relative to their interspecific effects. Importantly, this is simply an alternative formulation of the same general mechanisms for coexistence that emerge in the standard Lotka-Volterra competition equations, as well as in consumer-resource modules (Chase and Leibold 2003, Letten et al. 2017). However, simply having intraspecific effects greater than interspecific effects does not guarantee coexis-tence. If one species' overall fitness is so high that it overwhelms the other species, the per capita effects of one species on another becomes irrelevant for coexistence (Chesson 2000). This too is borne out of standard Lotka-Volterra (Vandermeer 1975) and consumer-resource models (Chase and Leibold 2003, Letten et al. 2017) of coexistence. Thus, in addition to stabilizing factors that result from dif-ferential usage of limiting factors (e.g., niche differences), equalizing factors that reduce the differences in the fitnesses of species are also required for coexistence (see Fig. 2.4 for an example). While equalizing and stabilizing factors are often considered in the context of local coexistence, Chesson's generalized framework is agnostic about the spatial scale on which stabilizing and equalizing processes (and associated trade-offs) take place and thus is relevant for coexistence at any spatial scale.

Adler et al.'s (2007) important insight was that the equalizing processes inher-ent to NT are actually key components of traditional coexistence theory, usually considered to be the realm of niche theory (incorporating aspects of SS, PD, and ME), and that these processes act in concert with stabilizing processes more tra-ditionally associated with niche-based theories. The NT is thus at one end of the extreme in coexistence theory: processes that are equalizing but not stabilizing. Coexistence in such a scenario is quite fragile, and any deviations in equalization (fitness differences) without a concomitant shift in stabilization (niche differences) will lead to the fairly rapid exclusion of all but the species with the highest fitness; hence Hubbell (2006)'s search for mechanisms that favor equalization.

An important insight that emerges from this perspective is the inherent tension between equalizing and stabilizing processes. Increases in the relative importance of equalizing processes (species becoming more similar) decreases the critical level of stabilizing processes (degree of niche differentiation) needed for species to coexist (and vice versa). Thus, in highly diverse systems such as tropical rainforests and coral reefs, as long as equalizing processes are strong, stabilizing processes need not be very strong, and the system can appear relatively neutral even though some degree of SS is operating. Indeed, Zhou and Zhang (2008) devised a "nearly neutral" model that largely anticipates these results. An important adjunct to the dichotomy between stabilizing and equalizing effects, however, is that a species' functional traits that influence stabilizing processes often also influence equalizing processes, complicating the strict separation of components that lead to coexistence in terms of ecological traits (Leibold 1998, Letten et al. 2017). For example, traits that influence a species' fitness in a given environment (equalizing processes) also influence its interactions with the resources that influence its frequency-dependent influence on other species (stabilizing processes). It is also likely that evolutionary processes can influence the relative importance of stabilizing versus equalizing effects (Hubbell 2006), which we will explore in more detail in Chapter 8.

Finally, it is worth noting that most of the ways we understand the various archetypes make the most sense when we consider species pairs one at a time. The various archetypes imply that the mechanisms apply to entire groups (more than pairs) of species. This, however, is only a convenient approximation, and it could be that different subsets of a community conform to different mechanisms of coexistence. For example, the ecological drift inherent to the NT is a more important process in smaller, relative to larger, communities (Orrock and Fletcher 2005, Orrock and Watling 2010) and may be more likely to act on the rarer members of a given metacommunity, whereas the more common members might be more likely to conform to SS. Likewise, stabilizing processes might be important in differentiating among functionally or phylogenetically disparate species, whereas in the same system, functionally or phylogenetically more similar species might coexist largely by equalizing mechanisms (Siepielski et al. 2010, but see Kelly et al. 2008 for a counterexample; see Chap. 6 for more discussion of this idea).

2.3.2 The SS-PD Interface

Basic PD models assume environmental homogeneity, whereby coexistence due to niche selection can occur only via spatial mechanisms, such as competition-colonization and competition-fecundity trade-offs, and/or through core-satellite dynamics. These all involve dispersal mechanisms and coexistence as a result of locally unstable (dynamic) mechanisms. SS models, on the other hand, assume that niche selection results from different species responses to environmental heterogeneity and that dispersal has minimal effects (other than to allow the assembly process to move forward). As a first step toward combining spatial coexistence mediated by dispersal trade-offs in PD with coexistence mediated by habitat heterogeneity trade-offs in SS, Horn and MacArthur (1972) devised a model that expanded the PD formulation by Levins and Culver (1971) to examine how patch heterogeneity and differential responses of species to that heterogeneity could alter coexistence among species at a regional scale. They found that coexistence (or lack thereof) in SS due to heterogeneity could be overridden by dispersal under certain conditions and that coexistence (or lack thereof) in PD due to dispersal trade-offs could be overridden by environmental heterogeneity. Similar model formulations incorporating habitat heterogeneity with dispersal heterogeneity have been examined for a variety of questions (Holt 1997, Shurin et al. 2004, Gross 2008, Nagelkerke and Menken 2013).

Leibold and Loeuille (2015) expanded upon the Horn and MacArthur (1972) formulation to evaluate how the SS-PD interface influences the relative importance of environmental variation in community composition. Overall, the results showed that the strength of environmental filtering (SS) decreased at higher extinction or lower colonization rates due to both increases in the prevalence of empty patches and decreases in the frequency patches inhabited by the competitive dominant (Fig. 2.6).

2.3.3 The SS-ME Interface

Most ME models include some degree of niche selection due to habitat heterogeneity and ask how this can interact with the dispersal of individuals who might migrate into suboptimal habitats. The model by Mouquet and Loreau (2003) is a good example of the application of ME principles at the interface of an SS archetype. Mouquet and Loreau considered environmentally heterogeneous local communities within a broader metacommunity. Each local community consist of numerous "micropatches" that can be occupied by a single individual of any of the species. Using a simulation model framework, Mouquet and Loreau assumed that niche selection favors different species in each of many heterogeneous habitat types, such that without dispersal (starting with a model in which all species have access to their favored habitat), the model collapses on a simple SS approach.

The results from the Mouquet and Loreau (2003) simulation provide highly testable predictions about how a variety of community-level variables change with

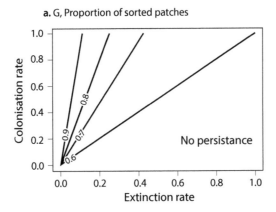

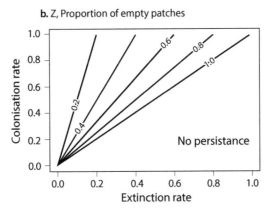

FIGURE 2.6. Proportion of patches that are occupied by competitively dominant species (a) and proportion of patches that are empty (b) as a function of the extinction and colonization rates (identical across species) in a metacommunity with two patch types that each favor a distinct species. These results indicate that high species sorting occurs when colonization is high and extinctions low and declines with extinction and is enhanced by high colonization. Modified from Leibold and Loeuille 2015.

changes in the rates of dispersal among localities within the metacommunity. Most well-known and tested is how local (α-) diversity and regional (γ-) diversity, as well as the variation in species composition from site to site (i.e., β-diversity), should vary as the rates of dispersal within a metacommunity vary (Fig. 2.7). Typically, slight increases in dispersal rate will increase α-diversity because ME allow species to coexist in localities where they are inferior competitors. Higher rates of dispersal, however, can lead to declines in α-diversity when the regionally dominant competitor homogenizes the metacommunity, driving weaker regional

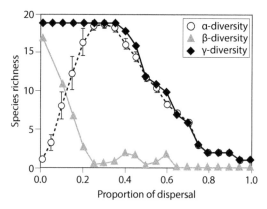

FIGURE 2.7. α- (local), γ- (regional or metacommunity-wide) diversity, and β- diversity (a measure of spatial differentiation or turnover) as a function of dispersal (proportion of individuals emigrating from natal patches) in a metacommunity with patches that each favor a distinct species. In the absence of dispersal each patch is occupied by its local dominant. As dispersal increases, β-diversity decreases as each patch is increasingly occupied by local subdominant species immigrating from other patches. This initially increases α-local diversity until it approaches γ-diversity. Further increases in dispersal lead to reduced γ-diversity because some species do better at the regional level due to average fitness differences. This constrains α-local diversity which decreases along with the decrease in γ-diversity, producing a hump-shaped relationship between α-diversity and dispersal. Modified from Mouquet and Loreau 2003.

competitors extinct (Mouquet and Loreau 2003). At the regional scale, γ- and β-diversity are maximized with the lowest rates of dispersal, when the role of SS is highest, and decline monotonically as dispersal increases; β-diversity declines with moderate increases in dispersal as ME-induced coexistence homogenizes localities, and then γ-diversity declines with higher rates of dispersal as regionally competitive (dispersive) species dominate and drive locally competitive species extinct from the metacommunity.

2.4 IS THERE AN "UBERMODEL"?

If the various models can be tweaked to show most of the interfaces among them, it seems it should be possible to develop a comprehensive model that includes all of the aspects that could be important in metacommunities. This is, however, more difficult than one might think, the reason being that different models are formulated in very different ways. In particular the PD models are formulated as "occupancy models," in which the presence or absence of species is modeled without regard to the abundance of those organisms in any communities in which they occur, whereas abundance is fundamental to the ways the source-sink relationships occur in ME models and demographic stochasticity in NT models.

Shoemaker and Melbourne (2016) took the approach of reformulating a range of metacommunity models within the framework for modeling spatial coexistence that had been developed by Chesson and colleagues (Chesson 2000, Snyder and Chesson 2003, Snyder et al. 2005). This framework partitions the ability of species to increase when they are rare in the metacommunity into components that can be attributed to the mutually exclusive categories of (1) nonspatial mechanisms, (2) storage effects, (3) fitness-density covariances, and (4) nonlinear competitive variances. The nonspatial mechanism quantifies the fitness in a landscape if spatial structure were absent. The spatial storage effect accounts for the ability of species to benefit from different high-quality habitats without losing individuals to poor-quality habitats. Fitness-density covariance quantifies the correlation between the fitness of a species and its population size within patches and accounts for the effects of variable population size such as might happen if there are disturbances of population fluctuations within patches. Finally, nonlinear competitive variance is a mechanism by which spatial variation in competition causes net increases or decreases in overall fitness (this mechanism has yet to be included in the metacommunity theory we have so far discussed).

Figure 2.8 shows representative results from Shoemaker and Melbourne (2016) for simple cases that illustrate the pure NT, pure SS, pure competition-colonization as an instantiation of PD, a representative ME model, a hybrid ME model (wherein the two species differ in dispersal rate), and a hybrid PD-SS model (wherein both species have preferred habitat space but they also differ in colonization rates following disturbances). A couple of interesting findings emerge. First, the overall strength of stabilizing coexistence (the magnitude of both species to increase when rare overall) is high and symmetrical for the SS case and zero for the NT case, as we might have expected. Second, in the ME scenario, in which there is maladaptive dispersal, the fitness-density covariance is reduced relative to the SS case (with a similar but small decrease in the storage effect). Here the effect is not particularly strong but would increase as the dispersal rate increases. Third, the competition-colonization scenario shows that stabilizing coexistence strength is much weaker than for the other cases and that the contributions of the different factors are different for the better colonizer than for the better competitor. Shoemaker and Melbourne (2016) also examined the effects of stochasticity resulting from a small number of patches and stochastic demographic, dispersal, and extinction processes and found that they tended to boost overall coexistence by enhancing most of the individual components in somewhat complex ways.

While this study shows that it is possible to find a common framework for looking more comprehensively at coexistence at the metacommunity scale, it tells us

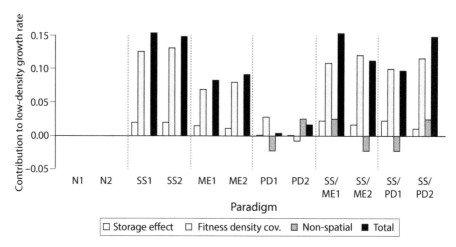

FIGURE 2.8. The contribution to total low-density growth rate from each of the three spatial coexistence mechanisms in different metacommunity archetypes (nonlinear effects are absent and not shown) shown in various shades of gray as well as their sum, the total low-density growth rate (shown in black) for two species in a metacommunity. N1 and N2 show results for two species interacting neutrally and show no contribution to coexistence (as would be expected from a neutral model). SS1 and SS2 show two species coexisting by specialization to two distinct stable habitat patch types and show that spatial storage and fitness-density differences are important (as expected). ME1 and ME2 show two species with distinct habitat preferences (as in SS1 and SS2) but with higher maladaptive dispersal and show that fitness-density covariance is lowered for both species with a net decrease in coexistence strength. PD1 and PD2 show two species coexisting wherein one (Species 1) is the better colonizer and the other (Species 2) the better competitor. N = neutral, SS = species sorting, ME = mass effects, PD = patch dynamic. Modified from Shoemaker and Melbourne 2016.

less about patterns at the patch scale and about interactions in communities that have more than two species. Fournier et al. (2017) took a different approach to explore the model frameworks in a more comprehensive manner. They did so by developing a general model that includes elements of all four archetypes in an occupancy model (albeit with some limitation on just how this happens). In addition, this approach allowed for the analysis of much more complex and realistic landscapes of patches that varied in their connectance and the spatial configuration of their heterogeneity (Fig. 2.9).

Using dynamic simulations, they then compared how the full model, including all of the processes, compares with each of the four simplified archetypes (Fig. 2.10). In all of the simulated landscape configurations they used, they found that the neutral archetype (NT) was consistently the least likely model to match the observations, whereas the other three archetypes (SS, PD, ME) contributed significantly more for at least one or more of the landscapes. SS and PD

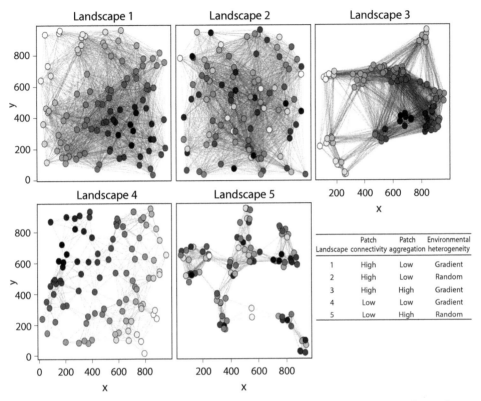

FIGURE 2.9. Five landscapes (on an *x-y* spatial plane) with different patches, aggregation, and connectivity and different distribution of environmental conditions. In each panel, patches have local environmental conditions that vary along and environmental gradient (indicated by their degree of shading). Additionally, depending on overall connectivity and the physical arrangement of patches, the actual dispersal connections among patches vary as indicated by lines connecting patches. Fournier et al. (2016) studied five different configurations as shown here that varied in connectivity, aggregation, and spatial autocorrelation in environmental conditions as highlighted in the inserted table. Modified from Fournier et al. 2017.

contributed most, especially SS in the first three landscapes, where connectivity was high.

While promising, this more complete exploration of the full complement of metacommunity processes still has a long way to go. Currently it is only implemented in patches using an occupancy framework, and this may constrain how ME are modeled in the landscape. Additionally, the approach on trait distributions in the full model is crude, and it is not quite clear how more realistic trait distributions might alter the outcomes. It nevertheless shows a way to simultaneously study the elements of all the core assumptions of the divergent metacommunity archetypes.

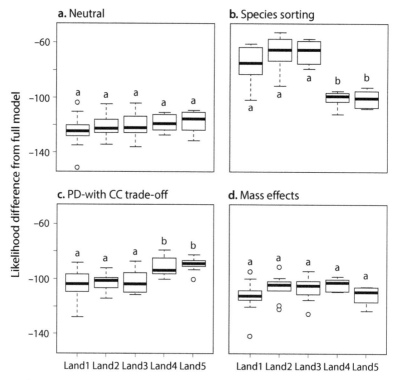

FIGURE 2.10. Likelihood differences between the probability distributions of the full model and each of the four archetypes for each of the five landscapes shown in Figure 2.9. In each case the full (complex) model was compared with modifications which "pretended" that each of the different archetypes was at play. The likelihood difference is plotted to compare how much the simplified model deviated from the full model (note that the values are negative so that differences of higher magnitude are plotted at lower position on the graph). The differences depend on the model (e.g., neutral theory is generally less likely to fit the full model than other archetypes) and landscape (e.g., species sorting is more likely to fit the full model in landscapes 1–3 than in landscapes 4 and 5). Modified from Fournier et al. 2017.

2.5 THE INFLUENCE OF LOCAL COEXISTENCE MECHANISMS

Much of the work on metacommunity theory has largely focused on how scale, heterogeneity, and dispersal, along with various aspects of species traits and stochasticity, combine to influence patterns of species coexistence and diversity at local scales. However, we have not yet really discussed the processes that influence coexistence at finer (local) scales in the absence of these metacommunity processes. This is largely deliberate because our interest is in exposing how taking a metacommunity approach can enhance our understanding of some of the

larger-scale patterns. Additionally, theoretical investigations that examine the role
of spatial processes and place a premium on dispersal and stochasticity (PD, NT,
ME) tend to make highly simplistic assumptions about the regulation of local diversity in deference to understanding the role of spatial processes; specifically,
they usually assume that a small number of species (usually one or two) can coexist in any given locality in the absence of metacommunity processes.

Clearly, these spatially based metacommunity models do not capture the reality of local coexistence. In fact, there are a multitude of mechanisms whereby species can coexist locally, including resource partitioning (MacArthur and Levins
1967, Schoener 1974, Tilman 1982), differential response to extrinsic temporal fluctuations (Levin and Paine 1974, Hastings 1980, Chesson and Warner 1981), intrinsic temporal fluctuations (Armstrong and McGehee 1980, Huisman and Weissing
1999), differential shared enemies (Paine 1966, Holt et al. 1994, Leibold 1996),
differential use of mutualists, and many others. While dispersal and stochasticity
can certainly influence the likelihood that this local coexistence can be realized,
a more complete theory of metacommunities will be needed that can incorporate
local and spatial coexistence mechanisms in a more complete, realistic way.

2.6 CONCLUSIONS

The theories of metacommunities have come a long way since Leibold et al. (2004)
synthesized the four archetypes and discussed their similarities and differences
in assumptions and predictions. While a fully satisfying ubermodel of metacommunities may still be a ways off, many insights can be gained by recognizing that
the four archetypes should not be viewed as categorical, but rather as representing
a continuum of approaches to represent spatial processes in an inherently complex
world. Every metacommunity likely experiences aspects of niche selection, stochastic drift, and dispersal and thus contains elements of SS, NT, PD, and ME. We
need to devise a more synthetic way to identify and understand when the different
processes play a stronger or weaker role in the structuring of a given community.

Despite the emerging synthesis of current metacommunity theories, there are
still many limitations to this effort, which we hope to identify in this book with
the goal of inspiring future work. For example, the majority of our discussion of
the metacommunities in this chapter, as well as the empirical tests of pattern and
process in the following three chapters, focuses on metacommunities where
interspecific competition for limiting resources is the primary interaction among
species. However, just as the last two decades have seen a move away from competition being the primary determinant of coexistence at local scales, metacommunity
ecology also needs to move away from a focus on competition to more seamlessly

integrate other types of species interactions, such as natural enemies and mutualists. Indeed, some of this is already taking place, particularly in the context of enemies in food webs, and to a lesser degree with mutualisms, which we discuss in more detail in Chapter 11. Likewise, there has been increased recognition that small-scale microevolutionary responses that occur on ecological timescales (eco-evolutionary dynamics), as well as large-scale macroevolutionary drivers, can strongly influence our theoretical expectations of coexistence and diversity in metacommunities, which we discuss in Chapters 8 and 9, respectively. Finally, the theories that we have discussed in this chapter have largely focused on patterns of species coexistence, with less emphasis on macroecological and biogeographic patterns, which we discuss in Chapter 10, or ecosystem-level patterns, which we discuss in Chapters 12 and 13.

CHAPTER THREE

Processes in Metacommunities

Prospectus

1. Each of the core metacommunity archetypes places different emphasis on which core mechanisms (including the role of dispersal, habitat heterogeneity, and ecological drift) are at play (or not). Evidence from experimental and observational studies provides evidence for almost all these.

2. There is substantial experimental evidence that locally coexisting species often show stabilizing and equalizing niche relations that allow them to coexist locally as a result of niche differentiation. Nevertheless, there are notable exceptions in which no such stabilizing dynamics have been found despite careful experiments, suggesting that locally neutral co-occurrence may also happen, albeit perhaps less frequently.

3. Stochastic extinctions are difficult to study because it is not always easy to identify the consistency and causes of extinctions when they occur. Although more careful quantitative studies are needed, such extinctions have been identified in a number of studies.

4. Demographic stochasticity between competing species has frequently been found in experimental studies, especially during the early phases of community and population growth. These can lead to long-term deterministic differences that result from such stochastic effects, but it is less clear how important long-term demographic stochasticity in larger populations (as assumed by NT) is.

5. Dispersal limitation has been repeatedly shown, especially in terrestrial plants, but also in other systems. However, effect sizes of dispersal on the size of the population of dispersers have generally been small, and many species show no evidence of dispersal limitation at the spatial scales that have been studied.

6. Variation in dispersal limitation among species has not been particularly well studied, but limited evidence suggests that such variation may be associated with different aspects of the ecology of species, including local density and regional occupancy, habitat specialization, and demographic and life-history traits.

7. Trade-offs between ecological traits, including dispersal and other traits that matter at local scales, are probably important but remain understudied. More extensive data indicates that trade-offs between traits that affect performance in different habitats (such as those associated with SS) are common and can lead to environmental segregation.

8. Although present, evidence for ME, or source-sink relations, among distinct habitats is surprisingly scarce.

9. Regional and local effects should not be viewed as alternatives but rather as an interacting set of factors that regulates metacommunity assembly at different spatiotemporal scales.

The large majority of empirical tests in metacommunity ecology have focused on testing predicted patterns that emerge, such as the shape of the species-abundance distribution or the degree to which variation in species composition among localities within a metacommunity correlates with environmental or spatial distances (Logue et al. 2011). We will describe these in depth in Chapters 4 and 5. However, the general rule that correlations do not necessarily imply causation is particularly pertinent in metacommunity ecology, especially because many of the patterns predicted by the metacommunity archetypes are more or less indistinguishable from one another without further information. Before we embark on extended discussions about patterns, it is useful to dissect and evaluate the core underlying assumptions and processes that occur within a metacommunity.

Each of the four archetypes of metacommunity structure described in Chapter 2 make different assumptions about the importance of two types of processes that act on local scales (Table 2.1)—(1) resource utilization and response and (2) demographic stochasticity—and two types of processes that act on regional scales—(3) environmental heterogeneity and (4) dispersal rates. At one end of the conceptual continuum, the traditional niche-based approach (SS) emphasizes local resource utilization and response and regional environmental heterogeneity, whereas the neutral approach (NT) emphasizes local demographic stochasticity and regional dispersal; the ME and PD models are intermediate. It is important to note that for the sake of simplicity, here we are primarily discussing processes that act within the context of competitive interactions within trophic levels, and thus we ignore many important types of processes (e.g., food-web and mutualistic effects) that will be discussed in later chapters. The processes that we overview here strongly match Vellend's (2010, 2016) synthesis, which separates the primary factors of importance into those dealing with selection (which we have separated into local resource utilization and regional environmental heterogeneity), dispersal, and drift; Vellend also discussed speciation as essential for "priming the pump" of biodiversity, and we deal with this issue in Chapter 9. Vellend (2016) also devoted substantial effort to evaluating the premises and predictions of his synthesis, an evaluation that overlaps somewhat with our review; we, however, couch these more specifically within the framework of metacommunity ecology.

We separate our discussion into processes that occur primarily at the local scale (resource utilization and response, demographic stochasticity) and those that occur primarily at the regional scale (environmental heterogeneity, dispersal). Throughout, we will emphasize the critical nature of the interactions between these processes and scales; that is, local-scale processes are modified by regional-scale processes (and vice versa), necessitating the consideration of both scales simultaneously in order to fully understand the system. We next overview the critical assumptions of each of the four metacommunity archetypes for each of these processes.

Resource utilization and response. NT assumes that all species are identical with respect to their interactions with each other and with resources and other limiting factors (hence "neutrality"), although "nearly neutral" models relax this somewhat. PD, SS, and ME allow for species to differ in their local resource utilization and interspecific interactions (i.e., relative competitive abilities depending possibly on the local environment).

Demographic stochasticity and drift. NT assumes that individual birth-death rates are completely stochastic in any given time and place, whereas PD assumes that population-level extinctions can be stochastic and, if so, that this property inherently emerges from individual-level stochasticity. SS and ME assume that birth-death rates are functions of the local environmental conditions and largely deterministic.

Environmental heterogeneity (in space and time). NT assumes that all species respond as if the environment is completely homogeneous. Simple PD models also assume homogenous environments, but more complex "interface" models with PD allow for spatial heterogeneity in response to the environment. SS and ME models explicitly assume that species respond strongly to heterogeneous habitats and their growth and that competitive abilities are greatly influenced by these heterogeneities.

Dispersal. There are three qualitatively distinct ways by which dispersal can influence metacommunity dynamics—dispersal limitation, dispersal sufficiency, and dispersal surplus. NT and PD assume that dispersal is limiting so that species are not in every locality that they could otherwise persist. PD also allows for interspecific differences in species' abilities to colonize and establish in unoccupied patches. SS assumes dispersal sufficiency, such that each species has ample opportunity to establish in any locality where it can have positive population growth. ME assumes that dispersal rates are in surplus, so that species can persist in suboptimal habitats so long as there is sufficient immigration from nearby localities where population growth is high.

3.1 RESOURCE UTILIZATION AND RESPONSE

As we described in Chapter 1, the history of research on interspecific competitive interactions and coexistence based on trade-offs in resource utilization is long and nuanced and is beyond the scope of this book. Indeed, much of the foundation for modern (local-scale) community ecology is based on the principle that species must differ in some set of traits that allows them to stably coexist (Grinnell 1917; Lotka 1925; Elton 1927; Gause 1936; Hutchinson 1957, 1959; MacArthur 1958, 1972; MacArthur and Levins 1967; Levin 1970). More recently, this concept has been formalized by stating that for two or more species to coexist locally, they must have paired trade-offs in resource utilization and response (Tilman 1982, Leibold 1995, Chase and Leibold 2003). Said another way, local species coexistence is thought to be only possible when stabilizing mechanisms act to reduce *relative* fitness differences among species (e.g., trade-offs in resource use), but the strength of stabilizing mechanisms necessary for coexistence also depends on the *average* fitness differences of the species (i.e., equalizing mechanisms; Chesson 2000, Adler et al. 2007, HilleRisLambers et al. 2012, Letten et al. 2017). This formulation contrasts sharply with the assumptions of NT, which posits no differences among species in any fitness parameters (Hubbell 2001, Rosindell et al. 2011).

Given that niche differences and trade-offs have long been cornerstone assumptions for local species coexistence, one might think that there would be quite strong empirical evidence for this. And indeed, there is a huge volume of literature that has shown with experiments and strong observational approaches that coexisting species interact strongly and often differ in their abundance and propensity for exclusion when niche axes vary (Connell 1983, Schoener 1983, Gurevitch et al. 1992). However, as pointed out by Siepielski and McPeek (2010; see also HilleRisLambers et al. 2012), there are in fact only a very few studies that were explicitly designed to be able to tease apart the influence of differential resource utilization and response on local species coexistence. Here we are only considering cases in which stabilizing processes are more likely to be local, rather than regional, phenomena due to differential use of spatial habitat heterogeneity. However, the distinction between resource-utilization factors that are local and differential utilization of resources across heterogeneous habitats in a region can be a subjective one. And so it is important to keep in mind that differential use of resources can occur at many scales (space and time), some of which we discuss in this section and some of which we discuss below in the context of habitat heterogeneity.

Table 3.1 presents an inexhaustive list of studies that have sought, and found, strong evidence for local coexistence resulting from stabilizing processes. These studies used a number of different approaches, including directly measuring the traits of species and relating them to environmental conditions and their relative abundance and manipulating the densities of species themselves. Many other studies have taken different, typically more inferential approaches to suggest similar patterns. For example, despite its status as a bastion for NT, the high diversity of

TABLE 3.1. Some Empirical Examples in Support of Local Coexistence Due to Differential Resources Utilizations from a Variety of Different Approaches

Study System	Approach	Result	Reference
Rocky marine intertidal invertebrates (Washington State,)	Parameterized the neutral model from observations; experimentally removed the dominant species and compared predicted with observed abundances.	Neutral model poorly predicted the abundance of specific species in the experimentally altered plots.	Wootton 2005
Prairie plants (Minnesota and Kansas)	Measured the level to which monocultures of a species could reduce the primary limiting resource (nitrogen), taken as its competitive ranking (R^*); compared competitive ranking with abundance ranking; also compared with cases in which systems were disturbed or nitrogen was added.	Competitive ranks strongly correlated with abundance patterns, as well as responses to disturbance and nitrogen addition, thus emphasizing the importance of resource utilizations.	Harpole and Tilman 2006
Picocyanobacteria from the Baltic Sea	Pigments of different species create color variation that allows species to trade off their ability to photosynthesize in different types of light conditions; compared parameterized models with experimental results.	Species coexistence and ability to increase when rare fit the model predictions. Species coexistence changed when the color of light was altered.	Stomp et al. 2004
Annual plants in the desert (Arizona)	Long-term monitoring of the demographic parameters of plants (germination, fecundity) across years that differ in precipitation; combined with observations of the functional trade-off between species in their relative growth rates and water-use efficiency.	Demographic rates of species in response to variation in environmental conditions (precipitation) were strongly tied to the variation among species in their trade-off along the growth rate/water use continuum.	Angert et al. 2009

(continued)

TABLE 3.1. (*Continued*)

Study System	Approach	Result	Reference
Annual plants on serpentine soils (California)	Experimentally "removed" niche differences by normalizing seed numbers among years that was independent of their relative abundances in a given year; compared these with control treatments in which niche differences were left intact to create the numbers of seeds produced.	Treatments in which niches were experimentally removed lost a large amount of diversity and became dominated by a few species.	Levine and Hille-RisLambers 2009

tree species on Barro Colorado Island (BCI) in Panama appears to have a variety of coexistence mechanisms operating, including negative frequency dependence, whereby species are more strongly limited by conspecifics than by heterospecific neighbors (Uriarte et al. 2004, Comita et al. 2010, Mangan et al. 2010); differential responses of species to light availability (which is furthermore correlated with their relative abundance; Ruger et al. 2011); and significant correlations between tree species distributions and soil nutrients (John et al. 2007).

Not all studies that have sought explanations of local coexistence based on differential resource utilization have found them however, including those that suggest that some species may co-occur even though they have no differences in their resource utilization. Although there are other possible explanations (e.g., they may not be at or close to equilibrium), they appear neutral with respect to their demographic traits, which can lead to ecological drift and lottery-like dynamics, which we discuss next.

3.2 EVIDENCE FOR DEMOGRAPHIC STOCHASTICITY AND ECOLOGICAL DRIFT

Despite considerable theoretical interest in the influence of demographic stochasticity on coexistence and ecological drift, direct evidence for these processes occurring in natural communities is rare. Instead, many ecologists assume that such stochasticity plays a role if other factors (e.g., environmental conditions) explain little of the variation observed in species abundances (described in more detail in Chap. 4). The four demographic properties of populations—birth rates, death rates, immigration rates, and emigration rates—are often considered stochastic processes (Vellend et al. 2014), even though they need not be neutral and though they also have a strong deterministic component that is a function of the species' fitness in a given environment. Before we can describe the evidence for demographic stochasticity and ecological drift as a process in metacommunities, we must first be very clear about a few points.

First, we wish to clearly separate demographic stochasticity from environmental stochasticity. It is often convenient to think about environmental variation as a stochastic process. For example, precipitation and temperature are often modeled as stochastic processes within some probabilistic boundaries, and this environmental stochasticity can strongly influence the demographic responses of species (Boyce et al. 2006) and their likelihood of coexistence (Chesson and Warner 1981, Chesson 2000). Importantly, however, while the environment is assumed to be stochastic, the responses of the species to those environmental parameters is deterministic. Demographic stochasticity, on the other hand, ascribes the probabilities to species births and deaths, which have a random component in any moment in time.

Second, there is the question, perhaps more in the philosophical realm, as to whether there is truly ever any stochasticity—that random probabilities are simply assigned to a process that is deterministic and could be known if more information were available (Clark 2009). For example, we might be able to predict exactly which trees will blow down in a forest based on their locations, soil conditions, size, etc. However, we usually do not attempt this degree of precision and simply assign a stochastic probability that any tree in the forest will be blown down by wind. Likewise, with perfect information, we might be able to predict exactly which individual juvenile fish will settle in a spot on a coral reef when a space opens up, although we usually model this as a stochastic "lottery." While this question provides for interesting discussion, most ecologists agree that there is value to depicting some demographic processes as having a stochastic element, such as births and deaths (Rosindell et al. 2011, Vellend et al. 2014).

Third, demographic stochasticity leading to ecological drift does not necessarily imply neutrality in species traits per se. Stochasticity can act on individuals' birth and death rates because they happen as discrete events with given probabilities, potentially leading to drift, even if the demographic rates of those species are otherwise not neutral. Just as in population genetics, ecological drift is much more likely to occur when community size is small (Fukami 2004a, Orrock and Watling 2010) or selection is weak (Chase 2003a, 2007, 2010) or both, but the concepts of stochasticity and drift need not be synonymous with neutrality.

Finally, demographic stochasticity need not lead to ecological drift, in the sense of randomly varying species frequencies, at the metacommunity scale. For example, stochastic colonization-extinction dynamics are included in many metacommunity models (e.g., PD), but this will not necessarily lead to ecological drift in the metacommunity unless community size is reduced below a certain threshold (Yu et al. 2001).

In metacommunities, stochasticity can emerge at the scale of individuals—which we call demographic stochasticity; here, subtle variations in birth and death rates can lead to drift in the relative abundances of individuals through space and time. Additionally, stochasticity can emerge at the patch scale—which we will call *patch stochasticity*—when, for example, a species has finite probability of going extinct from a patch in any period of time, as well as having a finite probability of colonizing an unoccupied patch. These are usually modeled as stochastic extinction and colonization rates, for example, in the equilbrium theory of island biogeography (MacArthur and Wilson 1967), metapopulation models (Levins 1969), and PD metacommunity models. Although both types of stochasticity are more appropriately viewed as a continuum rather than as discrete categories, we deal with each sequentially below. However, it is important to remember that stochasticity per se need not imply either neutrality or a lack of determinism; for example, in the equilibrium theory of island biogeography, larger islands are assumed to have lower extinction rates than smaller islands because population sizes are larger (MacArthur and Wilson 1967).

3.2.1 Stochasticity Leading to Ecological Drift

There are a few examples of species appearing to co-occur without any obvious niche differences; that is, they appear to be "neutral." For example, Shinen and Navarrete (2014) performed a variety of experiments on two species of barnacles living on the rocky shores of Chile, including measurements of fitness across ecological gradients and experimental reductions of each species, and found no evidence for any differential response or effect; that is, they appeared to be coexisting due to lottery dynamics. Munday's (2004) observations and experiments with two species of coral dwelling goby (*Gobiodon histrio* and *G. erythrospilus*) on Australia's Great Barrier Reef also appear to show coexistence of (nearly) identical competitors with lottery-like dynamics.

In a more direct test of the idea of neutrality, Siepielski et al. (2010) experimentally manipulated frequency-dependent competition between two similar species of *Enallagma* damselflies that did not appear to differ in any measured performance traits (Stoks and McPeek 2006). They manipulated the relative abundances of the species and found that neither experienced an advantage when rare (Fig. 3.1), and that there were no clear signals between demographic performance or relative abundances across a gradient of lakes that varied in environmental conditions. These patterns seem to suggest these species may be coexisting neutrally. Interestingly, however, this pattern is not universal among coexisting damselflies. When Siepielski et al. (2011) examined the interactions between one of these *Enallagma* species (*E. vespersum*) and species in two other damselfly genera (*Ischnura* and *Lestes*), they found a strong signature of coexistence via niche differences in that the growth rates of *Enallagma* were strongly limited by increasing conspecific density but much less so with increasing densities of either *Lestes* or *Ischnura*.

The damselfly example provides a window into a process that might in fact be quite general. Niche differences appear to play a strong role in determining the coexistence among members of different genera that clearly differ in a number of important traits. Furthermore, niche differences also seem to play a role in driving the distributions of species that live in different types of habitats (e.g., fish vs. fishless lakes) (McPeek 1998). However, within a given locality, there can be several species within each of the different genera that co-occur with less obvious, or nonexistent, niche differences. Similar observations have been made on groups of amphipods within the genus *Hyalella* that co-occur in similar lakes. While some of these species, although quite similar, clearly partition habitats and resources (Wellborn 2002, Wellborn and Cothran 2007, Cothran et al. 2013, Smith 2013, Dionne 2015), others seem to co-occur without clear differences in traits or demographic processes (Smith 2013, Cothran et al. 2015, Dionne 2015). We will address this example in more detail in Chapter 7—where we discuss how patterns and processes at the taxonomic (i.e., species level) may give different results than patterns and processes that emerge at higher levels of organization, such as functional traits—and in Chapter 8, where we discuss how nearly neutral coexistence may emerge.

It is not necessary for species to be strictly "neutral" with respect to their interactions for demographic stochasticity to be important. Just as genetic drift can be important in the face of natural selection, ecological drift can be important even when species differ (Vellend 2010, 2016). For example, when initial populations are small, as is often the case during the early parts of community assembly, or when community size is small, as might be the case in small or isolated habitat patches, stochasticity in birth and death rates can create drift in the abundances of species, even when there are substantial nonneutral differences among species (Orrock and Fletcher 2005, Orrock and Watling 2010, Vellend 2010). A classic example is from the work on competitive interactions between *Tribolium* beetles in small microcosms by Thomas Park and his colleagues. Under some environmental conditions, one species or the other was the eventual winner and outcompeted the other, but this outcome appeared to result from differences in initial conditions (Park 1957, Leslie et al. 1968). Likewise, Fox and Smith (1997) examined the competitive interactions between *Paramecium multimicronucleatum* (a protist) and *Philodina* sp. (a rotifer) in experimental microcosms and found that even though they took great care to have identical environmental conditions and

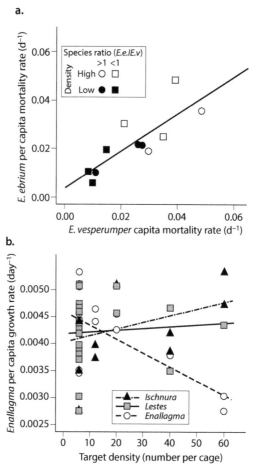

FIGURE 3.1. (a) Results from experimental manipulations of two similar species of *Enallagma* damselflies in a New Hampshire lake. Although both species had higher per capita mortality when the total densities were higher (open symbols), there was no influence of variations in their relative abundances (squares vs. circles), suggesting that they are interacting neutrally. Similar results were observed for growth (not shown). Modified from Siepielski et al 2010. (b) Per capita growth rates of *Enallagma vesperum* to experimental manipulations of damselflies from three different coexisting genera. *Enallagma vesperum* growth rates are plotted against densities of itself and members of two other damselfly genera, *Ischnura* and *Lestes*. The results show that *E. vesperum* has a strong influence on its own growth rate (steep negative slope) but is much less strongly influenced by the species in the other genera, suggesting a mechanism of niche partitioning leading to coexistence. Modified from Siepielski et al. 2011.

initiated their cultures with identical relative abundances of the two species, the replicates quickly diverged toward dominance of one species or the other.

In fact, there are many experiments that elucidate how stochasticity initiated during the early stages of colonization can lead to very different community outcomes and that a number of features of the environment appear to influence the relative strength this effect. These include habitat size (Drake 1991, Fukami 2004a), dispersal rates (Robinson and Dickerson 1987, Forbes and Chase 2002), disturbances (Chase 2007, Jiang and Patel 2008), and productivity (Chase 2010). These observations can be readily placed within the theoretical construct that the importance of drift should be highest when population or community size is small, dispersal is low, and selection is weak (Chase 2003a, Vellend 2010, 2016). An important issue, however, is that although these studies provide strong evidence for the importance of initial conditions leading to drift, they are typically performed in controlled experiments in which the overall habitat size is orders of magnitude smaller than the communities in which organisms typically interact. It is unclear how important these sorts of processes will be in larger communities where larger population sizes may be able to override any initial effects of stochasticity. To address this, we need to explore ways to try to infer the role of stochasticity and drift from more complex natural environments, which often involves analyses of spatial and temporal patterns, which we discuss in Chapters 4 and 5, respectively.

Finally, it is important to recognize that even though stochastic processes might lead to drift in the early parts of a community-assembly trajectory, other more deterministic processes can take over later in the trajectory. For example, stochasticity in early initial conditions can determine the priority effects that determine which species in a community will grow to dominate through priority effects (e.g., positive frequency dependence) and which will be rare or driven extinct from that location. Thus, once established, variation in initial conditions leads to multiple stable equilibria (reviewed in Beisner et al. 2003 and Fukami 2015), in which species maintain their relative abundances through deterministic interactions and no longer drift. We discuss this sort of interaction between temporal and spatial processes in more depth in Chapter 5.

3.2.2 Stochastic Extinction and Colonization Rates

Metapopulation models, and by extension, PD metacommunity models, assume that in any given population, a species has some finite chance of going extinct at any point in time and a finite probability of colonizing an unoccupied patch at any point in time. While theoretically elegant, evidence for this sort of dynamic equilibrial colonization-extinction pattern in natural systems has not been

particularly forthcoming (Harrison 1991). Nevertheless, a few notable examples show extinction (and recolonization) patterns in accordance with the expectations of demographic stochasticity (even if, in some instances, the cause for a
population's local extinction might be more deterministic in a given time period). Such examples include butterflies whose larval host plants are patchily
distributed (Harrison et al. 1988, Hanski et al. 1995), zooplankton living within
isolated rock pools (Bengtsson 1989, 1993; Hugueny et al. 2007), amphibians in
breeding ponds (Sjogren 1991, Carlson and Edenhamn 2000), and plant populations in patchy scrub habitat (Miller et al. 2012).

A second way that local stochastic extinctions have been observed in natural
communities is in the context of the theory of island biogeography (MacArthur and
Wilson 1967), which predicts that species richness on an island or patch emerges
as a dynamic equilibrium between stochastic colonizations and extinctions. We
will discuss the issue of colonizations in more detail in Section 3.3, on dispersal and
dispersal limitation, but for now we focus on extinctions as a potentially stochastic
local process. Although extinctions have often been observed in these sorts of studies, there are several limitations that make it unclear whether they are evidence for
the types of stochasticity assumed in metacommunity models.

First, a number of tests of the theory are incomplete in that they do not observe
the colonization-extinction rates per se but simply compare numbers of species
on islands of different sizes and isolations. There are many other mechanisms that
can lead to similar richness patterns without the stochastic element of colonizations and extinctions envisioned by MacArthur and Wilson (1967). For example,
larger islands are typically more heterogeneous, which would create a pattern of
increasing species numbers with island area similar to the predictions of island
biogeography, but without stochasticity (Connor and McCoy 1979, Ricklefs and
Lovette 1999).

Second, a number of tests have included estimates of species colonizations and
extinctions and show that these tend to balance each other and thus indicate that
local species richness was in a dynamic equilibrium. Nevertheless, observed patterns of turnover in many of these communities have been questioned because it
is hard to know when or if extinctions and colonizations happened (i.e., how thoroughly must we sample to know a species is truly extinct from a patch?; Lynch
and Johnson 1974, Simberloff 1976, Nilsson and Nilsson 1983).

Third, it is important to consider at which level any observed stochasticity in a
population's extinction probability is actually occurring. Although we often model
extinction probability as inherent to a given population, extinctions can often result from a very deterministic response to an external pulsed perturbation, such
as a disturbance, which itself may be stochastic. While these sorts of extinctions
might be reasonably modeled as stochastic in a higher-level metacommunity model,
the mechanisms that create the extinctions are not technically stochastic and can

strongly influence our understanding of and predictability in a system. For example, on a well-studied group of small islands in the Caribbean, several groups of organisms, including lizards (Schoener et al. 2001a, 2001b), spiders (Spiller et al. 1998, Schoener and Spiller 2006), and plants (Morrison and Spiller 2008, Morrison 2010), experience high rates of local extinctions as a result of periodic hurricanes, but these rates depend on the intensity of the hurricane, the size of the island (and its topography), and the population size of the organisms. In the interim between hurricanes, recolonizations of the same or different species occur, creating what appears to be a largely stochastically driven system, but one in which the stochasticity is at the level of the disturbance imposed on the system, not in the extinction rates themselves. The distinction is particularly important when considering the patterns that emerge when a colonization-competition trade-off scenario drives patterns in a PD system (Calcagno et al. 2006). If extinction can result only from externally imposed disturbances, changing disturbance rates will typically lead to a strong shift in coexistence patterns. Alternatively, if extinction is inherently stochastic, changing external disturbance rates will have a smaller influence on patterns of coexistence.

Fourth, when we observe extinctions, it is often quite difficult to discern whether these are truly due to demographic stochasticity or whether some process (e.g., an environmental change) has shifted the system to a position at which a species is no longer expected to persist in a system. For example, the numbers of extinctions exceeds that of colonizations in some island systems, leading to an overall decline in species richness that probably has something to do with a deterministic response to changing environmental conditions (Burns and Neufeld 2009, Morrison 2010). Likewise, local extinctions are a nearly ubiquitous observation at sites that have been isolated from larger, more continuous habitats due to habitat fragmentation (Newmark 1987, Gonzalez et al. 1998, Gibson et al. 2013, Haddad et al. 2015). However, what is less clear is whether these extinctions result because there are lower rates of colonization into local patches from the region, as would be predicted from a PD model with stochastic extinctions. Instead, these extinctions might result because the environmental conditions and/or biotic interactions within the remaining habitat have fundamentally changed in a deterministic way, for example, as a result of edge effects.

Despite these caveats, there appear to be some cases in which extinctions in a locality result from some degree of stochasticity, albeit often with a deterministic component. In an exemplary case, Alonso et al. (2015) collected data on the recovery of coral-reef fish communities in the northern Indian Ocean following a major El Niño–Southern Oscillation event that reduced coral cover to less than 10% of the benthic area in 1998. They developed a stochastic colonization-extinction model based on the principles of MacArthur and Wilson (1967) and parameterized it with the observations of the presence or absence of more than 150 species of fish

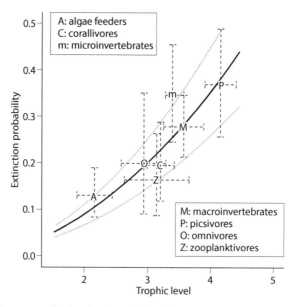

FIGURE 3.2. A parameterized extinction-colonization model based on observations of coral reef fish near the Lakshadweep Islands coral atolls in the northern Indian Ocean. Reefs were monitored from 2000 to 2011 after a massive mortality event caused by an El Niño–Southern Oscillation event in 1998. Colonizations and extinctions of 156 species were followed. Within a given guild, extinction probabilities were largely similar, but there were significant differences among guilds; higher trophic levels (Piscivores [P], micro- [m] and macro-invertivores [M]) had higher rates of extinction than the lowest trophic level (algae feeders [A]); intermediate trophic levels (zooplanktivores [Z], corallivores [C] and omnivores [O]) had intermediate extinction rates. Modified from Alonso et al. 2015.

thoroughly surveyed more than six times over the course of 11 years. From this, they estimated extinction and colonization rates of each species. First, they found that a general balance between colonization rates and extinction rates was obtained after only a few years, suggesting that the system reached a dynamic equilibrium in line with the predictions of the theory of island biogeography. Second, they found that (1) extinction probabilities were similar among species within a trophic guild, whereas (2) they differed among trophic guilds, with higher trophic levels having higher rates than lower ones (Fig. 3.2).

3.3 DISPERSAL LIMITATION

One of the principal distinctions between metacommunity ecology and more traditional "local" community ecology is that localities in the former are connected via dispersal. So how does dispersal influence the composition of communities?

One of the most obvious influences occurs when species are dispersal limited in their distributions in the landscape. If SS were perfect, we would expect that each species would be present in any locality where its expected population growth rate is sufficiently greater than zero. Alternatively, if dispersal limitation is frequent, we might expect metacommunity processes to be more closely aligned with the NT/PD archetype, in which species may be absent from many localities where they otherwise could have positive population growth. In contrast, ME processes are associated with the other end of the dispersal continuum, where dispersal rates are so high as to allow species to persist in habitats to which they are otherwise poorly suited; we discuss the issue of "surplus" dispersal below.

A number of empirical studies have explored the various aspects of dispersal limitation, far too many to adequately review here. We therefore give a simple overview of key results that show that, indeed, dispersal limitation is a frequent occurrence in natural metacommunities and that it often appears to be a highly context-dependent phenomenon. First, within a given metacommunity, dispersal limitation appears to be more prevalent in some types of environments than in others. Second, across metacommunities, the degree to which habitat patches are isolated from one another (relative to the dispersal capacity of the species of interest) strongly influences the degree to which dispersal is limited. Third, the scale of observation is critical for interpreting the degree to which dispersal is limiting.

3.3.1 Evidence for Dispersal Limitation within a Metacommunity

Are species present in every locality that provides favorable environmental and biotic conditions? This simple question is at the heart of the issue of the relative roles of niche versus dispersal in the assembly of communities. Most ideas of niche limitation, SS models included, assume that species have adequate ability to colonize any locality where their expected fitness is greater than zero. Even Hutchinson's (1957) classic definition of niche included a *fundamental niche*, which represented all of the possible resources and habitats a species could use and a *realized niche*, which was constrained only by biotic limitations. If dispersal limitation is prevalent, it might be necessary to think of two components of the realized niche— the realized niche due to biotic limitations and the realized niche because of dispersal limitation (Sax et al. 2013). Furthermore, if the answer to the above question about whether species occupy all of their possible habitat is frequently no, it would also undermine the critical assumption of many who use climate-envelope models (or niche models), which take the observed distributions of a species in different environment and habitat types as evidence of a species' possible distribution (Araújo and Peterson 2012, Sax et al. 2013). However, the degree to which dispersal, and in particular dispersal limitation, actually mediates metacommunity structure is less clear (Levine and Murrell 2003).

The easiest way to address the question of whether species are present in all of the habitats that they could occupy is a rather simple experiment—move individuals of a species to areas that are not occupied and see if they can establish. This experiment, or variants of it, have now been performed many times (Turnbull et al. 2000, Clark et al. 2007, Myers and Harms 2009).[1]

An early experiment by Tilman (1997) that examined the role of niche-based factors (environmental factors, species interactions) versus regional/dispersal-based factors in controlling community-level diversity remains one of the most robust. Therein, seeds of species that were present nearby, but not in the experimental plots, were introduced and monitored for several years (treatments ranged from 5 to 54 added species). This was one of the first experiments to test if the importance of dispersal limitation versus niche limitation could be detected by comparing the size of the regional species pool with the number of species occurring locally (Terborgh and Faaborg 1980, Ricklefs 1987). Although tests comparing patterns of local/regional diversity have been shown to have little explanatory power (see Chap. 4), Tilman's (1997) experimental test found strong evidence for dispersal limitation in this grassland community. After four years, species richness increased by more than 80% in plots where the maximum number of seeds were introduced. These "new" species, however, typically remained quite rare in the plots. At the same time, there was still a strong role for local factors, including local soil conditions and the species already present in the plots, in determining the likelihood that seeded species could establish. A similar experiment by Burke and Grime (1996) found largely congruent results, with dispersal limitation playing an important role but modified by environmental conditions (productivity, disturbance) and species interactions (diversity).

Results from Burke and Grime (1996) and Tilman (1997) provided a peek into what soon emerged as a general phenomenon—dispersal limitation and local environmental and biotic interactions combine to determine patterns of local communities. Results from one of the few studies that directly examined dispersal limitation for animals outside of the context of habitat isolation/connectivity, by Shurin (2000), epitomized the perspective that local and regional processes interact. Shurin introduced a diverse "slurry" of pond water that included many species of zooplankton that were present in the region into enclosures within several small freshwater ponds. When the resident community was left intact, this

---

[1] Here we are not talking about moving species from outside of their biogeographic boundaries, as in the case of exotic/invasive species, but rather within their ranges. We will discuss some of the biogeographic implications of dispersal limitation and invasions in Chapter 10.

enhanced propagule treatment had minimal effect—on average, "invaded" enclosures gained about 0.5 species relative to controls. However, when the resident community was experimentally perturbed by removing more than half of the individuals, establishment success of the introduced species was much greater and the invaded enclosures gained more than two species on average.

These early results emphasized that the local perspective, which held that patterns of diversity and community composition could be explained entirely by species interactions and environmental filters, was incomplete and that dispersal from the regional pool can also play an important role. At the same time, the regional perspective, which posited that patterns of local communities could be understood simply as a reflection of regional/biogeographic patterns was also incomplete because there was also a strong role for local factors influencing local diversity. Instead, an intermediate metacommunity perspective, with interacting local and regional processes, seemed to be more appropriate.

Tens to hundreds of experiments have now been performed that have examined dispersal limitation and its interactions with local factors, with the vast majority of these studies being done on plants. Early reviews of these experiments indicated that the majority of plant species were seed-dispersal limited (Turnbull et al. 2000, Moles and Westoby 2002), but Clark et al. (2007) used meta-analysis to show that while this was true in that the effect sizes of seed additions were on average greater than zero (seed additions led to more seedlings), the effect sizes were in fact quite low. Instead, the majority of the individuals added during experiments died during the seed-to-seedling transition, indicating a strong role for local filters (biotic or abiotic) in limiting establishment. Clark et al. (2007), however, did find that seed additions were much more likely to lead to more recruitment when local environments were disturbed (Fig. 3.3a). Likewise, in a meta-analysis of 62 community-level seed-limitation experiments, Myers and Harms (2009) found that more than 70% of the studies showed enhanced species richness when seeds were added (an overall positive effect size) and that the effect size was significantly stronger (more species established) when local environmental conditions were disturbed (Fig. 3.3b). Both of these meta-analytic results again support the notion of an interaction between regional and local factors (i.e., the metacommunity perspective).

One important limitation of the majority of dispersal-limitation experiments, particularly those performed with plants (where the majority of data are available), is the timescale on which the experiments took place. For example, the meta-analysis by Clark et al. (2007) focused only on emergence in the seedling stage, whereas demographic bottlenecks are known to strongly limit plant populations, such that seed-enhancement studies typically have weaker effects as the time of observation increases (Turnbull et al. 2000, Ehrlén et al. 2006). Likewise, at the community level, the positive effect of dispersal limitation on community richness

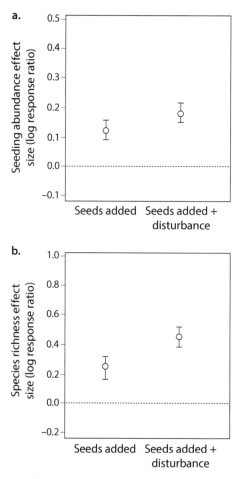

FIGURE 3.3. (a) Log-ratio effect sizes of studies that compared seedling numbers of plants in control plots and those with augmented seed numbers ($N = 102$ comparisons) and studies that compared disturbed plots with and without augmented seed numbers ($N = 83$ comparisons) (only studies on native species were used). Data from Clark et al. 2007. (b) Log-ratio effect sizes of studies that compared species numbers of plants in control plots and those with additions of species not present locally (but present in the regional pool) ($N = 32$ comparisons) and studies that compared disturbed plots with and without additions of new species by seed ($N = 17$ studies) (only studies in grasslands were used). Data from Myers and Harms 2009.

typically declines with length of observation in a given community. For example, the strong positive effects of seeding 54 species native to the region, but not site, on local species richness observed by Burke and Grime (1996), diminished over five years (Thompson et al. 2001). Tilman (1997) observed similar declines at a different study site in a four-year study of seed additions of species from the regional (but not local) community.

Many fewer dispersal-limitation studies have been conducted on animals. Two notable studies on zooplankton in freshwater communities have shown somewhat opposite results to those on plants. In experimental enclosures within a large lake, Lukaszewski et al. (1999) found that increasing propagule availability from nearby lakes did not influence local diversity. Likewise, using similar methods, Shurin (2000) found that few species from the broader region were able to invade local communities of zooplankton in ponds. However, perturbing the local community, which reduced the intensity of biotic interactions, lead to more invasions and higher local diversity.

Although we tend to think about dispersal limitation (or lack thereof) as a metacommunity-wide phenomenon, it is important to remember that not all species are equally dispersal-limited within a metacommunity, nor are all metacommunities of a group of species equally dispersal limited. We briefly overview each of these cases in the following sections.

3.3.2 Variation among Species in Their Degree of Dispersal Limitation within a Metacommunity

A number of factors can influence variation among species in a community with regard to their likelihood of experiencing dispersal limitation. Moore and Elmendorf (2006) used niche models to estimate whether several species of serpentine grassland plants were likely to be present in most of their availability habitat and then experimentally added seeds to unoccupied sites to determine whether they were dispersal limited. They found that the four species whose distributions matched their estimated niche envelopes tended to fare more poorly when transplanted to sites where they were not initially present; that is, they were not particularly dispersal limited. Alternatively, two species that appeared less likely to be distributed in all of their available habitats were equally successful in sites where they were present and absent; that is, they seemed to be more dispersal limited (Fig. 3.4).

Although not yet synthesized, a number of factors (e.g., relative abundances and traits of species) have been suggested to correlate, at least on occasion, with the degree to which a species experiences dispersal limitation. These include factors such as local abundance and regional occupancy, as well as demographic and life-history traits (Table 3.2). Of course, many of the factors listed in Table 3.2 are highly correlated. For example, dispersal limitation is expected to be more frequent in species that have relatively low regional occupancy and in species that are habitat specialists, but these are linked (Svenning et al. 2008). These low-occupancy species also tend to have traits associated with dispersal limitation, such as shorter dispersal distances and larger and fewer seeds produced (Van der Veken et al. 2007, Riibak et al. 2015).

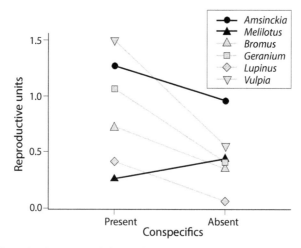

FIGURE 3.4. Reproductive output of six species of serpentine grassland when planted in sites that were occupied by the species (present) and sites that were unoccupied (absent). Species with gray symbols and lines were those that niche models predicted were well matched to their environments and all had significantly lower reproductive success in unoccupied sites, indicating a lack of dispersal limitation. Species with black symbols and lines were those that niche models predicted were not well matched to their environments and experienced no significant declines in reproductive output in sites that were unoccupied, indicating likely dispersal limitation. Modified from Moore and Elmendorf 2006.

3.3.3 The Influence of Species Interactions on Dispersal Limitation

Aspects of the biotic community itself can influence the degree to which species are dispersal limited. For example, many animals are seed predators and can strongly reduce the realized fecundity of plants. In some circumstances, the abundances of seed predators in natural communities has increased greatly, due, for example, to decreases in their own predators or increases in shelter. As a result, this higher abundance of seed predators can create patterns of dispersal limitation in plants that would otherwise be able to persist in more sites (Orrock et al. 2008, Dangremond et al. 2010). Alternatively, a number of animals can reduce the likelihood of dispersal limitation by serving as seed dispersers. Seed dispersal by animals occurs either as part of a mutualistic interaction (seeds that make edible fruits or other structures), inadvertently as part of their foraging activities (e.g., small mammals or birds that make caches of seeds, some of which germinate and grow), or as a hapless dispersal vector (e.g., seeds that stick to the fur of animals). As a result, alterations to the community of animal seed dispersers might strongly influence the degree to which plants are dispersal limited. For example, in many tropical forests, reductions in seed-dispersing birds and mammals through hunting and other habitat-degrading factors can greatly reduce seed dispersal rates (see

TABLE 3.2. Factors That Have Been Implicated in Determining the Degree to Which Different Types of Species Are Limited by Dispersal

Category	Factor	Description	Representative References
Abundance/ distribution	Local abundance / rarity	Species that are rare locally are more likely to be dispersal limited than more common species.	Ehrlén and Eriksson 2000, Mabry 2004
	Regional occupancy	Species that occupy fewer sites in the landscape are more likely to be dispersal limited than more widely distributed species.	Ehrlén and Eriksson 2000, Mabry 2004
	Habitat specialization	Species that are more specialized for certain habitats are more likely to be dispersal limited than more generalized species.	Bossuyt et al. 2004, Svenning et al. 2008
Demography/ life history	Seed mass	Species with larger seeds are more likely to be dispersal limited than species with smaller seeds.	Ehrlén and Eriksson 2000, Moles and Westoby 2002, Mabry 2004, Riibak et al. 2015
	Dispersal ability	Species with a smaller dispersal distance are more likely to be dispersal limited than species that disperse farther.	Ozinga et al. 2005, Van der Veken et al. 2007, Marini et al. 2012, Riibak et al. 2015
	Seed bank	Species with a shorter-lived seed bank are more likely to be dispersal limited than species with a longer-lived seed bank.	Ozinga et al. 2005, Van der Veken et al. 2007
	Longevity	Species that are shorter-lived are more likely to be dispersal limited than species that are longer-lived.	Ozinga et al. 2005, Van der Veken et al. 2007, Marini et al. 2012

the meta-analysis by Markl et al. 2012), and thus can sometimes leave a strong signature on the population abundances and spatial aggregations of plants within the community (reviewed in Stoner et al. 2007).

One clear example of the importance of seed dispersers on plant dispersal limitation and distribution comes from the long-term forest dynamics plot in Lambir,

Borneo (Harrison et al. 2013), shown in Figure 3.5. In 1992, a 52-ha plot was established where all trees greater than 1 cm in diameter were identified, counted,
and mapped spatially. Importantly, this plot was in an area that been relatively
untouched by humans until the late 1980s, when the road through the park was
expanded and bushmeat hunting increased. Thus, at the same time as the first
censuses in the plot were taking place, many of the large vertebrates that serve
as important seed dispersers in the system, such as hornbills, sunbears, and flying
foxes, were on their way to extirpation from the park. At the start of the study, species of trees that used nonanimal seed-dispersal mechanisms (e.g., ballistic, wind)
were more tightly clustered in space than were animal-dispersed seeds, suggested
that animal dispersal tended to reduce the degree of dispersal limitation of plants.
However, following the extirpations of mammalian seed dispersers, the animal-
dispersed seeds became more and more clustered in space, indicating a greater
degree of dispersal limitation through time (Fig. 3.5).

3.3.4 The Influence of Habitat Isolation and Dispersal
Rates on Dispersal Limitation

One of the fundamental observations that led MacArthur and Wilson (1963, 1967)
to devise the theory of island biogeography—the precursor to many of the basic
metacommunity theories described in Chapter 2—was that islands that were more
distant from a mainland source were more depauperate in their species richness
than islands that were nearer the mainland (see Fig. 3.6a for an example data set
of bird species on oceanic islands used in MacArthur and Wilson 1963 and Fig.
3.6b for a more recent global analysis of island floras from Kreft et al. 2008). The
primary mechanism thought to underlie this negative effect of isolation on patterns
of species richness is that more distant islands have lower rates of colonization.
Of course, for dispersal limitation to lead to fewer species in the long term, rather
than just a slower approach to an equilibrium number of species, another mechanism is needed to act in concert with colonization rates—namely, there must be
some extinction rate that may or may not be related to habitat isolation and other
features of the island (e.g., its size).

Even before MacArthur and Wilson (1963, 1967), Maguire (1963) tested the key
prediction of the influence of distance from the "mainland" (i.e., isolation) by
placing small buckets at different distances from a source pond and found progressively decreasing diversity with distance as a result of increasing dispersal
limitation. Many experiments have now been performed directly manipulating
the dispersal rate of species among localities within a metacommunity. Most of these
experiments have been performed in experimentally tractable systems of aquatic
microcosms or mesocosms consisting of small zooplankton, protists, algae, and

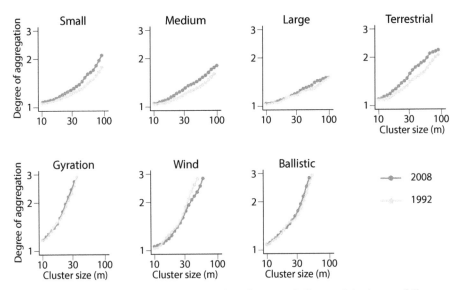

FIGURE 3.5. Results of the degree of spatial clustering, as an indicator of the degree of dispersal limitation, of trees in the long-term study of the 52-ha forest dynamics plot at Lambir, Borneo. The y-axis indicates the degree to which a given cluster of trees is aggregated, and the x-axis is the size of the cluster. Curves that are further to the left indicate a population that is more clustered than curves that are to the right. Different lines indicate surveys conducted in 1992 (gray) and 2008 (black). The three groups of species on the bottom panels represent nonanimal seed-dispersal modes, and these are in general more tightly clustered, than the panels on the top, which are all animal dispersed. The results suggest that initially, nonanimal-dispersed species had greater dispersal limitation. Through time, following the defaunation of the island's seed-dispersing community, plants that were more animal dispersed became more clustered, particularly the larger-seeded species, suggesting that they became more dispersal limited as a result of the loss of seed dispersers. Modified from Harrison et al. 2013.

bacteria, although some were in natural aquatic systems and a few in terrestrial systems (reviewed in Cadotte 2006, Logue et al. 2011, and Grainger and Gilbert 2016).

In a meta-analysis of 50 studies that experimentally manipulated dispersal rates, Grainger and Gilbert (2016) showed that, in general, increased rates of dispersal increased local (α-) diversity, though several studies found the hump-shaped pattern predicted by Mouquet and Loreau (2003) and some found negative or no effects of dispersal (Fig. 3.7). In addition, most studies showed that β-diversity declined with dispersal rates, consistent with the ideas of biotic homogenization. Finally, at the regional (γ-) diversity scale, more than half of the studies showed no influence of dispersal rates, while the other proportion showed about equal numbers of studies that showed positive or negative effects of dispersal.

While interesting, these experimental studies are limited to a few types of tractable systems. A number of observational studies have examined the influence of

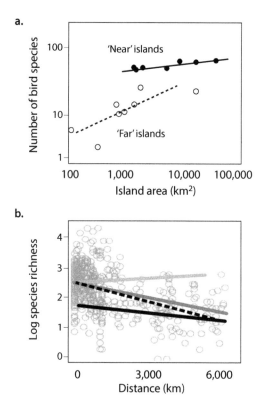

FIGURE 3.6. (a) Pattern of bird species richness on select Pacific islands in Oceania (Melanesia, Micronesia, and Polynesia) illustrating the influence of island isolation (and presumably dispersal limitation). Islands shown are grouped as "near" the mainland of New Guinea (less than 800 km) or "far" from the mainland of New Guinea (more than 3200 km). Modified from MacArthur and Wilson 1963. (b) Pattern of plant species richness on 488 islands distributed across the globe in relation to the distance of those islands to the nearest continent. Although considerable variation exists in this relationship, it is important to note that this includes islands with a number of different characteristics (e.g., size, climate, geological origin). The black hatched line shows the overall fit; the lightest-gray (nonnegative) line is through islands of continental origin; the lowest black line is through atolls; the intermediate-gray line is through islands of volcanic origin. Modified from Kreft et al. 2008.

dispersal on species diversity using isolation of islands and habitat patches as a proxy for dispersal rates. In fact, such studies are so frequent that a comprehensive review would be beyond the scope of what we can accomplish here. However, some recent examples include the lower diversity of mycorrhizal fungi on tree islands (Peay et al. 2010), of snakes on oceanic islands (Pyron and Burbrink 2014), of marine organisms at deep-sea hydrothermal vents (Rogers et al. 2012),

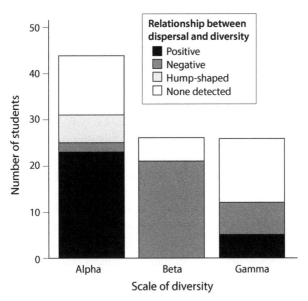

FIGURE 3.7. Results from a meta-analysis examining 50 studies that manipulated dispersal rates and measured patterns of diversity across spatial scales. The y-axis represents the numbers of studies that found the qualitative patterns listed in the legend (positive, negative, hump-shaped, or no response) for α- β-, and γ-diversity. Modified from Grainger and Gilbert 2016.

and even of human parasites (Jean et al. 2016) and languages (Gavin and Sibanda 2012) on oceanic islands.

Key exceptions to the generalization that increased dispersal rates and decreased isolation lead to more richness do exist, but these are largely explained when isolation and dispersal influence species that have a disproportionate influence on the rest of the community (e.g., strong competitors or predators; Kneitel and Miller 2003, Chase et al. 2010).

Even microbes are now recognized to be highly dispersal limited, at least in some cases. While microbes were historically thought to adhere strongly to the Baas Becking (1934) hypothesis—"Everything is everywhere, but the environment selects"—this is probably because early evidence tended to focus on morphological or crude molecular tools to distinguish among microbial taxa, leading to the supposition that similar species were actually the same and thus not dispersal limited. However, with more sophisticated sequencing technologies, evidence is rapidly amassing that individual microbial species can sometimes, though not always, be quite dispersal limited, even if functionally (and phylogenetically) similar species occur more broadly (Martiny et al. 2006, Hanson et al. 2012, van der Gast 2015). Though much more work needs to be done, van der Gast's (2013) revision of the

Baas Becking hypothesis to "some things are everywhere and some things are not. Sometimes the environment selects and sometimes it doesn't" probably applies just as equally to microorganisms as it does to larger organisms.

As a corollary to the reduction in richness typically seen when habitats are more isolated, habitat corridors between isolated habitat fragments can increase dispersal rates among localities and influence both composition and diversity of local habitats compared with otherwise similar localities without corridors. A meta-analysis has shown that corridors most typically have a strong positive influence on target organisms (Gilbert-Norton et al. 2010). While few of these studies specifically were able to examine influences on patterns of species diversity, those that did typically found increases in local diversity with the presence of corridors (Gilbert et al. 1998, Gonzalez et al. 1998, Damschen et al. 2006). Likewise, a meta-analysis by Haddad et al. (2015) of experiments on habitat fragmentation in which both habitat isolation and area are reduced has shown a general loss of species diversity and a variety of other community attributes consistent with the idea of dispersal limitation.

Again, notable exceptions exist where corridors have less influence on diversity patterns. For example, Collinge (2000) observed only a moderate influence of habitat corridors on species diversity, and this was largely driven by results from a particularly dry year when corridors increased diversity. Hoyle and Gilbert (2004) found that dispersal among moss patches was less influential on patterns of micro-arthropod diversity when conditions were wetter relative to results from an earlier experiment in which conditions were drier and corridors were more influential on species diversity (Gilbert et al. 1998). Both of these studies suggest that dispersal limitation might play a stronger role when communities experience stressful conditions, perhaps increasing the likelihood of random extinctions and rescue effects (see also Östman et al. 2006).

3.4 INTERACTIONS BETWEEN STOCHASTICITY, DISPERSAL, AND INTERSPECIFIC EFFECTS

The processes we have discussed so far capture the core elements of the NT (stochasticity and dispersal) but not the other metacommunity archetypes. PD incorporates interspecific trait variation along with stochasticity and dispersal. In the most prevalent PD model, the colonization-competition trade-off allows coexistence at the regional scale because some species are more adept at colonizing empty patches, whereas others are superior competitors within patches. Similar coexistence patterns emerge with trade-offs, colonization rate, and persistence in patches where species are already present (lower extinction rate).

There are a number of examples of systems that appear to have mechanisms akin to the colonization-competition assumption. Tilman (1994) formalized the well-known trajectory of successional patterns in Minnesota prairies into the colonization-competition trade-off by showing that there was a negative relationship between the ability of plant species to survive at low concentrations of nitrogen (R^*) and their ability to establish in old fields following abandonment (Fig. 3.8a). He then related these to the approximate trajectories of dominance of the different species in fields of different ages (Fig. 3.8b). Likewise, plant traits, such as differences in seed mass, are often taken as prima facie evidence for a competition-colonization trade-off, with smaller-seeded species being better dispersers and larger-seeded species being better competitors (Rees 1995, Turnbull et al. 1999). Other systems with trait distributions that are similar to the colonization-competition trade-off include carrion flies (Hanski 1987), *Drosophila* (Shorrocks 1991), laboratory protists (Cadotte et al. 2006), freshwater snails (Chase et al. 2001), ants that live on acacias (Stanton et al. 2002), and forest birds (Rodriguez et al. 2007).

Despite the popularity and intuitive appeal of the PD model based on competition-colonization processes and despite observations of traits associated with this trade-off, it is less clear how often the mechanism itself leads to patterns of metacommunity structure and coexistence in nature. For example, Lei and Hanski (1998) suggested that two species of parasitoid could coexist on one larval host (the butterfly *Melitaea cinxia*) because one species was a better local competitor and the other was a better disperser/colonizer, allowing them to coexist. However, later evidence from van Nouhuys and Punju (2010) suggested that in fact the superior competitor was also a better colonizer. The only way that the inferior competitor/colonizer was able to persist in the landscape was because a small isolated proportion of the butterfly larvae was left untouched by the superior competitor/colonizer. In a similar way, the trade-off between food discovery and interspecific dominance has been a popular mechanism to explain the high diversity of ants that can sometimes coexist at small scales (Fellers 1987, Davidson 1998, Pearce-Duvet and Feener 2010). However, Adler et al. (2007) used a combined modeling and empirical work to suggest that it was unlikely that coexistence could be explained only by this trade-off and that the addition of other limiting factors, such as parasitoids and multiple resource types, was necessary to explain the high diversity of ants in these systems.

In fact, we know of no fully complete case showing that coexistence is mediated purely by the competition-colonization trade-off with no further trait variation in other dimensions. One possible explanation is suggested by the work of Livingston et al. (2012). They constructed a model system of two strains of *Pseudomonas*: the wild type (the stronger competitor) and the less-competitive mutant type. Within this system, they experimentally manipulated the dispersal probabilities of the two

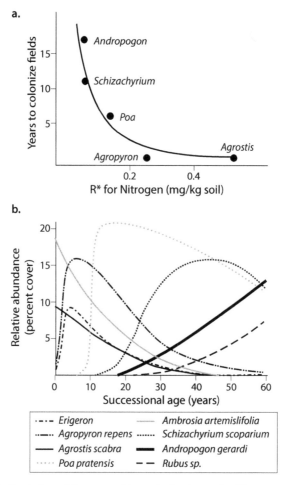

FIGURE 3.8. Idealized view of the competition-colonization trade-off among prairie plants in Minnesota. (a) The trade-off between a species ability to compete for nitrogen (its R^*) and the time it takes for it to colonize old fields following abandonment from agriculture. (b) A general sketch of the relative abundance trajectories of species at different times along the successional gradient. Modified from Tilman 1994.

species in an experimental landscape (96-microwell plates) so that they could match the predictions of the PD theory as closely as possible. However, rather than coexistence between the two strains when the colonization rate of the mutant was high enough to offset the competitive superiority of the wild type, one or the other species dominated in a seemingly stochastic way. This observation emphasizes that although colonization-extinction trade-offs can theoretically lead to a number of metacommunity patterns, they probably occur amid a more complex array of mechanisms based on environmental and resource constraints. This result is consistent with the modeling work of Shoemaker and Melbourne (2016; see Fig. 2.8 in Chap. 2), who found that competition-colonization models were much less stabilizing at the metacommunity scale than other archetypal models that had a role for SS dynamics.

3.5 THE INFLUENCE OF HABITAT HETEROGENEITY

The influence of habitat heterogeneity on species distributions, which forms the basis of the SS archetype, is probably one of the best studied of the metacommunity processes, and clearly our treatment of it can scratch only the surface of a voluminous literature. First, there is no doubt that habitat heterogeneity is typically strongly correlated with increased species richness, because communities with more heterogeneity allow the coexistence of species with varied habitat requirements, tolerances, and interactions. MacArthur and MacArthur (1961) were among the first to articulate and quantify this principle, and despite some dissention and confusion, recent meta-analyses have upheld it as a rather general phenomenon (Hortal et al. 2009, Stein et al. 2014). The results from the meta-analysis by Stein et al. (2014; Fig. 3.9) show that heterogeneity in a number of different landscape features strongly leads to increases in species richness.

Second, there are multitudes of studies that show how species abundances and performance vary across important habitat gradients, a tiny proportion of which are overviewed in Table 3.3.

Third, at moderate to large spatial scales, it is axiomatic that environmental conditions, such as climate and edaphic factors, determine species associations. For example, Whittaker's (1975) classic illustration (Fig. 3.10) shows how the world's major terrestrial biomes are distributed along axes of temperature and precipitation.

Given this ubiquity of evidence for its importance to the distribution, abundance, and diversity of species, it is clear that environmental conditions and their heterogeneity are essential variables in the development of a generalized perspective on metacommunities. This does not mean, however, that heterogeneity must always be an important driver of metacommunity variation.

The examples above clearly show that habitat heterogeneity is an important process, but it is also biased; researchers are more likely to investigate heterogeneity in environmental conditions and its importance to species diversity and distributions where they see important changes in species distributions. There is heterogeneity out there that some species do not "see" and thus that we as researchers may also not notice. Instead, the degree to which environmental heterogeneity influences

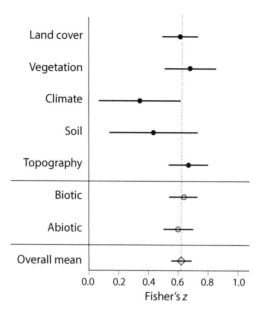

FIGURE 3.9. Results from the meta-analysis on the heterogeneity-diversity relationship by Stein et al. (2014). Fisher's z is the "effect size" calculated from correlations between heterogeneity and species richness; points represent mean effect size, and lines are the 95% confidence interval. The mean effect size is shown at the bottom, and the dashed line shows this value for other comparisons. All of the effect sizes are significantly positive. Modified from Stein et al. 2014.

species distributions will depend on how that habitat heterogeneity influences both the absolute and relative fitnesses of the species.

In Figure 3.11, we illustrate the four possible outcomes in terms of absolute and relative fitness of two species across two habitat types. First, the habitat gradient measured could have no influence on the fitness of either species (Fig. 3.11a). This is consistent with the NT, which assumes that there is no ecologically meaningful habitat variation and that species have equivalent fitness. Although not universally true, there will be a large number of cases in which the fitness of species across some habitat gradient are uninfluenced by that gradient; that is, even though we may measure heterogeneity, the species do not "see" it. Second, the habitat gradient could change the total fitness of each species but not their relative fitness (Fig. 3.11b). In this case, we might observe a positive habitat association for both species along the habitat gradient, although they are still "neutral" with respect to their traits. Third, the habitat gradient could change the relative fitness of the two species, such that one is favored while the other is disfavored along the gradient (Fig. 3.11c). This is a classic habitat trade-off, which will allow species to coexist

TABLE 3.3. Some Examples of Habitat Associations and Trade-Offs in Performance

System	Environmental Gradient	Presumed Performance Trade-Off	References
Amphibians and invertebrates in freshwater lakes	Lakes that have fish in them compared with lakes without fish	Trade-offs between species ability to compete for resources and species ability to tolerate high levels of predation by fish	Werner and McPeek 1994, Wellborn et al. 1996, McPeek 1998, Wellborn 2002
Plants in European wet meadows and South African fynbos	Soil moisture content	Trade-off between tolerance to aeration stress or tolerance to drought stress	Silvertown et al. 1999, Araya et al. 2011
Ants along elevational gradients in the Smoky Mountains (USA)	Elevation and temperature	Trade-off between competitive dominance and thermal tolerances; dominant species occurred only over a narrow range of warmer temperatures, whereas subordinate species were temperature generalists, including living in cooler areas.	Lessard et al. 2009
Zooplankton in freshwater lakes	Resource quantity and quality	Trade-off between growth rate and minimum resource requirements	Tessier et al. 2000
Trees in in lowland Amazonian rainforests	White-sand and clay soils	Trade-off between growth and defense against herbivores in different habitat types	Fine et al. 2004, 2006
Rodents in the Israel desert	Stabilized and unstable sand dunes	Trade-offs in habitat preferences, foraging efficiency, predation risk	Abramsky et al. 1991, 1994; Ziv et al. 1995, Wasserberg et al. 2005
Barnacles in the British intertidal zone	Tidal height	Trade-off in dessication resistance (upper zone), competitive ability (intermediate zone), and ability to withstand predation (lower zone)	Connell 1961a, 1961b

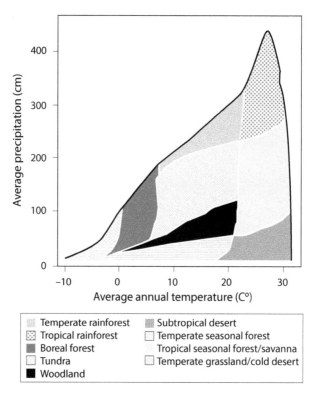

F<small>IGURE</small> 3.10. Illustration showing how temperature and precipitation interact to create a number of the world's terrestrial biome types. Modified from Whittaker 1975.

and strong opposing habitat associations to be observed. Finally, the habitat gradient could change the absolute and relative fitness of the species, where both are favored along the gradient, but one more so (Fig. 3.11d). Here, positive habitat associations would be observed for both species, but it would be stronger for one species than the other.

Given the many ways in which habitat heterogeneity *could* influence species distributions, a common practice is to correlate species abundances across habitats that differ in key characteristics in order to ask the organisms which degrees of heterogeneity they "see" with respect to differences in fitness (which is presumed to correlate with differences in abundances). This has perhaps been done most frequently in the context of the mapped forest plots that are part of the Center for Tropical Forest Science–Forest Global Earth Observatory (CTFS-Forest GEO) network (http://www.forestgeo.si.edu/). This network was initiated with the iconic 50-ha plot at BCI in Panama, the proverbial "birthplace" of the NT, and has since been replicated across the world. It provides some of the best data available to

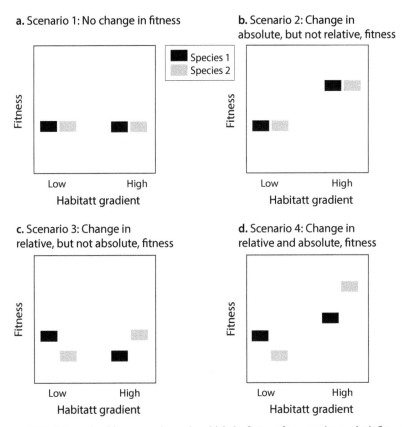

FIGURE 3.11. Schematic of four general ways by which the fitness of two species can be influenced by some environmental gradient. (a) No influence on either species. (b) An increase in absolute fitness of both species but no change in relative fitness. (c) A change in relative fitness of each species but no change in absolute fitness. (d) Change in both the relative and absolute fitness.

clearly test how species abundances correlate with a number of spatial and environmental variables. So it is not surprising that many of the strongest tests of habitat associations—and thus by association, the relative importance of niche-based SS—come from this network of forest plots.

A formal meta-analysis of the many studies that have asked versions of the question, to what degree do species abundances respond to environmental variation? in these forest plots is beyond the scope of what we can do here and would in fact require the raw data, since there are many ways that investigators have approached this question. However, there are some interesting patterns to observe, even if they may tell us at least as much about the human nature of bias in interpretation than about any of the real patterns.

A wide variety of statistical approaches have been employed to disentangle habitat associations from more "random" distributions. However, we can gain some insight by simply comparing the results of a number of these studies.

Table 3.4 lists several studies that have addressed this problem. From this list, we can see that between 30% and 90% of the species are either significantly positively or negatively associated with at least one habitat variable, with an average somewhere around 50%–60%. There are interesting and biological relevant patterns in this table, but it is clear that many species have strong habitat associations, while many others do not (at least within the context of the observed plot).

Given the schism between the niche and neutralist perspectives, one might look at these same data and come to very different conclusions. If you are a "niche believer," the observation that more than half of the species, and in fact many more than half of the most common species, are strongly associated with habitats suggests that the idea that the world can be understood just through a completely neutral model is absurd. However, if you are a "neutralist," the fact that the traditional niche-based perspective fails to provide any insight about the distributions of more than half of the species that could be analyzed in these studies is rather damning. Additionally, these analyses can be performed only if there are enough individuals of a species in the plot, which in the high-diversity systems can eliminate a further 50% or more of the species, further narrowing the niche theory's explanatory ability to around 25% or less of the species in a region.

This dual perspective on the same information is exactly what seems to have emerged from the literature on this topic. For example, the results from Harms et al. (2001) on the habitat associations of species at BCI have been cited a number of times in the literature for comparing those results with the results from other studies. However, the value used to discuss the BCI results when comparing other studies has itself varied. Some have reported BCI as having as few as 33% of the species associated with habitat variation (Gunatilleke et al. 2006), whereas others reported the same study as showing 64% of the variation (Chuyong et al. 2011). Of course, both are correct, because Harms et al. (2001) presented a number of different analyses, each of which told a rather different story depending on the numbers of species used, how conservative the test was, etc. And the concluding remarks in Harms et al. (2001, 957) state, "Our results do not support the hypothesis that habitat specialization is among the principle mechanisms of coexistence maintaining a large fraction of the alpha diversity within communities of tropical trees," whereas, Chuyong et al. (2011, 1372), citing the same study, conclude, "Similar to patterns reported for tropical forests in Asia and the Americas, . . . the results support the idea that niche partitioning contributes to the maintenance of

TABLE 3.4. Studies in Stem-Mapped Forest Plots That Have Tested the Degree
to Which Species Are Either Positively or Negatively Correlated
with Habitat Variables and Their Key Results

Study Site	Percentage of Species Significantly Associated with Habitat Variables	References
Barro Colorado Island, Panama	33%–64%, depending on analysis	Harms et al. 2001
Lambir, Borneo	87%	Davies et al. 2005
Korup National Park, Cameroon	63%	Chuyong et al. 2011
Sinharaja, Sri Lanka	79%	Gunatilleke et al. 2006
Yasuni, Ecuador	50%–75%, depending on analysis	Valencia et al. 2004
Gutianshan National Nature Reserve, China	55%–72%, depending on life stage	Lai et al. 2009
Dinghushan Biosphere Reserve, China	83%	Wang et al. 2009a

diversity in species rich plant communities." Clearly, much of the niche-neutral schism is in the eye of the beholder!

In addition to simple bias in perception regarding the importance of habitat associations in driving species distributions, other clear signals influencing these interpretations are the degree of habitat heterogeneity present in a given sampling scheme and the spatial scale on which species-by-environment analyses are performed (see also Pinto and MacDougall 2010). For example, it has often been noted that the plot at BCI was deliberately placed in an area that was relatively homogeneous in its degree of environmental conditions, whereas other forest plots are much more heterogeneous. Simple variation in the degree of heterogeneity among forest plots might explain much of the variation observed in Table 3.4 (we revisit this issue in Chap. 4).

Likewise, the spatial scale on which the data are collected will necessarily include different amounts of heterogeneity in the analysis and thus make different conclusions about the relative importance of habitat heterogeneity in driving species distributions. Garzon-Lopez et al. (2014) nicely showed this effect for several species on BCI itself. For these species, the strength of environmental associations was weakest at the smallest measured spatial scale (50 ha), but as spatial scale increased,

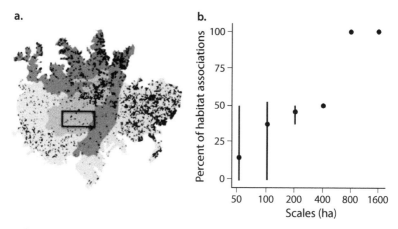

FIGURE 3.12. (a) A map of Barro Colorado Island, with the 50-ha plot outlined in the center of the island. Different habitat types are depicted in gray, and the distribution of one particular tree species, *Attalea butyracea*, is illustrated with black dots. The main conclusion is that this species may appear to have a somewhat "random" distribution when examined at the relatively small scale (such as within the 50-ha plot), but at the larger scale, there is clearly a strong habitat association of the species. (b) Variation in habitat associations with different sampling scales for *A. butyracea*. Error bars indicate the SE of significant associations among plots. Modified from Garzon-Lopez et al. 2014.

the degree to which habitat associations were significant and strong increased. Figure 3.12a shows the distribution of one species, *Attalea butyracea*, on a geological map of the entire BCI with the location of the 50-ha plot shown as a box near the middle of the island. Within the plot, this species appears to show a fairly random distribution of individuals; however, from the larger map, we can see that the species has strong habitat associations in other parts of the island. Figure 3.12b shows how the strength of species habitat associations increases with increasing spatial scale.

In sum, the answer to the question of whether habitat differences contribute to patterns of community composition, or whether instead, neutral processes are more important, as we argued in Chapter 1, is yes.

3.6 INTERACTIONS BETWEEN HABITAT HETEROGENEITY AND DISPERSAL

The most complex models that we described in Chapter 2 include interactions between habitat heterogeneity and dispersal. These include the ME models, which assume that dispersal is so high that there is spillover into adjacent habitats, as well as the "interface" models, which incorporate habitat heterogeneity in a PD framework and assume that dispersal is limiting. This interaction of heterogeneity and dispersal highlights the distinct roles that dispersal can play in metacommunities depending on whether it is on the low end of the continuum—where it serves to fuel community assembly and reduce dispersal limitation, and thus enhance the degree to which environmental filtering is important—or whether it is at the upper end of the continuum—where it can allow species to persist in local patches where they are not well suited, and thus reduce the degree to which environmental filtering is observed in accordance with ME theory. A cartoon for how increasing dispersal rates might influence patterns of species richness across a gradient of the different metacommunity archetypes is presented in Leibold and Miller (2004; Fig. 3.13).

3.6.1 Evidence for Mass Effects

Although the idea of ME provides intuitive appeal and compelling theoretical arguments, it is less clear how important they are in real systems. In fact, we might expect them to be relatively weak along shallow environmental gradients (assuming net dispersal is proportional to differences in population size) and also weak in the presence of abrupt environmental gradients where fitness difference are extreme. Thus, ME may be most likely not only with intermediate levels of environmental differences among habitats but also at high dispersal rates (Shmida and Ellner 1984; Pulliam 1988, 2000; Mouquet and Loreau 2003). And this might explain why there is little empirical evidence for them.

Kunin (1998) studied the degree to which ME occurred at the boundaries between different treatments in the Park Grass experimental plots at the Rothamsted Experimental Area. These plots were established more than 150 years ago and subjected to different fertilization treatments since then. By studying transects perpendicular to the boundaries of these treatments, Kunin documented only very weak ME (an increase of only about 0.5 species per transect), which dissipated quite abruptly.

Michels et al. (2001) studied the prevalence of zooplankton species in ponds connected by streamflows. They combined sampling of the downstream pond using drift nets with estimates of population immigration and found very weak ME even though thousands of individuals immigrated per hour (see also Cottenie et al. 2003). In a similar vein, Howeth and Leibold (2010b) manipulated dispersal of zooplankton among mesocosms and found weak evidence for ME even at extremely high dispersal rates (20% per day, roughly equivalent to three times per generation).

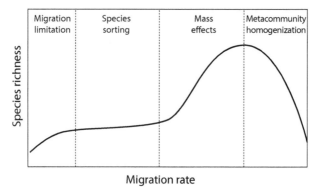

FIGURE 3.13. A hypothetical relationship between migration rate and species richness illustrating the major mechanisms by which migration affects community structure. Modified from Leibold and Miller 2004.

These rates are much higher than those observed in the study by Michels et al. (2001) and are probably rare in natural systems.

Experimental work in natural communities has similarly failed to find a strong influence of ME in a number of other situations. For example, in an experimental removal experiment of annual plants in the Israeli desert, Kadmon and Tielborger (1999) found no evidence for ME. Likewise, Harrison et al. (2010) examined the distribution of three species of annual tarweeds (Asteraceae) across habitat types. They found that the species dominant in the most common habitat type (*Holocarpha virgata*) was unable to establish populations in less prevalent habitats characterized by distinct plant communities with related annual tarweeds, even after large-scale bulk additions of seeds.

Despite the general lack of evidence for ME in most metacommunity experiments, this does not mean that the mechanism is universally unimportant. Indeed, there is strong evidence for population-level ME in a few systems. One of the most compelling is the case of *Calanus finmarchicus*, one of the most prevalent copepods in the North Atlantic, which serves as a key resource for numerous fish populations throughout the area. Amazingly, detailed life-history studies conducted throughout the North Atlantic suggest that it has a negative net population growth throughout its range except for a subset of "hot spots," where its population growth is extremely high (Miller et al. 1998).

3.6.2 Evidence for Combined Effects of Heterogeneity
and Patch Dynamics

Our discussion of the simplest PD models above, with colonization-competition trade-offs, emphasized the need to incorporate habitat heterogeneity or other niche factors in order to better explain natural processes. As a test of the interface model incorporating habitat heterogeneity into the PD model discussed in Chapter 2, Fukumori et al. (2015) used microcosms with two protist species that each strongly (but not completely) dominated microcosms subjected either to darkness or to light and thus supported either bacteria or algae as resources for the protists. They found that as the rates of extinction increased (both experimentally and stochastically imposed), the degree to which the environment explained variation in the relative abundance of the two species declined. This result corresponded with the predictions of the model by Leibold and Loeuille (2016) described in Chapter 2. Likewise, Mordecai et al. (2016) showed that coexistence and diversity of snail-inhabiting trematodes resulted from a combination of colonization-competition trade-offs and spatial heterogeneity.

3.7 IMPLICATIONS FOR LOCAL VERSUS REGIONAL
CONTROLS ON DIVERSITY

When the local environment determines the distribution and diversity of species, it is usually assumed that local processes dominate, such as those inherent to the SS archetype. Alternatively, when dispersal of species from the regional pool plays an important role, regional processes are implicated. This dichotomy has promoted the view that local communities are structured either by local *or* regional processes (Ricklefs 1987, 2004). In the metacommunity context, this perspective would imply that if SS predominates, dispersal should not be important, whereas the NT, PD, and ME modularities imply that dispersal plays a stronger role, negating the role of local processes. Unfortunately, this assumption is overly simplistic and does not capture the nuanced interaction between local and regional factors in metacommunities.

Instead, strong environmental controls on community structure driven by SS are not incompatible with the strong dispersal limitation observed in experimental analyses, even within the same system. Pattern analyses and local experiments are usually performed on relatively small spatial scales, typically within a relatively fixed pool of species. Those species that are present then respond to variable environmental conditions. When a metacommunity consists of a larger pool of species, as is the case when patches have increased dispersal rates (e.g., through proximity, corridors, etc.), this is because the metacommunity has more opportunities for regional coexistence. While we might say that the metacommunity is dispersal limited, this dispersal limitation is at a different scale than the dispersal limitation that can emerge as a result of PD, NT, or ME at local scales.

Because dispersal limitation can occur at multiple spatial scales, it is important to remember that even when regional processes influence patterns of diversity owing to dispersal limitation, this does not mean that local processes (such as SS) are not important (see also Pinto and MacDougall 2010). In fact, it is quite the opposite: reducing dispersal limitation (increasing dispersal) should often increase the importance of local environmental factors in determining metacommunity structure. For example, Foster et al. (2011) showed that the match between environment and species composition was stronger in a grassland community that was enhanced by seed dispersal from the regional species pool than in control communities without seed additions. The explanation was that when dispersal was less limiting (seed-addition treatments), a fuller complement of the regional species pool was available to participate in the SS process, whereas species that could persist in the more dispersal-limited scenario were more likely to be better colonizers that responded more to space than to environment.

Thus, as with most areas of ecology, the approach we usually use to test between local and regional controls on diversity and species composition is generally false. Local and regional processes occur simultaneously, and dispersal can be a limiting factor at one scale (i.e., dispersal from outside the metacommunity of interest), while simultaneously, environment limits species distributions at the smaller scale. Likewise, biogeographic processes might influence the size of the regional species pool that is then filtered by local conditions to determine local diversity (Ricklefs 1987, 2004). This scenario, however, does not imply that local SS processes are not occurring within those biogeographic constraints (Chase and Myers 2011); we will revisit this issue in Chapter 10.

3.8 CONCLUSIONS AND SYNTHESIS

The theoretical approaches to metacommunity ecology are based variously on different assumptions about the importance of major ecological processes ranging from habitat specialization, the role of niche differences in affecting coexistence, differences in dispersal, and dispersal limitation. The approaches also depend on assumptions about covariances among the traits of species that affect these processes (and the consequent trade-offs among species that may differentially affect them). Experiment- and process-based observational studies indicate varying support for all of these possible assumptions. The most surprising, perhaps is the relatively weak evidence for dispersal-maintained source-sink relations among distinct habitats that might affect the prevalence of ME. Future work should investigate how different aspects of the traits that affect these processes might vary among species in a metacommunity (see Chap. 6). While we still have a long way to go in developing a full understanding of metacommunity ecology, there is extensive evidence that the basic tenets that underlie the theories are well supported.

CHAPTER FOUR

Metacommunity Patterns in Space

Prospectus

1. Although understanding metacommunity processes cannot be fully accomplished by observational studies of their patterns, pattern-based studies are useful for exploring evidence regarding the core assumptions and predictions of the metacommunity theories.

2. NT received much of its initial support and traction from its ability to predict the shape of commonness and rarity in the species-abundance distribution (SAD) and related metrics, but more recent work shows that other metacommunity theories can also predict these same patterns as well or better.

3. One approach for evaluating the importance of biotic interactions and SS examines the structure of species distributions and co-occurrences within a metacommunity. This was catalyzed by perspectives such as Diamond's assembly rules, as well as by counterperspectives that emphasized the development of adequate null models to distinguish whether any structure differed from random. Truly distinguishing which, if any, processes structure patterns of species distributions and co-occurrence remains controversial.

4. A multidimensional approach that extends comparisons of these incidence patterns—elements of metacommunity structure—seeks to identify the degree to which distributions among entire groups of organisms show distributional patterns consistent with different metacommunity predictions about species distributions, including checkerboard, nested, and random distributions, and different types of gradients (evenly spaced, Gleasoninan, and Clementsian). Comparisons of a number of different data sets show support for each pattern, but there seems to be substantial evidence for Clementsian distributions of species along spatial gradients.

5. A variation partitioning approach, which attempts to disentangle the relative contributions of environmental and spatial predictors of metacommunity structure, has become prevalent. Again, there is considerable variation in the literature, but much evidence supports at least a reasonably strong influence of environmental variation on species compositional variation (consistent with SS, ME, and PD with environmental heterogeneity). However, there are also substantial influences of spatial effects that are unrelated to measured environmental differences (consistent with NT, ME, and PD). Finally, much of the variation in species distributions within

a metacommunity is unexplained, which might be indicative of stochasticity (consistent with NT and PD).

6. These findings suggest that species distributions are likely to result from a complicated mixture of processes that incorporates assumptions of all the metacommunity archetypes described in Chapter 2. Different systems, depending on their biology, the spatiotemporal scale of the study, and other elements related to regional context of the system, may lie at different points along the resulting spectra of possible patterns.

7. Overall, the study of distributional patterns of species within a metacommunity indicates that the idealized archetypes generally interact strongly with each other. Refinements and expansion of metacommunity theory to incorporate multiple processes are helping motivate and refine our approaches and improve the way we understand and interpret how species distributions are influenced by the interaction of local and regional processes.

If the whole point of metacommunity ecology is to account for spatial factors such as dispersal and heterogeneity, then a core provision of the metacommunity perspective should be to help us understand the relative abundances and occurrences of species both within and among localities. Chapter 2 described some theories that incorporate different assumptions about metacommunity assembly processes (e.g., the importance of dispersal, stochasticity, etc.), and Chapter 3 reviewed data on the underlying *processes* that influence metacommunity dynamics. Here, we explore just how well the theories and their underlying assumptions do at predicting critical *patterns* in the distributions and abundances of species (as well as their composition and diversity within and among localities).

Precursors of the various metacommunity theories were initially evaluated by comparing their predictions against some sort of null expectation. For example, patterns in support of SS drew largely from studies of the distribution of species along univariate and multivariate environmental gradients (Whittaker 1960, 1967), with little recognition that alternative processes such as dispersal or stochasticity might explain them. The influence of these processes were thus essentially lumped into a catchall "error" term or potentially confounded with each other. Likewise, several data sets were compared with PD models, which predict "core" species that are persistent through time and "satellite" species that are more transient (Hanski 1982, Hanski and Gyllenberg 1997), but these studies ignored possible underlying environmental factors that could also be playing a role. Finally, biogeographic patterns of species distributions were examined to determine the possible roles of interspecific interactions and other structuring factors against null models that assume random distributions (Connor and Simberloff 1978, 1983;

Diamond and Gilpin 1982), but these analyses typically ignored the possible influence of environmental constraints or dispersal limitation.

Following the conceptual integration of the various archetypes on metacommunity ecology (Leibold et al. 2004), in particular in response to the ongoing debate regarding the primacy of SS versus that of NT, there ensued a large number of studies that compared metacommunity patterns with the predictions of these different archetypes (reviewed in Chase et al. 2005, Cottenie 2005, McGill et al. 2006a, Logue et al. 2011, Rosindell et al. 2011, Soininen 2014, Brown et al. 2017). In this chapter, we focus on how static patterns of species relative abundances and spatial co-occurrences (largely within a single trophic level) have been used to evaluate the different metacommunity archetypes. These static "snapshot" patterns have received the most attention in the literature because of the wealth of available data sets from floral and faunal surveys. In following chapters, we will build on this discussion by incorporating patterns of community variation through time (Chap. 5), patterns of co-occurrences and distributions of species including phylogenetic and functional trait information (Chaps. 6 and 7), patterns of biodiversity and biogeography (Chaps. 9 and 10), and patterns involving food webs (Chap. 11).

Empirical studies on static patterns of species abundances and distributions in a metacommunity can roughly be divided into three categories: (1) studies that statistically combine the compositional differences (e.g., co-occurrences) and relative abundances of specific species in a metacommunity into emergent patterns such as the species-abundance distribution (SAD) or the species-area relationship; (2) studies that compare the co-occurrences of species among habitats to look for potential patterns such as species that co-occur less (or more) than expected by chance; and (3), studies that compare the distributions of species along spatially and environmentally explicit gradients to examine the relative influence of space and environment on the distribution of species. Here, we review each of these approaches and their relative ability to differentiate among the various metacommunity processes. The last approach in particular has become increasingly important as a way to disentangle, in ways that can help evaluate and interpret metacommunity processes implied by the theories we described in Chapter 2, the roles of environment and space in driving metacommunity patterns.

4.1 PATTERNS OF SADS AND RELATED DIVERSITY METRICS

The SS archetype of metacommunity assembly has historically been the default view of how communities are structured (reviewed in Hubbell 2001 and Chase and Leibold 2003). However, one of its biggest limitations was that it was not immediately obvious how it could predict certain patterns of regularities typically observed in natural communities. For example, while SS models can predict the number of species that can persist in a given community or region and the shifts in species composition across environmental gradients, it was not always so clear how species would differ in their relative abundances or regional occupancies. For example, why should the SAD be generally characterized by a relatively few very dominant species and many rare species? There have been periodic attempts to use various niche models to describe how resources are apportioned among species (MacArthur 1957, Sugihara 1980, Tokeshi 1996), but these were generally complex and full of questionable assumptions; further, they did not always do a great job of predicting the observed patterns. One of the features of Hubbell's (2001) NT that made it so compelling was that it could simultaneously predict patterns of species diversity and realistically shaped SADs with relatively few assumptions (hence the use of the word "unified" in the title of Hubbell's book).

Given Hubbell's (2001) emphasis on the NT's success at such prediction, and especially the presence of an inflated tail of rare species (the zero-sum multinomial), it is not surprising that a majority of early tests, both for and against the NT, were based on the shape of the SAD (reviewed in McGill et al. 2006a and Rosindell et al. 2011). For example, there was an important back-and-forth about whether the inflated-tail SAD predicted by the NT actually fit the data on tree species from Hubbell's field site at BCI (Fig, 4.1). First, using a simulation-based approach, McGill (2003) found that the standard log-normal SAD, which was typically equated with a niche perspective, fit the BCI data better than the shape predicted by the NT. As a follow-up, Volkov et al. (2003, 2005) devised an analytical approach that was less subject to some of the limitations of McGill's (2003) simulation-based approach and found that the NT predictions fit the BCI data better than the log-normal. Differences between these models' predictions, however, are quite subtle and generally express largely similar patterns of commonness and rarity (Fig. 4.1). Furthermore, although the NT might be more parsimonious than more complex SS models, several different types of metacommunity models, including those based on strong niche differences, can predict SADs that are largely similar to those predicted by the NT (Chave et al. 2002, Mouquet and Loreau 2003, Wilson et al. 2003). In a more recent installment of this debate, Chisholm and Pacala (2010) argued that SS archetypes could determine the species richness in a community, even if NT determined their relative abundances, and they found good concordance with this idea in the data from BCI.

This debate illustrates an essential point: that many patterns may actually not be particularly helpful in discriminating among different hypotheses. Here, both SS and NT can predict very similar patterns, and it might be easy to imagine that other, perhaps idiosyncratic or minor, features could tip the balance in favor of either SS or NT models. Indeed, McGill (2010) showed that models based on

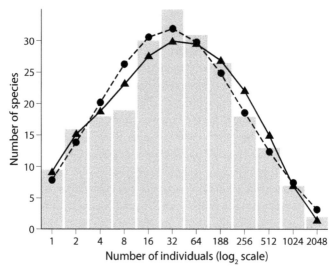

FIGURE 4.1. Comparison of the fit of an analytical version of the neutral theory's (NT) zero-sum multinomial (triangles, dashed line) with the more traditional log-normal relationship (circles, solid line) for the 21,457 individuals of 225 different species monitored in the 50-ha plot on Barro Colorado Island, Panama (gray bars). Volkov et al. (2003) argued that their version of the NT fit the data better than the log-normal, whereas a previous paper by McGill (2003) suggested the opposite with a simulation-based approach. Modified from Volkov et al. 2003.

processes inherent to SS, NT, and PD archetypes, as well as statistical mechanics models such as maximum entropy (Shipley 2010, Harte 2011), could all predict more or less the same SAD shape. This finding indicates that the shape of the SAD, as well as related patterns such as the shape of the species-area curve (Rosindell and Cornell 2007, 2009), is unlikely to be usefully diagnostic of the various metacommunity-level processes underlying this pattern (see Chap. 10 for more details on the macroecology perspective). While the success of NT in predicting SADs was thus supportive of the hypothesis, it turns out that this pattern is not particularly diagnostic for any of the proposed models.

4.2 NULL MODELS AND CO-OCCURRENCE
IN METACOMMUNITIES

A second approach that has been used as a diagnostic for understanding the processes underlying metacommunity structure has been to focus on the patterns of species distributions across space. Diamond (1975a) was among the first to use the biogeographic distribution of species among patches (islands) to infer ecological processes in his development of the assembly rules of bird distributions in the Bismarck Archipelago, in the Pacific Ocean near New Guinea. While the details of these assembly rules are not necessary for our purposes here, Diamond's (1975a) main assertion was that bird distributions among the islands were largely driven by interspecific competition, so that certain combinations of species were more often observed together than expected by chance (putatively owing to weaker competitive effects), whereas others were rarely observed together (putatively owing to strong competitive exclusion). Connor and Simberloff (1979) operationalized the analysis of these biogeographic patterns by placing the distributions into a site × species matrix,[1] which could then be subjected to a series of statistical analyses based on null models that compared the observed distributions of species to those that would be expected were species randomly distributed among islands. They argued that there was little evidence in support of Diamond's (1975a) arguments.

As more data became available and the null modeling approach became more sophisticated, researchers continued to analyze distributions of species to infer the role of species sorting and interspecific interactions from more random processes. More than twenty-five years after Diamond (1975a) first published his assembly rules, Gotelli and McCabe (2002) tested how general those rules were across a large number of data sets. Using improved algorithms and methods, they found support for the majority of the rules—fewer species combinations, more checkerboard distributions of species pairs, and fewer co-occurrences of species than would have been expected by random chance alone—suggesting that the distributions of species among patches was not random (Fig. 4.2).

Although Diamond (1975a) and others attributed nonrandom distributional patterns to interspecific competition, there are in fact, many possible explanations for such distributions, including environmental differences among the habitats or other types of interspecific interactions (e.g., predation, mutualism). Even the conclusion made by Gotelli and McCabe (2002) that either environmental variation or interspecific interactions—both quite deterministic—were probably driving nonrandom distributions is not necessarily true. Instead, it is quite plausible that the nonrandom patterns found using null models of species co-occurrences could by driven entirely by dispersal limitation and ecological drift inherent to the NT (Ulrich 2004).

Amazingly, thirty years later, the distributions of species at the core of this debate continues to be contentious, even among the original combatants. The assembly rules defined by Diamond (1975a), which were criticized by Connor and

---

[1] A matrix in which sites are in columns (or rows) and species names are in rows (or columns). Each combination of species × sites gives either a species' presence or absence or its abundance.

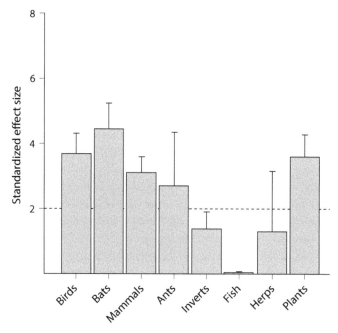

FIGURE 4.2. Results from a meta-analysis comparing the effect size (C-score, a measure of "checkerboardness") of deviations from the null expectation of co-occurrence patterns for several different groups of organisms. The dashed line with an effect size of 2.0 represents the 5% significance level. Modified from Gotelli and McCabe 2002.

Simberloff (1979, 1983) and defended by Diamond and Gilpin (1982), were revived by a new data set of bird distributions in the same biogeographic region as the original study (Mayr and Diamond 2001). Throwing the first punch of the millennium, Sanderson et al. (2009) used these data to suggest that a number of species pairs, especially among congeners, were less likely to co-occur than expected by chance, such that "the difference between chance and pattern can be unequivocally determined." Examples of congeneric species pairs that overlap considerably in traits and resource utilization but diverge in their distributions among islands to produce a checkerboard distribution included some of the same taxa that Diamond (1975a) discussed in his original study, such as the fruit pigeons (*Ptilinopus superbus* and *P. solominensis* in this case) and fruit doves (*Dulcula spilorrhoa* and *D. rubricera*), as shown in Figure 4.3. Such checkerboards among otherwise similar species were envisioned by Sanderson et al. (2009) to have probably emerged from a SS process; for example, if species compete locally and cannot coexist but partition environmental gradients such that some islands favor one species and other islands favor the other.

a.

Island	Ptilinopus superbus	Ptilinopus solomenesis
1	X	
2	X	
3	X	
4	X	
5	X	
6	X	
7	X	
8	X	X
9	X	X
10	X	X
11	X	X
12	X	X
13		X
14		X
15		X
16		X
17		X
18		X
19		X
20		X
21		X
22		X
23		X
24		X

b.

Island	Ducula spilorrhoa	Ducula rebricera
1	X	
2	X	
3	X	
4	X	
5	X	
6	X	
7	X	
8	X	
9	X	
10	X	X
11	X	X
12	X	X
13	X	X
14	X	X
15	X	X
16		X
17		X
18		X
19		X
20		X
21		X
22		X
23		X

FIGURE 4.3. Two representative sites (islands) × species matrices that show strong negative associations between pairs of congeneric species interpreted as "checkerboard" distributions that likely resulted from interspecific competition. Modified from Sanderson et al. 2009.

In lockstep with their skepticism from thirty years earlier, Collins et al. (2011) reanalyzed a subset of the data, noting that although they also found a preponderance of checkerboard-like patterns, analyses of presence-absence matrices "provide limited insight" because they cannot disentangle historical biogeography and dispersal limitation from interspecific competition in generating the observed patterns. Not to be deterred, Sanderson et al. (2011) countered that the analyses by Collins et al. (2011) were limited because they examined only a fraction of the available data and "their discussion adds nothing substantively new: it merely raises again, without the benefit of personal experience, the issues . . . that we had already

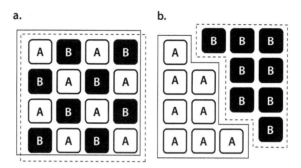

FɪɢURE 4.4. (a) A "true" checkerboard distribution as envisioned in Diamond's (1975a) original study. The distribution of Species A and B overlap throughout the archipelago, but they do not co-occur. This pattern could result from interspecific competition or simply differences among species among islands that have different environmental conditions, but either way would conform to a species sorting archetype. (b) A pattern in which species are distributed unevenly across the archipelago, which could result because of historical biogeography (i.e., allopatric speciation) and/or dispersal limitation. If geographic position is not explicitly considered, both of these distributions would lead to a conclusion of checkerboards. Modified from Connor et al. 2013.

discussed with knowledge of the species and archipelagoes involved." In the latest installment, Connor et al. (2013) accepted the challenge by Collins et al. (2011) and reanalyzed the data in Sanderson et al. (2009) from the Solomon and Bismarck Islands, as well as from Vanuatu. Crucially, this time they explicitly incorporated spatial geography into their analyses. They pointed out that the checkerboard analogy cannot be appropriately analyzed with binary site × species matrices unless geography is also explicitly incorporated. For example, if we were to compare the site × species matrices for the pairs of species in Figure 4.4a, which is a true checkerboard, with Figure 4.4b, in which species are distributed allopatrically on different locations within the archipelago, we would conclude that both had the same structure. Connor et al. (2013) accept culpability in prolonging what might have been a much-abbreviated debate had they initially realized the limitations of their original null models to detect checkerboards and other patterns of co-occurrences that did not explicitly include geographic structure. This seemingly endless debate continues (Connor et al. 2015, Diamond et al. 2015).

While the inability to definitively infer process from patten has been implicit in the discussion of null models on binary site × species matrices, the analyses by Connor et al. (2013) expose a more fundamental problem by not accounting specifically for the geographical structure of species distributions. As a result, they question the entire endeavor of trying to infer process from pattern. We agree with this concern, at least with low-dimensional patterns such as checkerboards or the SADs that we discussed in Section 4.1. Again, the problem is that multiple processes can generate patterns that are very similar. In the case of checkerboards, however, it does seem clear that there are real patterns that do deviate from those expected on the basis of unconstrained random sampling. One way forward is to modify null expectations (constrained random sampling) to account for other possible processes (e.g., historical biogeography), but this can make it very difficult to reject the null hypothesis unless the data sets are enormous or the patterns extremely strong. One can also imagine an endless sequence of invoked constraints that might be accounted for before claiming that any analysis is actually a test of a particular hypothesis. Another way forward is to increase the dimensionality of the patterns to include multiple descriptors.

4.3 ELEMENTS OF METACOMMUNITY STRUCTURE

We can take the analyses of species × site matrices a step further to dissect and discern the nature of any nonrandomness that might emerge. There can be many types of nonrandom structures in site × species matrices that a single descriptor and null model analysis will not be able to fully capture; for example, when a nonrandom distribution of species is as equally probable to result from spatial segregation, due to biogeography for example, as the checkerboard pattern often inferred from such analyses (Connor et al. 2013). Leibold and Mikkelson (2002) defined properties of a site × species matrix—which they called "elements of metacommunity structure" (EMS)—that allow some dissection of how those patterns relate to idealized types of metacommunities: coherence, turnover, and boundary clumping.

With EMS, patterns of metacommunity structure consistent with a number of idealized scenarios can be distinguished by using a hierarchical approach to examining EMS (Fig. 4.5). These include (1) *Clementsian gradients*, in which groups of species come and go along ecological gradients as cohesive units (Clements 1916); (2) *Gleasonian gradients*, in which groups of species come and go more independently along the gradient (Gleason 1926); (3) *checkerboard distributions*, in which pairs of species are less likely to be found together (but pairs are independent from other pairs, to differentiate from Clementsian), as would be expected by competitive exclusion (Diamond 1975a) or differential habitat preferences; (4) *evenly spaced gradients*, in which species ranges are more evenly spaced than expected by chance, which might be the case, for example, when species trade off their ability to persist along a continuous environmental gradient among sites (Tilman 1982); and (5) *nested subsets*, in which species in more diverse communities are proper subsets of those in less diverse communities (Patterson and Atmar 1986). Presley et al. (2010) improved upon the generalized EMS approach by

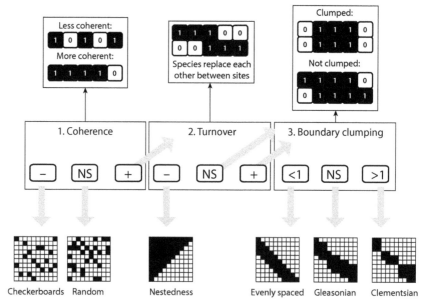

FIGURE 4.5. Conceptual flow diagram representing how the three elements of metacommunity structure are calculated and can allow the elucidation of several idealized metacommunity patterns. The first step looks at coherence. If it is negative, the pattern is classified as "checkerboard"; if not significant, the pattern is classified as "random"; if it is positive, a second test for turnover is conducted. If this is negative, the pattern is classified as "nested"; if not a third test is conducted for boundary clumping. If significantly less than one, the pattern reflects evenly spread niches along gradients; if it is greater than one, the pattern is classified as "Clementsian" because it implies species border clumping, and if not, it is classified as "Gleasonian." Modified from Henriques-Silva et al. 2013.

explicitly recognizing patterns that could emerge interior to the idealized structures (quasi-structures, such as quasi-Clementsian, etc.) and possible mechanisms that could lead to those patterns.

Coherence refers to the degree to which species in a metacommunity respond to the same gradient (spatial or environmental). Coherence can be analyzed using ordination (specifically correspondence analysis), which maximizes the positioning of sites on axes relative to the degree that communities share species and species share ranges. The number of absences of species embedded within the axis indicates the degree of coherence (Fig. 4.5). Embedded absences are then compared with a null distribution to determine whether coherence is nonsignificant, positive, or negative. If a metacommunity is not significantly coherent, we might infer that it is randomly assembled and stop there. If a metacommunity is negatively coherent, we infer that it has a checkerboard distribution and stop there. However, if a metacommunity is positively coherent, we need to examine one or both of the other elements—turnover and boundary clumping—in order to describe its full structure.

Turnover refers to the changes in species composition across sites and is measured by the numbers of times that one species is replaced by another across any two sites (Fig. 4.5). This is compared with a null expectation. If turnover is significantly less than expected by chance, we can say that the metacommunity has a significantly nested structure. If turnover is not different or higher than expected by chance, we next must examine the boundary-clumping patterns.

Boundary clumping refers to the degree of distinctiveness in groups (clumps) of species distributions (Fig. 4.5). It is measured using Morisita's index; values not different than the null expectation of 1 suggest that community boundaries are randomly distributed in the manner envisioned by Gleasonian gradients. If this index is significantly greater than 1, indicating clumped boundaries, the pattern is more akin to that envisioned by Clementsian gradients, whereas if the index is significantly less than 1, indicating overdispersed boundaries, the pattern is more evenly spaced along the gradient.

Leibold and Mikkelson (2002) applied this method to 35 haphazardly chosen data sets from the literature and found that each of the structures occurred in natural systems. However, only a small proportion was noncoherent, and nearly half showed Clementsian (positive coherence, positive turnover, and significant clumping) or quasi-Clementsian (positive coherence, nonsignificant turnover, and significant clumping) gradients. This analysis included some of the data sets that were used to support Gleason's over Clements's views (Whittaker 1967)!

Table 4.1 presents a nearly exhaustive review of the studies that have since used the EMS framework. While there is almost certainly bias in the data sets that have been applied to the framework, the patterns that have emerged are intriguing. None of the metacommunities examined appear to show true checkerboards; only a tiny fraction have no coherence (i.e., are random) or show nested patterns; Gleasonian gradients are rare; and, Clementsian and quasi-Clementsian gradients occur in the vast majority (17 of 19) studies and comparisons within studies.

The preponderance of Clementsian and quasi-Clementsian gradients in recent studies that have examined the EMS framework is particularly intriguing in light of the intense debates between Gleason and Clements nearly a century ago. Most ecologists have since tended to think that Gleason's view of individualistic responses of species to environmental gradients better approximated reality than Clements's superorganismic perspective (McIntosh 1998, Götzenberger et al. 2012). Part of the problem here may have arisen because Clements described several features of communities as organisms unto themselves, which flies in the face of our

TABLE 4.1. A Number of Studies That Have Applied the Elements of Metacommunity Structure Framework to Metacommunity Patterns

System	Coherence	Turnover	Boundary Clumping	Overall Pattern	Citation
Amphibians distributed among freshwater ponds in Michigan	Positive	Positive	Significant	Clementsian	Werner et al. 2007
Forest trees along elevation gradients in Puerto Rico	Positive	Positive, but variable	Significant, but variable	Clementsian (but variable)	Barone et al. 2008
Dragonflies distributed among freshwater ponds in Michigan	Positive	Non-significant	Significant	Quasi-Clementsian	McCauley et al. 2008
Bats across sites in Paraguay	Positive (variable among guilds)	Positive (variable among guilds)	Significant (variable among guilds)	Clementsian (variable among guilds)	Presley et al. 2009
Gastropods along an elevation gradient in Puerto Rico	Positive	Non-significant	Significant	Quasi-Clementsian	Presley et al. 2011b
Snails distributed among freshwater ponds in Michigan	Positive	Positive	Non-significant	Gleasonian	Hoverman et al. 2011
Rodents, bats and passerine Birds along elevation gradient in Peruvian Andes	Positive (all groups)	Rodents—positive Birds—non-significant Bats—negative	Significant (all groups)	Rodents—Clementsian Birds—quasi-Clementsian Bats—nested	Presley et al. 2011a

Terrestrial gastropods along distinct elevation gradients in Puerto Rico	Positive (both forest types)	Mixed forest—positive Palm forest—non-significant	Mixed forest—significant Palm forest—non-significant	Mixed forest—Clementsian Palm Forest—quasi-Gleasonian	Willig et al. 2011
Plants in English woodland. Measured twice across 70 years	Positive	Positive	Significant	Clementsian	Keith et al. 2011, Newton et al. 2012
Bats across sites in Mexico	Positive	Positive	Significant	Clementsian	López-González et al. 2012
Wetlands along the Front Range of the Rocky Mountains, Colorado	Positive	Positive	Significant	Clementsian	Johnson et al. 2013
Fish in Ontario, Canada lakes (85 different metacommunities analyzed)	82 positive 3 non-significant	42 negative 40 positive	35 significant 5 non-significant	3 random 42 nested (lower energy lakes) 35 Clementsian (higher energy lakes) 5 Gleasonian	Henriques-Silva et al. 2013
Grassland plants in the French Alps	Positive (across scales)	Positive (across scales)	Significant (across scales)	Clementsian (across scales)	Meynard et al. 2013
Trematode parasites within *Helisoma* snails in freshwater ponds	Positive	Non-significant	Non-significant	Quasi-nested	Richgels et al. 2013

(continued)

TABLE 4.1. (*Continued*)

System	Coherence	Turnover	Boundary Clumping	Overall Pattern	Citation
Plant and soil organisms in southern Appalachian Mountains	All groups but collembolans—positive Collembolans—random	Positive (fungi, collembolan); non-significant (plants, mesostigmatids)	Significant (all groups)	Clementsian or quasi-Clementsian for all groups but collembolans (random)	Shevtsov et al. 2013
Parasites in Sonoran Desert rodents	Positive (variable among groups)	Non-significant (variable among groups)	Significant (variable among groups)	Quasi-Clementsian (variable among groups)	Dallas and Presley 2014
Small mammals in South American Atlantic Forest fragments	Positive	Positive	Significant (when parsed out, marsupials not significant)	Clementsian (marsupials Gleasonian)	de la Sancha et al. 2014
Fish in seasonally flooded Pantanal wetlands, Brazil	Positive (all seasons)	Early season—negative Late season—largely non-significant	Significant (all seasons)	Early season—nested Late season—quasi-Clementsian	Fernandes et al. 2014
Fish in Hungarian streams	Positive (all seasons)	Non-significant (mostly)	Significant (all seasons)	Quasi-Clementsian	Eros et al. 2014
Bacteria in the East China Sea	Non-significant (December, August) Positive (May)	Non-significant (December, August) Significant (May)	Significant (May)	Random (December, August) Clementsian (May)	Yeh et al. 2015
Diatoms, bryophytes and invertebrates in headwater streams of Finland	Positive (all groups and locations)	Mostly non-significant (southern watersheds) Significant (northern-most watershed)	Mostly non-significant (southern watersheds) Significant (northern-most watershed)	Mostly Gleasonian or quasi-Gleasonian (southern watersheds) Clementsian (northern-most watershed	Heino et al. 2015

modern conceptions of evolution (Tansley 1935, Egler 1954). Instead, at least on the scales that these EMS studies have taken place, groups of species appear to respond somewhat in concert with environmental gradients, though perhaps not as tightly as envisioned by Clements. For example, a majority of studies along elevational gradients showed Clementsian or quasi-Clementsian patterns (Barone et al. 2008; Presley et al. 2011a, 2011b; Willig et al. 2011; Meynard et al. 2013), which might make some sense as groups of species replace one another across relatively short distances because of the strong physiological constraints imposed by changing temperatures and rainfall amounts.

While Table 4.1 provides some intriguing patterns from a variety of natural communities, we again caution that identification of these patterns is not necessarily straightforward (Ulrich and Gotelli 2013) and that the underlying processes that lead to patterns of EMS cannot be fully resolved. For example, it might be tempting to conclude that the preponderance of Clementsian and quasi-Clementsian patterns (and the scarceness of Gleasonian, nested patterns, and random patterns) in Table 4.1 means that metacommunities are much more strongly structured by niche-based processes such as SS (and also PD or ME) rather than by the randomness expected from NT. Instead, while NT is often equated with random species distributions in a metacommunity, implying no coherence, there are circumstances where an NT framework can predict Gleasonian patterns and nested subsets among more and less diverse sites due to dispersal limitation. In some cases, depending on the geographical distributions of sites, NT could even lead to checkerboard, Clementsian, or evenly spaced patterns.

Two other issues need to be addressed when considering the patterns described in these EMS studies. First, there may be some "publication bias" (or more appropriately "analysis bias") inherent to these patterns. Most of the studies describe comparisons of distributions of species within a specific group (e.g., snails, frogs, bats) and include significant environmental variation among the sites included. It might not be surprising, then, that patterns consistent with Clementsian gradients seem to emerge more often than, say, random or nested patterns.

Second, spatial scale (number of sites) and taxonomic scale (number of species) certainly play predominant roles in what sorts of conclusions we can deduce from an EMS analysis (as it does throughout this book). Random patterns at one scale can look structured at another. Likewise, patterns might look Gleasonian at one scale (say within a given biome, where species respond more individualistically), but Clementsian at another scale (say across biomes, where wholesale shifts in species composition turnover occur). Furthermore, the different structures that emerge across spatial scales (Münkemüller et al. 2014) may be reflective of shifts in the relative importance of local and regional processes (Presley et al. 2010, Henriques-Silva et al. 2013, Meynard et al. 2013), as well as of biogeography (Leibold et al. 2010, Pillar and Duarte 2010, Collins et al. 2011).

4.4 USING VARIATION PARTITIONING TO DIAGNOSE SPATIAL, ENVIRONMENTAL, AND RANDOM EFFECTS

While each of the approaches described above has the potential to identify some component predictions of different metacommunity processes, none of them relate these patterns to attributes of the sites themselves (i.e., whether they relate to environmental differences, spatial effects, or both). One way to disentangle these that has become quite popular has been to include multivariate environmental and spatial information into analyses of patterns of metacommunity structure. The idea is that the different metacommunity models place different emphasis on the importance of environmental factors (SS, ME) and spatial factors (PD, NT), and so examining the relative influence of space and environment on the similarity of species composition among sites (i.e., β-diversity) may help resolve the possible underlying processes. Figure 4.6a presents a simplistic scenario comparing the predictions of the SS and NT models in terms of the influence of environmental distance on community similarity: more different environments are more different compositionally in SS metacommunities, but there is no influence of the environment in NT metacommunities. Alternatively, Figure 4.6b compares SS and NT predictions as spatial distance among sites increases: more distant sites are more different compositionally in NT, but there is no influence of space in SS metacommunities.

The first studies to address the roles of space and environment in determining species composition in the context of the metacommunity categorizations typically treated them as discrete alternatives—one, space, was more consistent with the SS archetype, and the other, environment, was more consistent with NT. For example, Condit et al. (2002) set out to examine similarity of species composition with increasing distance among 1-ha forest plots in Panama, Ecuador, and Peru and found good concordance with NT predictions at least over small to moderate distances (0.2 to 50 km; Fig. 4.7a). At yet shorter distances, they found a steeper decline in compositional similarity than NT predicted, a result they attributed to high conspecific aggregations and habitat heterogeneity; at much larger scales, the decay of similarity was less than expected according to NT, a finding they attributed to the prevalence of geographically widespread species. Although consistent with aspects of NT at certain scales, an obvious criticism of the interpretations of Condit et al. (2002) is that both environment and spatial distances are often highly

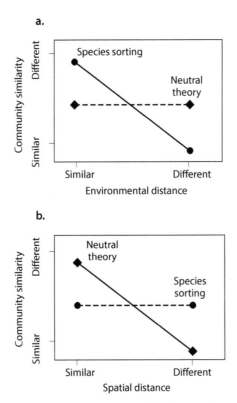

FIGURE 4.6. Predictions of how community composition (measured as a pairwise similarity) between sites that vary in their environmental (a) or spatial (b) distance are expected to differ in the species sorting and neutral theory archetypes that represent opposite ends of the continuum of the influence of environment and space, respectively. Modified from Chase et al. 2005.

correlated; that is, the more spatially distant plots are from one another, the more likely they are to differ environmentally (Duivenvoorden et al. 2002, Ruokolainen and Tuomisto 2002). As a counterpoint to Condit et al.'s (2002) study, Tuomisto et al. (2003) examined spatial patterns among forests in Amazonia and found a strikingly different result. Most likely because their surveys included forest types with very different edaphic (soil) conditions, Tuomisto et al. found a much stronger relationship between environmental distances among sites and species compositional variation (Fig. 4.7b) than between spatial distance and species compositional variation (Fig. 4.7c).

Although the studies by Condit et al. (2002) and Tuomisto et al. (2003) were instrumental in bringing patterns of species compositional along spatial and environmental gradients to the question of metacommunity patterns, a more sophisticated

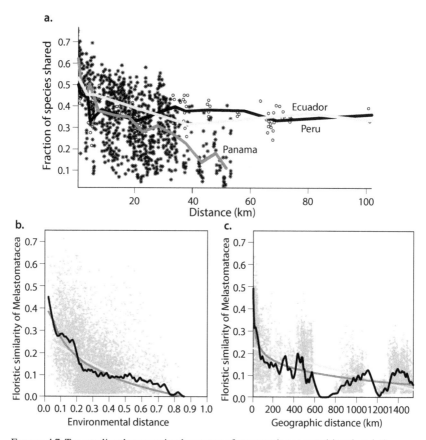

FIGURE 4.7. Two studies that examined patterns of community compositional variation among sites as a means to test predictions of dispersal-based neutral theory versus niche-based species sorting ideas of metacommunity structure. (a) Comparison of tree compositional similarity among 1-ha plots that differed in their spatial distances from one another in three different locations (Ecuador, Panama, Peru). Modified from Condit et al. 2002. (b) Comparison of species composition relative to environmental distance (a composite of climatic and soil variables) among sites in a family of shrubs and trees (Melastomataceae) from transects in western Amazonia. (c) Comparison of species composition relative to spatial distance from the same data in (b). Both (b) and (c) were modified from Tuomisto et al. 2003.

approach has since become popular to resolve the roles of environment and space in the patterning of metacommunity structure. This approach, initially developed by Borcard et al. (1992) and since refined in various ways, uses multivariate ordination techniques to decompose the amount of variation in the distribution of species in a metacommunity that could be explained by spatial or environmental differences among sites and how much variation was unexplained.

Cottenie et al. (2003) used this variation partitioning approach to evaluate the role of environmental and spatial components in determining zooplankton composition in a highly interconnected metacommunity of 35 Belgian ponds. They found that environmental differences among ponds explained a high proportion of the variation in communities even though they also measured high rates of dispersal between the ponds. Likewise, Gilbert and Lechowicz (2004) used this approach to examine patterns in the abundance and distribution of understory species in a temperate forest. They purposefully sampled the metacommunity so as to decouple possible spatial and environmental correlations and, again, found strong correlations between species distributions and environmental effects, but no spatial ones.

Cottenie (2005) was the first to propose that this variation partitioning method could provide a test of the various metacommunity archetypes (SS versus ME versus NT and PD), and the use of this compositional variation partition has skyrocketed. A recent review by Soininen (2014) identified more than 50 studies, and more than 300 comparisons within those, in which the relative influence of environment and space were partitioned; dozens more have been published since.

We briefly overview the mechanics behind the approach here in the context of interpreted metacommunity patterns, but we note that there are many complexities in the analyses (Peres-Neto et al. 2006, Blanchet et al. 2008, Peres-Neto and Legendre 2010, Dray et al. 2012, Legendre and Legendre 2012) and interpretations (Gilbert and Bennett 2010, Smith and Lundholm 2010, Stegen and Hurlbert 2011, Tuomisto et al. 2012) that will need to be carefully considered for anyone wishing to use the analyses themselves.[2]

As shown in Figure 4.8, variation in the site × species matrix is examined in relation to both the site × environment matrix and the site × space matrix. From this, we can calculate the fraction of the total variation that is explained by seven different components (after Peres-Neto et al. 2006): (1) the fraction explained by

---

[2] We are aware of issues that have been raised about the ability of variation partitioning to adequately address how environmental gradients affect distributions (references in text). We believe the issue is due largely to the fact that spatial factors are modeled using powerful nonlinear pattern-detection methods (such as principle coordinates on neighbor matrices [PCNM], Moran eigenvector maps [MEM], and asymmetric eigenvector maps [AEM]), whereas environmental variation is currently implemented primarily using linear methods such as reduced dimensional analysis (RDA) and is much less sophisticated. This means that nonlinear environmental patterns (e.g., the effects of species that occur in the middle of gradients but not at the endpoints) are not adequately accounted for in the environmental component of the variation partitioning. If these aspects of environmental patterning are spatially structured (as might often be the case), they would consequently be "captured" by the spatial component of variation partitioning instead and lead to difficulties in interpretation. In principle this can be solved by using nonlinear modeling frameworks for the environmental components, but they have yet to be implemented (Peres-Neto, pers. comm.).

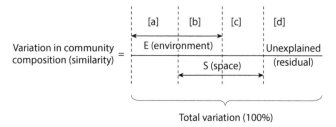

FIGURE. 4.8. Conceptual diagram of the variation partition of compositional variation among sites in response to environmental [E] factors, spatial [S] factors, and unexplained variation (calculated as adjusted R^2 statistics). Total variation can be disentangled using the following partition: (1) calculate fraction explained from both E and S predictor matrices (a+b+c); (2) calculate fraction based on the E matrix (a+b); (3) calculate fraction based on the S matrix (b+c); (4) the unique fraction of variation explained by E (a=[a+b+c] − [b+c]); (5) the unique fraction of variation explained by S (c=[a+b+c] − [a+b]); (6) the common fraction of variation shared by E and S (b=[a+b+c] − [a] − [c]); and (7) the residual fraction of variation not explained by E or S (d=1−[a+b+c]). Modified from Peres-Neto et al. 2006.

both environment and space, [E+S]; (2) the fraction explained by environment, [E]; (3) the fraction explained by space, [S]; (4) the fraction explained by environment when space is controlled, [E|S]; (5) the fraction explained by space when environment is controlled, [S|E]; (6) the fraction explained by both environment and space together, [E∩S], and (7) the fraction of variation not explained by the measured environmental or spatial variables, [Resid].

In the context of metacommunities, [E] is typically assumed to indicate the degree to which SS processes control community structure and [S] is assumed to indicate the degree to which the dispersal characteristics of either NT or PD (which are not distinguishable in this context) controls community structure (Cottenie 2005, Soininen 2014). Some authors have also suggested that ME, or a combination of mechanisms, is indicated when both [E] and [S] explain community variation (Cottenie 2005). Furthermore, although many authors ignore the amount of unexplained variation [Resid], it in fact can contain important information about the structuring of a metacommunity. Although both NT and PD predict spatial patterning (high [S]), they also predict a considerable amount of stochasticity in community composition, which could lead to high [Resid] (Legendre et al. 2009). Unfortunately, the interesting component of [Resid] is often embedded within a large component of artefactual unexplained variation, such as that driven by unsampled environmental components (Chang et al. 2013) and probabilistic sampling effects (Legendre et al. 2009, Myers et al. 2013), making interpretations of this component tenuous without a more explicit attempt to disentangle these effects.

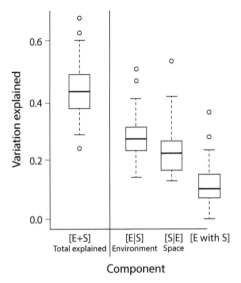

FIGURE 4.9. Results from a meta-analysis of 158 data sets with site × species data, along with spatial and environmental variables. The proportion of variation in community composition explained by environmental [E] and spatial [S] factors is shown. On the left of the vertical line is the total variation explained by both factors [E+S] and on the right of the vertical line is the breakdown of that variation into environmental [E], pure spatial [S], and spatial with environmental [E∩S] variation. The boxes represent 25, 50, and 75 quartiles, the whiskers are the minimum and maximum values, and the outliers are outside of 1.5 times the interquartile range. Modified from Cottenie 2005.

Finally, the component [E∩S], which consists of variation explained by confounded effects of environment and space, is important even though it is hard to dissect.

As an early attempt at the synthesis of the disparate perspectives, Cottenie (2005) used this variation partitioning approach to analyze 158 data sets in which site × species data were available along with environmental and spatial correlations. Across all data sets, he found that almost half (48%) of the variation in community structure could be explained by environment and space ([E + S]), with the remaining half unexplained. Within the proportion explained, 22% was explained by environment when space was controlled ([E|S]), 16% by space when environment is controlled ([S|E]), and 10% by spatially structured environmental variation ([E∩S]) (Fig. 4.9). Interestingly, Soininen (2014) found that [E] explained a very similar proportion of the variation (26%) using a different meta-analytic approach that included 326 data sets in all (including those in Cottenie 2005).

In addition to this overall approach, Cottenie (2005) was interested in more specifically partitioning the different data sets into categorizes on the basis of which model best fit a given metacommunity. He defined an SS metacommunity as one in which [E|S] was significant but [S|E] was not significant, an NT or PD meta-community as one in which [S|E] was significant but [E|S] was not significant, and a combination SS+ME metacommunity as one in which both [E|S] and [S|E] were significant; other metacommunities were either unable to be categorized or had no significant variation explained. With these distinctions, Cottenie found that of the 158 data sets, 69 (44%) were most consistent with SS, 46 (29%) were either ME or SS+ME, and only 13 (8%) were consistent with the NT/PD archetypes of strong spatial structuring; the remaining data sets were not defined. Taking these on face value, then, we would say that environmental factors appear to be much more important in determining community composition (73% of studies) than spa-tial factors (8% of studies). However, it is important to note that Cottenie assumed that residual variation was uninformative despite the fact that some of it could re-flect real stochastic processes operating in a metacommunity. Depending on how much of the residual variation emerged from real stochastic processes (e.g., colonization-extinction dynamics in PD or ecological drift in NT), the conclusions of the predominance of SS metacommunities might need to be tempered.

4.5 VARIATION IN THE RELATIVE IMPORTANCE OF METACOMMUNITY-STRUCTURING PROCESSES

For a few years, metacommunity ecologists were in the trenches, pitting one the-ory categorically against another (most predominantly SS vs. NT) with each side claiming victory in battles but with no clear winner of the war. We soon realized, however, that rather than being opposing views, the various metacommunity archetypes might better be thought of as a continuum rather than as discrete alter-natives (Alonso et al. 2006, Gravel et al. 2006, Leibold and McPeek 2006, Adler et al. 2007, Chase 2007).

Rather than asking whether SS or NT (or PD, or ME, or any mixture) explains a given metacommunity, the focus has turned toward quantifying the degree to which the different core processes (e.g., environmental filtering, dispersal limitation, stochasticity) explain patterns in a given metacommunity and what factors might influence their relative strengths. For example, properties of the metacommunity itself can influence the relative importance of metacommunity processes; [E] is more likely to leave a signature on a metacommunity with high environmental heterogeneity, whereas [S] can be more important in metacommunities with more isolated patches. Likewise, traits of the species being examined can influence the importance of metacommunity processes: smaller species might be able to partition environmental heterogeneities more finely than larger species and be structured more by [E], whereas species with weaker dispersal abilities are more likely to be structured by [S].

To make the idea of viewing metacommunities along a continuum of environmental and spatial properties more concrete, we contrast two very different metacommunities each of which has had a significant part in the development of disparate metacommunity theories: (1) metacommunities of small ponds that differ in a number of abiotic and biotic variables in Michigan and have played an important role in the development of the SS metacommunity archetype; and (2) metacommunities of tropical rainforest trees, most notably the 50-ha long-term research plot on BCI, in Panama, which had a fundamental part in the development of the NT metacommunity archetype.

Freshwater ponds have played an unusually strong role in the development of the niche theory that forms the basis of the SS archetypes. Notable highlights of the long history of an SS view in ponds includes Hutchinson's (1959, 1961) foundational essays regarding the coexistence among competing species and patterns of diversity, Tilman's (1976, 1977) initial empirical tests of consumer resource and competition models, and Werner and Hall's (1977, 1979) explorations of competition and coexistence among competing sunfishes. In metacommunities of small ponds in Michigan, a general finding is that there are strong compositional differences among groups of species in response to varying abiotic and biotic variables, including amphibians (Wilbur 1972, Werner and McPeek 1994, Skelly et al. 1999, Werner et al. 2007), dragonflies and damselflies (McPeek 1998, McCauley 2007, McCauley et al. 2008), amphipods (Wellborn 1994), zooplankton (Leibold 1998, Tessier et al. 2000), and snails (Chase 2003b, Hoverman et al. 2011). It was this system that both of us often relied on for insights as we worked to synthesize aspects of the SS theoretical archetypes in a metacommunity context (Leibold 1995, 1996, 1998; Chase and Leibold 2003).

In contrast to the SS archetypes, Hubbell's (2001) NT was developed very clearly within the backdrop of trees in highly diverse tropical rainforests. A main impetus for Hubbell's development of the NT was his dissatisfaction with traditional niche-based archetypes, which suggested that the numbers of species in a system was limited by the numbers of resources available to those species. Hubbell argued that this traditional niche-based view was inadequate to explain the exceptionally high diversity of species in tropical rainforests and other diverse types of metacommunities, so he took the opposite tack, devising a theory that could explain high diversity without considering niches at all; hence, Hubbell's focus on dispersal limitation and stochasticity (Hubbell 1979, 2001; Hubbell and Foster 1986). As he developed and refined his ideas, Hubbell and colleagues

TABLE 4.2. Results of a Variation Partition of Environmental and Spatial Effects
on Two Disparate Metacommunities

Metacommunity	Explained Variation [E+S]	Unexplained Variation [Resid]	Pure Space [S|E]	Pure Environment [E|S]	Environment with Space [E∩S]
Benthic invertebrates and amphibians in Michigan ponds[a]	0.59	0.41	0.03	0.56	0.0
Trees within Barro Colorado Island forest[b]	0.54	0.46	0.29	0.0	0.25

Note: E, environment; S, space.
[a]Traditionally associated with species sorting.
[b]Traditionally associated with neutral theory.

established a large-scale, long-term monitoring program in a diverse tropical rainforest on BCI starting in 1980. After 30 years of monitoring and more than 200 papers published with these data, Condit et al. (2012) concluded that the dispersal-limited view of NT provided a strong explanation of BCI's dynamics.

Here, we used variation partitioning to ground our discussion of the variation in the relative importance of spatial and environmental processes and its underlying cofactors. We specifically contrast a partition of the variation in the community composition of benthic invertebrates (mostly arthropods and mollusks) and larval amphibians among 29 ponds that vary in a number of biotic (e.g., presence/absence of fish) and abiotic (e.g., pond permanence, water chemistry) features in southwestern Michigan (data from Chase and Leibold 2002; Chase 2003a, 2003b; Chase and Ryberg 2004; unpublished data), with a partition of the variation in community composition of trees at BCI (from Baldeck et al. 2013b). Perhaps not surprisingly, the variation partitions largely conform to the divergent archetypes of these two metacommunities (Table 4.2). The Michigan pond metacommunity has a strong response to environmental variation but little or no influence of space, whereas the BCI trees have a strong response to spatial variation but little or no influence of environment. Both metacommunities have a considerable amount of unexplained variation, some of which may be due to unmeasured environmental variation and other sampling artifacts, but some of which is plausibly due to the sorts of drift processes envisioned in the NT. The environmental influence in the BCI metacommunity was entirely spatially structured ([E∩S]), indicating that the influence of environmental variation (measured and unmeasured) was itself spatially structured; this was not the case for the Michigan pond metacommunity, where important environmental variables, including pond permanence, canopy coverage, and water chemistry, were not spatially structured.

On the surface, the difference in the variation partition between these two metacommunities suggests that different processes underlie their dynamics—with SS underlying pond metacommunities with little influence of dispersal and NT/PD underlying the BCI metacommunity with little influence of environmental filtering. It also highlights a critical bias that probably enters into the psyche of metacommunity ecologists as they develop their worldviews. The Michigan pond metacommunity contained habitats that varied considerably in a number of variables known to strongly influence the distributions of a suite of important taxa in this region, including the presence or absence of fish that are important predators; permanence (i.e., whether they retain water throughout the year or are ephemeral); and physical (e.g., canopy cover) and chemical variables (e.g., nutrients, alkalinity) (Wellborn et al. 1996, McPeek 1998, Werner et al. 2007, McCauley et al. 2008, Hoverman et al. 2011). And even though dispersal limitation is sometimes implicated in experimental manipulations of these systems (Forbes and Chase 2002, Shulman and Chase 2007, Howeth and Leibold 2010b), these studies are often conducted in small mesocosms where dispersal might be more necessary to maintain diversity than in the larger natural ponds. Indeed, Shurin (2001) found little evidence for dispersal limitation influencing intact zooplankton communities in an experimental manipulation in natural ponds. Furthermore, many of these species can quite readily disperse long distances (Bohonak and Jenkins 2003), supporting the notion that dispersal limitation may often play less of a role in this system. This is not to say, however, that dispersal is always unimportant in this sort of system. For example, Cottenie and De Meester (2003) found a significant role for spatial processes controlling zooplankton community structure in a series of ponds that were interconnected by overflows and rivulets that were constructed within a large wetland in Belgium ([S|E] explained 14.8% of the variation), despite the fact that there was still an important environmental component ([E|S] explained 17.3% of the variation). Likewise, Chase and Ryberg (2004) showed that natural ponds that were less isolated tended to have reduced levels of residual (unexplained) variation in community composition.

At the same time, there are several good reasons why the BCI forest plot might be expected to show a much stronger influence of space, with little influence of environment. First and foremost, although there are some edaphic heterogeneities within the plot, it remains one of the most homogeneous in the growing network of similar forest plots (Brown et al. 2013). In addition, many species of trees in the BCI plot appear to be dispersal limited (Dalling et al. 2002, Seidler and Plotkin 2006), and experimental manipulations enhancing seed dispersal have indicated an important role for dispersal limitation among many species in this system

(Svenning and Wright 2005). In contrast to the results from BCI, other studies in tropical forests have found a stronger role for environmental variation in addition to spatial effects, such as the study by Gueze et al. (2013) in Bolivia ([E|S] explained ~18% of the variation, [S|E] explained ~25% of the variation, and [E∩S] explained ~12% of the variation).

Figure 4.10 allows us to visualize how these two disparate metacommunities (Michigan ponds and BCI) that played a role in the development of the divergent metacommunity archetypes (SS and NT, respectively) stack up against the 158 metacommunities compiled by Cottenie (2005) with respect to the relative roles of environment, space, and residual variation. In order to visualize these along three axes and deal with the [E∩S] component in an unbiased way, we have allocated the variation in [E∩S] to environmental and spatial variation in proportion to their "pure" contributions. More-sophisticated approaches for disentangling [E∩S], such as exploring variations across scales (Diniz et al. 2012) or traits (Peres-Neto et al. 2012), are available but beyond the scope of what we can accomplish here. Two observations are immediately apparent. First, the BCI and pond metacommunities appear to be relatively extreme cases. BCI has a much smaller influence of environment than any other metacommunity, although the influence of space is comparable to that in a few others. The pond metacommunity has a much smaller influence of space than any other metacommunity, although the influence of environment is comparable to that in several others. Second, the total unexplained variation in each metacommunity is a bit more than 50%, which is right in the middle of the range for that of the other metacommunities. Although we cannot know their relative influence for any given metacommunity, a combination of unmeasured variables, sampling effects, and stochasticity due to ecological drift contribute to this residual component and are clearly quite important (on average, 51% ± 12% of the variation in any given metacommunity was not explained by space nor environment.

If anything, Figure 4.10 seems to show that the environmental component is generally stronger than spatial patterning (there are more points in the lower left than upper right of the triangle), as also concluded by Cottenie (2005) and Soininen (2014). The way we have allocated the [E∩S] variation should also be kept in mind. Because we assumed proportionality, almost all of that variation in BCI is allocated to spatial effects because the pure environmental factors [E|S] are so small. However, even in BCI's relatively homogenous landscape, there are important topographic and soil variables that have a significant influence on tree-species distributions (Harms et al. 2001, John et al. 2007, De Caceres et al. 2012), even though these factors are themselves strongly spatially structured.

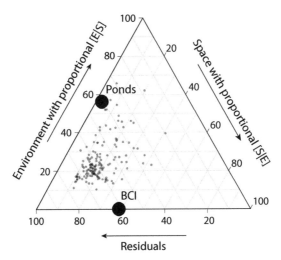

FIGURE 4.10. Visualization of the proportion of variation explained by space, environment, and unexplained variation from each of the 158 metacommunities presented by Cottenie (2005). To simplify the figure, the variation attributable to joint environment and space ($[E \cap S]$) was allocated to environment and space in proportion to their pure forms ($[E|S]$ and $[S|E]$); other ways of presenting the data showed similar patterns. The two extreme data sets discussed in text—tropical forest trees from Barro Colorado Island and benthic invertebrates and amphibians from Michigan ponds—are represented as large dots.

4.6 FACTORS THAT INFLUENCE VARIATION IN METACOMMUNITY PATTERNS

Can we explain or understand the disparate variation illustrated in Figure 4.10? The most important thing to recognize is that even though metacommunities are differentially structured, the broader universe of metacommunities can contain all combinations of the relative importance of environment, space, and unexplained factors. Thus, any synthetic treatment will require understanding the properties of a metacommunity that would influence the relative importance of these factors (though Fig. 4.10 suggests that some combinations are less likely, such as high environment and space but low residuals).

4.6.1 Effects of the Number of Measured Variables

One of the most widely recognized limitations of variation partitioning is that the degree to which environmental variables influence composition depends on having measures of the variables that are most relevant to the metacommunity of interest.

If an unmeasured environmental variable is spatially structured, it would enter into the [S] component; if it is not spatially structured, it would enter into the [Resid] component. This problem not only influences quantitative conclusions, such as those of Soininen (2014), who attempted to examine factors creating variation in the degree to which metacommunities were structured by [E], but also can lead to a fundamental "flip-flop" in conclusions regarding whether a metacommunity better conforms to a dispersal-limited system characteristic of NT/PD or to an environment-limited system of SS. For example, initial analyses of the tropical forest plots using this variation partitioning approach used only topographical variables in its description of [E] and concluded that a great deal of the variation in these metacommunities was due to [S] and [Resid] (Legendre et al. 2009, De Caceres et al. 2012). However, more recent studies, which included soil characteristics as well, led to conclusions that suggested a much stronger role for environmental factors (Baldeck et al. 2013a, Chang et al. 2013).

A simple illustration of this principle is shown in Figure 4.11, which uses the data set from Michigan ponds, indicating that the amount of variation explained by [E] declines and [Resid] increases as the number of environmental variables considered declines. Note that [S] remains unchanged because there was no [E∩S] in this metacommunity. However, if there were a strong [E∩S] component, reducing the measured environmental parameters would lead to an increase in the observed [S] component even if dispersal limitation itself was not important.

4.6.2 Effects of Habitat Heterogeneity

The BCI metacommunity was deliberately established across a relatively homogeneous environmental template, whereas the metacommunity from Michigan ponds was deliberately collected across distinct habitat types known to influence species distributions. The conclusion that [E] played a weaker role in BCI than in the Michigan ponds, then, becomes almost self-evident. In this context, the debate regarding the relative importance of SS versus NT as it is manifest in the role of environmental, spatial, and stochastic factors becomes much more a question of context (the degree of heterogeneity included, in this case) rather than substance (Chase 2014).

Imagine the probable scenario in which dispersal limitation and stochastic drift create community-level variation within a given habitat type, but habitat heterogeneity in environmental conditions plays an important role in determining the distributions among habitat types. If we were to sample only the metacommunity from within one part of the heterogeneous environment, there would simply not be enough variation in environmental conditions for there to be much of an influence

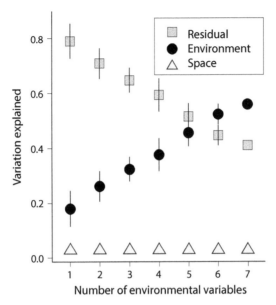

FIGURE 4.11. Results showing how sequential removal of environmental variables from the metacommunity of Michigan ponds leads to decreased proportions of the variation explained by environmental variables [E|S] and increased the unexplained [Resid] component. In this case, the proportion of variation explained by space was initially low and did not change with a reduction in the environmental variables included because there was no [E∩S] component. There is no variation for the estimate for all seven variables, there are six possible combinations of six variables, and seven combinations of one variable; for all other environmental variables, the average and standard deviation is given from 20 combinations of the given number of variables.

on community compositional variation (i.e., [E] will necessarily be low), and instead, the observed variation would necessarily be partitioned between [S] and [Resid]. Alternatively, if we were to sample across multiple heterogeneous habitat types, there would be much more opportunity for [E] to play an important role, which would necessarily drive the relative importance of [S] and/or [Resid] down (see also Steinbauer et al. 2012, Tuomisto et al. 2012, Brown et al. 2017 for general discussions about how spatial scale and sampling can influence these patterns).

Figure 4.12a presents a schematic for how the relative influence of [E], [S], and [Resid] might differ if we were vary the degree to which the different proportions of heterogeneity were sampled; at one extreme, where little heterogeneity is sampled within the metacommunity, results show a small influence of [E] (similar to what is observed at BCI), whereas at the other extreme, where much heterogeneity is sampled, a stronger influence of [E] may be observed (similar to that observed among Michigan ponds). Figure 4.12b illustrates this principle using the data from the Michigan pond metacommunity, which included at least four distinct habitat types along axes known to strongly influence species compositional distributions (fish vs. no fish; open vs. closed canopy; permanent vs. temporary water). When qualitatively distinct habitats are removed from the analysis, the relative importance of [E] goes down, [S] increases slightly, and [Resid] increases dramatically (this remains true whether the numbers of local communities are held constant in the analysis or allowed to vary). This result emphasizes that the extent of habitat heterogeneity included in an analysis is a critically important factor that must be considered when interpreting the relative importance of environmental and spatial controls of metacommunity structure. Indeed, despite the known strong influence of [E] in driving species distributions across intense habitat filters such as pond permanence or presence of fish, a more nuanced view emerges when examining variation within these habitat types, emphasizing an important role for dispersal limitation and stochasticity within the habitat-type boundaries (Chase and Leibold 2002, Chase 2003a, 2007, 2010). Similarly, strong habitat filters are known to play important roles in driving the distributions of tropical trees, but only a few of those filters are likely to occur within the 50-ha plot at BCI due to the way the site was selected (Garzon-Lopez et al. 2014).

Given their differential incorporation of habitat heterogeneity, we argue that the major reason for the divergent results and perspectives among metacommunities such as between BCI and the Michigan ponds is because the scale of BCI, and the amount of habitat heterogeneity it encompasses, is simply too small to be able to detect much influence of [E]. While it may sound absurd to suggest that the 50-ha plot of BCI and similar forest plots are small, while at the same time suggesting that the small ponds within a single site in Michigan are large, we have to scale this perspective to the size of the individuals and the amount of heterogeneity they experience. The approximately 240,000 individual trees on the BCI plot may seem like a lot, but it is orders of magnitude less than the millions to billions of invertebrates and amphibians in the Michigan pond metacommunity. On a per-individual basis, the BCI plot is probably equivalent to studying the distributions of organisms within a part of one of the 29 ponds in our survey. The environmental gradients faced by the organisms in the pond metacommunity, including variation in the presence or absence of important consumers, resource availability, and major disturbances (ponds drying), would perhaps be more like studying the distributions of trees across Central America, including sites with and without important browsing herbivores, sites impacted by hurricane or fire disturbances, and sites with the full range of variation in rainfall and soil chemistry (such as in the study by Condit et al. [2002], who showed deviations from NT predictions at this scale).

While habitat heterogeneity can clearly play an important role in mediating the relative importance of spatial and environmental factors, the degree of

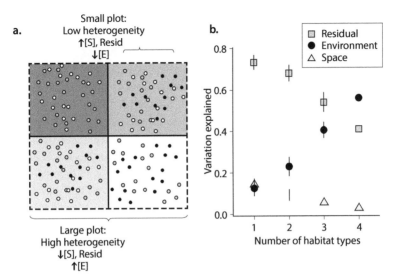

FIGURE 4.12. (a) A schematic illustrating the principle of the role of habitat heterogeneity in detecting the influence of the different components explaining variation in community composition. There are four species, denoted by different shades of gray, each of which dominates one quadrant of the entire landscape but also stochastically appears in the other habitat. The small box in the top right represents a sample that comes from within a single habitat type and includes little environmental variation and a random assortment of three of the four species; it will have a small proportion of the variation in community composition explained by environment (low [E]). The large box encompasses the entire landscape and its constituent habitat heterogeneity; it will have a larger proportion of the variation in community composition explained by environment (high [E]). (b) An illustration of this principle using the data from the metacommunity of invertebrates and amphibians in Michigan ponds. Four key habitat types emerge along important orthogonal environmental axes (permanent ponds with fish, fishless permanent ponds with open canopy, fishless permanent ponds with closed canopy, and fishless temporary ponds). When the number of habitat types considered was reduced, the amount of variation explained by [E] declined, the amount of variation explained by [S] increased slightly, and the amount of variation unexplained [Resid] increased greatly.

heterogeneity is defined by what the organism experiences. Some groups of species may respond to much finer levels of habitat heterogeneity than others, and it is the scale at which these organisms respond relative to the scale at which environmental variables are measured that will determine the degree to which environmental variables predict a species' distribution (de Knegt et al. 2010). For example, small-statured plants (e.g., herbs and shrubs) can respond to smaller-scale variation in soil and topographic conditions than can larger plants like trees. Likewise, the degree to which species respond to heterogeneity will depend on the scale at which the heterogeneity occurs and the movement rates of species across heterogeneity boundaries.

As a result, body size and dispersal rates are often found to mediate the relative importance of environmental versus spatial factors. Even within the same metacommunity, larger and less dispersing species may be less able to respond to habitat heterogeneity and are thus are often more structured by spatial processes, whereas smaller and better dispersers are often more structured by environmental process (Hajek et al. 2011, De Bie et al. 2012, Heino 2013a). Although Soininen (2014) did not find this pattern when comparing all data sets in a meta-analysis, we suspect that it was probably due to the fact that that study compared data sets that encompassed different degrees of habitat heterogeneity and thus could not disentangle dispersal mode effects from differences in the extent of heterogeneity included in the metacommunity. Likewise, Farjalla et al. (2012) found that environmental differences among water-filled bromeliad leaves explained almost none of the variation in bacterial communities, but a much larger proportion of the variation was explained by environmental features for larger organisms, such as zooplankton and macroinvertebrates. Farjalla et al. attributed these patterns to variation in dispersal rates and niche plasticity, but we note that a much more parsimonious explanation is simply that the spatial scale at which environmental variables were measured for all groups of species—the bromeliad level—were much less relevant to the bacteria (which probably respond to environmental heterogeneity at much finer-scale levels) than to the larger organisms.

In addition to the organisms within different metacommunities having different dispersal modes, and thus experiencing different relative proportions of [E] versus [S], metacommunities themselves can vary in their connectance and the degree to which organisms can disperse among patches.

4.6.3 Effects of Metacommunity Connectance and Dispersal Rates

In the context of metacommunity structuring processes, varying dispersal rates can have a complex relationship with the relative importance of environmental and spatially structuring processes, one example of which we present in Figure 4.13. At the lowest rates of dispersal, dispersal limitation inherent to NT and PD may play a more important role, leading to metacommunities more structured by [S] than [E]. At intermediate rates of dispersal, dispersal limitation is less likely and SS processes can predominate, leading to metacommunities more structured by [E] than [S]. Finally, at the higher rates of dispersal, ME can become important, leading to a reduction in [E] and an increase in [S]. Though other patterns can emerge, this complex dependence of patterns on scale and dispersal illustrates the difficulty in ascribing mechanism to patterns such as the role of [E] and [S] in explaining compositional variation. While some authors (e.g., Cottenie 2005) have

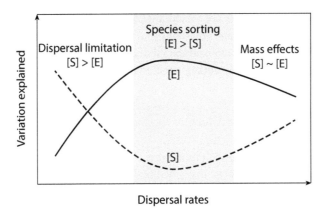

FIGURE 4.13. A hypothetical relationship between dispersal rates within a metacommunity and the relative influence of environmental [E] and spatial [S] factors. At the lowest dispersal rates, dispersal limitation inherent to the patch dynamics and neutral theory models predominate and environmental factors are less important. At intermediate dispersal rates, environmental factors such as those associated with species sorting are most important. And at the highest dispersal rates, environmental factors are important, but so are spatial factors that result from the high rates of dispersal into adjacent less favorable habitats as a result of mass effects.

suggested that ME are indicated if both [E] and [S] are significant, this need not be the case because this pattern can also indicate a transition between PD or NT and SS processes. Additionally, some authors have equated important spatial processes [S] as indicative of PD or NT, whereas others have suggested that they are indicative of ME. While the pattern of [S] may be similar, the underlying mechanisms behind the spatial signal observed in dispersal-limited systems (species not present in all of the habitats they could live in) is very different from that observed in ME systems (species present in habitats where they could not otherwise persist.

Although it is impossible to definitively identify the specific processes underlying these patterns, one possible way to distinguish among the degree of [S] that results from dispersal limitation (i.e., PD or NT) from ME is to examine the component of variation in which space and environment cannot be disentangled, [E∩S]. This component is usually assumed to result from autocorrelations between spatial and environmental distances (i.e., more spatially distant habitats also differ in environmental conditions). However, this need not be the case, for example, if ME processes are operating. Here, [E∩S] can emerge if adjacent habitats that differ in environmental conditions provide a filter for species that nevertheless also "spill over" into adjacent, less favorable habitats due to ME. As a result of the two very different mechanisms that might lead to [E∩S], one way these might be disentangled is to compare [E∩S] with the degree to which environmental and spatial distances are actually correlated. If [E∩S] is high but there is only a weak correlation between spatial and environmental distances, we might suspect ME are playing a role. If environmental and spatial distances are strongly correlated, however, it is more difficult to disentangle the relative importance of these drivers.

4.6.4 Effects of Regional Pool Size

Another important thing to keep in mind is that the amount of variation in community composition that can be explained by [E], [S], or other components in any given metacommunity is constrained because the magnitude of variation in community composition depends on the number of species present in the regional pool of species (Chase and Myers 2011, Kraft et al. 2011, Myers et al. 2013). Just as species-pool effects can have a big influence on α-diversity and interpretations of local coexistence (Pärtel et al. 1996, Zobel 1997), β-diversity necessarily increases with the size of the regional species pool (Kraft et al. 2011, Bennett and Gilbert 2016). All else being equal (e.g., the same numbers of environmental variables are measured, the species partition those variables the same), we would expect by random chance alone that a metacommunity with a larger species pool would have proportionately less of its compositional variation explained by [E] or [S] than a metacommunity with a smaller species pool and it would thus have higher [Resid].

The principle of the influence of the species-pool size on the amount of unexplained variation in species composition [Resid] is nicely illustrated by a comparative study of forest plots across a latitudinal gradient (and thus gradient in species-pool size) by De Caceres et al. (2012; Fig. 4.14a). Because the variation partition must add up to 1, any increase in one component, [Resid] in this case, will necessarily lead to reductions in the others ([E] and [S]). Although there could be a biological component that also influences this pattern (e.g., higher stochasticity in larger species pools), it is necessary to partial out the purely probabilistic effect in order to determine whether any differences in the underlying processes may be occurring. Myers et al. (2013) show one way this can be done using a null model that first accounts for the compositional variation expected for any given species pool and then compares the relative influence of [E], [S], and [Resid] on the value of the deviations from the null expectation. They compared the patterns of compositional variation in forest plots in a moderate-diversity temperate forest in Missouri (46 species) to that found from a similar sampling regime in a high-diversity tropical forest in Bolivia (607 species; Fig. 4.14b). When they compared raw compositional variation, they found that only around 15% of the variation was explained in the tropical forest, whereas almost 30% was explained in the

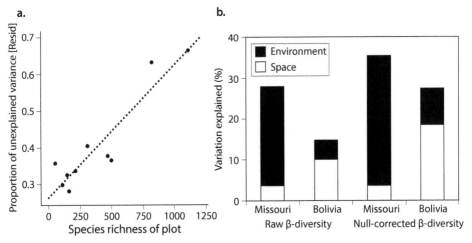

FIGURE 4.14. (a) The positive relationship between the proportion of variation unexplained by the variation partitioning approach and the species richness of forest plots from a network of plots across the globe ($N=10$, $R^2=0.8595$, P < 0.001; $y=0.2650+0.0003x$) Data from the 10-×10-m plots in De Caceres et al. 2012. (b) A comparison of the proportion of variation explained by [E] and [S] among a low-diversity temperate forest in Missouri and a high-diversity tropical forest in Bolivia when raw compositional variation (β-diversity) was used (left side) and when corrected for the size of the species pool (right side). Modified from Myers et al. 2013.

temperate forest. However, when they controlled for differences in regional pool size, the explained variation was much more similar between tropical and temperate sites (28% vs. 34% explained in tropical and temperate sites, respectively). After controlling for the null-expected differences in variation explained, Myers et al. were able to distinguish that the diverse tropical forests in Bolivia appeared to be more spatially structured (higher [S]), whereas the temperate Missouri forest was more environmentally structured (higher [E]), indicating potential differences in the importance of dispersal limitation in these different metacommunities. Nevertheless, it is important to point out that the form of the null model and interpretation of deviations from the expectation remains under debate (Tucker et al. 2016, Ulrich et al. 2017).

Thus, caution must be used when comparing the amount of variation explained by [E], [S], or other components among metacommunities that differ in the size of their regional pools. This is true both for studies that compare the influence of the different components, such as the role of [E] among different metacommunities within a meta-analysis (Soininen 2014), as well as for studies that compare and contrast the influence of the different components on groups of species that have different-sized regional pools (Hajek et al. 2011).

Returning to our comparison of the tropical forest and the pond, the metacommunity in the Michigan ponds examined above consisted of 51 species, whereas the BCI metacommunity has more than 300 species. Thus, even if BCI was as strongly influenced by SS as the Michigan ponds, the raw value of [E] would almost certainly be less at BCI simply because of probabilistic constraints.

4.7 CONTRIBUTIONS OF DIFFERENT SPECIES AND DIFFERENT LOCALITIES TO THE OVERALL METACOMMUNITY PATTERN

One important feature of the variation partitioning analysis that we have avoided so far is that it includes all of the species in a given community (or at least those surveyed), as well as all of the habitats within that community, and assumes that they all respond similarly. Clearly, however, this is not true, and it is possible that we can dissect a bit more information about a given metacommunity by training our perspective more finely on the details of the species and the localities within the metacommunity. A few features of metacommunities to consider include the following:

1. Not all species in a metacommunity have the same dispersal rate. If the dispersal rates among species are reasonably similar, this may not matter, and in fact, this is expected in the context of PD models that assume trade-offs in dispersal ability that are nevertheless highly constrained (Calcagano et al. 2006). However, if species within a metacommunity greatly differ in their dispersal rates or distances, they may respond to space and environment quite differently. Thus, a single metacommunity could include some species that are more spatially structured and others that are less so.
2. Not all species are likely to respond to the same levels or types of environmental heterogeneity (e.g., habitat specialists vs. generalists). Thus, a metacommunity could have species that are more environmentally structured and others that are less so.
3. Patches within a metacommunity itself can also differ. For example, not all patches in a metacommunity are similarly connected (or conversely, isolated), suggesting that patches that differ in isolation may differ in the degree to which they are inhabited by different species. One possibility is that more-connected patches in a metacommunity might experience stronger spatial effects, whereas less connected ones might incur greater environmental effects.

Flinn et al. (2010) explored some of these effects by separating the responses of strongly versus weakly dispersing plants within wetland communities. As expected, weaker-dispersing species were more influenced by spatial processes (higher [S]), whereas stronger-dispersing species were more influenced by environmental processes (higher [E]). Likewise, Alexander et al. (2012) found that spatial patterning was more prominent early during the successional processes, whereas environmental structuring became more important later in succession.

A more general approach to quantifying the contribution of species and localities can be gained, for example, using the model by Fournier et al. (2017) that we described in Chapter 2. If a given empirical landscape with a given set of species and their traits is provided to the model, it is possible to calculate how they contribute to the overall distribution pattern using likelihood methods. Additionally, the model provides statistical tests that can evaluate the contributions of individual species and patches to observed patterns. Future work along these lines should help better resolve how such site and species influences affect patterns in metacommunities.

4.8 HOW WELL CAN WE HOPE TO DO IN EXPLAINING METACOMMUNITY STRUCTURE?

Among the different methods for discerning process from pattern, the variation partitioning methods are the most sophisticated in capturing elements of metacommunity processes that might affect patterns in natural metacommunities. At a minimum they reveal the potential influence of environment (at least for those components that we measure in our studies) and the influence of spatial patterning (at least for those components that are revealed by the principle coordinates on neighbor matrices [PCNM] and related methods used). They also help us resolve how to account for spatially patterned environmental factors. Overall, the summary statistics from Cottenie (2005; Fig. 4.9) and more recent syntheses (Soininen 2014, 2016) indicate that environment by itself often accounts for about 20%–30% of the variation in metacommunity composition and that adding spatial patterning accounts for about an extra 20% or so. This leaves 50% or more of the variation in metacommunity structure unexplained by either environment or space. This somewhat low explanatory power together with other statistics-based critiques of the variation partitioning approach (Gilbert and Bennett 2010, Smith and Lundholm 2010, Stegen and Hurlbert 2011, Tuomisto et al. 2012, Brown et al. 2017) might make one wonder whether the whole endeavor of trying to explain metacommunity pattern from core metacommunity processes is worth it.

How might we explain so much residual variation? First, there are a number of sampling issues, some of which were described above, including unmeasured environmental variables, undersampling rare species, etc. Second, there are many types of "true" random effects in metacommunities, such as demographic stochasticity and colonization-extinction stochasticity, that result in ecological drift at the metacommunity level that is neither spatially nor environmentally structured (Hubbell 2001, Vellend 2010, 2016). Third, metacommunity processes can be much more complex than simply responses to environmental filtering (influencing the environmental component of the variation partition) and dispersal (influencing the spatial component of the variation partition). Interspecific interactions and the interactions between environment and space lead to a highly complex and multidimensional context that influences species distributions that may appear "random" or otherwise "stochastic" but actually result from a quite deterministic underlying structure (Clark et al. 2007, Clark 2009). Frequency dependence and the interactions between space and time will similarly create complex metacommunity structures that may appear "random" without additional pieces of information (see Chap. 5 for a deeper discussion of these issues).

While we will never be able to improve the variation partitioning approach to be able to fully dissect process from pattern, there are ways that we can improve the overall approach in addition to simply recognizing some of its limitations. What can we do with the approximately 50% or so of the variation in species distributions that are not explained by environment and space? Is there is any way to disentangle the potential different reasons for the unexplained variation (e.g., due to sampling, complexity, etc.)?

We are working with colleagues to see if there is a way forward, which we will revisit in Chapter 14. Although preliminary, one approach is to see if there are patterns in the unexplained residual variation itself. After all, the only thing we have done with variation partitioning is pull out the parts that relate to environment and space. However, there might still be much structure left that is not related to these two static predictors. If present, these patterns would show up as correlations among the species themselves, at least in part. Thus, the correlation structure among species in the residual variation (not explained by either environment or space) tells us something about the degree to which species are distributed independently of each other.

As an example of the utility of this approach, Pedro Peres-Neto and Mathew Leibold (pers. comm.) looked at the correlation structure of this residual variation in a large data set of fish communities among approximately 3000 lakes in Canada and found that these correlations could account for a surprising amount of the structure of these fish communities (Table 4.3). The standard variation partitioning showed that 25.2% of the variation was explained by the combination of environmental [E] and spatial [S] components, leaving almost 75% of the variation as apparently unexplained. An analysis of latent variables, however, showed that correlations among the residuals account for more than half of the entire variation even though we do not necessarily know how to explain these correlations. Unstructured residual variation is less than 25% of the total variation however.

TABLE 4.3. Variation Partitioning and Subsequent Analysis of Residuals Using Latent Variables in Fish Communities in Canada

Component	Proportion of Variation Explained
Pure environment [E|S]	0.092
Pure space [S|E]	0.082
Environment with space [E∩S]	0.078
Apparent residuals	0.748
Latent variation in residuals due to correlations	0.507
Remaining residuals	0.241

Note: E, environment; S, space.

These results mean that much of the residual variation in this data set is not really random but instead indicate that there are patterns in the correlation structure. The factors that could be involved in this structure could be any of a number of possible mechanisms, including the sorts of ecological drift envisioned by NT (increases in one species due to drift are associated with decreases in other species due to the zero-sum expectation; see Chap. 3), a multitude of spatiotemporal processes, including multiple stable equilibria, assembly cycles, positive and negative (e.g., Janzen-Connell) frequency-dependence (see Chap. 5), and even unmeasured environmental variables (with no spatial structure) or spatial patterns at scales that could not have been detected with the methods used.

Remarkably, this sort of analysis can also be used to reveal the dimensionality of the correlation pattern. Do these correlations involve lots of more or less independent correlations between pairs of species (high dimensionality), or are these correlations themselves organized into more aggregated patterns along a few axes (low dimensionality)? We do not yet have a good idea of what we will find, but it seems that the dimensionality of correlation structure is much lower than the number of species or patches. That assumption means that whatever factors drive these correlations, they might not be hopelessly complicated. Nevertheless, we still have a challenge ahead to be able to relate these correlation structures to their underlying causes.

4.9 CAVEATS AND CONCLUSIONS

Of the various pattern analyses that we have examined here, ranging from the simplest (e.g., form of the SAD) to the most complex variation partitioning method (and their various diversification of methods), none really provide a fully satisfying way to quantify and interpret metacommunity processes from patterns. At this point, it seems like variation partitioning is most likely to furnish insights, at least into the various ways that different species and patch attributes interact with species interactions and dispersal in a landscape. However, as we have discussed above, interpretations of these patterns need to be very cautiously made and considered in the context of the degree of heterogeneity, the numbers of variables measured, the size of the species pool, and many other factors.

It is also useful to remind ourselves of the constraints that come from trying to interpret processes from static snapshots of patterns. Metacommunity processes can be very dynamic in time as well as in space, and the approaches based on patterns that do not evaluate temporal change can be misleading. One important issue is whether SS and niche-based community assembly (related to deterministic components of species interactions) always result in predictable correlations between environment and composition. If they do, then snapshots can describe the SS process because we assume that everything has come to equilibrium. However, what would we expect if there is no fixed-point equilibrium for local community assembly, or if it is a dynamic rather than static equilibrium? In Chapter 5, we describe how temporal patterns may also inform our understanding of metacommunity processes.

In this chapter we have also focused exclusively on species-level patterns; that is, patterns based on taxonomy and abundances or distributions. These have had the longest historical basis and have the advantage that they connect population-level processes (birth, death, immigration, and emigration as well as speciation and extinction) to metacommunity dynamics. However, recent efforts have also argued that much can be gained by taking a step back from this approach and shifting the emphasis to functional trait distributions or to *community phylogenetic* approaches (see Chap. 6). These approaches have the advantage that they may more directly allow us to evaluate how mechanisms related to such traits affect patterns, but they also remove us from the population-level processes of ecology. For example, how do we evaluate NT if we ignore taxonomy? Trait- and phylogeny-based approaches are thus probably complementary to the taxon-based approaches, which we discuss further in Chapter 7.

CHAPTER FIVE

Interactions between Time
and Space in Metacommunities

Prospectus

1. Temporal patterns of change in communities and metacommunities are
often closely related to spatial ones. Such patterns can thus provide ancillary
and often complementary information about metacommunity dynamics
that can help discern the relative importance of metacommunity assembly
processes.

2. The amount of temporal turnover within localities can depend on the
dynamics involved. Within the simple metacommunity archetypes, for
example, SS and ME are expected to show much less turnover than
PD and NT.

3. There are, however, some important issues that influence our interpretations
of temporal turnover patterns. Some of these involve sampling and
scaling issues, as well as appropriate statistical tests necessary for comparisons
(e.g., null models). And some of these involve consideration of
environmental changes that could create temporal turnover in community
composition even in metacommunities that conform to SS or ME archetypes
as the species "track" such changes.

4. Temporal turnover can involve either transient dynamics (e.g., tracking)
or a dynamic equilibrium that entails a balance between colonization and
extinction, resulting in relatively stable local diversity. The simplest predictions
are derived from the theory of island biogeography, but more complex
situations can also involve many other types of dynamics.

5. Temporal turnover seems to be a fairly ubiquitous feature of natural communities,
but it varies in a multitude of ways with mean local diversity,
latitude, and other factors.

6. There are several mechanisms that operate within local communities and
are related to deterministic features of metacommunity assembly that can
result in spatial and temporal turnover. These mechanisms include temporal
priority effects leading to multiple stable equilibria (MSEs) and endpoint
assembly cycles (EACs); in the latter, nontransitive interactions among
species affect the dynamics of metacommunity assembly. MSEs are associated
with much less temporal turnover than EACs, but both of these
reduce the degree of environmental regulation seen in metacommunities

because of a spatiotemporal component (as evaluated for example by the E component of variation partitioning). Both MSEs and EACs may be more likely in highly productive environments.

7. In spatially continuous metacommunities (our metacommunity sensu lato from Chapter 1), frequency dependence (both positive and negative) can create dynamic landscapes that experience considerable temporal turnover within localities despite being largely deterministically determined.

Because metacommunity ecology specifically addresses the role of dispersal and habitat heterogeneity, it is easy to see that spatial patterns are likely to be affected (Chap. 4). And yet spatial effects are not the only features that might be affected by metacommunity dynamics. Temporal dynamics can also be a way to gain insights into metacommunities. For example, the classic theory of island biogeography (MacArthur and Wilson 1963, 1967) predicts that species richness in a given patch is stable and results from a balance between colonization and extinctions, whereas species compositional changes through time is a dynamic process with continual turnover of species. Although less well developed than theoretical and analytical approaches for detecting metacommunity patterns in space, the increased availability of data and tools for analyzing fluctuations of metacommunity composition through time has renewed researchers' interest in patterns of temporal turnover in metacommunities (Korhonen et al. 2010, Dornelas et al. 2013, 2014). It is also easy to see that temporal turnover can interact with spatial turnover since both are likely to have reciprocal effects on each other (Adler et al. 2005, Soininen 2010, White et al. 2010, Stegen et al. 2013).

The four metacommunity archetypes from Chapter 2 place different emphases on the importance of temporal fluctuations and their underlying causes. PD and NT assume that temporal processes, including dispersal limitation, ecological drift, and colonization-extinction dynamics determine long-term regional coexistence and compositional variation. Here, temporal fluctuations in species composition in any given locality is expected regardless of any changes in environmental conditions. On the other hand, SS predicts that species composition in any locality is determined by a combination of habitat tolerances and interspecific interactions and should only fluctuate through time when the environmental conditions themselves vary (Chesson 2000, Chase and Leibold 2003).

Examining temporal fluctuations in species composition might help disentangle the relative roles of dispersal and drift versus environmental change factors via transient dynamics (Adler 2004, Soininen 2010, Stegen et al. 2013, Dornelas et al. 2014). In addition, PD, NT, and ME each predict that we would observe a spatial signal in the species composition of local communities in a metacommunity, but in the simple case, ME would not predict temporal change (assuming source-sink effects in relation to environments and dispersal are stable), whereas PD and NT would predict some degree of continuous temporal change within localities. So here, the interaction between spatial and temporal patterns can indeed help us better understand metacommunity dynamics.

In this chapter, we first discuss some of the temporal patterns that exist in metacommunities and explore ways that these can help provide some inferences about possible mechanisms. We then discuss some limitations of this simplistic perspective by showing how more complex scenarios for community assembly can generate temporal turnover even when deterministic processes predominate (i.e., without stochastic drift or colonization-extinction dynamics as in NT and PD).

5.1 TEMPORAL TURNOVER: WHAT DOES THEORY PREDICT?

While PD and NT assume that colonization-extinction dynamics are frequent, leading to considerable temporal turnover (sometimes referred to as temporal β-diversity, by analogy with spatial turnover), SS and ME assume much less temporal turnover in community composition (Chase et al. 2005). Building on these simple models, several authors have suggested that rates and patterns of temporal turnover (often combined with other information) might be taken as indicative of which metacommunity archetype is most appropriate for a given system and that variation in temporal turnover might indicate variation in the predominant mechanisms underlying metacommunities (Ellis et al. 2006, van der Gast et al. 2008, Pandit and Kolasa 2012, Dornelas et al. 2014).

Nevertheless, just as using spatial patterns to make inferences about metacommunities can be tricky, the same is true when using temporal turnover:

First, estimates of species turnover rates are typically measured with a community similarity metric or with the scaling exponent of the species-time relationship (Soininen 2010, White et al. 2010). As we described in Chapter 4 with spatial turnover (i.e., β-diversity), these measures are strongly sample biased and influenced by the size of the regional species pool (and sample size; see, e.g., Chase et al. 2011). Thus, simply examining raw temporal turnover measures and comparing them among metacommunities might reflect sampling probability even if the underlying mechanism of metacommunity assembly is the same, and appropriate null models are needed for proper comparisons of temporal turnover rates (see, e.g., Stegen et al. 2013 and Dornelas et al. 2014 for exemplary analyses in this context).

Second, even highly deterministic SS models predict strong temporal species turnover when environmental conditions are also highly variable; that is, species composition should track changes in environmental conditions through time as a

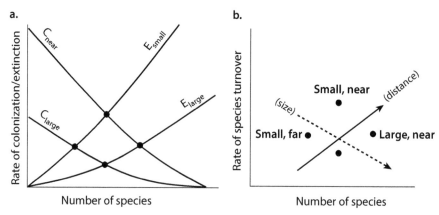

FIGURE 5.1. (a) A familiar application of MacArthur and Wilson's (1963, 1967) equilibrium theory of island biogeography. Colonization rates are higher on islands near (C_{near}) the mainland species pool than on islands far (C_{far}) from the species pool. Extinction rates are lower on larger islands (E_{large}) than smaller islands (E_{small}). Colonization is assumed to be unaffected by island size, and extinction is assumed to be unaffected by island distance. The point where the colonization and extinction lines cross is the dynamic equilibrium (denoted by a solid circle), and the expected species richness and turnover can be projected to the x- and y-axes, respectively (not shown). (b) is the same as (a) but with the colonization and extinction rates removed to emphasize the predicted relationship between local richness (x-axis value) and temporal turnover rate (y-axis value). Each point is labelled with respect to its size (large, small) and isolation (near, far). Arrows show that the relationship between local richness and turnover rate can be positive if higher richness results from higher colonization rates from far to near islands (solid arrow pointing to the upper right), or negative if higher richness results from lower extinction rates from small to large islands (dashed arrow pointing to the lower right).

transient dynamic. Here, temporal turnover is likely to scale with the magnitude of environmental change. Furthermore, deterministic processes of metacommunity assembly can sometimes also predict temporal turnover dynamics; for example, when they involve deterministic assembly cycles (discussed below) rather than single endpoint communities.

Third, completely stochastic models, like the theory of island biogeography (MacArthur and Wilson 1963, 1967), can also predict variable rates of species turnover, from very low to very high. For example, smaller islands are predicted to have higher rates of turnover than larger islands as a result of their assumed higher extinction rates (Fig. 5.1a). Likewise, islands that are further from the mainland source pool should have higher rates of turnover than islands closer to the source pool, owing to their lower assumed colonization rates. Here, the magnitude of community turnover differs between islands with different scenarios (it depends on size and isolation), but the mechanisms underlying metacommunity assembly are the same (Fig. 5.1b).

5.2 PATTERNS OF TEMPORAL TURNOVER

One way to study temporal turnover, which is closely related to the *species-area relationship* (SAR; which we discuss in depth in Chapter 10), is the *species-time relationship* (STR) (Preston 1960, White et al. 2006, 2010). It turns out that, like the SAR, the STR is quite ubiquitous in communities that have been observed for a sufficient length of time (White et al. 2006, Dornelas et al. 2014). Some of this ubiquity is almost certainly due to sampling issues, because the probabilities of observing rare species increase cumulatively through time. However, much of this turnover represents real compositional change through time.

Figure 5.2a shows some empirical evidence for the STR from some classic studies showing the consistent rise in cumulative species numbers with an increasing observation window (years), including data from Lepidoptera from Rothamsted (UK) collated by Fisher et al. (1943), data from birds on a 31-ha plot in Ohio described by Preston (1960), and data from plants in 1-m^2 quadrats at the Konza Prairie (Kansas) analyzed by Adler and Lauenroth (2003). Figure 5.2b shows the results from the analyses of more than 1000 STRs of a number of different ecosystem types and organisms (White et al. 2006, Shade et al. 2013), illustrating the ubiquity and rough consistency of STRs among groups or ecosystems.

One important factor that influences the shape of the STR is the amount of area encompassed by the sample; with more area sampled, and thus a larger proportion of the species pool, fewer species will be available to replace each other in the context of temporal turnover. This leads to a more generalized *species-time-area relationship* (STAR; Adler and Lauenroth 2003, Adler et al. 2005).

The average magnitude of STRs shown in Figure 5.2 is relatively uniform (the exponent is around 0.3) and largely consistent among groups of organisms and ecosystems (White et al. 2006), although microbes appear to have slightly higher rates of turnover (Shade et al. 2013). Nevertheless, there is a fair amount of variation across the more detailed studies. For example, White et al. (2006) found that communities with greater local species richness tended to have shallower STRs than those with less. Interestingly, Shurin et al. (2007) found that both species richness and temporal turnover decreased with latitude (suggesting the inverse of the pattern found by White et al. [2006]); a meta-analysis of temporal turnover in aquatic ecosystems confirmed this pattern (Korhonen et al. 2010). Clearly, much more needs to be done before we can really begin to understand these patterns and their variation, including development of appropriate scaling relationships and null models, as well as a clearer understanding of what metacommunity theories predict. We view this as an area ripe for future work.

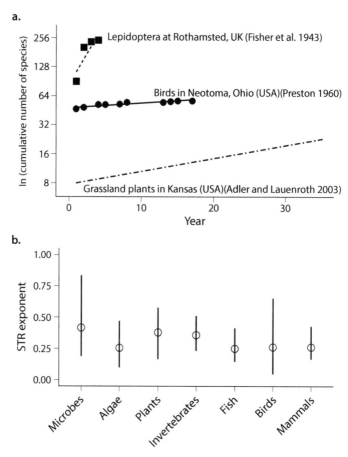

FIGURE 5.2. (a) Shows examples of three species-time relationships (STRs) from three classic studies, including Lepidoptera species observed from the Rothamsted experimental site over 4 years (squares; Fisher et al. 1943), birds observed in a 31-ha forest in Neotoma, Ohio, over 17 years (dots; Preston 1960), and plants observed in 1-m² plots at Konza Prairie, in Kansas, over 35 years (triangles; Adler and Lauenroth 2003; data from only one habitat type presented here). (b) Results from meta-analyses of STR analyses from a number of ecosystem types and across a number of organism types (circles are means; lines are ranges). While there are no systematic differences within groups of microbes (Shade et al. 2013) or macroscopic organisms (White et al. 2006), it appears that microbes have higher rates of turnover (exponent of about 0.4 on average) than macroscopic organisms (exponent around 0.3 on average). Modified from White et al. 2006 and Shade et al. 2013.

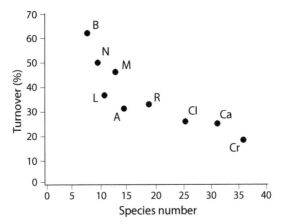

FIGURE 5.3. The percentage of turnover of bird species composition from 1917 to 1968 on several of California's Channel Islands. Letters refer to names of the different islands. Modified from Diamond 1969.

Despite the substantial turnover typically seen in the STR, the average numbers of species in a given locality is often unvarying. One might first be tempted to interpret such a pattern as being consistent with the theory of island biogeography (MacArthur and Wilson 1963, 1967). Indeed, in one of the first explicit tests of the theory, Diamond (1969) compared data on the diversity and composition of bird species on nine islands within the Channel Island archipelago off the coast of California from surveys conducted in 1917 and 1968 and found that islands that were more depauperate in species richness experienced much higher rates of turnover than those that were more species rich (similar to the meta-analytic results of White et al. 2006; Fig. 5.3). However, he observed that the patterns were more complex than the simplest predictions of island biogeography; turnover was not a result of island size or isolation but instead appeared to result from habitat quality, which influenced the equilibrium number of species. Nevertheless, Diamond ascribed the turnover patterns to colonization-extinction dynamics. Because Diamond derived his estimates of turnover from observations at only two time points and did not study possible shifts in environmental conditions that may have occurred in the time interval between samples, this study cannot eliminate the alternative hypothesis that environmental changes caused the shifts in species composition through time.

It is not just islands that show that turnover can occur even when local diversity is constant. Brown et al. (2001) demonstrated that similar patterns were observed in several long-term data sets from observations in mainland regions, some of which are shown in Figure 5.4. Although Brown et al. also could not discern if

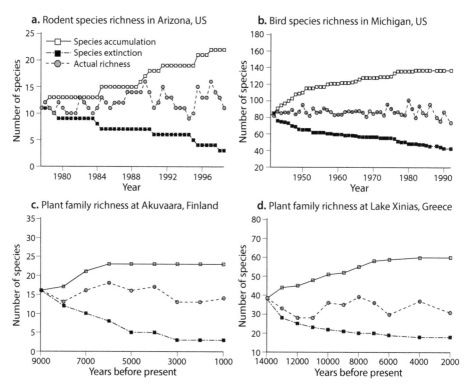

FIGURE 5.4. Results from Brown et al.'s (2001) comparison of colonizations and extinctions (indicating turnover) and average richness from multidecade observations of rodents in Arizona (a) and birds in Michigan (b), and multi-thousand-year observations of plant-family colonizations and extinctions from subfossils in Finland (c) and Greece (d).

the mechanisms involved were stochastic or deterministic (or some mixture), they argued that colonizations and extinctions that result from shifting responses of species to fluctuating environmental conditions, also involving a balance between colonizations and extinctions, could explain their results. Unfortunately, they also did not have data available on the shifting environmental conditions nor on the species responses to those conditions. Consequently, these results cannot be used to rule out the alternative hypothesis that stochastic drift and colonization-extinction dynamics associated with NT-like models caused the shifts in species composition through time.

Echoing these case studies, a recent meta-analysis of species richness and compositional change among 100 study sites showed that there were no general trends in local diversity patterns, although particular studies showed decreases and others showed increases in local diversity (Fig. 5.5a; Dornelas et al. 2014). However,

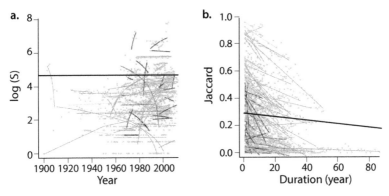

FIGURE 5.5. Results from Dornelas et al.'s (2014) meta-analysis of 100 long-term studies that monitored local richness (a) and compositional turnover measured by the Jaccard's similarity index, comparing the composition of a given community to itself through time (b). The thin lines reflect trends in individual time series, the bolder line shows the overall trend.

there was an overall substantial change in species composition through time, shown by a smaller Jaccard's similarity and therefore greater divergence of community composition from an initial condition through time (Fig. 5.5b; Dornelas et al. 2014).

Although many studies that have examined temporal turnover patterns have not been able to discern the nature of the possible mechanisms, turnover probably emerges from a combination of stochastic (e.g., colonization-extinction, drift) and deterministic processes (e.g., environmental change). However, some compelling evidence suggests that deterministic processes play a strong role in many of these patterns. For example, Dornelas et al. (2014) compared their observed results of species turnover to a realistic NT model with only demographic stochasticity and found that observed temporal turnover rates were an order of magnitude higher than those predicted by NT. Likewise, Hatosy et al. (2013) found that demographic stochasticity could explain some degree of community turnover but that environmental variation (both within and among seasons) was of overriding importance in surveys of marine microbes. Finally, Kalyuzhny et al. (2014) examined long-term patterns of compositional change in two well-analyzed data sets—the trees on BCI associated with Hubbell's (2001) studies on NT and the North American Breeding Bird Survey. By borrowing a technique of fluctuation analysis from statistical physics, they showed that the patterns of compositional change were most consistent with a system driven primarily by environmental stochasticity rather than by compositional drift.

Overall, there is often substantial temporal turnover in local communities (and often at larger scales as well) that cannot generally be explained by drift and colonization-extinction alone. The magnitude and tempo of this turnover often seems to be due to temporal fluctuations in environmental conditions, even though species richness stays similar and even though the nature of these fluctuations is not clear. Virtually all of these studies, however, seem to interpret the effects of environmental fluctuations as the result of external factors (e.g., gradual or cyclical changes in climate, disturbances, etc.). Is this necessarily the only possibility?

5.3 HOW TIME CAN INFLUENCE DETERMINISTIC COMMUNITY ASSEMBLY

In Chapter 4, we argued that one of the most obvious ways to distinguish SS from other possibilities was that it predicts a strong association between environmental conditions and community composition, whereas this is not so (or at least a lot less so) when other processes dominate (Tuomisto et al. 2003, Gilbert and Lechowicz 2004, Cottenie 2005, Legendre et al. 2009, Soininen 2014). While this is an intuitively appealing dichotomy, it is not necessarily true, the reason being that there are several niche-based deterministic mechanisms that can paradoxically lead to a strong temporal influence on metacommunity structure. These mechanisms, if they are not synchronized in space, can also produce spatial patterning. In the context of variation partitioning that we discussed in Chapter 4, such mechanisms can produce a strong influence of space [S] or random factors [Resid] and weaker effects of environment [E], even when environmental filtering and deterministic niche-based processes (which we normally associate with SS) are important.

There are at least three such temporally mediated processes. First, differences in the timing by which species arrive in a local community can lead to priority effects and multiple stable equilibria under identical environmental conditions. Second, community composition can be maintained regionally but fluctuate locally as a result of frequency-dependent interactions and/or nontransitive interactions between species (e.g., "rock-paper-scissors" scenarios). Third, frequency dependence (positive or negative) can create spatially structured metacommunities that fluctuate through time owing to variable local interactions among species at different frequencies in the landscape. For each case, the patterns of spatiotemporal variation in species composition would, on the face of it, support dispersal-limited and/or stochastic models of metacommunity organization even though the mechanism is primarily one involving deterministic processes of metacommunity assembly. Comparing patterns of turnover in space and time can provide important clues about the role of these more complex effects of deterministic community assembly.

5.4 PRIORITY EFFECTS AND MULTIPLE STABLE
EQUILIBRIA IN METACOMMUNITIES

Multiple stable equilibria (MSEs) emerge in a local community when the initial conditions of the species entering the community determine the final community structure even though there is no important difference in environmental conditions.

MSEs can occur in the simple Lotka-Volterra two-species competition model when the strengths of interspecific effects exceed the strength of intraspecific effects, allowing whichever of the two species colonizes the habitat and is first to grow past a threshold to eventually dominate the community. In fact, this is a generalizable effect that can emerge in any number of modules with interacting species (Law and Morton 1993, 1996; Chase and Leibold 2003); when the sum of interspecific effects of a species on the other members of the community exceeds its intraspecific effects, it will have positive frequency-dependent interactions leading to MSE. When a species is initially relatively rare in an MSE scenario, it will not be able to increase, whereas when it is initially relatively common, it can increase and dominate species that are initially less common.

Even though explicit evidence for MSEs in natural communities remains somewhat equivocal (Schroder et al. 2005), there are many reasons to expect that they can often occur among localities in natural communities. For example, by assuming that community assembly proceeds by a series of rare or slow colonization events and fast dynamics within interaction webs that characterize a locality's community structure, Law and Morton (1993, 1996) found that a broad parameter space of typical species interactions can lead to MSEs among communities (as well as cycles and other dynamic assembly phenomena that we discuss below).

MSEs are also frequently expected in simple models when species traits that are associated with its being able to persist on low amounts of a resource (e.g., a low R^*) also confer strong interspecific effects on neighboring species. For example, a plant with large leaves that allow it to capture even low levels of light (and thus a low R^* for light) simultaneously strongly reduces the amount of light available for other species that are more light limited, thus causing interspecific effects to exceed intraspecific effects (Tilman 1988, Reynolds and Pacala 1993). Likewise, in more complex communities with soil feedbacks, interactions between plant and soil microbes can result in both negative feedbacks, creating conditions for species coexistence, and positive feedbacks, creating conditions for priority effects (Klironomos 2002, Ehrenfeld et al. 2005, Kardol et al. 2007, Bever et al. 2010). Although usually examined at the population level, demographic stochasticity and Allee effects (positive density dependence), in which a population has to exceed a certain threshold to be able to establish and maintain viability in a given patch can lead to MSEs at the community level (Amarasekare 2000a, Keitt et al. 2001, Ferdy and Molofsky 2002, Lutscher and Iljon 2013, Gerla and Mooij 2014). In Chapter 8, we show how local adaptive evolution can also lead to priority effects in metacommunities.

Priority effects leading to MSEs have been suggested to be more common in some types of environments than in others, because environmental conditions influence community size and the strength and nature of interspecific interactions (reviewed in Chase 2003a and Fukami 2015). For example, priority effects and MSEs have been suggested to be more frequent in environments that are more productive (Chase 2003a, 2010; Ejrnæs et al. 2006; Chalcraft et al. 2008), less frequently disturbed (Chase 2003a, 2007; Jiang and Patel 2008), or smaller or less interconnected (Chase 2003a, Orrock and Watling 2010). One repercussion of this would be that metacommunities in environments with more frequent MSEs would be expected to have a lower environmental match (lower [E] and higher [S] or [Resid]) relative to environments with less frequent MSEs. However, it would be inaccurate to characterize the higher [S] or [Resid] as having resulted simply from dispersal limitation or ecological drift without temporal information. Dispersal limitation and ecological drift imply that localities within metacommunities should vary in species composition through time (i.e., high temporal turnover), whereas if MSEs were common, we might expect this lower [E] but nevertheless expect no relationship to turnover through time.

For example, in a long-term experimental study, Chase (2010) varied the initial colonization histories in mesocosms during an initial period of two years and then monitored the dynamics of the resulting communities for several more years. Low-productivity mesocosms were quite similar to one another regardless of the initial colonization histories and through time (i.e., they generally seemed to converge on a single stable equilibrium). At high productivity, however, there was much more variation among mesocosms. This variation appeared to be clustered into four groups, which might be expected as a result of MSE (Fig. 5.6a). Because environmental conditions were more or less controlled in this experiment (same water, same nutrients, same sunlight, etc.), we might take a single temporal snapshot (one sample year) and assume that because [E] seems to explain little of the variation observed, then dispersal limitation, ecological drift, or both must have been important. However, this need not be the case. If dispersal limitation or drift was responsible for the results, we would expect temporal turnover that would eventually break down this pattern. Instead, there was little turnover in species composition following the initial colonization period (Fig. 5.6b), suggesting that the results were more likely due to MSEs.

MSEs are less likely to occur in metacommunities with extinction-colonization dynamics (Shurin et al. 2004). When patches in a metacommunity are homogenous,

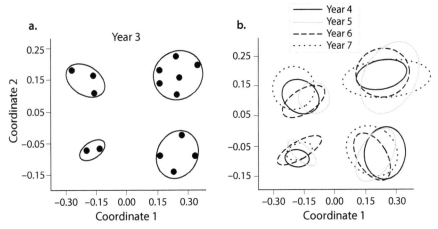

FIGURE 5.6. Nonmetric multidimensional scaling plots showing dispersion of community composition of macroinvertebrates and amphibians among the 15 experimental mesocosms that were identical in environmental conditions but experienced different assembly histories for the first two years (measured with the Bray-Curtis abundance-based similarity metric). (a) The relative position of each mesocosm's composition (denoted by black dots) in the third year of the experiment, after the assembly history treatments were complete. Circles are drawn around groups of mesocosms that appeared to cluster into four qualitatively distinct community types. (b) The relatively static composition of these community types through the final four years of the experiment. Only the circles that surround the entire cluster of communities are shown for simplicity. Data from Chase 2010.

once one of the species becomes relatively common, it will tend to increase the frequency of times it becomes the first colonist in empty or disturbed patches. This effect leads to positive feedback at the metacommunity scale, where this species increasingly dominates the metacommunity as a whole even though there is no change in the strength of its local interactions with other species. One way that MSE dynamics can be maintained under such conditions is if the component species also have spatial refuges where they are deterministically favored regardless of priority effects (i.e., they each have patches with different environmental conditions that do not have MSE dynamics; Shurin et al. 2004). Another way that this can happen is if the MSE is not complete, so that priority effects determine relative dominance rather than complete exclusion (Calcagno et al. 2006).

In sum, MSEs can be a mechanism that decouples the correlation between community composition and environment and generates spatially correlated and residual variation. Whether it does so frequently in nature depends on the prevalence of MSEs. In contrast with many of the other mechanisms that enhance spatial turnover among patches with identical environmental conditions, MSEs are temporally mediated (via priority effects) but predict little temporal turnover. Because studies have rarely compared spatial and temporal variation simultaneously, especially at the resolution needed to make such inferences, we unfortunately cannot make much headway in addressing this question until better data are available.

5.5 ENDPOINT ASSEMBLY CYCLES

A second way by which spatiotemporal variation can lead to a decoupling between community composition and environmental conditions under deterministic niche-based community assembly is temporal variability of community composition due to the internal dynamics of the interacting species—we term these *endpoint assembly cycles* (EACs). A classic way EACs can emerge is if species compete in a nontransitive way, such as a "rock-paper-scissors" interaction (e.g., species A outcompetes species B, B outcompetes species C, and C outcompetes A; Fig. 5.7a). All three species coexist in the region, but in any given locality the identity of species will change from one time period to the next as each species is sequentially replaced by its superior competitor (Durrett and Levin 1998, Law and Leibold 2005, Rojas-Echenique and Allesina 2011, Vandermeer and Yitbarek 2012, Schreiber and Killingback 2013; Fig. 5.7b). Any snapshot in time of a system like this will indicate much variation in community composition that is unrelated to environment (Fig. 5.7c) and would emerge in the [Resid] or spatial [S] component of a variation partition. Importantly, however, this pattern, which might otherwise be attributed to NT or PD without more explicit temporal information, is in fact very deterministic.

While the simple nontransitive case leading to EACs might seem somewhat contrived, there are in fact documented scenarios in which processes similar to these might occur, for example, in microbial communities where species produce antibiotics that differentially impact some but not other species (Kerr et al. 2002, Hibbing et al. 2010, Prasad et al. 2011), in sessile marine invertebrates where species can overgrow or undercut each other (Buss and Jackson 1979), and in plant communities that interact via allelopathy and plant-soil feedbacks (Lankau and Strauss 2007, Lankau 2011). This simple nontransitive scenario can be viewed as a limiting case of a much broader set of interactions that lead to EACs among species that coexist regionally despite local instability. For example, when species interact with multiple limiting factors (e.g., more than two resources, predators), the likelihood of nontransitivities increases, allowing the regional coexistence of large numbers of species, but with considerable local turnover in species composition (Allesina and Levine 2011). Likewise, many other types of community assembly dynamics in food webs can lead to EACs (Law and Morton 1996, Steiner

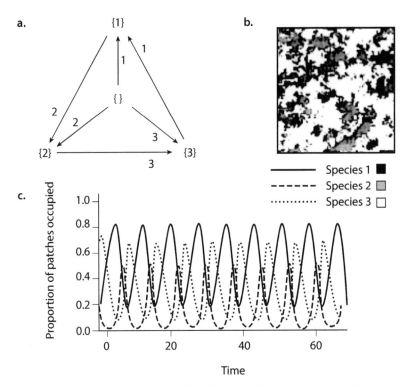

FIGURE 5.7. (a) A simple nontransitive (rock–scissor–paper) competition scenario shown as an "assembly graph." Each node in the graph shows one of the possible community configurations; the community can be empty (shown in the center) or occupied by one of the three species; invasions by particular species (labels associated with each arrow) will lead to changes as shown by the arrows. (b) Snapshot of the spatial configuration of these species as they interact on a grid during one time period. (c) Numerical simulation of a set of parameters in which colonization rates of all species are the same. Modified from Law and Leibold 2005.

and Leibold 2004), at least in some circumstances (see Chap. 11 for more details on some of these in cases involving food-web dynamics). Steiner (2014) demonstrated how sequential assembly (that facilitates the dynamics of EACs) of plankton communities generated temporal and spatial turnover, whereas the simultaneous introduction of all the species did not and instead led to a single stable community (Fig. 5.8). However, as with MSEs discussed above, this effect depended on productivity, and the joint effects on spatial and temporal turnover were absent in low-productivity conditions.

The prevalence of EACs can also critically depend on the regional context in which localities are embedded; that is to say, EACs cannot occur without an appropriate metacommunity context. The reason is that the spatially mediated coexistence

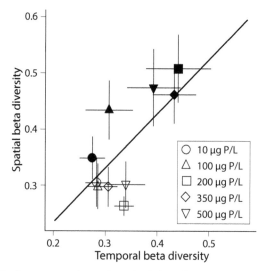

FIGURE 5.8. Results from an experimental manipulation of assembly history and environmental productivity in communities of six species of freshwater cladoceran zooplankton. Treatments that led to higher levels of temporal β-diversity associated with endpoint assembly cycles included those that were higher in productivity (depending on phosphorous levels as shown in the legend) and had sequential assembly histories (filled symbols). Low productivity and simultaneous introduction of species (open symbols) did not. This situation also created higher observed spatial β-diversity in sequential high productivity ecosystems. Modified from Steiner 2014.

of nontransitively interacting species and other types of EACs require the metacommunity to experience some degree of compartmentalization and dispersal limitation. If rates of dispersal throughout the region are too high, the EACs will not be out of phase, the numbers of species in the region will collapse to a few or single species (Law and Leibold 2005, Reichenbach et al. 2007, Schreiber and Killingback 2013), and the EAC will disappear over time.

As with MSEs, EACs could be important for understanding the spatiotemporal dynamics of metacommunities and interpreting their patterns if they are common. Although there are examples of EACs involving simple competition akin to the "rock-scissors-paper" model we described above (Kerr et al. 2002), many ecologists would guess that such situations are not the rule because it requires what might be an unlikely combination of interspecific competition patterns. However, Soliveres et al. (2015) analyzed grassland and arid-land plant communities and found a surprisingly high prevalence of nontransitive interactions between species. In Chapter 11, we argue that they may in fact be very common when we consider community assembly involving more than one trophic level.

5.6 FREQUENCY-DEPENDENT COEXISTENCE IN SPATIALLY CONTINUOUS METACOMMUNITIES

Most of our models of metacommunities are framed within the context of very discrete separation of spatial scales (the sensu stricto definition we described in Chap. 1) with discrete patches connected by dispersal into a metacommunity with a completely distinct spatial scale. On the other hand, there are many situations in which this separation of spatial scales is less distinct (the sensu lato definition in Chap. 1). In such habitats, space itself, as well as the interaction between space and time, can allow species to coexist. These mechanisms include those that result from intrinsically (endogenously) generated temporal and spatial variation in resource levels, as well as from frequency-dependent interactions among individuals. Although continuous instead of discrete, these models often result in dynamics that are very similar to those we discussed for community assembly with MSEs or EACs above; namely, that a purely SS process can lead to spatiotemporal variation in community composition even under initially uniform conditions and that this decouples the relationship between environmental variation and community variation.

The processes of frequency dependence in continuous habitats are quite similar to those described above for the discrete cases that result in EACs (Law and Leibold 2005). Intrinsically generated temporal variation can lead to local coexistence, for example, if species differ in their ability to respond to resources when they are at different abundances (e.g., different shapes of their Type II functional responses) but simultaneously influence the abundance of those resources, thus generating resource cycles whereby each species is favored at different times in the cycle (Armstrong and McGehee 1976, Abrams and Holt 2002). If multiple resources are considered, many more species than resources can coexist through these intrinsically generated cycles (Huisman and Weissing 1999). Such intrinsically generated temporal fluctuations can lead to more regional coexistence when placed in a homogenous regional context (Wilson and Abrams 2005) and to more regional coexistence when there is also spatial heterogeneity (Allesina and Levine 2011). An important factor that determines how much this is so is the degree of synchrony of these fluctuations across the metacommunity (Gouhier et al. 2010).

Spatial heterogeneity can also be intrinsically generated by the actions of the organisms themselves, allowing species to coexist regionally due to spatial processes even with initially uniform environmental conditions. For example, species can coexist spatially in a manner similar to the colonization-competition mechanism in PD if actively foraging species vary in their foraging traits: when some are slow to move through space but are effective at reducing resources in those patches (i.e., "crumb pickers"), whereas other species move more rapidly but are more superficial in their foraging ability (i.e. "cream skimmers") in any given area (Wilson et al. 1999, Chase et al. 2001). Likewise, similar mechanisms can work for sessile species, whereby coexistence is achieved by spatial structure that emerges when species have limited dispersal ranges and the size of the interaction neighborhood differs between conspecific and heterospecific species; the species can thus coexist via *heteromyopia*, whereby intraspecific effects are larger than interspecific effects on average due to spatial aggregations that emerge when a species has greater impacts on heterospecifics at short distances and on conspecifics at greater distances (Murrell and Law 2003). This could occur, for example, if species have neighborhood-level allelopathic effects on heterospecifics that operate at smaller scales than the resource competition among conspecifics, or if the interspecific effects of resource competition occur across short distances, whereas intraspecific effects, say via shared specialist enemies, occur more distantly.

When habitats are modeled in a spatially explicit way, a variety of frequency-dependent processes can facilitate coexistence in which the relative intra- and interspecific effects of species are nonlinear, allowing spatial structure to emerge and species to coexist in a dynamic mosaic. Nevertheless, this result requires appropriate trade-offs among species in their responses to the environment and to each other. One of the first frequency-dependent mechanisms invoked to explain coexistence in otherwise homogeneous environments was what we now know as the Janzen-Connell mechanism, which was independently introduced by Janzen (1970) and Connell (1971) in an attempt to describe the coexistence of a large diversity of plant species in tropical areas despite seemingly few limiting factors. If specialist enemies of a plant (herbivores, seed predators, or pathogens) are more likely to recruit to areas where the frequency of their host species is higher, they can subsequently limit subpopulations of that plant in locations where it is locally abundant. In areas where that plant is locally rare, it will be proportionately less influenced by enemies and grow more rapidly. If each plant has its own suite of specialist enemies, a species experiences proportionally greater limitation when it is locally common and will be at a relative advantage when it is locally rare, thereby allowing many species to coexist in a regional mosaic (see Caswell 1978 and others for theoretical explorations of this effect).

The Janzen-Connell effect and related frequency-dependent mechanisms that emerge indirectly through enemies (e.g., predator learning; Ishii and Shimada 2012) are negatively frequency dependent, giving a proportional advantage to species when they are locally rare and allowing regionally stable coexistence with locally variable dynamics. Other types of negative frequency dependence can also occur and lead to similar spatiotemporal mosaics of species coexistence, including plant-soil feedbacks, in which individuals of the same species are inhibited by the soil microbes conditioned by their conspecifics (e.g., enemies or mutualists) relative to individuals of different species (Bever 1999, Molofsky et al. 2002).

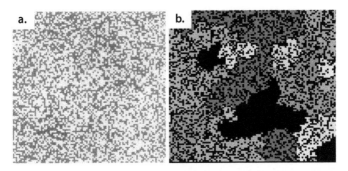

FIGURE 5.9. (a) Results from simulations by Molofsky et al. (2002) of two species with weak negative frequency dependence. The two species are dispersed throughout the space and have a lot of variation in their distribution that is not explained by environment or space. (b) Results from the simulations by Molofsky and Bever (2002) with strong positive frequency dependence allowing the coexistence of several species (each a different shade) but also with some space that is unoccupied (denoted by black dots). Individuals of each species are spatially clumped, which would create conditions explained by space even though there is no environmental variation and dispersal is not particularly limited. Modified from Molofsky et al. 2002 and Molofsky and Bever 2002.

Likewise, positive frequency-dependent effects can increase the possibilities of species coexistence as a result of the interaction between space and time in an otherwise environmentally homogenous region. Priority effects and Allee effects, for example, shift the per capita effects of a species as its density increases; if multiple species in a community experience these, they can (co)exist, once a lower threshold of abundance is reached through local aggregation (Britton 1989, Molofsky et al. 2001, Molofsky and Bever 2002). Both of these types of frequency-dependent interactions, two examples of which are presented in Figure 5.9, can decouple the environment-species composition relationship ([E]), and increase [S] or [Resid] in community composition depending on the strength and direction of frequency dependence, even if dispersal is not particularly limiting and there is little demographic stochasticity.

5.7 CONCLUSIONS

If local communities are closed, temporal changes in diversity and composition are very limited; for the most part, species can either go extinct and never come back (unless they have dormant stages), or they can speciate in situ via sympatric speciation. Studies of the temporal dynamics of communities indicate instead that temporal turnover of species in communities is a rather frequent (even ubiquitous) outcome—depicted as STR. The simple fact that the STR and the SAR are two of the most frequently observed patterns emphasizes the necessity of metacommunity thinking (e.g., dispersal, spatial heterogeneity) in order to understand pattern and process of species interactions, coexistence, and biodiversity. Echoing others (e.g., White et al. 2010), we suggest that a deeper look into the interrelationships between space and time (STAR) will be an important area for future research.

Although species composition turnover in metacommunities is frequent, the view that higher temporal turnover implies the overriding importance of stochastic processes associated with NT and PD is far too simplistic to be of use in testing the relative importance of metacommunity assembly mechanisms. Even though temporal turnover can point toward the importance of some stochastic processes (e.g., dispersal, extinction, drift), a number of other mechanisms can create temporal turnover patterns. There is a long way to go in developing the conceptual and statistical tools, as well as in collecting the necessary observational and experimental data, for addressing these questions. Two issues are particularly important.

First, it seems that the vast majority of systems show temporal turnover rates that are much higher than would be expected under a simple demographic drift scenario in NT (Dornelas et al. 2014). This observation implies that species composition shifts may often be tracking environmental variation more in accord with an SS perspective or may be changing due to a variety of other possible reasons, such as EACs. We currently lack strong analytical tools that are able to tease apart the relative influence of environmental variability and spatial variability on species compositional variation in both space and time from empirical systems (e.g., a variation partition that can include both spatial and temporal β-diversity).

Second, multiple mechanisms can lead to temporal turnover in species composition even when they emerge from niche-based processes; that is, emergent patterns may "look like" they emerged from stochastic processes such as ecological drift, but in fact they result from more deterministic processes (although dispersal and time are necessary to create these patterns). These mechanisms include MSEs, EACs, and other frequency-dependent processes that act in a spatiotemporal context. We also need theoretical and statistical tools that can help us to understand and tease apart the influence of spatiotemporal dynamics that emerge from these mechanisms relative to those that emerge from drift and dispersal limitation associated with NT and PD.

Overall, the conclusion is that the simplistic SS, NT, PD, ME archetypes are too crude to adequately address metacommunity dynamics in a realistic manner. To date, the temporal dynamics of metacommunities have not been sufficiently studied to make definitive conclusions, but they represent an important component of metacommunity assembly that must be implicitly incorporated into our collective research program.

CHAPTER SIX

What Can Functional Traits and Phylogenies Tell Us about Coexistence in Metacommunities?

Prospectus

1. Functional trait and phylogenetic comparisons among species that co-occur (or not) within metacommunities can provide additional insights into the study of metacommunity pattern and process.
2. Trait-based approaches assume that the traits used in the analysis are strongly related to the ecological interactions that affect the organisms. Likewise, phylogenetic approaches typically assume that traits that affect ecological interactions are evolutionarily conserved. However, both of these assumptions are often unfulfilled, suggesting that more caution is needed in interpreting patterns (e.g., whether co-occurring species are more similar, or more different, than expected by chance).
3. One of the most prevalent approaches uses patterns in the similarity in traits and relatedness in phylogenies of co-occurring species to resolve the relative roles of environmental filtering (i.e., co-occurring species are more similar than expected by chance, suggesting that they have adaptations to similar environmental conditions) and interspecific competition (i.e., co-occurring species are more different than expected, suggesting that competition with similar or related taxa prevents coexistence). The validity of this approach assumes a strong SS perspective that does not adequately address other elements of metacommunity assembly. More important, it does not adequately address niche differences that may be destabilizing as a result of multiple stable equilibria and other forms of positive feedback, including endpoint assembly cycles.
4. Explicitly taking a scale-dependent approach may improve the insights from the study of dispersion in traits or phylogeny because the relative importance of different mechanisms may often be scale dependent under different metacommunity scenarios.
5. More carefully identifying traits that may be differently informative about aspects of species interactions—such as habitat-related traits, dispersal traits, and resource-use traits—may also improve the way we use them to understand metacommunities.

6. Another way to evaluate metacommunity dynamics is to compare taxonomic patterns in β-diversity to functional-trait or phylogenetic patterns. Typically, we find that trait-based patterns appear to be more predictable than taxonomic-based approaches, which suggests an increase in the role of drift affecting β-diversity patterns among similar species.
7. Temporal changes in functional β-diversity can also be informative about mechanisms, especially when compared with changes at the taxonomic level.
8. It can also be informative to examine which traits covary with which environmental variables. This is especially true when traits are selected that have close links with alternate possible mechanisms for SS.
9. We conclude that trait-based approaches can be an important complementary way to understand metacommunities. This is especially true when the traits involved can be linked to mechanistic aspects of metacommunity dynamics. Nevertheless, ignoring taxonomic patterns ignores the fundamental issues that regulate metacommunities, since these involve contrasting processes that affect individual species.

Species traits are inherent to most of the metacommunity processes we have discussed (other than the NT, which assumes that all species are identical); that is, in metacommunity models, species have traits that produce differential competitive abilities, preferences of environmental conditions, and dispersal abilities. These trait differences among species are critical to determining exactly how species persist and grow (or not) in a given locality and how that persistence varies within the metacommunity. Furthermore, species traits are formed over millions of years of evolution, and so species phylogenetic (evolutionary) relationships can often have a large influence on species-trait differences, which in turn should influence the patterns of abundance and distribution of species within a metacommunity.

While these statements seem rather self-evident, it is interesting to note that we spent the majority of our discussion about both the processes (Chap. 3) and spatiotemporal patterns (Chaps. 4 and 5) and included very little discussion about traits or evolutionary relationships among species. We might argue, then, that one can make a great deal of progress in understanding patterns of metacommunity organization without an explicit focus on traits and phylogenies—often referred to as a *taxonomic* focus.

However, a closer look at the traits of species (the *trait-based* approach) and their phylogenetic relationships (the *phylogeny-based* approach) might allow us deeper additional insights into the patterns and processes that structure metacommunities. In Chapters 4 and 5, we noted that a number of approaches that have purported to be able differentiate between key metacommunity processes (e.g., SS vs. NT), such as the shape of the SAD or the partitioning of variation in community composition due to environment and space, are in fact rather equivocal. And when we add evolutionary processes into the mix (such as in Chaps. 8 and 9), these patterns are even more difficult to discern. Are we reconciled to accepting the fact that even when environmental filtering and SS processes may be quite strong, we are often unable to disentangle this effect from patterns in nature and differentiate it from other model frameworks such as PD or NT?

Perhaps not. A number of authors have argued that a focus on functional or phylogenetic information not only provides added information, but also may often be necessary in order to fully evaluate and differentiate the mechanisms driving metacommunity structure (Webb et al. 2002, 2010; McGill et al. 2006b; Cavender-Bares et al. 2009; Pavoine and Bonsall 2011; Weiher et al. 2011; Mouquet et al. 2012; Münkemüller et al. 2012; Spasojevic et al. 2014). For example, in a test comparing the SS and the NT perspectives, Harpole and Tilman (2006) noted that the shape of the SAD in grassland communities was indistinguishable from both SS and NT predictions, but that the traits of the organisms (their R^* for limiting nutrients in this case) was highly correlated with their relative abundances in a way that was inconsistent with NT, putting more weight on the SS perspective.

This recent surge of interest in functional and phylogenetic information is due in part to the increase in the availability of information on traits and phylogenies, new analytical tools to examine them, and frustration with the "taxonomy only" approach for attributing structure (or lack thereof) to patterns of coexistence in local communities and metacommunities. However, there remains much variation in how these tools are applied and what inferences can be made from them. For example, there are at least 70 different phylogenetic metrics that integrate divergent properties of phylogeny to address different aspects of community structure and divergence (Tucker et al. 2016). Likewise, trait-based approaches can focus on traits that are considered to have loosely specified or indirect functional importance for an organism's performance (e.g., body size, leaf area)—which we call the *functional-trait* approach (McGill et al. 2006b). Or, less frequently (but perhaps more usefully), trait-based approaches can focus on factors that are directly relevant to an organism's performance in a given environment, such as its demographic responses, consumption of resources, etc.—we call this the *niche-trait* approach.

If we start with a simple SS view, we expect a strong relationship between spatial variation in an environmental factor and spatial variation in the traits of species in habitats associated with that environmental factor. For example, the composition of plant species with traits associated with water relations is typically influenced by variation in water availability (Cornwell and Ackerly 2009, Craine et al. 2013, Silvertown et al. 2015), and the morphology of fishes is typically linked to their success in different habitat types (Werner 1977, Ingram and Shurin 2009, Brind'Amour et al. 2011).

Continuing this highly simplified thought experiment, we might still expect a signal between environmental variation and trait variation of the species present when ME are important, but we would expect this signal to be weakened because of the spillover of less well adapted species in adjacent habitats. In the simplest PD models, we would expect covariation among species with different traits in different localities, but these traits would be involved with the competition-colonization trade-off that allows coexistence in these sorts of systems. For example, there might be negative covariation between species with traits that allow long-distance dispersal (e.g., seed size for plants) and species with traits that allow stronger local competition (e.g., traits associated with the tolerance of low light levels in plants). Finally, in its purest form, NT would predict no structured variation in species with various functional traits, because NT assumes that traits are not "functional" with respect to any variation in species performance (i.e., "functional traits" are unrelated to "niche traits").

Although individuals of species (and their traits) are ultimately the currency that matters for metacommunity patterns and processes, their phylogenetic history might also provide insight into the possible underlying mechanisms structuring metacommunities. Only the NT makes specific predictions about how species phylogenetic relationships should be distributed in and among metacommunities (Jabot and Chave 2009, Kembel 2009, Davies et al. 2011b, Pigot and Etienne 2015), though the predictions are often falsified (Ricklefs and Renner 2012).

At the outset, it is important to remember that phylogenetic perspectives on metacommunity assembly involve at least several distinct approaches (see, e.g., Webb et al. 2002), which can sometimes be confused and confounded. First, phylogenetic information is often thought to provide useful insights into the mechanisms that influence metacommunity assembly from a regional species pool (traits are assumed to be phylogenetically conserved, such that more closely related species are also more similar in their functional traits). Second, phylogenetic information can provide important insights at the biogeographic scale, when dispersal limitation is more important because of vicariance and large-distance dispersal events. Finally, it can reveal the interplay between metacommunity interactions and the diversification process itself. We discuss the interaction between metacommunity and macroevolutionary processes in Chapter 9 and here only focus on the first approach.

6.1 A BRIEF HISTORY OF TRAIT- AND PHYLOGENY-BASED "ASSEMBLY RULES"

An early bifurcation in taxonomy- versus trait-based approaches to metacommunity assembly followed from Diamond's (1975a) treatise. As described in Chapter 4, Diamond suggested that there might be nonrandom structures in the distributions of species among sites (using site × species matrices), and these structures (e.g., checkerboard distributions that suggest competitive exclusion) might tell us something about the nature by which metacommunities are structured. The ensuing debate about null models focused in part on this taxonomic approach to metacommunity assembly (see Strong et al. 1984, Diamond and Case 1986). However, as we discussed in Chapter 4, simply identifying patterns in the structure (or lack thereof) using the taxonomic approach to examine patterns of species distributions in a metacommunity can go only so far for detecting possible mechanism.

However, Diamond (1975a) also took an explicitly trait-based perspective. He discussed the traits of bird species as a primary mechanism for the types of islands in the Bismarck Archipelago on which certain bird species would be found, as well as the likely co-occurrence (or not) of different bird species. For example, Diamond described a number of cases in which bird body size or feeding traits seemed to be too similar for them to coexist, which he suggested led to checkboard species distributions (in which islands have one or the other of two competing species, but not both). Thus, Diamond was implicitly suggesting that there was a limit to how similar species could be in their traits to coexist locally, as a result of the competitive exclusion principle. This is not surprising, as Diamond's chapter was in a volume dedicated to Robert MacArthur (Cody and Diamond 1975), who had played a prominent role in developing and testing the idea that coexisting species could not be too similar in their traits (i.e., they needed to have some niche partitioning; MacArthur 1958, 1972; MacArthur and Levins 1967).

Trait-based assembly rules such as those in Diamond's (1975a) study originated from the perspective of community-wide character displacement, in which species need to be sufficiently different in their traits in order to co-occur locally. The basic concept has deep historical roots going back to Charles Darwin and Alfred Russel Wallace, if not before, and has been examined using a number of different approaches over the years, including trait similarities among coexisting pairs of species, genus-species ratios, and other similar lines of evidence (Grinnell 1904, Elton 1946, Brown and Wilson 1956, Hutchinson 1959, Williams 1964, MacArthur and Levins 1967, Simberloff 1970, Grant 1972, Schoener 1974). However, as with discussions of species co-occurrences, the use of trait-distribution patterns to infer ecological processes, such as competitive coexistence and exclusion, was wrought with debate and the expression of the need for rigorous statistical tests (e.g., null models) (Strong et al. 1979, Simberloff and Boecklen 1981, Schluter and Grant 1984). Some of the angst that accompanied this debate is symbolized (and satirized) in the Christmas card by Shahid Naeem that is reproduced in the preface. This early work continues to influence studies on patterns of coexistence in relation to phylogenetic relatedness (Webb et al. 2002, Cavender-Bares et al. 2009, Mouquet et al. 2012) and trait-based similarity (McGill et al. 2006b, Weiher et al. 2011, Enquist et al. 2015).

These more recent concepts of trait-based assembly explicitly include more-rigorous null model tests (though the specific forms of the null models are still debated). For example, in small mammal communities, Fox (1987) described an assembly rule based on lumping species into large groups based on their functional similarity (i.e., species with similar traits, diets, etc., placed into discrete functional groups, guilds, or higher taxonomic groupings, like genus). The idea here was that species should sort themselves into local communities according to a "rule" based on functional-group membership. Among local communities that varied in their species richness, Fox argued that one member of each functional group should be present in the community before a second species of one of the groups is added, and so on. Fox (1987) and Fox and Brown (1993) found good concordance of this basic rule among rodent communities from several hundred sites in Australia and North America.

Although the basic idea of Fox's assembly rule seems sound if interspecific competition is strong, and similar patterns have been found in other groups of organisms, there are many exceptions and it has a number of limitations (Weiher and Keddy 1995). First, one must be able to apportion species into discrete functional groups, which is not always easy when species trait variation is more continuous. More important, there are a number of problems with the null modeling approach even if the idea is intuitively appealing. One of the most important occurs when the regional species pool from which random local species communities are drawn is itself influenced by the interaction of interest (this is known as the *Narcissus effect*; Harvey et al. 1983, Stone et al. 1996). Another occurs when certain groups of species are overrepresented in the species pool for some reason other than assembly (Wilson 1995, Feeley 2003).

A more general perspective on the ideas of assembly rules is simply that when competitive exclusion is operating, species that co-occur together should be more different from one another (in traits or phylogenies) than expected by chance. With appropriate null models, larger trait data sets, and robust phylogenies, there is reasonable evidence for the idea that at least some local communities are overdispersed with respect to both (at least some) species traits and phylogenies (Cavender-Bares et al. 2004, Rabosky et al. 2007, Cooper et al. 2008, Ingram and Shurin 2009). However, as often as not, local communities do not differ from random expectations and may even be underdispersed (more similar than expected by chance) with respect to phylogenetic or trait structure (reviewed in Mayfield and Levine 2010 and HilleRisLambers et al. 2012; Fig. 6.1). This is evident even in Darwin's (1859) writings, which suggested a role for both possibilities and was formalized as an alternative hypothesis to the strong competitor-induced overdispersion of traits among co-occurring species. Using the metaphor of an environmental filter, the idea leading to underdispersion is that in a given environment (especially a

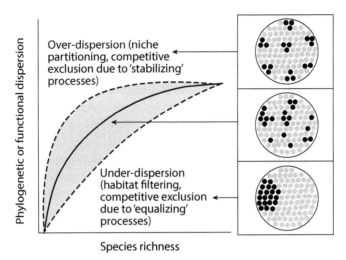

FIGURE 6.1. Illustration of the concepts of underdispersion and overdispersion of phylogenies or functional traits in a local community in relation to a "null expectation" of random co-occurrences. The large circles on the right depict the regional species pool, and the small circles represent individual species. Black circles represent individual species that are co-occurring together in a given locality, and gray circles are species that do not occur in that locality. The top panel represents a case of overdispersion, in which species co-occurring are more different from one another in traits or phylogeny than would have been expected by chance. The middle panel represents a case in which the dispersion is random (null expectation). The bottom panel represents a case of underdispersion, in which species are more similar to each other than would have been expected by chance. Modified from Cisneros et al. 2014.

"harsh" one), species with similar traits that allow them to tolerate those environmental conditions should pass through the filter. Thus, locally co-occurring species here should share traits (or phylogenetic relationships) that influence their environmental tolerances, and they should be more similar than expected by chance (Keddy 1992, Weiher and Keddy 1995).

Today, the dominant perspective seems to be that the degree of dispersion of traits and phylogeny depends on the tension between the strength of competition within localities and the importance of environmental filtering among localities (Leibold 1998; Chesson 2000; Kraft et al. 2015a, 2015b; Letten et al. 2017). However, there are some important nuances to consider in the way we think about this tension.

First, phylogenetic and trait differences and similarities have long been thought to be interchangeable because of an expected correlation between traits and phylogenies (i.e., niche conservatism in the broad sense), leading to the expectation that the strengths of competitive interactions between pairs of species are directly correlated with their phylogenetic relatedness (Webb et al. 2002). However, this expectation is often not realized, as we will discuss in Section 6.2. Second, while competition can lead to overdispersion by eliminating species with some types of trait values that are too similar to those of competitively dominant species as a result of stabilizing niche differences among species, competition can also create situations in which some types of traits are underdispersed because species that are too different in their equalizing fitness differences will be eliminated (see, e.g., Mayfield and Levine 2010, HilleRisLambers et al. 2012, Adler et al. 2013); we discuss this issue in more depth in Section 6.3. Third, and perhaps most important, the question of trait dispersions and co-occurrence is a metacommunity-level, scale-dependent problem; we discuss scale-explicit predictions of trait dispersions and coexistence in SS metacommunities in Section 8.4 and non-SS metacommunities in Section 6.5.

6.2 THE CORRELATION BETWEEN PHYLOGENETIC AND TRAIT-BASED INFORMATION AND "REAL" METACOMMUNITY PROCESSES

Before we dive too deeply into a synthesis of the utility of trait- and phylogeny-based approaches for understanding patterns in metacommunities, we need to lay out a few caveats. In our theoretical depictions of metacommunities (Chap. 2), traits of interest involve variables such as consumption rates, dispersal rates, death rates, and so on; these are often context dependent (e.g., they vary across environments, densities, etc.). Clearly, these sorts of niche-based traits are not often ones that can be easily measured in the field or museum and instead require detailed experiments or observations to quantify. Consequently, the use of phylogenetic and functional-trait information has been touted as "short cuts" that allow us to understand the responses of species to the environment and interspecific interactions without the necessity of estimating the parameters of the dynamical equations (Webb et al. 2002, McGill et al. 2006b).

The use of phylogenetic information to infer mechanisms of community assembly was largely inspired by the idea that factors associated with species niches were, at least to some degree, conserved within the phylogeny (Webb et al. 2002, Wiens et al. 2010) and goes back at least to Darwin (1859, 73), who suggested that "as species of the same genus have usually . . . some similarity in habits and constitution . . . the struggle will generally be more severe between species of the same genus . . . than between species of distinct genera."

Early studies often examined this idea—for example, by measuring and comparing the ratios of species to genera or related patterns to infer levels of competition and coexistence (Elton 1946, MacArthur and Wilson 1967), though it was quickly realized that interpretations of these patterns required null models (Simberloff 1970). The use of more sophisticated phylogenetic information as a proxy for niche relations to understand patterns of metacommunity assembly was reinvigorated by a number of studies synthesized by Webb et al. (2002).

There are many good examples in which traits important for community assembly are phylogenetically conserved. In such cases, we would expect patterns of the distributions of species to result from the balance of two interrelated patterns. First, more closely related species, with similar traits, should respond similarly to strong environmental gradients. For example, among terrestrial plants, nitrogen fixation is largely within the purview of the legume family (or more precisely the microbes in the legumes' root nodules). Phylogenetic structure in the representation of legumes versus non-legumes along a nitrogen gradient would be indicative of an environmental filtering process. Second, more closely related species, with similar traits, should compete more strongly with one another and thus be potentially less likely to coexist locally. For example, in a survey of forest birds, we might expect only a few of the possible perching/insect-gleaning species (e.g., warblers, sparrows) to coexist in a given site among a few of the possible forest-floor-feeding species (e.g., thrushes), diurnal predators (e.g., hawks), and nocturnal predators (e.g., owls). Other members of these groups would exist in other types of habitats and thus be overdispersed with respect to phylogeny.

Unfortunately, however, there are few such clear-cut examples of strong niche conservativism for traits that are important to community assembly patterns. Examples of convergent evolution abound where multiple, distantly related types of species utilize the same habitats and resources, interacting accordingly. Indeed, although there are some traits that are deeply conserved within plant phylogenies, some of the most important functional traits for plants and their responses to environmental gradients and species interactions, including seed size, leaf and wood morphology, and water and nutrient use efficiency, are not well conserved and instead are well mixed throughout the plant phylogeny (Cornwell et al. 2014, Garnier et al. 2016). Likewise, common functional traits for animals include life history, body size, and diet, which often have little phylogenetic signal (Losos 2008, Khalig et al. 2015).

Even the supposition that more closely related species should compete more strongly with one another than more distantly related species has garnered only mixed support. Some studies have supported this assumption, including experiments with protists and yeasts in microcosms (Violle et al. 2011, Peay et al. 2012) and grassland plants (Burns and Strauss 2011, Allan et al. 2013). However, many more studies have suggested that phylogenetic information is a poor predictor of competitive interactions, including studies of terrestrial plants (Cahill et al. 2008, Bennett et al. 2013, Godoy et al. 2014), freshwater algae (Narwani et al. 2013, Fritschie et al. 2014, Venail et al. 2014, Alexandrou et al. 2015, Naughton et al. 2015), and marine amphipods (Best et al. 2013). A common explanation for this poor predictive ability is that traits related to competitive abilities are not well conserved within phylogenies and that when traits, rather than phylogenies, are used, the expected positive relationship between trait similarity and competitive intensity are recovered (Best et al. 2013, Narwani et al. 2013, 2015).

As a result of the frequent lack of correlation between the phylogenetic relatedness of species and the features of their performance (i.e., their niche), enthusiasm for the phylogenetic approach to metacommunity assembly has abated more recently (Cavender-Bares et al. 2009, Kembel 2009, Pavoine et al. 2010, Mouquet et al. 2012, Gerhold et al. 2015). Even if there are correlations between phylogenies and traits, such correlations do not necessarily mean that both phylogenetic and trait information will equally predict species distributions (Pavoine et al. 2010). For example, although Pavoine et al. (2014) found positive (albeit weak) relationships between species traits and phylogenies among a group of Belgian butterflies, they found that only traits, but not phylogenies, could predict species distributions in the landscape.

Despite concerns regarding the use of phylogenetic information to infer mechanisms of metacommunity assembly, there remain good reasons to continue refining the phylogenetic approach in the context of community ecology (Mouquet et al. 2012, Cadotte et al. 2013, Gerhold et al. 2015, Cadotte and Davies 2016). First, a majority of analyses using phylogenetic similarity as a proxy for trait similarity have implicitly assumed that the difference in traits between species is linearly related to the difference in phylogenetic divergence time between those species. However, it may be more appropriate to compare trait differences to the square root of the phylogenetic distance, which may greatly improve the trait-by-phylogenetic relationship (Letten and Cornwell 2015). Second, Cadotte et al. (2013) have suggested a method taking both trait and phylogenetic information into account that provides a deeper understanding of the community assembly process (see also Pavoine and Bonsall 2011, Münkemüller et al. 2012, Chalmandrier et al. 2015, Cadotte and Davies 2016). Finally, as mentioned above, phylogenetic information can still provide quite useful information in a biogeographic and macroevolutionary context, as we discuss in Chapter 9 (see also Leibold et al. 2010, Peres-Neto et al. 2012, Gerhold et al. 2015).

So, the link between phylogeny, functional traits, and species performance is not as strong as is often assumed. How well do functional traits alone predict species performance? The answer seems to be, better than phylogenies but perhaps still not well enough to be the panacea that many have suggested. For example, a number of studies have found strong correlations between plant functional traits and performance in different environmental and competitive contexts (Goldberg and Landa 1991; Poorter et al. 2006, 2008, 2015; Angert et al. 2009; Sterck et al. 2011; Adler et al. 2014; Kraft et al. 2014, 2015b). For animals, there are fewer, but still several, studies that find correlations between functional traits and aspects of performance, with examples from fishes (Sirot et al. 2015), beetles (Davies et al. 2000), birds (Newbold et al. 2013), bees (Williams et al. 2010), bats (Farneda et al. 2015), ants (Arnan et al. 2012), stream invertebrates (Menezes et al. 2010), and amphibians (Smith et al. 2009).

Nevertheless, the direct link between functional traits and ecological performance is not always as strong as we might hope. For example, Pilière et al. (2015) examined stream invertebrates from a large-scale survey across Ohio and found no correlations between trait similarity among species and the responses of those species to broad-scale environmental gradients. Likewise, Kunstler et al. (2012) estimated interaction strengths among tree species in the French Alps, finding little correlation between the strength of interaction and the similarity in functional traits among species (rather than absolute differences, hierarchies of trait differences explained competitive differences). And in a global-scale survey, Kunstler et al. (2016) found a similar lack of correlation between functional-trait similarity and competitive intensity. There are a number of interrelated reasons why species' functional traits (and trait similarities or differences) might not always predict species performance (i.e., niche traits), including unmeasured or "hidden" traits or lack of a one-to-one relationship between a measured trait and a specific fitness outcome.

To sum up, while there is clearly value in both the phylogenetic and functional-trait approaches to understanding patterns of metacommunity assembly, it is important to recognize that the full cycle of associations is often (in the case of phylogenies) or sometimes (in the case of functional traits) poorly linked to species performance and response to environmental conditions (i.e., niche traits). Currently, the application of trait- and phylogeny-based methods is focused on using them to understand patterns of species coexistence, often separating the roles of fitness-equalizing and niche-stabilizing effects (Chesson 2000, Mayfield and Levine 2010, Adler et al. 2013, Kraft et al. 2015a, 2015b).

In the remainder of this chapter, we overview this literature with a critical eye to how well it addresses the full scope of metacommunity ecology. Here, we discuss how trait and phylogenetic information is used to explain why species that coexist actually do so (which was the original intent of the theory), as well as how well these ideas can be extended to address metacommunity-level processes that are not often well captured by the somewhat simplistic coexistence paradigm.

6.3 TRAIT AND PHYLOGENETIC OVER- OR UNDERDISPERSION: WHAT DOES COEXISTENCE THEORY PREDICT?

The reality from modern coexistence theory is unfortunately more nuanced than the simple perspective that patterns of overdispersed traits or phylogenies equals competitive exclusion, while patterns of underdispersed traits or phylogenies equals environmental filtering (Leibold 1998, Chase and Leibold 2003, Mayfield and Levine 2010, Adler et al. 2013, Kraft et al. 2015a). For example, using consumer-resource models, Leibold (1998) pointed out that coexistence would be most likely when certain types of traits—those involved in resource requirements—were more similar than expected by chance, whereas other types of traits—those involved in resource impacts—were more different than expected by chance. Chase and Leibold (2003) provided some evidence for this proposition from a number of data sources. Using a slightly different perspective based on the view that coexistence results from the balance between stabilizing and equalizing factors (Chesson 2000), Mayfield and Levine (2010) noted that overdispersion would be expected only for some types of traits (those more involved with stabilizing niche differences), whereas underdispersion would be expected for other types of traits (those more involved with equalizing fitness difference; see also HilleRisLambers et al. 2012 and Adler et al. 2013).

An important, but virtually ignored, issue in coexistence theory involves traits (or phylogenetic effects) that mediate destabilizing niche relations, leading to multiple stable equilibria (competitive exclusion by one species or the other depending on priority effects) rather than coexistence (see Chap. 5). Here, there are niche differences in effects and responses (sensu Chase and Leibold 2003), but these lead to interspecific effects being *greater* than intraspecific effects. Under these conditions, trait differences that contribute to this effect should be positively correlated with the difference between inter- and intraspecific competition and it is *less* likely that species with these sorts of niche differences would coexist. In the face of other possible mechanisms that would enhance coexistence (e.g., drift, dispersal, etc.), the relationship between trait difference and coexistence is the reverse of that assumed by the approaches that pit filtering against stabilizing effects! Nevertheless, despite some important documented cases of these sorts of destabilizing effects in natural communities (overviewed in Chap. 5), it remains unknown just how prevalent they are.

Kraft et al. (2015b) examined Mayfield and Levine's (2010) predictions (ignoring the possibility of destabilizing niche differences) by measuring demographic responses of 18 species of annual plants from California grasslands along a density gradient of each of its pairwise competitors. For each pairwise competitive interaction, they examined the average differences in equalizing fitnesses and stabilizing niche differences, finding that most species pairs were so different in their relative fitnesses that they were unlikely to be able to coexist. Instead, only 12 of the possible species pairs had stabilizing niche differences that were large enough, and equalizing fitness differences that were small enough, to be able to coexist locally (Fig. 6.2a). They identified a number of functional traits related to fitness differences but found no simple correlations between any individual functional trait and the stabilizing niche differences they measured. A more complex multi-trait analysis, however, suggested that some trait combinations could correlate with stabilizing effects; these included some, but not all, of the same traits involved in the fitness differences.

Narwani et al. (2013) performed a similar analysis on phytoplankton in microcosms and found ranges of stabilizing niche differences and equalizing fitness differences in which species pairs were expected to coexist and in which competitive exclusion was expected (Fig. 6.2b). In this study, there was a much smaller range of variation in equalizing fitness differences (compare the range along the y-axis in each study). Not surprisingly, they found that stabilizing niche differences were more important than fitness differences in explaining species coexistence, at least in the lab. Interestingly, and unlike the results from Kraft et al. (2015b), Narwani et al.'s (2013) results indicate that there seems to be a rather strong correlation between stabilizing and equalizing trait differences among species ($R^2 = 0.503$), again emphasizing the difficulty of fully separating functional traits associated with stabilizing or equalizing processes, because both involve traits involved in resource acquisition, conversion, and mortality (but see Leibold 1998, Chase and Leibold 2003, HilleRisLambers et al. 2012, Kraft et al. 2015b, and Letten et al. 2017 for some discussion of this possibility for some types of traits).

There are two other interesting observations from the in-depth analyses of traits and coexistence among species pairs by Narwani et al. (2013) and Kraft et al. (2015b), as well as a number of other detailed analyses that have evaluated aspects of species coexistence based on trait differences (Tessier et al. 2000, Angert et al. 2009, Kraft et al. 2014, Fritschie et al. 2014). First, a large number of species pairs used in the studies are not expected to coexist according to the criteria given (e.g., species pairs that fall above the lines in Fig. 6.3). Importantly, in Narwani et al. 2013, coexistence was measured as a very local phenomenon (i.e., microwells), whereas in Kraft et al. 2015b, coexistence was studied in 1-m^2 plots, and only one set of environmental conditions was used to test niche differentiation. Of course, the differences measured between species are likely to be context dependent, and these parameters will change as spatial heterogeneity and other spatial mechanisms of coexistence are considered, allowing many more species pairs to meet the conditions for coexistence in some places and times. Coexistence

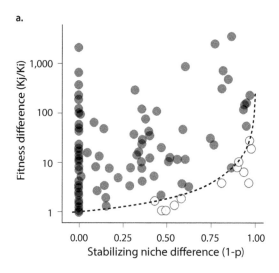

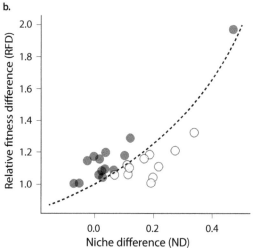

FIGURE 6.2. Experimentally determined relationships between relative niche fitness differences (RFD) and stabilizing niche differences (ND) in grassland plants from California and in freshwater algae from culture collections. In each case, the region below the dashed line indicates the region where coexistence is met (also illustrated with data-point shading). Note the differences in scales between (a) and (b). (a) Pairwise estimates of RFD and ND among 18 grassland plants. Modified from Kraft et al. 2015b. (b) Pairwise estimates of RFD and ND among eight species of freshwater green algae. Modified from Narwani et al. 2013.

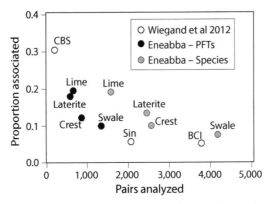

FIGURE 6.3. The relationship between species richness (shown as the numbers of species pairs analyzed) relative to the proportion of species pairs that show small-scale associational patterns (either positive or negative). For the data from the Eneabba region, pairwise associations are shown between species (gray circles) and plant functional types (PFTs; black circles). Labels on the data points refer to the habitat types in which data were collected. For the data from the Center for Tropical Forest Science forest plots, data come from associations between species (open circles) for a lower diversity (52 species) temperate forest in Changbaishan, China (CBS); a diverse (300+ species) neotropical forest at Barro Colorado Island (BCI), Panama; and an Asian tropical forest (200+ species) in Sinharaja, Sri Lanka (Sin). Modified from Perry et al. 2014.

is thus more likely at larger spatial and temporal scales, where this heterogeneity is present, than predicted by these studies of coexistence at very small scales.

Second, there are many species pairs that are quite similar in both stabilizing and equalizing factors (i.e., their points are not far away in either dimension from the line through the graph in Fig. 6.3), at least under the conditions under which these parameters were measured. Though not "neutral" in the purest sense, these species might be more prone to the sorts of demographic stochasticity that lead to a signature of ecological drift in a metacommunity (see Chaps. 4 and 5).

When evaluating the co-occurrence of species relative to the similarity of their traits, it is important to recognize that there are at least two mechanisms that could create species distributions that appear to be independent with respect to traits. First is the NT, whereby species traits are irrelevant, at least in the context of traits that influence an individual's demography (Hubbell 1979, 2001; Hubbell and Foster 1986; Bell 2001). Second, species distributions can be independent from one another even in the face of strongly deterministic competitive interactions and niche variation, especially in highly diverse communities. For example, in diverse communities, the demographic responses of a species to its neighbors will often be heterogeneous if there is some degree of dispersal limitation such that the most competitive species in a given site are often missing (Hurtt and Pacala 1995). Even if the underlying process of species interactions is highly deterministic, dispersal limitation and other processes can mask the signal of competitive interactions and lead to unpredictable biotic neighborhoods that cannot be distinguished from largely stochastic assembly—a phenomenon known as the *stochastic dilution effect* (McGill 2010, Wiegand et al. 2012).

Although interesting, the stochastic dilution effect cannot necessarily be empirically tested and disentangled from the predictions of a purely neutral model without detailed study (and probably experimentation). However, there is some recent intriguing evidence from comparisons of fully stem-mapped tree plots that vary in species richness. For example, Wiegand et al. (2012) found that pairwise species associations (either positive or negative) were most likely in a temperate forest plot in China with only 54 species, intermediate in a tropical Chinese forest with more than 200 species, and least likely in the BCI plot with more than 300 species. Using a similar analysis from four stem-mapped plots in shrublands in the Eneabba region of Western Australia, Perry et al. (2014) found a similar pattern of decreasing pairwise associations of functional types as species richness of the plots increased. Results from both of these studies are shown in Figure 6.3. Finally, Wang et al. (2016) used five stem-mapped forest plots to examine both functional and phylogenetic similarities among the species, finding largely similar results to those of Weigand et al.; the lower-diversity temperate plots showed higher (positive or negative) pairwise associations with functional similarity of species pairs than the more diverse tropical plots. From these observations, we might tentatively conclude, then, that even if SS-type forces are acting similarly in each habitat type, the stochastic dilution effect might mask the signature of species associations and trait dispersions in more diverse communities due solely to geometric constraints, making it difficult to discern the underlying mechanisms without further information.

Despite the continued appeal of the idea, modern coexistence theory cannot easily be distilled into the simple dichotomy in which trait underdispersion equivalent to abiotic filters predominate while trait overdispersion equivalent to biotic interactions predominate (Mayfield and Levine 2010, Adler et al. 2013, Kraft et al. 2015a). More important from our perspective, conspicuously missing from this discussion is the lack of consideration of metacommunity processes and the role of spatial scale despite the implicit role of spatial scale in discussions regarding the "regional species pool" (i.e., γ-diversity) and local coexistence (i.e., α-diversity). Most approaches to coexistence theory are largely agnostic about scale—they simply state that in order for species to coexist (at any scale), the balance of stabilizing and equalizing differences must be met (Chesson 2000, Adler et al. 2007, Hille-RisLambers et al. 2012). Likewise, the geometric constraints of the stochastic dilution effect are determined by neighborhood size—larger neighborhoods are more likely to show associations than smaller neighborhoods (Wiegand et al. 2012, Wang et al. 2016)—emphasizing a need to explore these patterns across scales (i.e., in a metacommunity).

6.4. PHYLOGENETIC AND FUNCTIONAL-TRAIT DISPERSIONS IN A SIMPLE SS METACOMMUNITY

Weiher and Keddy (1995) were among the first to explicitly consider the issue of scale for functional-trait dispersions (Fig. 6.4). To paraphrase Weiher and Keddy (see also Swenson et al. 2007 and Weiher et al. 2011), if we were to consider an entire pool of species in a reasonably heterogeneous metacommunity, we would find that environmental filtering would lead to species with similar traits being more closely associated with each other and that traits of locally co-occurring species would be significantly underdispersed. However, if the pool of species was restricted only to those that could persist within certain environmental conditions, we might find stronger competitive interactions and patterns of trait overdispersion.

Several authors have suggested that this simple scaling perspective—based on what is essentially an SS view of metacommunities—will allow us to achieve synthesis in the question of functional-trait and phylogenetic community assembly patterns (Weiher and Keddy 1995, Cavender-Bares et al. 2006, Swenson et al. 2006, Weiher et al. 2011). One of the first strong examples of scale dependence in trait dispersions comes from Moulton and Pimm's (1987) study of passerine birds introduced to the Hawaiian Islands after European colonization. They documented the success and failure of documented introductions of 49 nonnative bird species to the main Hawaiian Islands and compared those with null expectations based on morphological traits (wing length, leg length, bill size, etc.). At the scale of the entire island, Moulton and Pimm found no patterns with respect to the similarities or differences in the traits of species that were successfully established versus those that went extinct following their introduction. However, when they looked at patterns of co-occurrence within forests (a more local scale), they found that the morphological traits of surviving species were overdispersed and regularly spaced, suggesting the likelihood of competitive interactions determining which species could co-occur. This study thus shows that much better insights can be gained by an explicit attention to scale rather than by studies that ignore it.

Table 6.1 lists a number of studies that have explicitly examined functional and phylogenetic dispersion patterns at multiple spatial scales (based on a literature search of papers citing the seminal paper by Weiher and Keddy [1995]). Even a cursory look at the table shows that scale dependence in the degree and direction

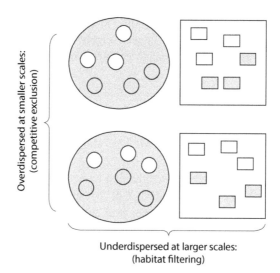

FIGURE 6.4. Cartoon depiction of Weiher and Keddy's (1995) idea that overdispersion versus underdispersion in species traits (or phylogenies) might be scale dependent. At smaller scales within a patch (large circles and squares), species (small circles and squares) are overdispersed due to competitive exclusion. However, because there is habitat heterogeneity, indicated by the shape (circle or square) of the habitat and because there are habitat preferences of the species, indicated by the shape (circle or square) of the species, there is underdispersion at the larger spatial scale due to habitat filtering.

of dispersions is a typical occurrence. Only a few studies found no evidence for shifts in the patterns of dispersion across scales. One in the BCI tropical forest (Kembel and Hubbell 2006) involves a situation in which there was no evidence for phylogenetic structuring at any scale, which is perhaps unsurprising for this notoriously neutral forest plot, and it contrasts with scale-dependent results from similar analyses in a number of other tropical forest plots (Swenson et al. 2007, Kraft and Ackerly 2010, Parmentier et al. 2014). And two studies—one on ants (Donoso 2014) and one on stream fishes (Troia and Gido 2015)—found that assemblages were underdispersed at all spatial scales. Further, a number of studies found scale dependence when some types of metrics and null model approaches were used, but no scale dependence when other approaches were taken.

Table 6.1 indicates that scale dependence in the trait dispersions does indeed seem to be a relatively general phenomenon. Nevertheless the exact nature of the scale dependence in functional-trait and phylogenetic relationships is not always consistent. In many cases, trait and phylogenetic relationships were underdispersed at larger scales and overdispersed at smaller scales, as suggested by the scheme discussed above (Weiher and Keddy 1995, Cavender-Bares et al. 2006, Swenson et al. 2007, Kraft and Ackerly 2010, Parmentier et al. 2014). However, all other permutations of overdispersion, underdispersion, and random dispersion at smaller and larger scales were also observed.

There are several possible reasons why these studies do not show consistent results. First, the simple expectations (filtering vs. competition) are not necessarily well connected to modern metacommunity theory. The conceptual model in Figure 6.4 is a reasonable "first cut" as to what an SS model might predict, but it has a number of implicit assumptions that are not always upheld. As discussed above, modern coexistence theory (Chesson 2000, Chase and Leibold 2003, HilleRisLambers et al. 2012) is more nuanced than the classic view that coexistence requires species traits to differ; they need to differ in some ways but cannot be too different in other ways (Leibold 1998, Mayfield and Levine 2010, Letten et al. 2017). Our discussion above suggested that it might be reasonable to approximate this version of coexistence theory in an SS metacommunity where equalizing factors are related to environmental filtering at larger scales (leading to underdispersion) and where stabilizing factors are related to species interactions and resource consumption at smaller scales (leading to overdispersion, but only if their connection to destabilizing effects is ignored); however, this connection has not been fully explored and it may not apply under other metacommunity scenarios.

Second, the conceptual view in Figure 6.4 explicitly assumes that habitat heterogeneity is reasonably well distributed across the landscape of interest. This assumption leads to the idea of local overdispersion (because of competitive interactions) but underdispersion at the metacommunity scale (because of environmental filtering). Imagine, however, a scenario in which environmental conditions are autocorrelated spatially along a more continuous gradient rather than being independent. Here, due to the SS process, we would expect species with similar traits (e.g., traits for water use or temperature response) to sort more closely to one another—leading to underdispersion at smaller scales, because their favored habitats are closer to one another—and species with more different traits further apart—leading to overdispersion at larger scales.

Third, the degree of heterogeneity of habitat types and the magnitude of their environmental differences will create different expectations regarding trait dispersions at different scales. Less heterogeneous metacommunities should have less underdispersion at larger spatial scales than more heterogeneous metacommunities. This view suggests that overdispersion in traits should occur primarily at the smaller spatial scales at which competitive interactions take place. However, there are several mechanisms that can allow species that have similar traits to coexist in metacommunities at larger spatial scales due to multiple stable equilibria (discussed above), endpoint assembly cycles, and related positive (locally destabilizing) frequency-dependent interactions (e.g., Janzen-Connell effects) (see, e.g., Chaps.

TABLE 6.1. Studies That Have Examined Scale Dependence in the Dispersion of Traits among Co-occurring Species

Study	System	Functional or Phylogenetic	Small Scale	Large Scale	Notes
Kraft and Ackerly 2010	Tropical forest trees (Ecuador)	Functional and phylogenetic	Overdispersed	Underdispersed	
Parmentier et al. 2014	Tropical forest trees (Cameroon)	phylogenetic	Overdispersed	Underdispersed	
Cavender-Bares et al. 2006	Oak trees (Florida)	phylogenetic	Overdispersed	Underdispersed	
Goncalves-Souza et al. 2014	Spiders (Espírito Santo, Brazil)	Functional and phylogenetic	Overdispersed	Underdispersed	
Swenson et al. 2007	5 tropical forest plots	Phylogenetic	3 plots overdispersed; 2 plots neutral	Neutral	Results depend on which phylogenetic metrics used
Luza et al. 2015	Small mammals (Brazil)	Functional and phylogenetic	Overdispersed in one habitat type; others underdispersed	Underdispersed	
Trisos et al. 2014	Insectivorous birds (Peru)	Functional	Overdispersed	Neutral	When multiple traits combined, pattern disappears
Algar et al. 2011	Tree frogs (Hylidae)	Functional	Temperate—overdispersed Tropical—random	Temperate—underdispersed Tropical—random	

Results largely consistent with Weiher and Keddy's (1995) "scale" hypothesis

Results partially consistent with Weiher and Keddy's (1995) "scale" hypothesis

Reference	System	Data type	Pattern 1	Pattern 2	Notes
Hubert et al. 2011	Fish (Indo-Pacific)	Phylogenetic	Random	Overdispersed	Suggests that allopatric speciation leads to large-scale patterns
Gonzalez-Caro et al. 2012	Hummingbirds (Andes)	Phylogenetic	Random	Overdispersed	Patterns vary with elevation
Münkemüller et al. 2014	Plants (European Alps)	Phylogenetic	Variable	Variable	Depends on null models and phylogenetic measures
Belmaker and Jetz 2013	Bird and mammal assemblages globally	Functional	Variable	Variable	Depends on how local and regional are defined, and how null models are established
Yang et al. 2014	Tropical forest trees (China)	Functional and phylogenetic	Phylogenetic—overdispersed (lower magnitude) Functional—random	Phylogenetic—overdispersed (higher magnitude) Functional—random	Patterns vary with size class examined
Kembel and Hubbell 2006	Tropical forest trees (Barro Colorado Island, Panama)	Phylogenetic	Random	Random	Some under- and overdispersion in local habitats, but in opposite directions
Donoso 2014	Ants (Barro Colorado Island, Panama)	Functional and phylogenetic	Underdispersed	Underdispersed	
Troia and Gido 2015	Stream fish (Kansas)	Functional	Underdispersed	Underdispersed	Degree of underdispersion higher upstream
Sobral and Cianciaruso 2015	Forest and savannah birds (Cerrado, South America)	Functional and phylogenetic	No pattern	No pattern	Depends on structural form of null model used

Scale matters, but results different from Weiher and Keddy's (1995) "scale" hypothesis

Scale does not seem to matter

5 and 11). They would, nevertheless, extend the scale at which overdispersion patterns are expected, especially when metacommunities are otherwise relatively homogeneous in their environmental conditions.

Fourth, even though different sorts of traits are often treated interchangeably in analyses of trait dispersions, different traits can often be quite distinct in their patterns of distribution and coexistence, and this observation can make much sense from a metacommunity perspective. For example, the SS metacommunity perspective suggests that traits associated with habitat tolerances would be more likely to show patterns of underdispersion due to habitat filtering at the larger scale, whereas traits associated with resource consumption abilities might be more likely to show patterns of overdispersion due to competitive interactions at smaller scales. To describe this phenomenon, others have used the terms α-niche (by analogy with within habitat, α-diversity) to describe traits associated with local species interactions, and β-niche (by analogy with across habitat, β-diversity) to describe traits associated with among-habitat filtering (Pickett and Bazzaz 1978, Wilson et al. 1999, Silvertown et al. 2006, Ackerly and Cornwell 2007). For example, Ingram and Shurin (2009) showed that among coexisting rockfish (*Sebastes* spp.) assemblages, traits associated with local feeding performance (body size, gill raker morphology) were more overdispersed, whereas a trait associated with habitat type (eye size, correlated with depth) was more underdispersed. Unfortunately, many of the traits that we might measure among natural communities, such as plant leaf size or shape or bird bill morphology, are not easily categorized by how they influence either environmental tolerances or interaction strengths, and indeed, they often probably influence both (Leibold 1998, Chase and Leibold 2003, HilleRisLambers et al. 2012, Adler et al. 2013, Kraft et al. 2015a, 2015b). Likewise, approaches that combine multiple traits into multivariate metrics (e.g., functional diversity or dispersion) can confound traits important for environmental tolerances versus those important for consumption and interspecific interactions (Weiher et al. 2011, Spasojevic and Suding 2012, Trisos et al. 2014); phylogenetic measures are likely to be particularly sensitive to this problem, even if traits are phylogenetically conserved (Swenson and Enquist 2009). Solutions to this problem involve comparisons of traits specifically expected to influence the factor of interest (e.g., competitive interactions; Trisos et al. 2014, Kraft et al. 2015b), but such comparisons are not always straightforward or possible without much more detailed information (e.g., species growth rates or resource reductions when isolated compared with when they are together with other species).

Fifth, the separation of analyses into different scales (i.e., local vs. regional) is often arbitrary, such that what one investigator considers to be relatively local can at the same time be regional to another investigator. And the results that emerge are often strongly influenced by the choice of such scales (Swenson et al. 2007,

Belmaker and Jetz 2013). A possible solution to this approach is to treat scale more continuously. For example, Smith et al. (2013) proposed that one could calculate a functional-diversity–area relationship (FAR) (which is analogous to SAR), in which phenotypic trait diversity increases with increasing sampling scale but in a decelerating way that depends on the degree of functional dispersion versus redundancy in a metacommunity. The shape of the FAR, compared with null expectations, can be used to evaluate whether species traits are over- or underdispersed and how such deviations do (or do not) vary with increasing sampling scale. Morlon et al. (2011) similarly discussed the utility of a phylogenetic-diversity–area relationship (PDAR) to detect patterns of phylogenetic metacommunity structure and to compare it with SAR patterns to infer possible mechanisms of community assembly (see also Helmus and Ives 2012 and Mazel et al. 2015).

Sixth, there are other methodological issues that can confound analyses of trait dispersions and other inferred community assembly mechanisms that should be carefully considered in the development and interpretations of these sorts of tests. For example, the taxonomic scale of the organisms used for comparison can greatly influence estimates of trait dispersions (Cavender-Bares et al. 2006, Swenson et al. 2006): patterns of dispersion within a single genus or functional group might give one result, whereas patterns within a family or entire community might give a different result. The form of the null model used and the definition of the scale of the species pool can strongly influence conclusions (Hardy 2008, Kembel 2009); one solution to this effect might be to develop process-based null models and definitions of a species pool (Lessard et al. 2012, 2016; Pigot and Etienne 2015).

Finally, patterns of over- versus underdispersion are usually associated with SS-based perspectives and tested against random expectations with a null model. However, as we have emphasized throughout this book, SS processes are but one mechanism for how spatial processes influence species coexistence at local and regional scales in a metacommunity. In Section 6.5, we discuss predictions for patterns of phylogenetic and trait associations expected when other metacommunity-level processes are important, such as those inherent to the other metacommunity archetypes.

6.5 PHYLOGENETIC AND TRAIT DISPERSIONS IN NON-SS METACOMMUNITY ARCHETYPES

Thinking about phylogenetic and traits dispersions has thus focused only on cases in which there are two deterministic processes (environmental filtering and stabilizing niche differences) that work against each other. This view generally lines up with an SS approach and ignores how processes associated with the other metacommunity archetypes (including NT, PD, and ME, as well as the mix of other possibilities we have so far discussed) might affect our expectations. Although a more robust exploration of the theoretical expectations of patterns of trait and phylogenetic variation is greatly needed, we can make some speculations.

If some degree of stochasticity and dispersal limitation is included within what is otherwise a standard SS archetype, patterns of associations within local neighborhoods of species are likely to be less predictable, especially in high-diversity systems (Hurtt and Pacala 1995), leading to a stochastic dilution effect caused by geometric constraints at smaller spatial scales (McGill 2010, Wiegand et al. 2012). Here we would expect no patterns of species or trait dispersions at smaller scales, although this effect would decrease with scale (Wiegand et al. 2012, Wang et al. 2016), and we might expect trait underdispersion at larger scales.

Instead, if the metacommunity is governed more by processes associated with PD, in which coexistence occurs via spatial processes such as competition-colonization trade-offs (Tilman 1994, Hanski and Gyllenberg 1997), we might predict that species occurring more closely together (i.e., at local scales) would possess similar traits—colonization specialists with traits associated with rapid dispersal and high growth rates living together in recently disturbed areas, and competition specialists with traits associated with resource acquisition and competitive ability living together in less recently disturbed areas. When coexistence is via processes associated with PD, we might expect patterns opposite to those predicted from the SS metacommunity above, with underdispersion at smaller scales (species with similar traits living together) and overdispersion at larger scales. ME metacommunity models (Amarasekare and Nisbet 2001, Mouquet and Loreau 2003) might make similar predictions, but because of "spillover" of species among localities, the expectations of trait dispersions would become diluted.

The NT metacommunity archetype (Hubbell 2001) is agnostic about traits, and thus we might expect random trait dispersions at all spatial scales, as well as perhaps random phylogenetic dispersions, when this archetype predominates. However, there is an important caveat with this expectation: most tests of phylogenetic and trait dispersions use a traditional null modeling approach, comparing observed to "random" communities. It is important to remember that "random" in the statistical sense does not necessarily translate into "random" in the mechanistic sense of the NT (Gotelli and McGill 2006).

There are several types of random processes under the guise of NT (e.g., speciation, extinction, birth, death) that can lead to significant patterning in natural communities (Ulrich 2004, Hardy 2008, Kembel 2009, Pigot and Etienne 2015). For example, Pigot and Etienne (2015) showed that even when processes were completely neutral, patterns of allopatric speciation, combined with colonization and extinction dynamics, are expected to lead to patterns of phylogenetic overdispersion of communities. They applied this model as a dynamic null hypothesis to data sets of bird and primate assemblages in the Manu Biosphere Reserve in Peru and found that these groups showed phylogenetic overdispersion consistent with the random expectation from NT even though these taxa had distributions that had previously been interpreted as an outcome of competitive interactions (Cooper et al. 2008, Graham et al. 2012). Likewise, there can be traits that are neutral with respect to demographic differences and speciation potential, and these may evolve and diverge in concert with phylogenetic divergence, leading to the possibility that even neutral traits can be overdispersed (Hardy 2008, Kembel 2009, Pigot and Etienne 2015).

6.6 CONCLUSIONS

Although functional and phylogenetic approaches for understanding patterns and processes of metacommunity assembly have been around since Darwin and Wallace, the past fifteen-plus years have seen an explosion of interest in this area (reviewed in, e.g., Webb et al. 2002, 2010; McGill et al. 2006b; Enquist et al. 2015). Clearly, this interest has helped to advance the field and our understanding of metacommunity assembly processes.

At the same time, some of the concepts and analyses associated with the phylogenetic and trait-based approaches to metacommunity structure have not provided the purported panacea that we might have expected. Using phylogenies and functional traits to understand patterns of small-scale metacommunity assembly processes assumes a strong correlation between phylogenetic relatedness, functional-trait similarity, and demographically important processes (e.g., performance). These links however are often weak, meaning that phylogenetic and functional-trait analyses can sometimes give limited and even misleading answers.

Despite some clear limitations, functional-trait and phylogenetic information can allow a richer understanding of metacommunity assembly processes if scale, habitat heterogeneity, and the processes (beyond SS) by which metacommunity assembly occurs are more explicitly considered. A key aspect is to better understand how trait differences that influence "destabilizing" effects may complement our current understanding of how they influence co-occurrence when they are stabilizing. Using functional or phylogenetic measures together with taxonomic measures can almost certainly help to disentangle metacommunity assembly processes in a way that is not possible when only considering each of these facets in isolation. In Chapter 7, we describe a number of approaches by which the utility of comparing patterns' different facets (e.g., functional/phylogenetic and taxonomic) of metacommunity composition can provide important complementary information that can allow us to get a few steps closer to understanding some of the potential processes by which metacommunities assemble.

CHAPTER SEVEN

Combining Taxonomic and Functional-Trait Patterns to Disentangle Metacommunity Assembly Processes

Prospectus

1. Combining taxonomic and functional approaches may provide a way forward in disentangling some degree of metacommunity assembly process from pattern.
2. One way to evaluate metacommunity dynamics is to compare taxonomic patterns in β-diversity to both functional-trait or phylogenetic patterns. Typically, we find that trait-based patterns appear to be more predictable than taxonomy-based patterns, which might suggest an increase in the role of dispersal limitation and drift affecting β-diversity patterns among similar species.
3. Comparing temporal changes in functional-trait-based β-diversity and taxonomic β-diversity, especially in response to external perturbations (anthropogenic or otherwise), can provide important information regarding the underlying mechanisms of metacommunity assembly (e.g., drift, dispersal limitation, environmental filtering).
4. It can also be useful to examine how and which functional traits covary with which environmental variables. This is especially true when traits are selected that have close links with alternate possible mechanisms by which SS occurs.
5. We conclude that trait-based approaches can be quite useful, particularly when combined with taxonomic approaches, toward achieving a more synthetic view of metacommunity assembly. This is especially true when the traits involved can be linked to mechanistic aspects of metacommunity dynamics. Nevertheless, ignoring taxonomic patterns ignores the fundamental issues that regulate metacommunities, since these involve contrasting processes that affect individual species.

One thing is clear from our discussions so far: even when environmental filtering and species interactions are strong, and thus consistent with a simplistic SS worldview, a great deal of variation in metacommunity organization is usually left unexplained. For example, in Chapters 4 and 5, we discussed the variation partitioning approach for detecting the role of environmental, spatial, and random factors and for attributing the resulting correlations to the different metacommunity perspectives. However, even the best case studies are usually able to explain only a modest proportion of the variation in community structure (i.e., less than 50%, although maybe there is "hope" that we can do better; see Chap. 4), and the majority of variance in community structure remains unexplained by these analyses (see also Cottenie 2005, Soininen 2014, Heino et al. 2015).

One might conclude that this lack of determinism implies a pervasively strong influence of stochastic processes perhaps consistent with ecological drift that are predominant in the NT and PD archetypes. However, as we discussed in Chapters 4 and 5, even when deterministic forces such as environmental filtering and biotic interactions inherent to SS are operating, we are often not able to predict the exact taxonomic composition of a given locality within a metacommunity with great accuracy (e.g., due to the effects of multiple equilibria, assembly cycles, frequency-dependence, and geometric sampling constraints). And in Chapter 8, we show how microevolutionary processes can further erode the correlation between species composition and environmental variation. This high degree of apparent stochasticity can create high β-diversity even among localities that are otherwise similar in their environmental conditions and not spatially distant (i.e., normally ending up as unexplained variation). This high stochasticity is particularly likely when the regional pool is diverse and there is high functional redundancy (or at least functional similarity) (reviewed in Chase 2003a and Fukami 2015).

Because taxonomic measures of diversity (e.g., α-, β-, γ-diversity) are often not very reliable predictors of mechanisms of metacommunity assembly (see, e.g., Chap. 4), many researchers have instead espoused the value of using phylogenetic relatedness and functional traits as a means to understand metacommunity assembly processes. However, as we discussed in Chapter 6, this perspective comes with its own set of limitations, particularly when we make assumptions as to exactly which patterns the various metacommunity archetypes might predict and just how closely matched phylogenetic and functional information is to the actual fitness and demographic processes that influence species persistence and coexistence.

Many authors have more recently suggested that combining taxonomic information with functional or phylogenetic information might provide deeper understanding of the metacommunity assembly processes (Sokol et al. 2011, Münkemüller et al. 2012, Purschke et al. 2013, Spasojevic et al. 2014, van der Plas et al. 2015). Indeed, a number of measures have been developed that intend to capture taxonomic, phylogenetic, and functional measures of α-, β-, and γ-diversity in order to resolve how these facets of biodiversity might enhance our understanding of metacommunity pattern and process (Graham and Fine 2008, Swenson et al. 2011, Siefert et al. 2013).[1]

7.1 USING FUNCTIONAL INFORMATION TO ENHANCE TAXONOMIC PATTERN ANALYSIS

Separating community patterns related to functional traits from those that are not is very old, going back at least to Gleason (1926, 10–11), who suggested that "much of the structural variation in an association would disappear if these taxonomic units which have the same vegetational form and behavior could be considered as a single ecological unit."

An excellent example of this phenomenon comes from Fukami et al.'s (2005) analysis of community variability in experimental grassland plots in the Netherlands, the results of which are shown in Figure 7.1. The experiment consisted of plots sown with either high or low species richness or left open to natural colonization. After eight years of allowing natural colonization of the plots and changes in relative species abundances, Fukami et al. found that the initial differences in species composition remained and that pairwise differences in community composition remained high throughout the study when viewed at the taxonomic level. One could interpret this as a situation in which stochasticity in initial conditions left a lasting signature on community structure, reducing the predictability of the community through time and across space. However, at the same time, the functional similarity among plots became more and more similar through time regardless of the initial conditions.

Similar results were observed by Swenson et al. (2012) in their observations of in-species species and functional-level temporal change when comparing an old-growth tropical forest (BCI) with one that was undergoing succession following logging and hurricane-induced disturbance (Luquillo, in Puerto Rico). They found that both forest types exhibited considerable taxonomic turnover during the 20-or-more-year observation periods, possibly indicating a considerable amount of stochastic drift at the species level. However, the older-growth forest (BCI) showed little functional-trait change during that time period, whereas the successional forest (Luquillo) showed high functional-trait change, indicating that at the

---

[1] It is important to note here that we will remain largely agnostic on exactly how these measures of taxonomic, functional and phylogenetic α-, β-, and γ-diversity are actually calculated and compared, although this can be critically important for making inferences and will require much more thought and standardization before we are able to fully understand what these measures indicate and/or how to understand and synthesize them across studies (Anderson et al. 2011, Chao et al. 2014, Pavoine 2016, Tucker et al. 2016).

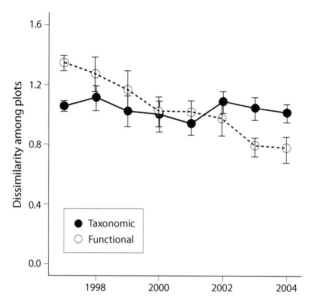

FIGURE 7.1. Dissimilarity in taxonomic (solid circles) and functional (open circles) composition among plots in a long-term grassland experiment in the Netherlands. Plots were initially different in treatments manipulating planted-seed diversity (initially high diversity, initially low diversity, or no added seeds). These plots differed substantially from one another, primarily owing to the seed-addition treatment. After eight years of natural colonization and growth, plots were still quite different from each other in taxonomic composition and did not show any evidence for convergence. Nevertheless further compositional change through time reduced the functional dissimilarity. Modified from Fukami et al. 2005.

functional level, patterns were more consistent with expectations from deterministic SS-based successional models.

We next extend the idea that stochastic and spatial processes might be more important at the taxonomic level, whereas more deterministic environmental processes might be more important at the functional level. For example, we can use the variation partitioning approach to compare results when β-diversity is calculated at the taxonomic level and when it is calculated at the functional level. We illustrate this approach using the same data set on invertebrates and amphibians in a metacommunity of ponds in Michigan that we explored in depth in Chapter 4. Here, functional traits were derived from independent observations of the species' responses to environmental characteristics of the ponds (e.g., presence of predatory fish, drought), their feeding relations, and other important traits (e.g., susceptibility to predators, dispersal ability, life history) (Chase 2003b, 2007; Chase et al. 2010; unpublished data). Table 7.1 shows that a much larger proportion of the variation in composition is explained by differences in functional traits than

TABLE 7.1. Variation Partition of Michigan Ponds at Taxa and Trait Levels

β-diversity facet	Unexplained [Resid]	Total Space [S]	Total Environment [E]
Taxonomic	0.41	0.03	0.56
Functional	0.12	0.00	0.88

taxonomic variation, and that this effect was driven almost entirely by the increase in variance explained by environmental differences among the ponds.

Intuitively, the relatively strong explanatory power of environment for species functional traits makes sense given our knowledge of the natural histories of species that live in different pond types (Wellborn et al. 1996, McPeek 1998, Tessier et al. 2000, Werner et al. 2007). Species that live in ponds that periodically dry out will be characterized by traits that allow them to persist in (e.g., resting stages) or to recolonize (e.g., complex life cycles, rapid dispersal) those habitats. Alternatively, species that live in more permanent ponds will have traits allowing them to be superior competitors or more resistant to the predators that live in these sorts of ponds. However, at the taxonomic level, most metacommunities have species with similar functional traits in the regional species pool (where traits are often sorted into functional groups or have high functional redundancy). Even if these species are not "neutral" per se, ecological drift and dispersal limitation can lead to much less predictability about exactly which species will be present in a given locality. In our Michigan ponds, for example, there are many functional groups that have several species in the metacommunity with high trait similarity (e.g., multiple species of actively swimming predaceous diving beetles, species of sit-and-wait predatory dragonflies, many types of snails, several frogs, etc.; Chase 2003b). While any given pond has only a subset of the species of each functional type, the ponds typically have representative species of the majority of functional types in that metacommunity. Thus, taxonomic β-diversity can be high and likely due to ecological drift (more multiple equilibria, or cycles, or drift), especially in certain types of metacommunities that have much functional redundancy in the species pool (Chase and Leibold 2002; Chase 2003a, 2007, 2010; Chase et al. 2009). However, at the same time, because many types of species are functionally quite similar, functional β-diversity is, almost by definition, much lower.

This idea is obviously quite reminiscent of our discussion of assembly rules in Chapter 6 (Diamond 1975a, Fox 1987, Fox and Brown 1993, Weiher and Keddy 1995). If this is typical, a lesson we can take by comparing patterns of taxonomic versus functional β-diversity might be that deterministic processes associated with environmental filtering can occur simultaneously with stochastic and dispersal-related processes within a metacommunity. However, the signatures of the processes will be differentially detected depending on which facets of community variation are measured (e.g., taxonomic versus functional, see also Sokol et al. 2011); that is, we might perceive a much larger role for ecological drift and dispersal limitation when variation is viewed at the taxonomic level, and at the same time, a strong role for environmental filtering when variation is examined at the scale of species functional traits.

This simple perspective comparing taxonomic and trait variation in metacommunities might help reconcile why the niche-versus-neutral debate has been so polarizing. Clearly, everywhere we look, including at the hyperdiverse tropical rainforests (Kraft et al. 2008, Shipley et al. 2012, Yang et al. 2014) where NT gained early support (Hubbell 1979, Rosindell et al. 2011), we see highly structured and deterministic patterns of species distributions based on their functional traits and environmental features. This observation makes global claims about NT as an explanation for diversity patterns seem unlikely. Alternatively, there can be much functional redundancy (or at least high similarity) in the traits of many subgroups of species that co-occur in metacommunities, especially in high-diversity systems (Leibold and McPeek 2006, Mouillot et al. 2013a, Pigot et al. 2016). The two different foci give very different pictures of the processes that influence metacommunity assembly, but they might be complementary rather than contradictory.

Several studies have found general support for the conjecture that SS-type patterns are more apparent at the functional and phylogenetic level, whereas ecological drift and dispersal limitation often predominate at the taxonomic level. For example, Spasojevic et al. (2014) examined the alpine plant communities of "sky islands" (mountain tops) in Colorado. The sky islands differed both in their spatial (isolation, area) and environmental (temperature, elevation) conditions. When analyses were done at the species level, they found significant effects of spatial variables on local (α-) and patch (β-) diversities, but no influence of environmental factors. However, when the analyses were conducted at the level of functional traits, the patterns reversed, and there was no influence of spatial factors, but a strong influence of environmental factors on functional α- and β-diversity. Likewise, comparing patterns among more than 200 assemblages of 126 species of hummingbirds in the South American Andes, Weinstein et al. (2014) found strong signals of both environmental and spatial distances for taxonomic β-diversity, but only environmental (and not spatial) signals influenced functional and phylogenetic β-diversity. Using data on the breeding-bird distributions in France, Meynard et al. (2011) showed similar patterns in which the relative importance of environmental structuring increased, and spatial structuring decreased, when comparing taxonomic with functional and phylogenetic β-diversity. Sokol et al. (2011) found larger proportions of variation explained by environment relative to space for functional relative to taxonomic β-diversity in mountain-stream invertebrates.

Finally, Wang et al. (2016) found a role for both environment and space in determining the functional and phylogenetic composition of tree neighborhoods in both
tropical and temperate forest plots, whereas there was only a signature of space
(but not environment) in a previous study by Wang et al. (2011) when considering
only taxonomic composition.

Other, less direct evidence also supports the idea that environmental controls
(in accordance with an SS perspective) might play a more important role in structuring communities at the functional and phylogenetic level than at the taxonomic
level. For example, meta-analyses of tropical-forest tree communities have shown
that even when significant, environmental variation explains a relatively small proportion of the variation in species composition (typically less than 10%–20%),
whereas the majority is unexplained (usually much more than 50%) (De Caceres
et al. 2012, Baldeck et al. 2013b). However, studies that have examined similar
partitions of variance on phylogenetic and functional diversity have found typically larger proportions of variance explained by environment, even if there was
still a lot of variation explained by space and unexplained (Swenson et al. 2011,
Liu et al. 2013, Yang et al. 2015).

There are, of course, exceptions to the possible generality that environment has
a stronger signature at the functional and phylogenetic levels relative to the taxonomic level; these exceptions require more in-depth consideration. For example,
Siefert et al. (2013) found that functional diversity metrics were able to explain
more of the variation in tree-community composition across the eastern United
States than taxonomic measures (~50% vs. 30%), but the larger amount of variation explained was largely due to an increase in the signature of space rather than
to environmental factors. This reason might have something to do with the much
larger scale of observation, which encompassed biogeographic processes, rather
than with the metacommunity perspective described above. Likewise, in a study
on a similar continental scale, Arnan et al. (2015) found spatial correlations among
ant communities across Europe at the taxonomic, functional, and phylogenetic
scales, with a smaller role for environment; environmental variation nevertheless
played a stronger role in describing functional diversity measures, suggesting a role
for SS dynamics. Saito et al. (2015) also found a stronger signal of space relative
to environment in their study of phylogenetic β-diversity of stream insects. They
attributed this result to the fact that flight capacity was phylogenetically conserved
(species with high flight capacity were more closely related to one another), leading
to a strong signal of space on the phylogenetic structure of the metacommunity.
Finally, da Silva and Hernández (2015) found that the relative importance of environment increased, while space declined, when they compared functional with
taxonomic β-diversity of dung beetles in Brazilian forests. However, the amount of
variation explained by all measures was exceedingly small (much less than 10%),

suggesting either a great deal of drift or the necessity to measure other types of functional traits.

7.2 USING FUNCTIONAL INFORMATION TO ENHANCE ANALYSES OF CHANGE THROUGH SPACE OR TIME

Another area where use of functional information has enhanced our understanding of taxonomic patterns of diversity and composition in metacommunities is in the exploration of how metacommunities may change across space or through time. For example, as we discussed in Chapter 5, a general expectation is that there can be considerable change in the composition of local communities through time (temporal β-diversity), but little change in the total number of coexisting species (α-diversity) in models that place high importance on stochasticity and ecological drift, such as the original equilibrium theory of island biogeography (MacArthur and Wilson 1963, 1967). This presumption also applies to modern metacommunity derivatives of this framework, such as NT and PD and might explain why contrary to expectations based on human impact, recent meta-analyses often do not find consistent declines of local species richness through time, even when they find substantial changes in species composition (Vellend et al. 2013, Dornelas et al. 2014, Supp and Ernest 2014).

One reason that there might be no changes in local species richness (α-diversity), even though we know there are patterns of extinctions at broader spatial scales, could be if the composition of local communities (β-diversity) became more similar over time, a form of biotic homogenization that could be the result of human activities. A recent example of this phenomenon comes from Magurran et al.'s (2015) analysis of marine groundfish communities off the coast of Scotland, which were monitored for nearly 30 years. They found no changes in local species richness (α-diversity) across the time series, but probably as a result of warming oceanic temperatures, they found that species from northern localities became increasingly more similar to those from warmer southern localities with time (thus decreasing β-diversity across the latitudinal gradient).

Decreased taxonomic β-diversity (biotic homogenization) as a result of anthropogenic activities has been of strong interest in applied ecology, with the idea that it could often lead to less diversity at larger spatial scales (McKinney and Lockwood 1999, Olden and Rooney 2006, Baiser et al. 2012). Although this may be the case, it is possible that other factors are involved, including (1) changes in factors related to dispersal limitation, (2) ecological drift, and (3) frequency-dependent processes or (4) changes in factors related to species environmental tolerances and interactions.

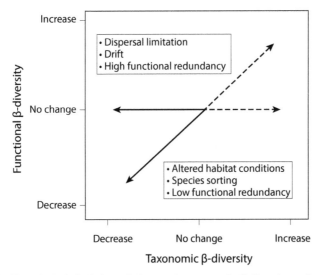

FIGURE 7.2. Hypothetical depiction of changes in taxonomic β-diversity and concomitant changes in functional β-diversity that are expected to occur if (1) changes are due to alterations to dispersal limitation and ecological drift, and/or there is high functional redundancy (horizontal line), or (2) if changes are due to alterations in habitat conditions and species sorting processes (diagonal line). Though we often view changes in β-diversity through the lens of expected decreases in response to anthropogenic activities (solid lines), these can also increase (dashed lines) due to other factors.

While changes in taxonomic β-diversity alone do not allow us to strongly infer which mechanisms might underlie these patterns, the relationships between changes in taxonomic and functional β-diversity can provide a stronger signature to infer something about the processes that influence metacommunity changes in space or time. Figure 7.2 outlines a conceptual framework by which to infer such changes. Specifically, if taxonomic β-diversity is driven mainly by the dispersal/ drift components of metacommunity assembly, when frequency dependence or priority effects are strong or when functional redundancy is high, then we would expect no correlated changes in patterns of taxonomic and functional β-diversity (see , e.g., Baiser and Lockwood 2011). Alternatively, if taxonomic β-diversity is driven mainly by processes altering the habitat filtering and species interactions components of metacommunity assembly or when functional redundancy is low, then we would expect correlated changes in taxonomic and functional β-diversity. These are not the only two possibilities, because there can be a combination of both processes (leading to intermediate cases), as well as cases in which functional β-diversity may change with little change in taxonomic β-diversity or cases in which taxonomic and functional β-diversity respond in opposite directions.

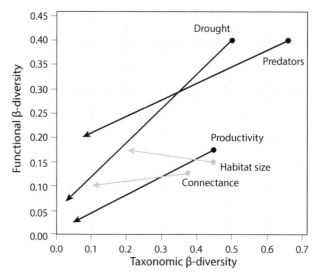

FIGURE 7.3. Comparison of taxonomic and functional β-diversity from several experimental studies that manipulated three environmental variables (shown with dark arrows)—drought, predators, and productivity, and two spatial variables (shown with gray arrows)—connectance and habitat size. The arrows each point from the more benign environmental conditions (low drought, no predators, high productivity) and small/isolated spatial conditions to the less benign condition. When environmental conditions were varied, both taxonomic and functional β-diversity declined (but to a lesser degree with productivity), but when spatial conditions were varied, only taxonomic, but not functional, β-diversity declined.

Nevertheless, we start with this as a simple hypothesis to examine how comparisons of changes in taxonomic and functional β-diversity can provide deeper insights into the mechanisms underlying metacommunity assembly.

As an illustration of the utility of comparing taxonomic and functional components of β-diversity simultaneously, we revisit some of the experiments and observations in pond metacommunities that we discussed in Chapter 4 about patterns of taxonomic β-diversity (Chase and Leibold 2002; Chase 2003a, 2007, 2010; Chase and Ryberg 2004; Chase et al. 2009, 2010). Although the patterns of taxonomic β-diversity were published in those studies, none of the functional β-diversity patterns have been published previously. (Experimental and sampling methods are available in the published papers.)

Figure 7.3 shows five different scenarios in which taxonomic β-diversity differed between experimental treatments. In three of the cases, the environment changed between treatments (drought vs. no drought, Chase 2007; predators vs. no predators, Chase et al. 2009; high vs. low productivity, Chase 2010), and in two cases, spatial processes changed between treatments (connectance, Chase et al. 2010;

habitat size, Chase, unpublished data). For the cases in which spatial processes changed, taxonomic β-diversity changed, but there was little influence on functional β-diversity. This result suggests that these two processes acted mostly by "culling" some of the functional redundancy in the species pool but retaining most of the functions, thus altering taxonomic, but not functional, β-diversity. Alternatively, for the cases in which environment changed, functional β-diversity changed concordantly with taxonomic β-diversity, suggesting that these treatments culled entire functions of species from the species pool. This idea is certainly consistent with our knowledge of the natural history of species that can persist with drought or species such as fish, and that certain types of species traits will be lost (Wellborn 1994). Likewise, we might expect only certain traits of species to be able to persist in low-productivity ponds, although, the effect on functional β-diversity was slightly less in low-productivity ponds.

It is currently not possible (or at least practical) to perform a true meta-analysis of other studies that have simultaneously examined taxonomic and functional β-diversity and its change through time and space. The reason is that there are almost as many metrics used to calculate both taxonomic and functional β-diversity as there are studies, and these different measures can give quite heterogeneous results (Tuomisto 2010a, 2010b; Anderson et al. 2011; Chiu et al. 2014; Cardoso et al. 2015). However, even a qualitative exploration of these can be illuminating to corroborate and refine the above synthesis.

Figure 7.4 presents a qualitative picture of the patterns observed from a literature search of studies that examined aspects of changes in taxonomic and functional β-diversity either through time or across experimental treatments. Although there are comparatively few such studies, some patterns emerge. First, many more studies show decreases rather than increases in either taxonomic or functional β-diversity through time (or across space using a space-for-time approach), though this could certainly be due to a bias in looking for homogenizing effects. Second, the majority of studies published showed either decreases in both taxonomic and functional β-diversity or no change in one and a decrease in the other. These results are consistent with those we presented above in the comparisons among freshwater ponds. Other analyses have suggested that patterns of declining functional β-diversity (i.e., increasing generalists and decreasing specialists) due to anthropogenic activities are quite general, which is consistent with these findings, but do not have the concomitant analysis of taxonomic β-diversity (Devictor et al. 2008, Ibarra and Martin 2015). Finally, it might not be surprising that we found no studies that showed negative correlations between taxonomic and functional β-diversity, because this is an unlikely outcome (Baiser and Lockwood 2011).

While it is clear that comparisons of different facets of biodiversity (e.g., taxonomic, functional, phylogenetic) across scales (α-, β-, and γ-diversity) can lend

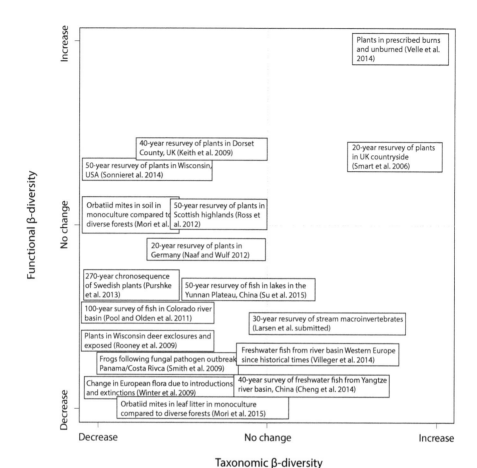

FIGURE 7.4. Qualitative review of studies that simultaneously compared changes in taxonomic and functional (or phylogenetic) β-diversity through time or in the face of experimental treatments intended to simulate an anthropogenic change. Studies are briefly described in each box by timespan or treatment, main taxa, and location. Locations of boxes are meant to capture qualitative changes in β-diversity for each facet, in lieu of a quantitative meta-analysis, which is impossible among these studies, which used different metrics to quantify β-diversity, thus rendering quantitative differences meaningless.

insight into the possible mechanisms underlying metacommunity assembly, we again note the problems of interpretation that can emerge when definitions of scales vary considerably among researchers (i.e., one person's α is another person's γ, and estimates of β-diversity by any measure are scale-dependent; see Chap. 10). As we discussed above, a promising avenue that might provide more direct comparability among studies is to explicitly examine how the different diversity facets change along a more continuous scaling relationship (e.g., the SAR, FAR, and PDAR; Morlon et al. 2011, Smith et al. 2013, Mazel et al. 2015).

7.3 TOWARD A TRAIT-BASED THEORY
OF METACOMMUNITY ASSEMBLY

As we described in Chapter 6, the recent years have seen an explosion of trait-based approaches to community ecology (McGill et al. 2006b, Cadotte et al. 2010, Webb et al. 2010, Weiher et al. 2011, Enquist et al. 2015). In fact, reading these articles might give the impression that the taxonomic-level perspective on biodiversity and community assembly is "old-school" and largely uninformative and that trait-based approaches are essential for moving forward. Clearly there is much value to the trait-based perspective; however, we disagree that approaches based purely on taxonomic, rather than functional, information are uninformative. Indeed, the mismatch between taxonomic and trait-based approaches can provide key insights into metacommunity assembly. In the extreme case, for example, NT is an important possible mechanism for metacommunity assembly, but it cannot be detected by a purely trait-based approach.

There are a number of approaches that identify how traits are related to patterns of species abundances and distributions. These approaches build on a long tradition of how traits vary with environmental conditions that go back hundreds of years and are evident in the writings of Darwin (1859) and Wallace (1876), their predecessors (e.g., Von Humboldt 1807) and followers (e.g., Schimper 1898), as well as in the works of early modern ecologists (Clements 1916, Gleason 1926, Elton 1927, Tansley 1935, Hutchinson 1959). More recent but still classic works on the expected correlations between species traits and environments include Grime's (1974, 1977) exposition of fundamental plant strategies across environmental gradients, Margalef's (1978) exploration of phytoplankton traits associated with environmental variation, and descriptions of the distributions of traits of stream organisms with respect to energy availability, temperature, and flow (Vannote and Sweeney 1980), among others.

Nevertheless, we are in a good position to state that a number of different types of functional traits of a large variety of organisms appear to predict their distributions and abundances across larger climatic regions, as well as among more local and regional environmental variation. Table 7.2 presents a few such examples. Often, such associations are taken as mechanistic evidence that metacommunity assembly is driven by environmental filtering in accord with an SS worldview and, as a corollary, as evidence against ecological drift and dispersal limitation (Westoby and Wright 2006, Weiher et al. 2011, Mouillot et al. 2013b, Enquist et al. 2015).

TABLE 7.2. Traits in Selected Groups of Organisms That Show Correlations
with Important Environmental Drivers of Their Ecosystems

Group	Select Traits	Environmental correlates	Select References
Terrestrial plants	Specific leaf area, wood density, root:shoot ratio, C:N:P content, leaf turnover, photosystem, seed size, seed dispersal	Precipitation, hydrology, soil nutrients, disturbance	Diaz et al. 1998, 2016; Reich et al. 2003; Westoby and Wright 2006; Lavorel et al. 2007
Phytoplankton	Cell volume, cell chlorophyll, mucilage, measured tolerances (nutrients, pH, temperature)	Nutrients, temperature, depth, light, herbivory	Reynolds 2002, Litchman and Klausmeier 2008, Edwards et al. 2013, Jamil et al. 2014
Mammals	Body size, life history	Latitude/temperature (lots of variation), seasonality, precipitation, island size, etc.	Badgley and Fox 2000, Blackburn and Hawkins 2004, Clauss et al. 2013, Tokolyi et al. 2014
Birds	Body size, life history	Latitude/temperature, seasonality, habitat type, resources, island size, etc.	Wiersma et al. 2007, Jetz et al. 2008, Olson et al. 2009
Stream fish	Body size, body shape, swimming speed, feeding niche	Hydrological and geomorphological features	Lamouroux et al. 2002, Ibanez et al. 2009, Mims and Olden 2012

There are two reasons, however, that we cannot simply use the fact that there are functional traits that co-vary with environmental conditions to conclude a preponderance of SS. First, simply quantifying the correlations between species traits and environmental variation is not always a straightforward task, and associating those traits with performance measures that actually influence a species' ability to persist in a given site even less so. Second, even if traits are strong predictors of the species distribution along some coarse-scale environmental gradients, the fact that there is a great deal of trait similarity (even nearly neutral redundancy) among species can still create considerable unpredictability at fine-scale resolutions, especially because there is rarely a one-to-one match between measured functional traits and a species' performance in any given habitat.

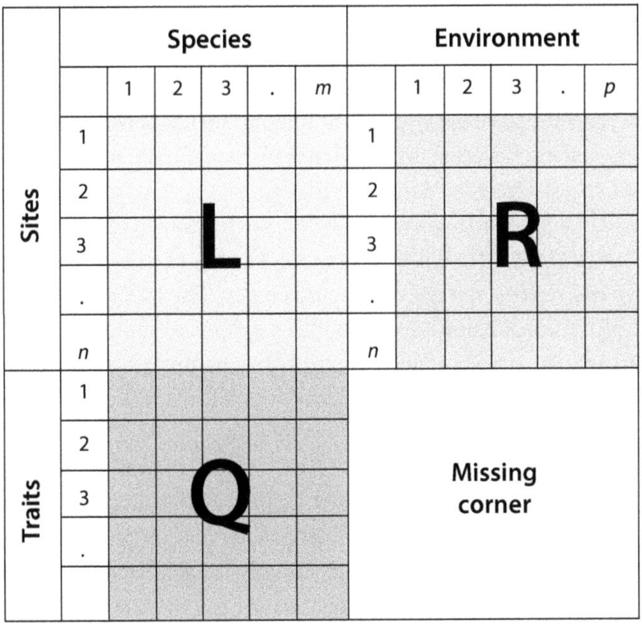

FIGURE 7.5. Diagram illustrating how the fourth-corner analysis works. In the upper left is the site × species matrix (L), the upper right is the site × environment matrix (R) and the lower left is the species × trait matrix (Q). The unknown matrix, which is the objective of the fourth-corner analysis, is the environment × traits matrix. Modified from Legendre et al. 1997.

Several statistical approaches have been devised to examine how species traits are correlated with environmental variation, many of which can be used to address slightly different questions (see, e.g., Kleyer et al. 2012). Here we primarily focus on the multivariate approaches that have been developed to examine species trait-by-environment relationships that are related to variation partitioning (described in Chap. 4 and discussed above) as presented in Figure 7.5. We define L as the classic species × site matrix (which is used to examine metacommunity structure as discussed in Chap. 4; see, e.g., Gotelli and McCabe 2002, Leibold and Mikkelson 2002, Presley et al. 2010); R is the environment × site matrix (used in the variation partitioning approach discussed in Chap. 4; see, e.g., Borcard et al. 1992, Peres-Neto et al. 2006); and Q is the species × trait matrix. Each of these matrices is readily collectable in natural metacommunities, but there is a missing matrix relating traits to the environment. Identifying what this unknown matrix looks like can provide key insights into a trait-based metacommunity framework.

Two of the most commonly used and complete approaches for exploring the structure of the trait-by-environment relationship are (1) the *RLQ analysis*, which is named after the three matrices used for the analysis and is a multivariate ordination analysis that scores elements of each matrix (species, sites, traits, environments) orthogonally (developed in Doledec et al. 1996); and (2) the *fourth-corner analysis*, so named because it is aimed at finding the relationship in the fourth matrix (environment × trait) when information from the other three matrices are known (developed in Legendre et al. 1997).

Both the RLQ and fourth-corner analyses use the same basic elements (the RLQ matrices) but provide different information; for this reason, many empirical examples have used both approaches simultaneously. The RLQ approach provides a global test of the associations and provides a graphical representation of the entire structure of the trait-by-environment relationship; however, it cannot tease apart which environmental factors influence which traits. The fourth-corner analysis allows one to test hypotheses about whether any given trait is significantly correlated (positively or negatively) to each environmental variable. Some limitations of this approach are that it cannot account for covariation among traits or environmental variables and that only the significance (or not) of the positive or negative associations between traits and environments can be calculated, not their strengths.

Unfortunately, despite the fact that there have been more than a hundred studies that have used the RLQ, fourth-corner, or combined approaches as of this writing (probably with many more on the way), it is not necessarily straightforward to review or synthesize these results. First, aside from a few notable exceptions with certain analyses (Brind'Amour et al. 2011, Ossola et al. 2015), the large majority of studies showed at least some significant relationships between environmental variation and species trait values. While we may often expect such significance, particularly when SS processes are important, there is also likely to be a strong publication (and analysis) bias here—one might not seek to analyze whether (and which) trait-by-environment relationships are present unless one suspects they exist.

Second, studies vary considerably in their choice of "functional" traits and "environmental" measurements, as well as the degree of environmental heterogeneity they encompass. Studies that include a large number of traits and environments that are known to strongly covary might inflate the perception of many strong trait-by-environment relationships. Alternatively, other studies may seem to show fewer trait-by-environment relationships but include a smaller degree of environmental heterogeneity or a larger number of traits that do not necessarily have a strong connection to that heterogeneity. Figure 7.6 shows a sample of the results from two studies that vary considerably in their choice of traits and environments. The study by Böhnke et al. (2012) examined the relationship between a number of traditionally measured wood traits and natural variation in environmental conditions in a subtropical Chinese forest (Fig. 7.6a). While they found a number of significant relationships, less than 30% of the comparisons were significant (note: many

a.

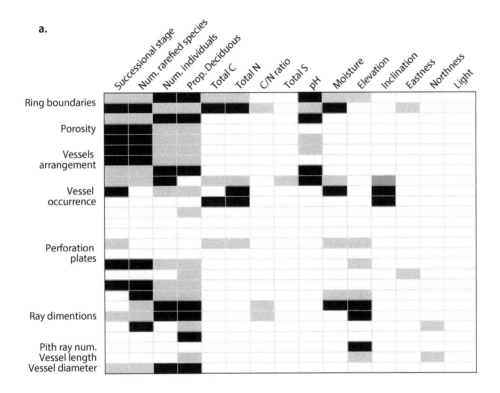

b.

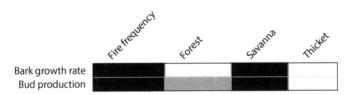

FIGURE 7.6. Two fourth-corner analyses. (a) An example from a Chinese subtropical forest examining the relationship between wood traits (rows) and environmental variables (columns). Black cells indicate a significant positive relationship between the trait and environmental factor, gray cells indicate a significant negative relationship, and white cells indicate no significant association. Modified from Böhnke et al. 2012. (b) An example from South African savannah where two traits related to bud types (rows) were measured as indicators of fire tolerance. Cell shading is as in (a). Modified from Charles-Dominique et al. 2015.

published studies have many fewer). The other study, by Charles-Dominique et al. (2015), examined how traits specifically related to fire tolerance (bud growth rate, protection) corresponded with fire frequency as a strong environmental gradient in a South African savannah, and the investigators not surprisingly found more than 60% of the trait-by-environment comparisons to be significant (Fig. 7.6b).

Recent developments have aimed at improved insights by combining the two approaches in order to facilitate global hypothesis tests (Dray et al. 2014) and developing a general linear method to allow predictions of the strength of the relationship between traits and the environment, rather than just whether or not they are "significant" (Brown et al. 2014). Broader use of these approaches will enhance our ability to achieve some synthesis but cannot overcome some of the limitations of comparability among studies—choice of traits, environmental variables, and magnitude. Not surprisingly, some of these are similar to the limitations that we described in Chapter 4 for the use of the variation partitioning approach at the taxonomic level.

Despite these limitations, it would be a shame if we could not take the multitude of studies that have evaluated trait-by-environment approaches and try to achieve some sort of generality or synthesis. To that end, we systematically reviewed the literature and created a database where we could explore some more general questions. We limited our quantitative analysis only to those studies that quantified trait-by-environment relationships using the fourth-corner method, which despite its limitations (described above) provides a quantitative test of hypotheses regarding the significant (or non-significant) relationship between individual traits and environmental features. We did not consider studies that used only the RLQ approach, which despite some advantages is less straightforward for summarizing the results in a quantitative comparison.

We searched the literature following citation threads of the most important fourth-corner methods papers (Legendre et al. 1997, Dray and Legendre 2008), ultimately finding 51 published papers containing 60 fourth-corner analyses (several studies contained more than one analysis). A full meta-analytic review is beyond the scope of our intent (and the analytical tools for such are not readily available); however, a few interesting patterns emerge.

We found that the number of significant trait-by-environment associations increased with the number of possible combinations in the analyses (Fig. 7.7a), suggesting a sort of "sampling effect." However, the proportion of the significant relationships to total possible relationships decreased as the total possible relationships increased (Fig. 7.7b). This observation suggests that the studies that included more environmental variables or traits may have been "fishing expeditions" and thus less likely to yield significant associations than tests that measured traits and environments that were more directed.

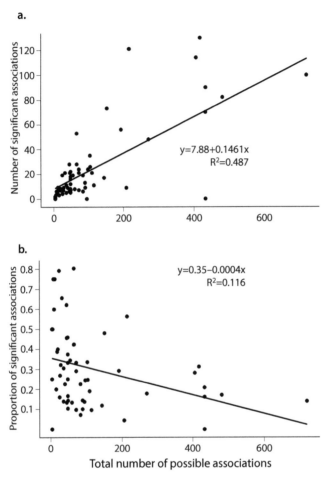

FIGURE 7.7. (a) Curve showing a significant positive association between the numbers of possible trait-by-environment associations in fourth-corner analyses and the total numbers of significant relationships. Similar results (not shown) were found comparing the numbers of traits or environments separately. (b) Curve showing how the proportion of significant associations declined with increasing numbers of possible trait-by-environment associations.

However, even amid the sorts of "sampling effects" from Figure 7.7, we found some interesting patterns. In Table 7.3, we show differences that were relatively large (more than a 10% difference in significant correlations among qualitatively distinct categories). We did not attempt to perform any significance analyses on these results.

First, we found that studies on producers (i.e., plants and algae) tended to show higher ratios of significant trait-by-environment correlations than those on animals,

TABLE 7.3. Cases in Which There Were >10% Differences in the Number
of Significant Fourth-Corner Trait-by-Environment
Relationships between Qualitative Comparisons

Comparison	Groups	Average ratio (+/–SD)
Taxonomic category	Animal (N = 32)	0.22 (0.16)
	vs.	
	producers (N = 28)	0.39 (0.22)
Ecosystem type	Terrestrial (N = 45)	0.21 (0.21)
	vs.	
	aquatic (N = 15)	0.33 (0.20)
Natural vs. disturbed	Natural (N = 33)	0.26 (0.21)
	vs.	
	disturbed (N = 27)	0.34 (0.20)
Size of environmental gradient	Small (N = 10)	0.16 (0.15)
	vs.	
	medium (N = 36)	0.26 (0.17)
	vs.	
	large (N = 14)	0.49 (0.20)

despite the fact that there were no systematic differences in the numbers of traits or environmental types measured. One possible reason for this finding might be that because producers are sessile, plant distributions and abundances are more directly influenced by environmental influences than those of animals, which are typically more mobile.

Second, we found that studies done in aquatic ecosystems (mostly freshwater) tended to show higher ratios of significant trait-by-environment correlations than studies done in terrestrial systems. This pattern is less easy to interpret but could have resulted because organisms in many types of aquatic systems (e.g., streams, lakes) might be more limited in their ability to disperse among localities that are typically interspersed within a harsher (terrestrial) matrix, whereas organisms in terrestrial metacommunities can more freely transit among localities.

Third, we found that studies where some sort of "disturbance" (e.g., fire, flooding, anthropogenic effects) was included among the environmental conditions tended to have higher ratios of significant trait-by-environment correlations than studies that did not include such environmental conditions. This finding is relatively straightforward to explain by the well-known relationship between many types of disturbances and the traits of species that are favored or disfavored by those disturbances (Grime 1974, 1977; Wellborn et al. 1996; Poff 1997; Clavel et al. 2011).

Finally, we found that studies that we categorized as encompassing "large" environmental gradients tended to have higher ratios of significant trait-by-environment correlations than studies that encompassed smaller gradients. This pattern is also relatively straightforward to understand since larger environmental gradients are more likely to include significant amounts of habitat heterogeneity on which species sort than smaller gradients. This finding mirrors the general pattern we showed in Chapter 4 in which studies across larger gradients tended to show a stronger degree of species compositional variation correlated with environmental variation.

We expect that future studies can provide even deeper insights into some of the variation structuring natural communities by (1) taking advantage of new statistical tools that will allow a more explicit quantification of the strengths of effects and account for multiple comparisons (Brown et al. 2014, Dray et al. 2014); and (2) recognizing the key role of the choice of the magnitude of environmental heterogeneity incorporated into the study design, the types of traits measured, and the numbers of species included in the analysis—all of which play a critical role in the observed trait-by-environment correlations and conclusions reached.

From this overview of the recent explosion of studies in this area, we can conclude that species traits often play an important role in driving how species distribute themselves along a metacommunity landscape. Of course, this is unsurprising, because even the staunchest advocates of NT do not really believe that traits have no influence on species distributions, particularly along broad environmental gradients. However, we can still learn a lot from deeper explorations of trait-by-environment relationships. When they are strong, trait-by-environment relationships provide deeper insights into the mechanisms by which SS operates, including the traits that are important for the sorting process along a given environmental gradient. When they are weak, they tell us that in the context of the metacommunity as defined, and of traits as measured, processes other than species sorting according to functional traits, such as dispersal and ecological drift, may be playing a stronger role.

7.4 CLOSING THE LOOP: PREDICTING SPECIES ABUNDANCE AND DISTRIBUTION FROM TRAITS

The use of species traits, rather than taxonomic names, has been suggested to be the holy grail not only for providing a mechanistic link between evolved species traits and the abundances and distributions of species, but also for understanding the functioning of ecosystems and responses to global change (Lavorel and Garnier 2002, Lavorel et al. 2007). In Section 7.3, we found that we often do pretty well to understand how species traits vary along important environmental gradients.

But traits are not species, and many of the types of traits measured and used in trait-by-environment relationships are often quite similar in a number of species in a metacommunity. And so trait-by-environment relationships, by themselves, cannot necessarily tell us whether species traits can actually predict the abundances and distributions of species. To address this, we need to go a step or two further.

To deal with the question of environmental distributions of species, Jamil et al. (2014) expanded on the fourth-corner approach using Bayesian methods and generalized linear mixed models to derive unimodal response curves of species to environmental tolerances. They calculated a number of functional traits from 60 species of phytoplankton distributed along very broad environmental gradients in a set of 203 lakes (ranging from tropical to arctic) and used them to predict the optima of species environmental distributions (as well as the range). As with most fourth-corner studies (discussed above), they found that some traits were much more predictive along environmental gradients (volume and the presence of flagella, mucilage, and siliceous exoskeletons) whereas others were not (surface area and maximum length). A particularly intriguing aspect of their work is that they were able to predict the response curves for each of these species along gradients in temperature, chlorophyll concentration, and a composite latent variable (akin to a principle axis of variation) that sorts quite nicely into two broad groupings of species that respond to each environments, at least at these large scales (Fig. 7.8). Using these trait optima, we can see strong clumping in the distribution of species along environmental gradients as in our discussion of the EMS method in Chapter 4. However, the patterns are fairly dramatic and we can only wonder how Gleason and Clements might have reacted to these results!

Edwards et al. (2013) used a related approach, also with phytoplankton, but rather than using only morphological traits, they gleaned performance traits for 25 species of freshwater phytoplankton that focuses on maximum growth, competitive ability for phosphorous (roughly equivalent to low R* for phosphorous), and competitive ability for light (low R* for light). Edwards et al. found a two-tiered result that at least superficially supports our hypothesis that trait-sorting associated with SS and ecological drift/dispersal limitation can occur simultaneously. Specifically, by comparing their trait-based results with a survey of phytoplankton abundances and distributions from more than 500 lakes in the United States (Taylor et al. 1979), they found that two important traits (maximum growth and competitive ability for light) were quite important in predicting the relative abundance of these species (when they were present) under varying light conditions. However, at the same time, they found that these traits were not very predictive in determining whether the species was present or absent among lakes varying in

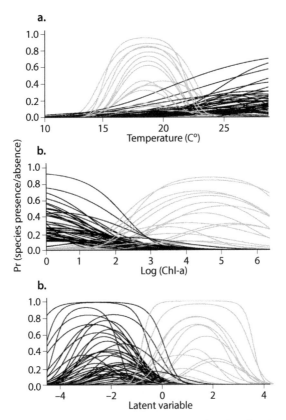

FIGURE. 7.8. Individual species response curves based on trait values for phytoplankton along (a) temperature and (b) chlorophyll (as a proxy for productivity) gradients, as well as a latent variable gradient (c) that explains much of the residual variation not accounted for by either temperature or chlorophyll. Modified from Jamil et al. 2014.

environmental conditions, which they argued may have resulted because of dispersal limitation.

Species traits are not always strongly associated with environmental filters, however, as we discussed above. In such cases, it is rather expected that species traits would be less predictive of species distributions and abundances than the cases described above. For example, in a survey of species traits, environmental conditions, and species distributions of butterflies near Prague (Czech Republic), Robinson et al. (2014) found only weak relationships between environmental variation and species trait variation, as well as in the distribution and diversity of species. In this case, the authors suggested that this lack of relationship may have resulted because of the relatively small range of environments sampled, but it could also emerge if species traits are decoupled from environmental filtering as a result of ecological drift or dispersal limitation.

7.5 CONCLUSIONS

Neither taxonomy-only (Chaps. 4 and 5) nor function/phylogeny-only (Chap. 6) pattern analyses can provide a full picture of the metacommunity assembly processes (e.g., environmental filtering, dispersal limitation, drift, etc.). Although we will probably never be able to fully discern metacommunity assembly processes solely from pattern analyses, our overview in this chapter shows that by combining taxonomic and functional approaches, we can get one step closer. Doing this should allow us to gain a more synthetic view of the metacommunity archetypes and of their possible action at different levels of organization.

Importantly, if functional and taxonomic patterns respond to environmental variation or some perturbation in the same direction, this is interesting and important information (i.e., such responses imply that SS processes are acting on both levels). If, however, taxonomic patterns respond differently from functional patterns, we gain different insights. For example, most communities are characterized by at least some, if not much, functional redundancy among species within certain groups (often referred to as functional groups, guilds, etc.), while at the same time there are clear functional differences among these groups of species. As a result, we might expect processes more closely aligned with NT (e.g., drift, dispersal limitation) to play an important role within the groupings where species traits are more similar, whereas we expect processes more closely aligned with SS to play a more important role among the groupings where species traits are more differentiated. Here, then, we would expect qualitatively different patterns to emerge when comparing taxonomy- to function-based analyses.

This contrast also reiterates the importance of recognizing the role of taxonomic scale when making inference about metacommunity assembly patterns (as discussed in Chap. 4). When analyses are performed within a phylogenetically constrained group of species (e.g., only within a particular family, such as hummingbirds [Trochilidae] or the Melastomataceae), or within a functionally constrained group of species (e.g., benthic grazers or woody species), we might expect processes associated with NT to play a larger role than when these same analyses are performed on larger groupings of species that necessarily will differentiate more, such that processes associated with SS appear more important.

To sum up, our evaluation of phylogeny- and function-based approaches in Chapter 6 errs a bit more on the side of caution in its interpretation of a field that seems to have skyrocketed in the past decade or so. However, in this chapter, we have emphasized that by combining and comparing taxonomic and functional information at the same time, we can find new ways to conceptualize and analyze metacommunities that are probably characterized by a combination of processes. For example, by considering both taxonomic and trait variation simultaneously, we suggest that processes that characterize NT and related metacommunity archetypes (PD) probably play a stronger role among species that are more functionally similar (e.g., within functional or phylogenetic groupings), whereas processes that characterize SS and related archetypes (ME) probably play a stronger role among species that are more functionally distinct (e.g., among functional or phylogenetic groupings).

We also think that much more thought needs to be given to which traits we consider in our studies. "Functional traits" come in a huge variety of flavors, ranging from those that are simply easy to measure and have very general effects on responses and effects of species in their environment, to those that are tightly focused on very specific features or those that are likely to be very context dependent. One profitable possibility would be to more carefully target traits that have strong connections to niche relations and are thus more meaningful from a metacommunity perspective (Letten et al. 2017).

CHAPTER EIGHT

Eco-evolutionary Dynamics
in Metacommunities

Prospectus

1. Recent findings show that adaptive evolution within populations can be rapid enough to interact with ecological processes. Furthermore, populations of a species can vary substantially in ecologically important trait distributions, even across surprisingly small distances, leading to important eco-evolutionary feedbacks in metacommunities.

2. These feedbacks may be particularly important where patch turnover is high or where there has been substantial local environmental change. Subsequent responses may be dominated either by the evolution of resident or early colonist species or by changes in community composition.

3. The *community monopolization hypothesis* predicts that local adaptation will prevail when dispersal is low and disturbances rare, whereas community change will prevail when dispersal is high and disturbances are common.

4. Local adaptive evolution can also in theory lead to local coexistence of species under some circumstances by favoring ecological-trait convergence akin to NT.

5. Evidence for both community monopolization and the evolution of neutrality comes from systems with plankton and other microorganisms, but there is also some evidence to support these ideas in other systems.

6. Destabilizing frequency-dependent selection can also create evolutionary dynamics that alter the diversity of organisms. Under high local background extinction, such frequency-dependent selection has little effect on landscape diversity. However, this sort of frequency-dependent selection can greatly enhance landscape diversity (even in the absence of environmental heterogeneity) when extinction rates are low and can lead to taxon cycles when extinction rates are intermediate.

7. A general model that examines interactions between local adaptation and disturbance in metacommunities reveals that this interaction can greatly enhance spatial patterning and reduce compositional correlations with the environment compared with models that ignore eco-evolutionary feedbacks. The resulting patterns resemble those predicted by NT but differ because they show little or no ecological equivalence at local scales.

Hutchinson (1965) entitled one of his more endearing books of essays *The Eco-logical Theatre and the Evolutionary Play*. However, the inverse phrase, "the ecological play in the evolutionary theatre," could make just as much sense, and authors often unintentionally misquote Hutchinson with this alternative perspective. The original title implies that ecological principles are fixed entities that serve as the structure within which the dynamics of evolutionary change plays out. There is certainly much to be said for such a view since ecological processes follow principles of thermodynamics, stoichiometry, and demography, and these have tremendous potential to influence fitness, as well as gene flow and genetic drift (i.e., the ingredients for evolutionary change). Alternatively, evolutionary change modifies how organisms interact with their environment and thus modifies numerous and sometimes dramatic aspects of ecological change. If both views are valid, then ecology (processes that affect population dynamics and distributions) and evolution (processes that influence genetic features of populations and their consequent traits) can affect each other reciprocally, a phenomenon called *eco-evolutionary feedback* (see Hendry 2016 for a broad overview of this rapidly emerging field).

The eco-evolutionary concept dates much farther back, including to Darwin's (1859) concluding chapter, which used the "entangled bank" metaphor to describe how intricate ecological interactions create selective forces on organisms. Nevertheless, a common recurrent theme is that evolutionary change is slow and gradual and thus of secondary importance in influencing rapid ecological processes, whereas longer-term evolutionary processes can assume ecological dynamics are at a "quasi-equilibrium" and effectively ignore those dynamics.

Although the separation of timescales (fast ecology, slow evolution) is convenient, several features of eco-evolutionary systems make it likely that ecology and evolution can often interplay in ways that dramatically (and even qualitatively) change expectations. For example, an increasing number of studies have shown that local adaptive evolution can be very fast, with substantial change of ecological-trait means occurring in one or two generations, and often comparable to demographic changes in affecting rates of population change (Tessier et al. 1992, Thompson 1998, Hendry and Kinnison 1999, Reznick and Ghalambor 2001, Yoshida et al. 2003, Hairston et al. 2005, Pelletier et al. 2009, Ellner et al. 2011, Schoener 2011). These eco-evolutionary feedbacks will depend on the genetic architecture of species, the nature of environmental change, and dynamic population feedback within populations (Ellner 2013), as well as on spatial processes that alter gene flow and dispersal (Urban et al. 2008, Vellend 2010, Malcom 2011, Richardson et al. 2014).

In this chapter, we examine some ways that eco-evolutionary feedbacks may affect the structure and dynamics of metacommunities. We focus especially on trait evolution under simple models of genetic evolution, recognizing that there are numerous complications that could arise that could depend strongly on how trait evolution relates to genetic evolution.

8.1 BUILDING AN EVOLUTIONARY ECOLOGY OF METACOMMUNITIES

To date, most of metacommunity ecology, and that which we have focused on in this book, has implicitly assumed that species traits are fixed in space and time. Clearly this is not true, and there is a lot of intraspecific variation both within and among populations (Bolnick et al. 2003, 2011). How does such variation alter the ways by which we understand how interactions among species shape metacommunity processes and resultant patterns?

Often, local populations diverge as a result of simple adaptation to local conditions—both abiotic and biotic. This feedback can alter patterns of distribution and abundance by altering interactions with other species. Perhaps the oldest idea in this vein is that of character displacement: pairs of species might show ecological divergence in their traits in parts of their distribution where they coexist (Brown and Wilson 1956, recently reviewed by Dayan and Simberloff 2005 and Pfennig and Pfennig 2009). Although early studies of character displacement were criticized for not taking into account alternate and simpler explanations (Grant 1972, Schluter and McPhail 1992), more recent work has confirmed that character displacement can occur, maybe quite commonly (Dayan and Simberloff 2005, Pfennig and Pfennig 2009, but see Stuart and Losos 2013). However, it is also possible that species might converge in areas of coexistence (Abrams 1987) and that evolutionary effects might be less directed by pairwise interactions and more responsive to diffuse competition with other species (Hubbell 2006).

More generally, Thompson's (2005) *geographical mosaic theory of coevolution* shows how the co-evolutionary give-and-take between species can vary spatially as a result of heterogeneity in both abiotic (environmental conditions) and biotic (interactions with other species) conditions. Individuals from populations that undergo varying strengths and directions of selection and drift in these different locations then migrate among habitats, creating variation in genetic structure among species and sites that are not that different from the patterns of species compositional differences we described in Chapter 4. Thompson argues that such context dependence in ecology can be a major feature of the evolution of population differentiation and codifferentiation (via coevolution) of species.

A related approach is embedded within the concept of *community genetics* (Whitham et al. 2006, Hughes et al. 2008, Bailey et al. 2009). This focus examines how evolutionary processes that have created genetic variation within a species might influence the distribution and abundance of species that interact with that species. For example, variation in the genetic diversity of foundational species, such as host plants for insects, can have a dramatic influence on the number and types of insects that live on those hosts (Whitham et al. 1999, Crutsinger et al. 2006, Barbour et al. 2016). This can feed back to the entire ecosystem; for example, when different microbial communities grow beneath different plant genotypes, altering the decomposition of organic matter and the cycling of nutrients (Schweitzer et al. 2004, Fitzpatrick et al. 2015, Jackrel et al. 2016). Related to this idea, *community selection* can occur when feedbacks occur between plants and their microbiota such that selection on plants influences the community structure of the microbes, which in turn feeds back to influence plant performance (Wilson 1992; Goodnight 2000, 2011; Whitham et al. 2006). While interesting, it is nevertheless likely that the conditions necessary for this sort of dynamic might be rare and likely that they are confined to situations in which the "partners" are highly specialized.

A third conceptual thread integrating aspects of spatial metacommunity ecology and eco-evolutionary processes is the *niche construction* paradigm (Odling-Smee et al. 2003, 2013). Here, via their impacts on the abiotic and biotic environment, species alter their own selective fate, which can create spatiotemporal legacies and feedbacks between species traits, species abundances, and ecosystem structure.

Each of these ideas has moved the fields of evolutionary biology and spatial (metacommunity) ecology closer together. However, they do not fully consider how these genetic differences arise and are maintained or what determines their spatial distribution. A more comprehensive evaluation about the role of evolution in metacommunities consists of looking at the reciprocal effects of evolutionary change on communities (akin to community genetics) and of community effects on evolutionary change (akin to mosaic coevolution and niche construction), combined within a spatial framework (Haloin and Strauss 2008, Urban et al. 2008). This approach has been called the *evolving metacommunity* approach (Urban and Skelly 2006, Urban et al. 2008), and although promising, it is complicated by the necessity for simultaneous attention to spatial processes that affect community structure (dispersal) and adaptive evolution (gene flow). In the following sections, we give some indication of how this can be achieved, keeping an eye on eco-evolutionary feedbacks that explicitly influence the interplay between environmental and spatial structuring of metacommunities.

8.2 ADAPTIVE EVOLUTION IN METAPOPULATIONS AND METACOMMUNITIES

Environmental change and environmental gradients are just as important for maintaining genetic variation and evolutionary potential in individual species as they are for the metacommunities in which they exist, and many of the processes and resulting patterns are analogous (Vellend 2010, 2016). Consider, for example, what might happen if an environmental change has negative impacts on the population growth rate of a species in a local patch. Assuming that the patch is sufficiently isolated, such change could drive a species toward extinction in that patch unless local adaptation to this new environment can act quickly enough to "rescue" this species before its extinction (Gomulkiewicz and Holt 1995; Fig. 8.1), a process we call *evolutionary rescue*. Such adaptive evolution will depend on several factors, including the influence of the environmental change on the demography of the population (the rate of decline) and several features of the species' ability to respond to selection (e.g., standing genetic variation, initial population size; Gomulkiewicz and Holt 1995).

The process of adaptive evolution in the face of environmental change becomes more complicated when populations are not isolated (i.e., they are in a metapopulation). This is because dispersal influences gene flow as well as the demography of the species and its genetic structure (Holt et al. 2004, Bell and Gonzalez 2011, Hanski 2012). Gene flow can either enhance local adaptation by increasing genetic variation and the adaptive capacity of the local population, or it can restrict local adaptation by homogenizing gene frequencies away from local optima even if there is strong local selection. The outcome depends substantially on the genetic architecture of the species involved, their mating systems, and the genetic and demographic structure of the metapopulation, but these details are beyond the scope of this book.

For our purposes, we primarily focus on the more general effect of gene flow, restricting local adaptation when it is too high and possibly enhancing adaptation at low gene flow (Urban et al. 2008). Likewise, dispersal can alter the demography of local populations experiencing environmental change. Through immigration, it can prevent the extinction of local populations via *rescue effects* (these are distinct from evolutionary rescue, described above; Brown and Kodric-Brown 1977), or through emigration, it can reduce local population size and enhance the probability of extinction. Thus, dispersal is likely to have a generally unimodal relationship, first enhancing and then reducing a population's ability to persist in the face of environmental change.

While useful, these simple ideas ignore how local adaptation affects, and is affected by, the interactions with other species in a spatially explicit context. Such feedbacks have the potential to substantially alter how we interpret pattern and process in metacommunities, as well as their outcomes, such as how biodiversity might respond to climate change (Urban et al. 2012, Norberg et al. 2012). For example, in a metapopulation, dispersal can influence the response of a species to environmental change through genetic and demographic effects. However, in a metacommunity with spatiotemporal heterogeneity and dispersal, environmental

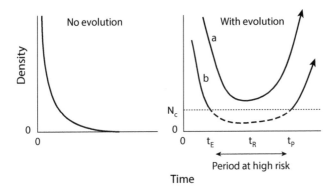

FIGURE 8.1. In the absence of evolution (left panel), a population that experiences environmental change leading to negative fitness shows an exponential decrease in population size leading to extinction. Adaptive evolutionary response to this change decreases the rate of decrease with time (as the population adapts) and, assuming it continues, would eventually allow the species to recover (right panel). If evolution is too slow (line b), it may not be sufficient to prevent the population from decreasing below a critical population size (N_c), at which there is a high likelihood of going extinct. However, sufficiently rapid adaptation (line a) can rescue the population from extinction. Figure modified from Gomulkiewicz and Holt 1995.

change can favor species that were previously excluded from the metacommunity and disfavor resident species (as an analog to SS in time). In an evolving metacommunity, such environmental change will also interact with the ability of species to respond evolutionarily to the environmental change and with each other. In the following sections, we will review three ways by which local evolutionary change can alter metacommunity processes and patterns—the community monopolization hypothesis, the evolution of neutrality, and frequency-dependent evolution—as well as interactions among these ways. Nevertheless, we have to admit that even if these ideas are intriguing and have some theoretical and empirical evidence, each of these foci are still relatively underexplored and the study of evolving metacommunities is far from providing a full synthesis.

8.3 THE COMMUNITY MONOPOLIZATION HYPOTHESIS

The *community monopolization hypothesis* has its roots in the *niche monopolization hypothesis* and includes many of the same underlying principles (De Meester et al. 2016). The niche monopolization hypothesis was initially developed to explain why local populations could be highly differentiated in both ecological traits and neutral genetic variation even among closely adjacent populations that were not likely to be dispersal limited (De Meester et al. 2002). Specifically, the hypothesis posits that as novel habitats are created or disturbances drive populations extinct, the initial colonizing genotypes are determined in large part by chance from a regional pool and that subsequent rapid local adaptation by these initial colonists acts to strongly prevent secondary colonizing genotypes.

The community monopolization hypothesis (Urban et al. 2008) links species interactions with local adaptation to changing environments in a community context. For example, if local adaptation is fast relative to the arrival of new species, the adaptation of extant populations to the changed environment may make invasions by previously absent species difficult, even if the new environmental conditions are now favorable to that species, thus impeding the process of SS. Alternatively, if local adaptation is slow relative to the arrival of new species, invasions could make it more likely that previous residents will go extinct rather than adapt. The degree to which one process or the other wins depends on several factors, but the degree of isolation among habitats is one of the most obvious (Urban et al. 2008; Fig. 8.2).

8.3.1 Community Monopolization Theory

To better understand the process of community monopolization, Urban and De Meester (2009) examined an idealized three-patch model. They started with two species that each exist in distinct but environmentally identical patches to which they are optimally adapted (these species are, at least initially, allopatric equivalent species). They then asked what happens if there is a new patch in the landscape with environmental conditions that differs from the initial two patches. If there is no evolution, the two species can co-occur in the new patch via neutral coexistence (their simulations show there is nevertheless a bias related to which species initially colonizes; Fig. 8.3a). If one of the species can adapt but the other cannot (or more generally, if one can adapt substantially faster than the other), the adaptive species will outcompete its competitor regardless of which species arrives first. However, if both species are equally capable of adapting, there is a strong priority effect by which the first species can reach the novel optimal trait value first and then outcompete the second arrival, leading to the community monopolization effect. Interestingly, if the two species arrive at roughly the same time, they may still co-occur via neutral coexistence, but there is nevertheless a bias toward the first arriving species (similar to what happens in the absence of evolution). The difference here is that without evolution, the two species co-occur with the same trait values they started with, whereas in the presence of evolution, the two species co-occur with trait values that are close to the optimal values in the new environment.

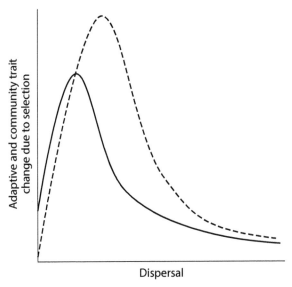

FIGURE 8.2. The relative rates of adaptive trait change (change in mean fitness due to trait evolution) as a function of dispersal by evolutionary change within a resident species (solid line) or by species replacements ("species tracking"; dashed line). In an isolated population (no dispersal), there is little or no species tracking (only extinctions of resident species can affect community trait change), but resident populations can show adaptive evolution depending on standing genetic variation and the rate of beneficial mutations; at very low dispersal, the solid line thus lies above the dashed line. At high dispersal rates, local trait change is swamped by dispersal (either gene flow within a species, or mass effects involving multiple species in the metacommunity), and homogenization increasingly constrains local trait differentiation from the metacommunity; and so, the dashed line lies above the solid line. There is a peak at intermediate levels of dispersal if trait variation in the metacommunity is unbiased with respect to local trait values because the rate of local adaptive trait change is enhanced by the resulting amount of trait variation. Trait variation is assumed to be higher among multiple species than within single species so the peak height is higher for species tracking than for local adaptation. In this scenario, adaptation by resident species is more efficient to the left of the intersections of the two lines and should favor evolutionary priority effects whereas the reverse is true to the right of the intersection. Modified from Urban et al. 2008.

Furthermore, when one species takes over the novel patch via the community monopolization effect, it can subsequently alter the distribution of species in their original patches (Fig. 8.4). This effect occurs because there is now maladaptive gene flow from the species that dominates in the new patch that alters the phenotypic distribution in its original patch and reduces its ability to compete with the other species (which has no such maladaptive gene flow).

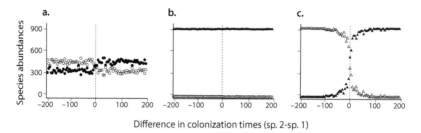

FIGURE 8.3. Abundances of two species (Species 2 shown with open symbols, Species 1 with filled symbols) in a novel patch as a function of the differences in colonization time from each of two source patches occupied by ecologically identical but allopatric species. Both species have identical initial trait values in their original patches that differ from the optimal trait value of the new patch. (a) If there is no trait evolution, both species co-occur in the new patch with somewhat of a bias toward the species that colonized earliest. (b) If evolution only occurs in Species 1, it evolves to conditions in the new patches whereas Species 2 cannot, and Species 1 consequently excludes Species 2 regardless of the differences in colonization time. (c) If both species can evolve (here with similar initial genetic variation and mutation rates), the species that arrives first has an evolution-mediated priority effect that allows it to exclude the competitor unless their arrival times are almost identical, in which case the earlier-arriving species still shows a bias in abundance. Modified from Urban and De Meester 2009.

Although this model is rather contrived, two interrelated points relevant to our interpretation of metacommunity structure can be gleaned from Urban and De Meester's (2009) analysis. First, even if species in a metacommunity were to start out "neutral" with respect to their traits, eco-evolutionary processes can eventually cause them to segregate by habitat type through habitat specialization. Second, adaptive evolution in this case enhances the degree to which species match their favored environments (i.e., SS), enhancing correspondence between composition and environment and decreasing spatial and residual variation that would otherwise have been predicted in the absence of evolution (at least starting from the initially neutral conditions modeled by Urban and De Meester). If this were to occur more broadly in a larger landscape of patches, the sorts of variation partitioning we discussed in Chapters 4 and 5 would more likely yield stronger environmental effects and weaker spatial and residual effects. Importantly, however, this is not the only outcome we might expect from community monopolization processes.

Rather than focusing on what happens when new patches are created, Loeuille and Leibold (2008) asked what happens when the environments in individual patches change through time. They started with a SS situation in which each of three species was specialized to the distinct environments of each of three patches. They then imposed environmental change to one of the patches. Like Urban and

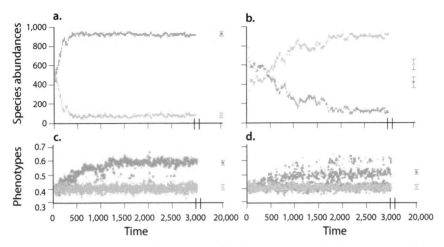

FIGURE 8.4. Dynamics through time of competition between two species in the presence of the community monopolization effect and substantial gene flow among patches. The top panels show the abundances of the two species in the novel patch (a) and in the ancestral patch (b). The lower panels show the phenotype distributions of the two species in the novel patch (c), where the optimal value is 0.6, and the ancestral patch (d), where the optimal value is 0.4. In the novel patch (a), Species 1 rapidly excludes Species 2 and more gradually evolves to have phenotype values close to the optimal value of 0.6. In its original patch (b), however, gene flow from the novel patch pushes mean phenotypic values away from the local optimum of 0.4 and allows Species 2 to invade and outcompete Species 1 from its original patch even though it was initially optimally adapted to that patch. Modified from Urban and De Meester 2009.

De Meester (2009), they found that the adaptive capacity of the resident species can prevent the other species from invading even when the conditions of the local environment are suitable for other species (Fig. 8.5b). This community monopolization effect occurs when dispersal is slow and adaptive capacity is high. Here, using a variation partitioning approach would indicate that the proportion of variation explained by environment is initially strong but weakens as the local environment changes and the species adapts. Alternatively, with high dispersal and low adaptive capacity, the resident species in the patch with the changing environment is driven extinct as the environment converges on one of the other patches due to colonization and competition by the residents of this other patch and subsequent changes involve community composition that tracks environment change (Fig. 8.5a). Finally, at high dispersal and high adaptive capacity, the evolving species in the changing environment can overwhelm the other two species and drive one or both extinct from their respective patches even though these other species are perfectly adapted to their local conditions (Fig. 8.5c). This happens because the adapting species can become equally adapted to the other environment and benefits from immigration

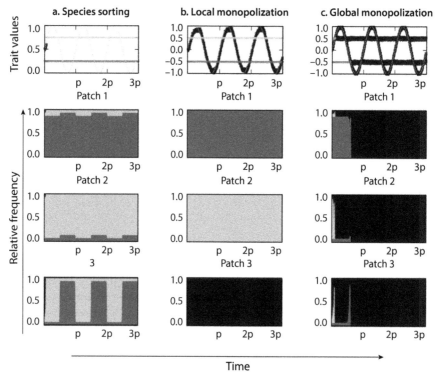

FIGURE 8.5. Changes in trait values (top row of panels) and community composition in Patch 1 (second row) through time with fixed environment at optimum of −0.5, Patch 2 (third row) with fixed environmental at optimum of 0.5, and Patch 3 (bottom row), which cycles between optimal values of −1.0 and 1.0 as shown in the top row of panels. Species 1 and 2 do not evolve in trait value, but Species 3 can evolve in response to local selection. Species 3 has trait values (mean and variance in the topmost row) shown in black. The other nine panels show the relative frequency of each species in each patch with Species 1 shown in light gray, Species 2 in dark gray, and Species 3 shown in black. Depending on parameters the outcomes can often be classified among three types. (a) *Species tracking* (left set of columns) occurs when Species 3 goes extinct because it is outcompeted by one of the other species when the optimal trait value in Patch 3 approaches that of one of the other patches; because this outcome happens at relatively high dispersal rates, there are also observable source-sink relations between patches. (b) *Local monopolization* (central panels) occurs when Species 3 can successfully prevent either of the other two species from successfully colonizing Patch 3 even when the optimal trait values in Patch 3 are equivalent to their own trait values; this happens when dispersal is low, so there are no observable source-sink relations in this scenario. (c) *Global monopolization* occurs when Species 3 dominates in Patch 3 (as it did in the community monopolization scenario) but also successfully invades and outcompetes Species 1 and 2 when conditions in Patch 3 resemble those in either of the fixed patches. Modified from Loeuille and Leibold 2008.

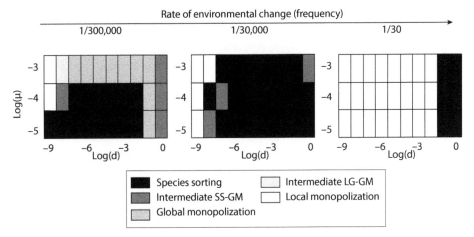

FIGURE 8.6. Distribution of outcomes as a function of three rate parameters used in the model. High dispersal (d) generally enhances species sorting, low dispersal enhances local community monopolization, and high mutation rate (μ) enhances evolutionary potential and in turn enhances global monopolization. This graph also includes outcomes that are intermediate between the three main types shown in Fig. 8.5. Modified from Loeuille and Leibold 2008.

from the other patches via ME. Loeuille and Leibold (2008) called this the *global monopolization hypothesis*. In this case, the strength of environmental filtering is weakened with environmental change. Which of these scenarios occurs will depend on three parameters in the model in a somewhat complicated way (Fig. 8.6).

The results from these two simple but distinct scenarios show that monopolization can emerge either from an initially neutral situation (as in Urban and De Meester 2009) or an initially SS situation (as in Loeuille and Leibold 2008), and that local adaptation can strongly influence the nature by which communities are environmentally or spatially structured depending on the amount of dispersal and environmental heterogeneity. However, these two studies just scratch the surface of what is likely to be a complex and dynamic process. Because the processes of metacommunity ecology and adaptive evolutionary change are largely structured by similar processes (drift, selection, dispersal; Vellend 2010, 2016), we might expect many such interactions between the processes that alter the way communities are environmentally or spatially structured.

8.3.2 Community Monopolization: Empirical Evidence

The above theories suggest that local adaptation in populations of interacting species can change interactions enough to alter the process of metacommunity assembly. Of course, such evolutionary effects depend on just how much scope there is for local differentiation to occur. In the models described above, this scope is determined by the parameters of the model in highly simplified ways (often in asexual organisms that evolve a single continuous trait). In nature this may not hold, either because the selection scenarios are more complex or because there are genetic and other constraints on the interaction of metacommunity assembly and evolution (Malcom 2011). One argument that could be leveled is that, in practice, niche differences among species are large compared with the entire possible scope for niche differentiation within species because of basic "architectural" or genetic constraints (i.e., the size of the "hump" for the local adaptation rate at intermediate dispersal in Fig. 8.2 is tiny compared with that of interspecific community change). If this constraint is sufficiently strong, the *qualitative* rules of community assembly might not differ even if there are more subtle processes that might change things *quantitatively*. How can we tell if adaptive evolution within species is sufficiently large to qualitatively affect metacommunity assembly?

To do so, adaptive evolution must alter at least one of the three key components of community assembly: (1) whether species can successfully invade a particular community, (2) whether species (or species combinations) can successfully resist invasion by other species, or (3) whether species influence the extinction of other species after they invade. Although sparse, there is some evidence that population differentiation can alter all three of these components in at least some cases. Studies on species in the genus *Daphnia* (an often dominant and important herbivore in freshwater lentic communities) have provided the most complete exploration of how population differentiation can alter community assembly.

Can population differentiation within a species alter if other species can invade? De Meester et al. (2007) used two ecologically and genetically differentiated populations of *Daphnia magna* to show that when each was placed into experimental mesocosms first, subsequent invasion by other zooplankton species differed between the populations (Fig. 8.7a). Can population differentiation within a species alter how well it can invade established communities? Pantel et al. (2011) found that two of three differentiated populations of *Daphnia pulex* could invade experimental mesocosms with resident zooplankton communities, whereas a third population could not (Fig. 8.7b). Can population differentiation within a species alter the likelihood that other species will go extinct? Mathew Leibold and Alan Tessier (unpublished data) showed that two populations of *Daphnia pulicaria* that originated under different amounts of predation by fish differed in their ability to cause the extinction of a congener, *D. dentifera* (although the extinction depended additionally on whether fish were present or not; Fig. 8.7c).

These studies all showed that genetic differentiation among populations of a single species could qualitatively alter different aspects of metacommunity assembly. To what degree could this depend on rapid adaptive evolution, as might be

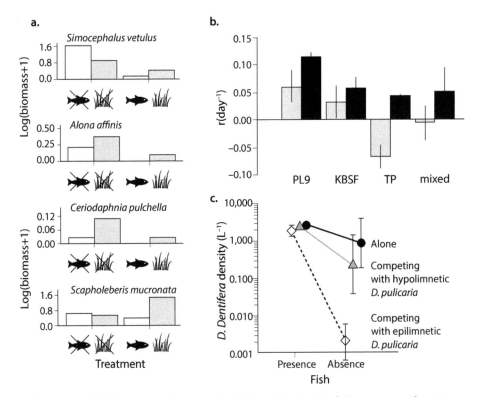

FIGURE 8.7. (a) Effects of genetic population differentiation by *Daphnia magna* on other zoo-plankton species. White bars denote the effects of *Daphnia* from clear-water ponds, gray bars the effects of *Daphnia* from turbid water. The effect is shown for two environmental conditions: on the left in mesocosms without fish or macrophytes; on the right in mesocosms with fish and macrophytes. There were significant effects of *Daphnia* source in each case and an interactive effect with environment for *Simocephalus* and *Scapholeberis*. (b) Different per capita growth rates of genetically different *Daphnia pulex* populations from three different ponds (labeled PL9, KBSF and TP) in the presence (gray bars) and absence (black) of previously established multispecies zooplankton communities (that did not have *D. pulex* in them). All three *D. pulex* populations could grow when rare in the absence of other species (black bars), but population TP could not do so in the presence of other species whereas the other two populations could (gray bars). Modified from Pantel et al. 2011. (c) Mean ($\pm$ 1 SE) density of *D. dentifera* as a function of fish presence or absence in response to three different manipulations of *D. pulicaria*. When *D. pulicaria* was absent (filled circles), *D. dentifera* was at high abundance. In the presence of genetically distinct populations of *D. pulicaria* that were "migratory" (showed diel vertical migration in their original lake), *D. dentifera* was essentially driven extinct (only one individual was present in one of the replicate mesocosms). This was not so when fish were present (open diamonds) in the presence of nonmigratory *D. pulicaria* (filled triangles). From unpublished data of A. J. Tessier and M. A. Leibold.

required for the community monopolization hypothesis? Pantel et al. (2015) exposed a single natural population of *D. magna* to four selection regimes (fish presence or absence × macrophyte presence or absence) in mesocosms for two months (typical of the seasonal active part of the cycle of this species). Consistent genetic differences among these five populations (the original population and the four populations exposed to different selection regimes) thus reflect the amount of evolution that happened during a single season from a naturally occurring amount of standing genetic variation. In the subsequent year, they used each of these five populations to inoculate different mesocosms. Each of the four "preselected" populations was used to inoculate mesocosms having the same environment in which they had been selected and were compared with control inoculations from the original pond population. Differences between the control and the preselected treatments thus reflect the consequences of a single season of exposure to "local" conditions (differing in the occurrence of fish and macrophytes). They then added in a diverse inoculum of other zooplankton species and asked how community structure was changed by the effects of preselection. They found that preselected populations of *D. magna* were substantially more abundant than the control populations (2–100 times more so!) in all environmental treatments (Fig. 8.8a). More important, they also showed that whether *D. magna* was preselected or not had a substantial effect on the community composition of the other zooplankton species (Fig. 8.8b).

Another line of evidence that suggests that evolutionary priority effects of the type hypothesized by the community monopolization hypothesis can occur is illustrated by the experiments of Fukami et al. (2007). They introduced different locally adapted ecotypic strains (largely equivalent to species) of *Pseudomonas fluorescens* into experimental microcosms at different times and found that this led to adaptive evolution and eventual dominance of the strain that was first introduced into the community, corresponding to the predictions of Urban and De Meester (2009). Interestingly, adaptation involved niche diversification within the community as well as changes in the mean trait evolution of the strains (Fig. 8.9). This finding suggests that community monopolization dynamics may also affect within-community differentiation in ways that can lead to adaptive radiations.

Although the most complete evidence for these effects comes from organisms with rapid dynamics, there is evidence showing at least some of the conditions for community monopolization effects from longer-lived species. For example, Crutsinger et al. (2008) found that genetically distinct populations of goldenrod (*Solidago altissima*) differed in their ability to resist invasion by competing old field-plant species. Likewise, Vellend et al. (2009) showed that genotypic variation in Kentucky bluegrass (*Poa pratensis*) influenced its ability to resist invasion by dandelions (*Taraxacum officinale*).

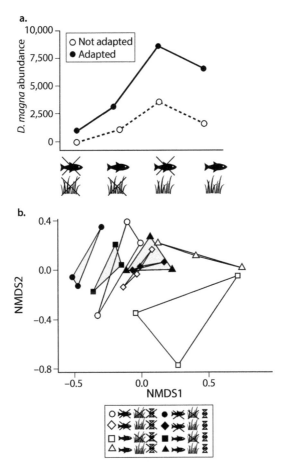

FIGURE 8.8. Effects of fast evolution in response to selection by *Daphnia magna* on its own abundance (a) and on community composition by other zooplankton (b). (a) Adapted populations of *D. magna* were more abundant than the ancestral population in each of four different habitat conditions (presence/absence of fish with presence/absence of macrophytes as shown in cartoon labels). (b) Composition of zooplankton shown in the first two axes of an Non-metric Multi-Dimensional Scaling (labeled NMDS1 and NMDS2) analysis for each of four habitat conditions (shown with different symbols) in the presence of adapted populations (filled symbols) and in its absence (open symbols). Modified from Pantel et al. 2015.

Overall, these studies suggest that the interpretation of metacommunity dynamics when there is local monopolization can differ substantially from our expectations in cases with purely ecological processes (see also De Meester et al. 2016). In cases with sufficiently high dispersal or connectivity, the predictions for metacommunity structure converge on nonevolutionary models since the degree of

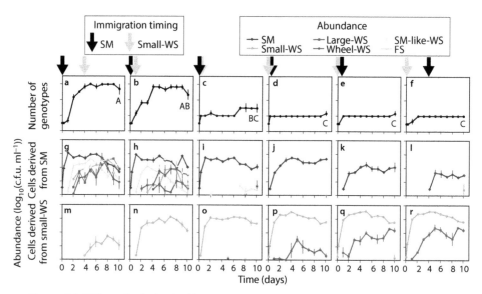

FIGURE 8.9. Effects of colonization history and rapid evolution on the assembly of *Pseudomonas fluorescens* strains. Columns indicate a progression of delays in the arrivals of two strains (SM = smooth morph, Small-WS = small wrinkled morph) ranging from 4-day advantage for SM (leftmost column) to 4-day advantage for Small-WS (rightmost column) as indicated by the location of the downward arrows. After inoculation, either morph subsequently differentiated into multiple ecomorphs (particularly so for SM). The abundances of each morph is shown in the lower two rows of panels, with those derived from SM in the middle row and those from Small-W in the lower row. The total number of genotypes (both ecotypes and ancestral strains) is shown in the top row of panels. When SM was introduced early, it differentiated and strongly suppressed the abundance and eco-differentiation of Small-WS. The reverse was true, albeit to a lesser extent when Small-WS was introduced early. When introduced simultaneously, there was little differentiation and little suppression by either strain. Modified from Fukami et al. 2007.

local adaptation does not substantially change the process of metacommunity assembly (i.e., assembly trumps local adaptation). However, when dispersal is low, local monopolization can alter metacommunity assembly by preventing successful colonization by species that could otherwise have done so (i.e., local adaptation can trump assembly, at least to some degree). The result is that isolated communities subject to disturbances or environmental change that can cause extinction (as in PD models) can nevertheless be resistant to invasions (contra the situation in PD). The outcome thus results in priority effects that are akin to those predicted for multiple stable equilibria (see, e.g., Chap. 5) but that occur via trait differentiation within a single species. Thus, from the point of view of community composition, which is based only on species identity, we would predict a relatively weak role for environmental filtering associated with SS and a stronger role for spatial and residual processes. Indeed, this could be one reason why even meta-analyses that suggest that there are strong correlations between species composition and environmental variation (which is consistent with SS) (Cottenie 2005, Soininen 2014) nevertheless are able to explain only about 20%–30% of the variation in species compositional differences due to environment (see Chap. 4). These ideas are quite conjectural, however, and more theoretical and empirical work is needed to disentangle the complex interactions between evolutionary and ecological processes, as well as to identify key signatures of these effects in natural metacommunities.

8.4 EVOLUTION TOWARD NEUTRALITY

Usually when we think of how evolutionary processes influence the coexistence of species, we think of things like character displacement, in which selection acts to differentiate species and reduce their competitive intensity (Brown and Wilson 1956). Typically, character displacement serves to enhance the stabilizing component of species coexistence, as we discussed in Chapter 2, and is a component of the community monopolization hypothesis, which allows species to adapt to multiple habitats (Urban and De Meester 2009). However, the other component of coexistence—equalizing factors—can also evolve when species are in competition. In this case, coexistence is more likely when species have more similar equalizing traits, which in many scenarios can lead to character convergence (Abrams 1987, Hubbell 2006, Scheffer and van Nes 2006, Fox and Vasseur 2008, terHorst et al. 2010). In some cases, convergence can lead to coexistence of species with identical traits (i.e., neutral), at least in the short term (Hubbell 2006, Scheffer and van Nes 2006, terHorst et al. 2010). This suggests that eco-evolutionary dynamics could, at least in some circumstances, lead to NT dynamics.

Riley (1963; see also Ghilarov 1984) was among the first to argue that species could coexist neutrally (see Chap. 1). One reason for this, he argued, was that under conditions of strong nutrient limitation and co-limitation, competing species would converge in their niches by natural selection. Abrams (1987) showed that such evolutionary convergence in resource use was especially likely when resources were "essential" (i.e., each represented a qualitatively different resource that affected growth, such as when different elemental nutrients limit growth due to stoichiometric needs of species). However, Fox and Vasseur (2008; Vasseur and Fox 2011) showed that this convergence in resource use did not generally result in neutral coexistence but instead led to frequency-dependent eco-evolutionary processes that could enhance coexistence (we discuss this general idea more below). Nevertheless, the idea that niche evolution via natural selection could favor niche convergence, at least under some conditions, is an intriguing one that has received renewed attention in the light of NT.

Hubbell (2006) argued that diffuse competition in a diverse metacommunity could lead to convergence of species-level traits toward a generalist ecotype. He imagined that in species-poor communities, two species would experience consistent selection to diverge because of consistency in the selection gradient they experience when in proximity to competitors (Fig. 8.10a-i), but that this would not be so in species-rich communities, where selection would differ wildly and idiosyncratically depending on which competitors were nearby (Fig. 8.10a-ii). The same principles are evident in the idea of the *stochastic dilution effect* in high-diversity communities, which we discussed in Chapter 6, whereby co-occurrences of species with similar or different functional traits cannot be disentangled from neutral expectations even if competitive exclusion is occurring (McGill 2010).

To explore this process, Hubbell (2006) simulated highly simplified scenarios. In the first scenario, he assumed the presence of an environmental gradient in a single important abiotic feature and asked if dispersal-limited, initially genetically similar, species that were nonrandomly "seeded" into this gradient would evolve to have different niches. This scenario is one that corresponds to environmental tracking by local adaptation, and he showed that this nonneutral result would occur under these conditions (Fig. 8.10b). In a second scenario, Hubbell imagined that dispersal-limited species were again introduced into a landscape of patches that were arranged randomly in space and had a much more limited distribution of environmental conditions. He asked how a diverse set of species with similar genetic variation of traits that were maladapted to this landscape would evolve and found that they all adapted in parallel to the overall distribution of environmental conditions. In this scenario, initially neutrally coexisting but identically maladapted species continued to coexist neutrally even though all of them adapted to the new

---

FIGURE 8.10. Hubbell's (2006) models of evolution in metacommunities. (a) Schematic diagram contrasting evolution of character displacement (i) between individuals of two species in response to their local coexistence, where each species evolves in a distinct direction, with evolution in a diverse set of species (ii), where each individual of each species is selected to favor a different direction of trait change depending on the identity of its neighbors. (b) Evolution of niche differentiation of 10 species along an environmental gradient (black bars) depending on random initial locations along an environmental gradient with very limited dispersal (open bars). (c) Evolution of 10 species with initially similar and maladapted trait distributions. Each species is shown with distinct stacked black or white squares and the optimal trait distribution curve is shown by the dotted line. Through time, the entire trait distribution shifts but none of the species go extinct (ii). (d) Evolution of 10 species with initial trait distributions as shown in (c) but with a strong spatial gradient in the environment (as in b). Here, the initial spatial distribution of each species was random along the gradient. Eventually, species evolve to have broad ecological niches along the environmental gradient but cluster at various locations along that gradient. Modified from Hubbell 2006.

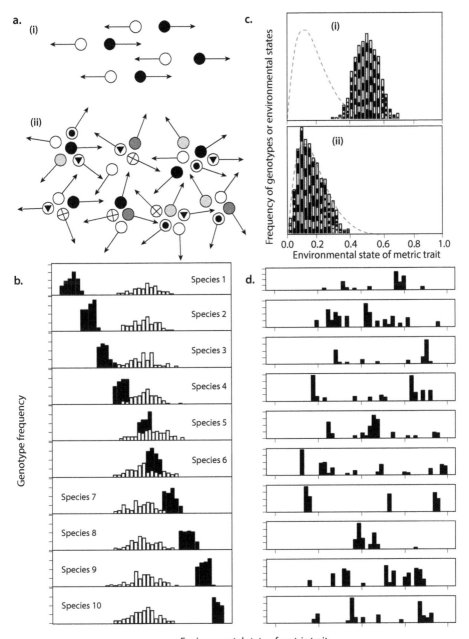

a.

(i)

(ii)

b.

Genotype frequency

Species 1
Species 2
Species 3
Species 4
Species 5
Species 6
Species 7
Species 8
Species 9
Species 10

Environmental state of metric trait

c.

Frequency of genotypes or environmental states

(i)

(ii)

0.0 0.2 0.4 0.6 0.8 1.0
Environmental state of metric trait

d.

set of conditions (Fig. 8.10c). Perhaps the most intriguing simulations, however, involved a third scenario. Here, environmental conditions varied along a spatial gradient (as in simulation 1) and species were randomly seeded into this spatial gradient, but now these species differed in initial trait values. Hubbell found that species evolved rapidly to their local conditions but any single species became locally dominant at numerous positions along the gradient. The result was that all the species had similar mean trait values when averaged across the entire gradient even though they differentiated and differentially dominated at multiple points along that gradient (Fig. 8.10d). He argued that this result illustrated how diffuse coevolution in a spatially structured metacommunity could lead to convergent evolution of generalists.

Others have also examined how evolution might tend toward convergence of competitors rather than to divergence. Although these models focus on evolution within a single community (i.e., they ignore spatial processes), they provide insights into the evolution of neutrality that shed light on what might happen in metacommunities. TerHorst et al. (2010) provided a highly simplified example by modelling a scenario in which a local community has two "niches" (e.g., substitutable resource types) whose exploitation depends on a single trait (specialization at either extreme values of the trait and generalization at intermediate trait values) (Fig. 8.11). They assumed that a single species in such a community will evolve to be a resource generalist (which is expected from many other formulations of niche evolution). If there are two species, they will diverge to be resource specialists on each of the two resources (also a common finding). However, the novelty of terHorst et al.'s analysis comes when they examined what happens when a sequence of generalized species are introduced in close temporal succession. When one or two species are introduced, terHorst et al. found the expected niche partitioning results described above. However, when more species are subsequently introduced, they can sometimes reach the extreme specialized trait values before being outcompeted and then form sets of equivalent species that can subsequently coexist neutrally. Similar processes could probably also lead to equivalent species and neutral coexistence when there is a single niche in the community, which would match the first scenario described by Hubbell (2006), above.

TerHorst et al. (2010) were at least partially motivated by their work on successional patterns of competition in the inquiline biota of the pitcher plant, *Sarracenia purpurea*, where they had found that resource competition among multiple coexisting protist consumers increased through time even though resource availability also appeared to increase (Miller and terHorst 2012). More recently, Miller et al. (2014) tested for the role of niche convergence by studying how the effect and response components of niche relations among four of the main protist species (standardized against intraspecific effects) changed during this process (Fig. 8.12).

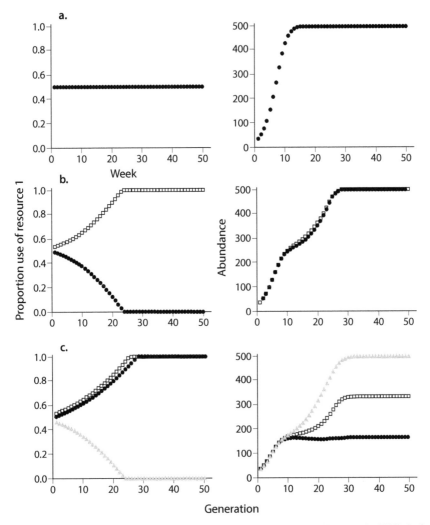

FIGURE 8.11. Ecological trait evolution through time in the model of terHorst et al. (2010). Left panels show mean trait values for each species; right panels indicate abundances of each species. The environment has two distinct substitutive resources (sensu Tilman 1980). All the species start at low densities with intermediate (i.e., generalist) trait values. (a) If a single species colonizes at time 0, it retains the generalist phenotype and reaches a high carrying capacity. (b) If two species are introduced simultaneously, they will diverge with each species becoming a specialist on the alternate niche types and both reach distinct (here symmetrical) carrying capacities. (c) If three species are introduced simultaneously (there are more species than available niches), two of the species will evolve toward one of the specialized niches, whereas the third will diverge. If both species reach the optimal trait value for the shared niche before either of them reaches carrying capacity, they will then coexist as neutrally equivalent species with reduced equilibrium densities (relative to the density observed in the second case). Modified from terHorst et al. 2010.

They found that initial species formed a clear hierarchy in which some species were initially strong suppressors of other species but only weakly affected by them (i.e., dominants), and others were weak suppressors that were also strongly suppressed by others (i.e., subordinates). Through time (and apparently in response to evolutionary dynamics), all the species converged to being moderately suppressive of others and moderately sensitive to others (i.e., the hierarchy disappeared and interspecific effects were more symmetrical). Nevertheless, the average strength of interspecific competition did not change (neither decreasing as might have been predicted by character displacement, nor increasing as predicted by convergence models), and the authors attributed this finding to the effects of diffuse competition. Although this finding may superficially resemble the evolution of neutrality, in fact strict neutrality would predict an overall increase in the interspecific effects when standardized against intraspecific ones. Clearly, the results are a bit more complicated than the models predict, but the effects nevertheless imply the evolution of enhanced coexistence, perhaps via a complex mixture of evolving niche partitioning (increased stabilizing components) and more symmetric overall fitness under competition.

Another complex scenario leading to a similar mixture of niche partitioning and equivalence (or at least nearly so) was studied by Scheffer and van Nes (2006). They studied niche evolution involving traits that have a mean value that determines fitness (as in the terHorst et al. 2010 model) and a variance within each species (as in Hubbell 2006) that describes the ability to exploit intermediate resources along a continuum of resource types. They found that the interaction between mean trait values and the variance in resource use led to "clumpy" trait distributions across large sets of species (Fig. 8.13a). Some studies indicate that such clumpy pattern of species trait distributions can be seen in natural communities (Fig. 8.13b, c). Barabás et al. (2013) questioned this interpretation, however, and argued that clumping along a single trait axis can be predicted by models that predict spread-out distributions of other traits within such clumps.

Scheffer and van Nes (2006) observed that each clump consists of multiple species with very similar mean trait values, but Scheffer et al. (2015) suggested that there could also be evolution of single species (instead of clumps of species) that space themselves out along similar trait values. An appealing example involving body size can be found in cave beetles in Australia (Fig. 8.13d; Scheffer et al. 2015). Indeed, although he did not necessarily invoke evolutionary processes, Hutchinson suggested the presence of such "community wide character displacement" back in 1959! And it is important to note that the existence of trait variation such as size ratios has a complex (and controversial) history (see Chap. 6), so that these examples should be taken with a grain of salt.

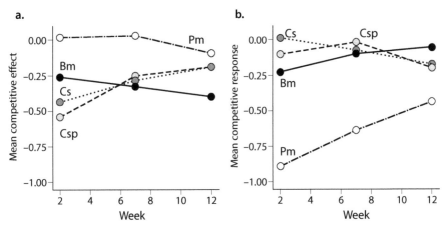

FIGURE 8.12. Evolution of competitive effect (a) and competition response (b) of four protist species through time in microcosms. Initially, there was a strong hierarchy of competitive relations with some species (e.g., Pm = *Poterioochromonas malhamensis*) being dominant over the others (Bm = *Bodo menges*, Cs = *Colpoda steinii*, and Csp = unidentified colpodid) by virtue of having strong interspecific effects on others (denoted by other initials) and weak responses to them. Over time, due to broad-scale heritable change in these species, the dominant species became less so, and previously subordinate species become more dominant. Modified from Miller et al. (2014).

What do these models of niche convergence mean for species compositional variation in metacommunities? Recall that character displacement among species so that they differ in key traits allowing them to exploit different parts of the landscape is the cornerstone for the importance of environmental controls on species composition in the SS perspective. If character convergence is important, leading toward (even if not full) neutrality, it can increase the likelihood of stochastic processes and dispersal limitation to play a stronger role in determining community composition across the landscape, decreasing the strength of the environmental effects and increasing the importance of spatial structuring and residual, apparently stochastic, patterns. Unfortunately, there is almost no empirical evidence for the evolution of character convergence in local communities (but see Muschick et al. 2012), and so it is not possible to know how important this process is in driving the patterns observed in natural systems. The diving beetles (illustrated in Fig. 8.13) may even represent striking counterexamples: even if they show regular spacing of body size, and even if this seems to result from convergent evolution at the regional scale, there is always only one species in each niche (body size) at each location (ditch or cave) in this example.

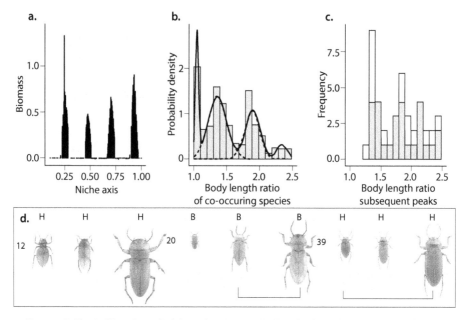

FIGURE 8.13. (a) Clumping of niche traits along a single axis through time starting from an even distribution. Selection, in response to diffuse competition along the niche trait results in clumping. Modified from Scheffer and van Nes 2006. (b) Body-length ratios of pairs of species found in 1507 samples of Dutch ditches where >2 species of diving beetles were found (yielding a total of 28,762 locally co-occurring species pairs). Dashed curves represent the fitted constituent distributions computed by means of latent class analysis. (c) Frequency distribution of the peak-to-peak body length ratio for diving beetles for each of 27 zoogeographical regions and of the northern and southern hemisphere aggregated ($n = 11$). (d) Illustrative communities consisting of triplets of blind diving beetle species found in three underground aquifers that became isolated ~5 million years ago when Australia became arid. Independent of the founder species, evolution led to a small, a medium and a large species in each aquifer. H, B, and C code the tribes Bidessini, Hydroporini, and Copelatini, respectively. Bars connect pairs of species that evolved from the same founder species (Fig. a modified from Scheffer and van Nes 2006, Figs. b-c, modified from Scheffer et al. 2015, Fig. d courtesy of Chris Watts and Howard Hamond).

8.5 FREQUENCY-DEPENDENT EVOLUTION

Just as frequency-dependent interspecific interactions can allow multiple species to coexist on few resources by a variety of mechanisms (e.g., cyclical assembly, predator-prey dynamics, species aggregations, and differential use of space; see Chaps. 2–5), frequency dependent coevolutionary processes can also influence species coexistence. For example, using a simple model with two species, Vasseur and Fox (2011) showed that species undergoing eco-evolutionary dynamics could coexist with, but not without, evolution as a result of frequency-dependent selection.

When a species is common, it experiences stronger selection from its neighbors (which are primarily of the same species; i.e., intraspecific competition), and
when it is rare, it experiences stronger selection from its neighbors (which are
primarily from a different species; i.e., interspecific competition). This neighbor-
dependent selection acts on the same timescale as ecological processes, allowing
intransitive competition among genotypes to lead to stable coexistence in some
circumstances.

As an empirical example of this sort of effect, Lankau and colleagues showed
that a mustard species trades off growth rates (enhancing intraspecific competition) with allelopathic root exudates (enhancing interspecific competition; Lankau
and Strauss 2007, Lankau 2011). When the mustard was relatively common in a
community, selection favored rapid growth at the expense of allelopathic production, whereas when it was relatively rare, selection favored allelopathic production
to the detriment of growth.

Frequency dependence can also act within a co-evolutionary context, influencing the relative frequencies of species and potentially allowing coexistence. For
example, many species of North American passerine bird species appear to be
largely competitively equivalent, at least with respect to their traits, and to have
abundances and occurrences that appear random with respect to ecological traits
and the biotic and abiotic environment, which is consistent with NT dynamics
(Ricklefs 2011). Ricklefs, however, conjectured that these same patterns could
instead result from a more cryptic case of ecological sorting resulting from
frequency-dependent co-evolutionary dynamics of the birds with important pathogens. Specifically, he envisioned a scenario in which species that are most common in a particular time and place are those that have evolved a "truce" with their
specialized pathogens: reduced virulence and high immunity. Rarer species are
those that have not yet achieved that truce and are strongly limited by pathogens
but are able to persist through a rare species advantage: escaping control by pathogens when they are at low densities. However, because the virulence of pathogens
and immunity of hosts are evolutionarily labile, these can change continuously
through time and space, leading to a mosaic of distributions and abundances of
species that would otherwise appear random with respect to environment and traits.
While speculative, this mechanism is reminiscent both of the Janzen-Connell (Janzen 1970, Connell 1971) mechanism for enemy-mediated frequency-dependent
coexistence in ecological time and of the geographic mosaic theory of coevolution
(Thompson 2005) and is a plausible scenario in which species co-occurrences
might appear to be largely random but instead are responding to a largely unseen
deterministic force. In such cases, the match between environment and species
composition would be low, and metacommunites would have large spatially structured or residual variation. However, the interpretation that these systems would

be dominated by NT or PD would be wrong because the co-evolutionary dynamic with pathogens represents a frequency-dependent stabilizing factor that is independent of environment but consistent with SS.

Loeuille and Leibold (2014) modeled dynamics involving such feedbacks in a simplified spatial context that allowed them to examine how these sorts of feedbacks might interact with adaptive evolution and metacommunity assembly. They found three qualitatively distinct outcomes in the phylogeographic "signature" and associated landscape-level patterns in trait variation (Fig. 8.14).

(1) *Permanent specialization.* This scenario (Fig. 8.14a) tends to occur when extinction rates are low. Here, the traits of the species in each local patch becomes increasingly distinct from those neighboring patches in an "arms race" between the effect of the trait on the environment (that excludes other species) and its ability to grow in that environment that reduces its ability to grow in patches occupied by other local populations. In this case, trait diversity can increase almost indefinitely in the metacommunity and is limited only by the total number of patches in the landscape.

(2) *Permanent generalism.* This scenario (Fig. 8.14b) tends to occur when extinction rates are high. Here, the effects of local feedbacks remain very small and temporary. Consequently, there is little evolution in response to them, and a single generalist phenotype is maintained and dominates the entire metacommunity.

(3)*Taxon pulses.* This scenario (Fig. 8.14c) tends to occur under intermediate extinction rates. Here the metacommunity tends to undergo alternating patterns of diversification and contraction, because the local "arms races" that happen in permanent specialization are always limited in spatial extent. As these areas of high specialization arise and increase in degree of specialization, they become more vulnerable to extinctions and thus collapse back to favoring the generalist species. When that happens, evolution favors the same generalists that are present in the case of permanent generalism, and the cycle starts over again.

Interestingly each of these three scenarios has a distinct effect on the landscape distribution of both environmental variables (driven by the feedback process on the environment) and on trait distributions that evolve across the landscape.

8.6 THE INTERACTION OF COMMUNITY MONOPOLIZATION AND NEUTRAL EVOLUTION

Both of the above eco-evolutionary processes—community monopolization and neutral evolution—act to reduce the association between community composition and environmental variation, but they do so in different ways. In the community monopolization hypothesis, any given species can be associated with a variety of

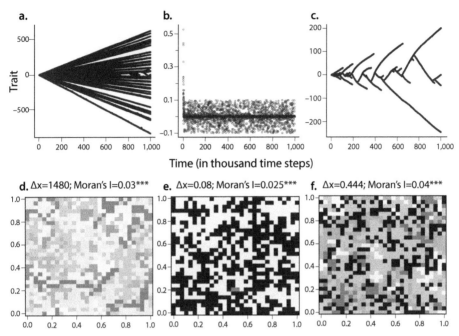

FIGURE 8.14. Eco-evolutionary dynamics represented as traits through time (a, b, c) and as landscape variation (d, e, f). There are three qualitative types of outcome under different extinction rates: a continuous divergence of traits that produce a continuous escalation of specialization of species in the landscape (a, d); no evolutionary divergence (b, e); and a divergence of traits with periodic extinctions of specialist morphs (c, f). Note the differences in scale between (a), (b), and (c). Each grid in each landscape is 30×30 patches. Lowest trait values are in dark gray and highest values in light gray. On (e) and (f), black corresponds to empty patches. Above each panel, Δx show the range (maximum-minimum) of environmental and trait variation, while Moran's I measures spatial autocorrelation. Stars indicate the degree of significance of the autocorrelation: $*p < 0.05$, $**p < 0.01$, $***p < 0.001$ compared with a random organization. Modified from Loeuille and Leibold 2014.

environmental conditions even though they still show strong niche differentiation with other species at the local scale. In the case of evolving neutrality, there is no such niche differentiation. We also note that the models above often show interactions between community monopolization (and niche differentiation) and the evolution of neutrality. For example, Urban and De Meester (2009; see also Vanoverbeke et al. 2015) observed that neutral coexistence occurred when colonization events were identical or almost so, and Hubbell (2006) showed that the outcome could involve either landscape-level neutrality or SS depending on initial conditions.

In an ongoing study, Joost Vanoverbeke, Luc DeMeester, Mark Urban, and Mathew Leibold are examining this question using simulations of a metacommunity of patches arrayed on a spatially explicit lattice landscape. They are employing a modeling framework that combines the main features of the model by Urban and De Meester (2009, see also Vanoverbeke et al. 2015) for evolution and population dynamics within patches with the framework of Leibold and Loeuille (2015), in which there is a spatially explicit landscape of patches and repeated random disturbances that cause extinctions. This feature means that community assembly will occur numerous times in each patch and results in a metacommunity that consists of a mosaic of patches at different stages of assembly. As an aggregate, the metacommunity can develop stable patterns at the landscape level even if each patch is continuously changing. However, in contrast with other models of assembly dynamics involving only species interactions (e.g., Leibold and Loeuille 2015), they model the additional effects of local adaptive evolution (as in Urban and De Meester 2009 and Loeuille and Leibold 2008).

Figure 8.15 shows the results of one of the simulations with two species and two patch types and compares a situation with rare disturbances and low dispersal and either no evolutionary change and one with high evolutionary rates, which promote community monopolization and the evolution of neutrality. Figure 8.15a shows a "snapshot" of what happens in the absence of evolution: that patches are occupied by the initially best adapted species and that there are strong correlations between species composition and environmental variation (high [E]) but lower spatial [S] and residual effects (Fig. 8.16). When evolution occurs (Fig. 8.15b), local adaptive evolution can reduce the importance of environmental correlations [E] and increase spatial [S] and residual effects (Fig. 8.16).

While the patterns that emerge from this model look similar to those that would result from NT, at least superficially (i.e., with the variation partition), they are not the same as would emerge from NT. For example, there is strong trait matching where each type has traits that match local conditions (not shown) and patches are resistant to invasion. Additionally, individual patches are sometimes occupied by a single species (consistent with community monopolization) and sometimes occupied by both (consistent with either *evolved neutrality* or transients). To identify which fraction of the co-occupied patches were consistent with evolved neutrality, Vanoverbeke and colleagues treat each species as an "allele," as in population genetics theory, to determine whether evolved differences were sufficient to overcome drift; if mean fitness differences are very small, this indicates local neutrality (or "near-neutrality"). In the absence of evolution, there are (trivially) no cases of community monopolization or neutrality (Fig. 8.15b), whereas with evolution, most patches conform to community monopolization, but there are also cases of evolved neutrality that occur at much lower frequency. The other

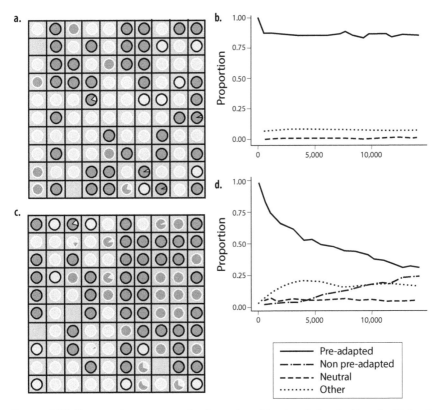

FIGURE 8.15. Contrasted metacommunity dynamics without evolution (a, b) and with local evolution (c, d) under relatively high mutation rate and low dispersal and extinction. (a, c) Two types of patches identified by the outline of the patch (Patch type A, white, vs. Patch type B, black) and two species shown by the fill shade (Species 1, light gray; Species 2, dark gray). When species coexist locally, their relative abundances are shown as a pie chart. Under initial conditions, Species 1 is highly adapted to white patches and Species 2 to black patches. (a) No evolution; there is a very strong correspondence between species presence and patch type corresponding to species sorting. (c) With evolution; there is virtually no such correspondence. (b, d) Temporal dynamics of patch occupancy types: (b) is without evolution, (d) is with evolution. Solid lines show patches occupied uniquely by the species best adapted to that patch time at the beginning of the simulation (i.e., Species 1 in A patches or Species 2 in B patches) with optimal trait values for each patch. Dashed lines show patches occupied by both species with trait values that are almost identical (i.e., effectively neutral); in (b) this value is zero, in (d) the value is very small. Alternating dot-dash lines show patches where niche dominance has switched species due to evolution (i.e., Species 2 occupies patch type A with a locally optimal trait value and vice-versa for species 1 in patch type B. Dotted lines show patches of other types including empty patches and patches occupied by species that do not have locally adapted trait values. From Vanoverbeke and collaborators, unpublished data.

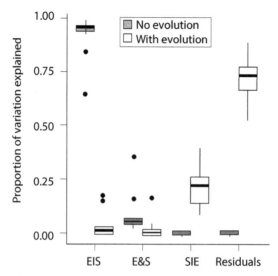

FIGURE 8.16. Variation partitioning in the presence (open bars) and absence (filled bars) for replicate runs of simulations with parameters with and without evolution. The variation components include pure environment (E|S), pure space (S|E), the mixed component (E&S) and residuals. In the absence of evolution (filled bars), the environmental component explained almost all the variation in community composition; in the presence of evolution (open bars) the environmental component explained almost none of the environmental part and purely spatial and residual variation explain much more of the variance. From Vanoverbeke and colleagues unpublished data.

interesting result is that the two species become essentially "equivalent" at the metacommunity scale (both have identical average fitness across the entire landscape) even though they are only very rarely so at the local scale.

For the cases we describe here, the resulting metacommunity has a structure that does not correspond very well to any of the four archetypes (SS, PD, NT, ME). First, like NT, species are equivalent, at least at the metacommunity scale, but unlike NT, species are very rarely equivalent at the local scale. Demographic stochasticity within local communities is thus rarely important and it does not enhance local diversity. Second, like SS, local communities are resistant to invasion by colonists and do not generally support sink populations, but unlike SS, there is little matching between environment variation and species compositional variation, but there is high spatial patterning and high unexplained variation (i.e., appearing like drift). Third, like PD, disturbances in the form of extinctions often lead to changes in species composition, but unlike PD, these do not produce consistent patterns and often lead to less compositional turnover because of the community monopolization process. Importantly, using the variation partitioning approach at a single point in time as a possible diagnostic tool (as in Chap. 4) suggests that the distributions would correspond more to NT or PD than to SS (much variation explained by space, little by environment). However, if we took a temporal turnover perspective (as in Chap. 5), the simulation models predict less turnover than either PD or NT and instead look more like SS.

8.7 THE INTERACTION BETWEEN COMMUNITY MONOPOLIZATION AND NEUTRAL EVOLUTION IN THE "REAL" WORLD?

Although the scenario discussed above is still highly simplified (i.e., a single niche per patch), we might speculate based on other evidence (e.g., Vanoverbeke et al. 2015) that this sort of dynamic would include metacommunities with the following features: (1) strong competition for very similar niche factors (e.g., resources); (2) evidence that there is sometimes neutral and nearly neutral co-occurrence; and (3) evidence that important niche relations are reversed across broader spatial scales in the landscape.

An intriguing example is the one of freshwater amphipods in the genus *Hyallela*. Here, there are three commonly co-occurring species that seem to coexist in virtually every fish-containing lake in postglacial parts of the northeast of North America. These three species (they are as yet unnamed and are known as Species A, B, and C) are almost impossible to tell apart morphologically and can only be reliably distinguished using molecular markers. In Michigan, Species B is slightly larger than the other two (Wellborn and Cothran 2004). These three species co-occur in three main habitat types in these lakes—shallow macrophyte beds, deep macrophyte beds, and the zone at the water's edge—but seem to differ in their relative use of the habitat types. In two lakes, Wellborn and Cothran (2007) found consistent differences in habitat use among the three species—that Species A specialized on shallow macrophyte beds, Species B specialized on nearshore habitats, and Species C specialized on deep macrophyte beds. More extensive work of 13 lakes in the same region by Smith (2013), however, showed that this was not always so (Fig. 8.17). While the species usually showed habitat segregation, their relative preference for the different habitat types switched between lakes, and in a couple of lakes, there were no discernable differences in habitat use between the species.

Experimental work in which typical populations (as in the leftmost figures in Fig. 8.17) were studied in mesocosms showed that habitat type had an effect on competition among the three species (Smith 2013). Interestingly, removal of the nearshore habitat led to the extinction of Species B, whereas removal of either the deep or shallow macrophyte habitat had no effect on the relative abundances of

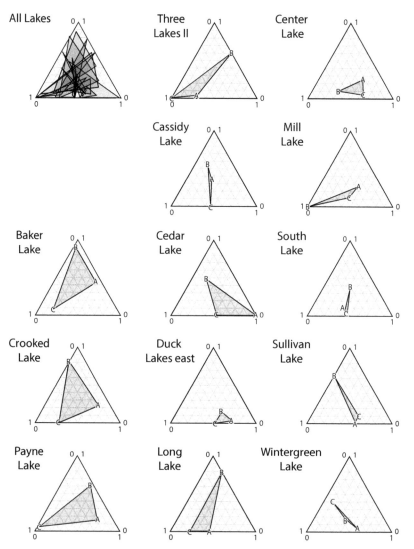

FIGURE 8.17. Proportional habitat use by species A, B, and C in thirteen different lakes in southern Michigan. Habitat use is shown using triplots (each species proportional use is shown as a single point that plots their proportions among the habitats, which have to sum to one). The pattern for a given lake can thus be shown as a triangle within the triplot. The top-left figure shows how these vary across all lakes. The most common pattern shows B in the nearshore, A in the surface, and C in the deep habitat. However, qualitatively different patterns occur in the five lakes shown in the rightmost column of plots. In many of the lakes, the differences are substantial; however, there are some lakes where these differences are essentially nil (e.g., Duck Lake East, Center Lake). Modified from Smith 2013.

the three species. This finding indicates that the two morphologically similar species (A and C) may coexist neutrally even if they show some tendencies to sometimes use different habitats, whereas the morphologically distinct species (B) may show habitat partitioning from the other two species.

While instructive about the possible role of neutral coexistence in a natural metacommunity, the study by Smith (2013) is not able to specifically discern what role eco-evolutionary processes played in driving the observed patterns. However, there is an intriguing contrast between the situation in Michigan studied by Smith (2013) and the situation involving these same three species in Quebec studied by Dionne (2015). Congruent with the findings by Smith, Dionne found that inferred diets of the three species differed from each other, but that these differences were not consistent from lake to lake. However, in contrast to the situation in Michigan, where Species B is largest and A and C may coexist neutrally, in Quebec it is Species A that is larger and Species B and C that are the same size and seem more likely to coexist without niche differentiation. If these differences in size are heritable, it may indicate that ecological-trait evolution has changed across the landscape from Michigan to Quebec.

Although these results are roughly consistent with the predictions we made above, there are many other possible explanations, and there are significant gaps in our knowledge of the system (for example, we don't know if the variation in habitat use among lakes is heritable). Nevertheless, the results suggest that niche relations need not be fixed, even in a qualitative way, among species within a metacommunity and that species can sometimes "switch niches" with each other. If these differences are heritable, they suggest a fascinating scenario for "evolving metacommunities" that can lead to variable niche partitioning across a landscape of patches.

A complementary example occurs with *Niphargus* amphipods in springs of the Istrian Peninsula in the Balkans (Fišer et al. 2015). Here, there were thought to be two pairs of species that showed ecological niche partitioning, with one pair in the northern half of the peninsula and a different pair in the south. In each pair, one species was somewhat more likely to be predacious and live deeper in the spring whereas the other tended to be more filter-feeding and lived at the surface. Subsequent molecular work, however, showed that there are really only two species but that the two species had switched niches in the north and south. While this observation is only preliminary, it might also suggest that the species differ at the local scale (within the north or south of the region) but have more symmetry in the peninsula overall. It would be intriguing to know of more examples that might suggest such effects, but we suspect that part of the problem is that we simply have not looked carefully enough to dissect patterns consistent with such scenarios. Alternatively, it may be that this sort of dynamic is not long lasting, since it could favor the evolution of reproductive segregation and possibly consequent speciation. We discuss this scenario in Chapter 9.

8.8 CONCLUSIONS

If some form of neutrality (be it at the metacommunity or the local scale) can evolve, how prevalent would we expect it to be? Here, we have to be almost entirely speculative. One possibility is that species with similar niches can evolve, but they do not do so in exactly the same ways, because they differ in the genetic architecture that determines the traits involved in niche evolution. It may be the case that species have traits that allow more fine-tuning to particular niches due to the genetic architecture (e.g., the number of loci and their epistatic and pleiotropic relations) that determines those traits. If so, niche monopolization (and local neutrality) would still be possible, but different species would have an evolutionary bias that would favor a particular niche at the expense of the others. Although species may have equivalent niches (due to historical events such as allopatric sexually driven speciation), niche evolution in sympatry toward neutrality would thus be unlikely, at least at the local level.

Another possibility is that any given patch type does not actually have a precise trait optimum. Instead, we might imagine that there is a large number of possible ways that evolution could improve local adaptation and that organisms in nature are still evolving to improve their adaptive fit to their habitats. If so, each species might have an initial advantage to habitats that always give it an edge over others that begin adapting later; a process closely related to "red queen" dynamics (Van Valen 1973). Again, we might find some degree of niche monopolization and local neutrality, but not so much that global neutrality would ever occur.

It is important, however, to keep in mind that many of the models that we have described here are highly simplified and the empirical evidence is often incomplete at best. Thus, these ideas are perhaps much more speculative than those in other parts of this book. Nevertheless, they are an important starting point for asking questions at the interface of metacommunity ecology and eco-evolutionary feedbacks at a scale that is probably much more appropriate than purely local scales.

An important feature of the eco-evolutionary metacommunity models described above (Louille and Leibold 2008, Urban and De Meester 2009, Vanoverbeke et al. 2015) is that they are developed within a framework similar to the more conventional PD archetypes we outlined in Chapter 2, but they have quite different consequences for how we understand metacommunity assembly. A key feature is how fast adaptive evolution might be. If it is slow (or if the metacommunity as a

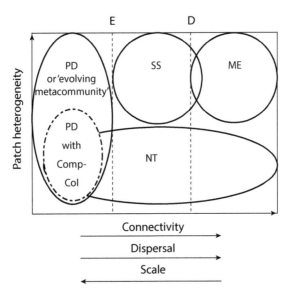

FIGURE 8.18. Modified scheme for metacommunity theory to incorporate the evolving meta-community theory outlined in Chapter 8 (Compare with Fig. 2.5). If local adaptation can be a powerful force, it can produce effects, such as community monopolization and emergent neutrality, that are distinct from those predicted by the patch dynamics theory. This would depend on potential rates of adaptive evolution (due to standing genetic variation or to mutability of ecology traits for example), dispersal (with associated possible effects on gene flow) and other assumptions (e.g., mating systems).

whole has not existed for very long), then the PD framework and associated theory is likely to capture much of the dynamics involved. On the other hand, if adaptive evolution is fast or if the metacommunities have existed for a long time, then it is possible that the predictions from the evolving metacommunity theory described in this chapter are more likely to hold. We might thus modify our initial view of the positioning of metacommunity archetypes along axes of heterogeneity and connectivity presented in Chapter 2 (Fig. 8.18). Here, we argue that the community monopolization hypothesis may result in the global neutrality dynamics under conditions that overlap broadly with the assumptions of the PD models (low dispersal and low disturbance). The two hypotheses have some predictions in common, such as high spatial variability and low environmental determinism of community composition, but they also differ, for example, in the amount of turnover they may show and in the degree of invasibility they may show.

So what should we make of this rich array of possible consequences of eco-evolutionary feedbacks in metacommunities? From one point of view, they complicate the issue now that we see that evolutionary processes which can happen rapidly even at the local scale (Ellner 2013) can also interact with slower ecological processes that happen at metacommunity scales. Whether these complications are worth the effort in exploring more deeply depends on how important they are expected to be and on what their influence on expected outcomes is. In this chapter we have highlighted some of the evidence that might indicate when and where we might anticipate these processes to be important and worth exploring more deeply, but there is still much to be known, and there is as yet very little evidence about the prevalence or importance of these effects in nature.

In this chapter, we have focused exclusively on how adaptive evolutionary change within species—that is, microevolutionary change—might alter our study of metacommunity dynamics. However, it is also quite clear that macroevolutionary and biogeographic processes can influence, and be influenced by, metacommunity processes and patterns. For example, the model described above predicts that global neutrality will be associated with fairly large scale spatial patterning; that is to say, we observe that there are large parts of the landscape where particular patch types (A or B in Fig. 8.15) are dominated by different species (Species 1 or 2 in Fig. 8.15). Such a pattern could certainly lead to macroevolutionary processes (e.g., parapatric ecological differentiation in this case). Thus, it is tempting to think that such a scenario could serve as a template for speciation if reinforcement and reproductive isolation of the ecotypes were also to evolve. In Chapter 9, we discuss this and other ways in which metacommunity ecology can inform, and be informed by, larger-scale evolutionary processes and patterns.

CHAPTER NINE

Macroevolution in Metacommunities

Prospectus

1. As the spatial scale at which we study community structure increases, there will be a transition between scales at which metacommunity processes dominate to those at which historical biogeography and phylogenetic processes dominate.

2. Metacommunity and historical biogeography can also interact, feeding back to each other such that metacommunity processes influence phylogenetic and biogeographic dynamics, and vice versa.

3. Metacommunity dynamics related to adaptive evolution, as discussed in Chapter 8, can influence ecological speciation. Such processes tend to enhance SS dynamics in the metacommunity.

4. Nonecological speciation, involving, for example, sexual selection, can also have important consequences in the distribution of organisms. Although these consequences may enhance NT dynamics, the degree to which they do may be limited as a result of reproductive interference.

5. Methods related to variation partitioning and to trait-by-environment correlations (described in more detail in Chap. 7) can reveal how phylogeny and biogeography interact with niche evolution to affect variation in community structure over large spatial scales. These methods show that clades can be differentially affected by historical versus environmental factors even when they involve organisms that interact with each other.

6. Novel interactive relations between metacommunity dynamics and historical biogeography and phylogeny will probably become more apparent as more work is done at the interface between these two disciplines.

In Chapter 8, we considered how microevolutionary adaptive change might influence metacommunity assembly. This interaction, however, ignores the very origin and spread of the species involved to begin with. We have already considered how the species pool in a metacommunity could affect metacommunity patterns (Chap. 4), but it is also possible (and indeed quite likely) that metacommunity processes can influence the species pool itself by affecting speciation and biogeographic dynamics (Mittelbach and Schmeske 2015). What can we say about the reciprocal influences of metacommunity and large-scale evolutionary processes such as speciation, phylogeny, and historical biogeography?

One perspective is that such large-scale evolutionary processes are largely independent of ecological ones. Ricklefs (2004, 2015) has provocatively suggested that biogeographic and phylogenetic processes that are largely driven by historical contingencies strongly determine the biota of regions, and that together with important regional or local influences (such as ME), they are of primary importance to determining coexistence and biodiversity patterns. Consequently, he relegates niche partitioning and other metacommunity assembly processes to a much smaller, secondary status.

An alternate perspective is that ecological processes play important roles in the ways that species are formed and how they spread through space within a biogeographic setting. Indeed, many descriptions of diversification suggest that interactions with competitors, enemies, and mutualists play important roles in determining the shape of phylogenies and their distributions (Simpson 1944, Sepkoski 1996, Schluter 2000, Thompson 2014). Nevertheless, many descriptions of the ways that species interactions shape biotas across relatively large spatial scales often seem highly simplistic. For example, they may extend ideas developed for purely local ecological processes (e.g., Lotka-Volterra competition equations) to the shape of regional (or even global) diversity-through-time relations. Or they may take an entirely neutral perspective on species interactions (Rosindell et al. 2015) that greatly simplifies the processes needed to generate and maintain biodiversity but necessarily obviates a majority of metacommunity assembly processes that we wish to consider. Because metacommunity ecology addresses ecological processes at larger spatial scales, it may serve as an important way to solidify the link between local ecological processes, dispersal, and the evolution and historical biogeography of biotas.

More important, metacommunity ecology may provide insights about the interactive effects of regional community assembly processes (via phylogenetic and historical biogeography) and the consequences of these effects on patterns of diversity and composition across different spatial and temporal scales (Mittelbach and Schemske 2015); that is, metacommunity ecology may help us understand the reciprocal relations between these two views that link local and regional biotas. Within this context, we seek to shed light on how dispersal and ecological interactions, on the one hand, and evolutionary processes that affect adaptation (or its absence) and population differentiation (or its absence), on the other, interact with each other within a historical biogeography context. A meaningful synthesis of these various processes is, of course, a challenge that goes beyond our current understanding. In this chapter, our discussions with therefore be more speculative than conclusive.

9.1 HOW METACOMMUNITY PROCESSES INFLUENCE
PHYLOGENY AND RADIATIONS

9.1.1 Ecological Speciation in Metacommunities

Ecologists often think about *ecological speciation* as being associated with changes in niche attributes of species (Schluter 2000, Rundle and Nosil 2005). We can start to think about the role of metacommunity dynamics on ecological speciation by building on ideas from the interaction of microevolution and metacommunity ecology that we developed in Chapter 8. For example, Urban and De Meester (2009) showed that population differentiation between local patches can be a powerful force for metacommunity assembly resulting in adaptive ecological differentiation. It is not hard to imagine that this could, under appropriate conditions, lead to allopatric speciation. In their model, there is no speciation, but gene flow between the patches occupied by the successful invader reduces its fitness in the ancestral habitat. One consequence of this reduction in fitness is that various mechanisms to prevent reproduction among habitat types might be favored (e.g., reinforcement), enhancing the probability of reproductive isolation and eventually leading to speciation.

In the very simple case modeled by Urban and De Meester (2009), reproductive isolation would probably occur in a single population and lead to the evolution of a species that, at least initially, would be a local endemic. Of course this species could then spread if there are suitable patches that have not been monopolized by others and become more widespread.

The overall process might be enhanced if there are frequent disturbances or if there is a rapid habitat expansion via the formation of new patches as described in Section 8.7 in Chapter 8. Here we note that species might form spatial clusters of populations with similar ecotypic traits (as evidenced by the fact that there is enhanced spatial structure in the population) in a landscape. This process could thus be associated with a more widespread speciation process in which speciation that is quasi-parapatric would occur without any obvious barriers to dispersal.

Vanoverbeke et al. (2015) modeled another interesting case in which community monopolization in isolated communities can occur but where it can be followed by niche differentiation within patches (as occurred, e.g., in Fukami et al.'s [2007] experiments). If such niche shifts are associated with speciation (and not just with population differentiation), the result can lead to patterns in adaptive radiations associated with the colonization of new patches or divergence into different ecological niches and habitats within patches. These patterns have intriguing parallels with a wide array of studies of adaptive radiation patterns, especially in island archipelagoes. These patterns are particularly apparent when inferences can be made from phylogenetic relationships (Mittelbach and Schemske 2015, Gillespie 2016). We further discuss this in Section 9.4.

9.1.2 Nonecological Speciation and Near Equivalence

Speciation can also occur with little or no differentiation of ecological traits, because speciation can also be driven by other processes, particularly by sexual selection (Gavrilets 2000, Panhuis et al. 2001). How are the distributions of closely related species that have speciated without niche differention shaped by metacommunity processes?

Inspired by his studies of evolution and ecology of damselflies in the genus *Enallagma*, McPeek (2008) has suggested that a contrast between ecological and nonecological speciation can often be seen at the clade level. He used a simple metacommunity model with speciation to show that diversification patterns may show some form of saturation (or at least declining diversification with greater diversity) if ecological speciation is high, whereas the reverse (accelerating diversification rates with diversity) occurs if nonecological speciation is dominant (Fig. 9.1a, b; McPeek 2008). Although most clades showed patterns associated with ecological speciation, a substantial number were more consistent with nonecological speciation (Fig. 9.1c).

More interestingly for our purposes is that the shape of these diversification curves in McPeek's (2008) study is associated with predictions about standing diversity and its dynamics within the metacommunity. Under the ecological speciation scenario, standing diversity flattens out with time to a relatively low level and species show strong environmental filtering (high [E] in the variation partitioning terminology). This saturation of diversity through time is associated with low local diversity and high turnover along environmental gradients. Under the nonecological speciation scenario, the reverse is true, and local diversity is high because species can co-occur via locally neutral coexistence.

A meta-analysis of differentiation rates in different clades shows that the majority (~80%) of them show declining differentiation rates with time (i.e., some degree of saturation), but that a substantial fraction (~20%) show the reverse (McPeek 2008). The consequence of this model reinforces our previous conclusions about the prevalence of processes associated with the NT in metacommunities. It suggests that neutral coexistence at both the local and metacommunity levels is possible but depends on the prevalence of nonecological speciation. This argument also suggests that neutral distribution patterns are likely to be clade specific (assuming speciation type is evolutionarily conserved) and involve intraclade patterns (e.g., patterns among *Enallagma* species only), rather than biome specific

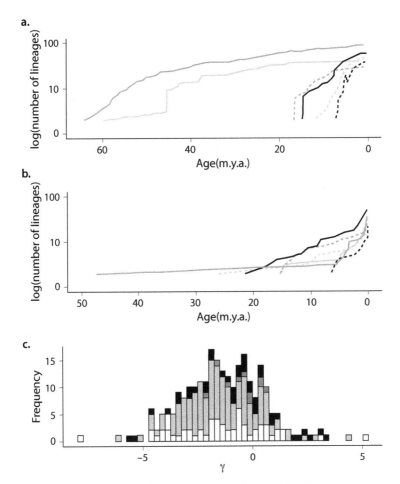

FIGURE 9.1. Plots of clade diversification through time. Some clades show saturating patterns of diversity (a), consistent with ecological speciation in McPeek's (2008) metacommunity model, whereas others show near exponential patterns (b), consistent with non-ecological speciation. The distribution of diversification rates can be described by the parameter γ (with values less than zero indicating saturating and those larger than zero indicating exponential curves). The distribution of values of γ for a wide array of organisms is shown in Panel c (different fill shades indicates different taxonomic groups: black = magnoliophyta, dark gray = chordate, white = arthropods. Modified from McPeek 2008.

(e.g., patterns involving all invertebrate predators), so that it is unlikely that entire biotas follow neutral metacommunity dynamics.

McPeek and Gavrilets (2006) proposed that an important mechanism for non-ecological speciation—reinforcement through mate recognition—may also involve metacommunity processes. They found that reinforcement through mate recognition can be a powerful force under conditions where species have very different relative abundances but very similar fitnesses. One scenario for this result could involve cases where there are numerous empty patches (either because of range expansions or because of disturbances) and where species with (near) equivalent niches arrive asynchronously into such patches. The first colonist would increase and achieve high abundance before the second colonist would colonize, and its relative abundance would thus be low (as in the model by terHorst et al. 2010). If they have similar fitnesses, the cost of mating inhibition on the second species due to interference from inappropriate mate selection by the first species could impose strong selection for modified mate recognition in the second species (costly mistakes would occur frequently) without much consequence to the first. The net result is the rapid evolution of species that differ strongly in mate-recognition traits (e.g., reproductive anatomy and behavior) but not ecology.

Nevertheless, there are still some open questions about how neutrality can be maintained and how it spreads in a metacommunity. In McPeek's (2008) simplified model, there is no spatial structure. Instead, dispersal is uniform among all patches in the metacommunity. A more spatially structured situation would reduce the likelihood that colonizations would be synchronous. There are also no evolutionary priority effects, such as those involving community monopolization (remember, in Chap. 8 we found these to be more likely than the maintenance of neutrality), nor any other spatial evolutionary processes (e.g., spatial clines). How might such effects influence the form of differentiation functions or the prevalence of actual neutral dynamics in metacommunities? Although speciation might not be associated with ecological shifts in some clades, this situation does not prevent local adaptation from happening as these species distribute themselves across a landscape of heterogeneous patches (as described in Chap. 8).

It is also useful to explore the speciation mechanism of mate recognition and reinforcement considered by McPeek and Gavrilets (2006). If the species involved are otherwise ecologically equivalent and relative abundance is low for a recent colonist, strong selection to reduce mating interference seems likely to also imply a substantial absolute fitness cost (Groening and Hochkirch 2008). This fitness cost would suggest that such colonists would have lower fitness than residents, even if they are otherwise ecological equivalents, which would make them unlikely to invade. Thum (2007) argued that this may have been an important mechanism constraining the biogeography of copepods in the northeastern United States. Similarly, Bourret et al. (2012) concluded that mating interference between ecologically equivalent species led to strong priority effects that lower coexistence and create strong spatial patterning in the metacommunity. However, the evolution of reinforcement suggested by McPeek and Gavrilets (2006) may be more likely when superior competitors arrive after less well adapted species have established. Population growth in the second colonist would thus be possible despite the costs of mate interference and Allee effects. It seems like this scenario probably does require a lack of local adaptation in ecological traits as hypothesized by McPeek and Gavrilets (2006), but it would not be associated as strongly with local equivalence as hypothesized by McPeek (2008).

9.2 HISTORICAL EFFECTS

Ricklefs's (2008) criticism of community ecology and his proposition that biogeography and phylogeny are more important drivers of community structure than local processes (Ricklefs 2004, 2015) may make some sense if we think that community processes are disjunct from larger-scale biogeography and phylogeny. However, the picture is more complicated if we think that ecology can play an important reciprocal role in the diversification of species. Nevertheless, Ricklefs (2008) is correct to point out that there are strong historical components that influence the species pool, often only loosely connected to ecological dynamics, and that these constraints will affect distribution patterns in metacommunities, at least over some spatial scales.

It is one thing to highlight that low connectivity can produce certain patterns in ecological metacommunities (e.g., PD or community monopolization), but it is another to recognize that historically unique events change large-scale biogeographic patterns (e.g., the continental exchange that occurred when the land bridge formed and connected North and South America) or that have constrained the biotas of faunas on different sides of the Wallace Line (Wallace 1876, Holt et al. 2013). Even within a less dramatic context, the assembly of regional species pools can be shaped by fairly unique historical events in ways that might have important consequences for metacommunity dynamics. For example, McPeek and Brown (2000; see also Turgeon et al. 2005) have argued that many features of the metacommunity involving damselflies (as well as numerous other likely taxa) in postglacial areas of North America can be understood as being the result of the reconstruction of the biota in this area from unconnected unglaciated areas.

While such historical events probably occur more often at larger spatial scales (and slower time scales) than the metacommunity processes we have discussed so far, the scale that separates the two may also often overlap and be rather fuzzy. As the spatial scale increases, the tools and ideas from metacommunity ecology will interact and become interwoven with the biogeographic ones (Fig. 9.2; Jenkins and

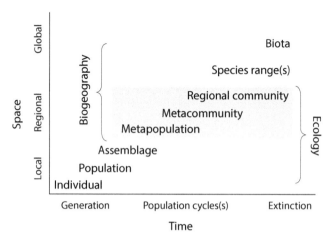

FIGURE 9.2. The general scaling of different biological processes in relation to time and space. Metacommunity ecology is at the center of overlap between ecological and biogeographic processes. Feedback mechanisms between the ecology and biogeographic of organisms are thus most likely to involve metacommunity effects. Modified from Jenkins and Ricklefs 2011.

Ricklefs 2011). These interactions are likely to be complicated, but resolving them may be the key to understanding some of the broader-scale patterns in spatially distributed metacommunities at large scales.

One way to study the interaction between metacommunity and biogeographic scales might be to incorporate spatial biogeography and phylogeny into the analysis of metacommunities. In this vein, Leibold et al. (2010) developed methods related to community phylogenetics that begin to identify how ecological and historical effects act in such large-scale metacommunities (Fig. 9.3). To do this, they used methods of variation partitioning on nodes in a phylogeny to identify the contributions of environmental filtering versus the spatial effects predicted by an a priori historical biogeographic hypothesis. Pillar and Duarte (2010) and Borregaard et al. (2014) have provided different, but closely related, methods. The method is similar to the fourth-corner method we discussed in a bit more depth in Chapter 7, but here we seek correlations between phylogenetic events (here coded as independent phylogenetic contrasts by identification of nodes) and either environmental variables (measured separately from the distribution of species) or the occurrence of biogeographic events (using independent data such as geological reconstruction of habitats).

The use of this method can be illustrated by applying it to a setting in which the timing and spatial extent of biogeographic events is relatively clear. For example, Christine Parent, Pedro Peres-Neto, and Mathew Leibold are using this method to study the distribution of land snails in the Galapagos Islands. These snails, in

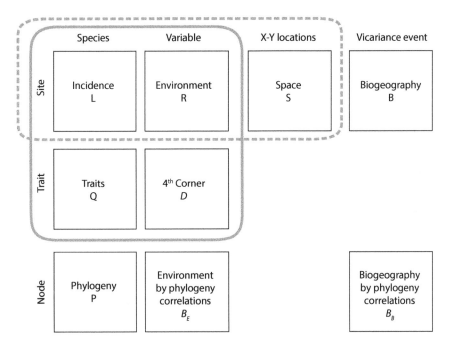

FIGURE 9.3. Matrices that evaluate the evolution of distributions of species in a metacommunity in relation to their evolutionary and biogeographic history and environments. L, R, S, B, Q, and P are data matrices obtained independently from each other. D, B_E, and B_B are derived association matrices. B_E evaluates how nodes are associated with distributional differences related to individual environmental variables, thus traits related to ecological phylogenesis; B_B evaluates how nodes are associated with distributional differences related to particular biogeographic events, thus the role of phylogeny in biogeography. The significance of these associations are evaluated using suitable permutation tests. The thick grey outline identifies the components used in fourth-corner and RLQ analyses (discussed in Chapter 7); the dashed gray outline identifies the components used in variation partitioning (discussed in Chapter 4). From Leibold et al. 2010.

the genus *Bulimulus*, have radiated in the Galapagos in a way that resembles (and is actually much more extensive) than Darwin's finches and have since served as a model system for ecological radiations in archipelagos. Most of these snails have current distributions that reflect a mixture of ecological processes (they are found in different habitats) and biogeographic ones (they are found on different islands that originated at different times). The geological history of these islands is known and it is thus possible to have predictions about the timing of biogeographic events and how they relate to phylogenetic events. In addition to the phylogenetic events that occur between islands and that may or may not have any environmental component, it is also obvious that there have been phylogenetic events within

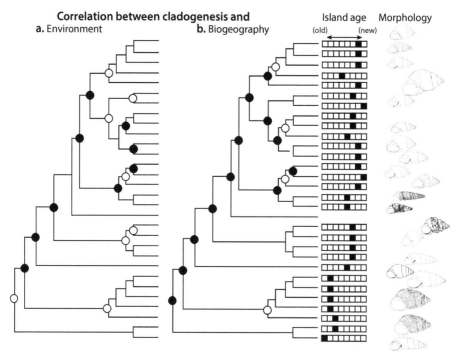

FIGURE 9.4. Node-by-node analysis of distributions of land snails in the Galapagos Islands. (a) Map of significant differentiation by environmental variables. (b) Map of significant differentiation in relation to cladogenesis between islands (notation same as in a). Nodes labelled with a solid symbol are significant at the 0.05 level even after correcting for multiple contrasts using a false detection method whereas the open symbols are significant at the 0.05 level, but not when imposing such a correction. Nodes that have no symbols had no significant association. The bars to the right of (b) indicate which island was colonized; islands are ranked from oldest to youngest. Figures to the right show the shell morphology of the snails. Parent et al., unpublished data.

islands that are largely associated with habitat shifts. Thus, the distribution of these snails reflects a mixture of biogeographic effects and ecologically driven effects that regulate the ways that these species distributions relate to environmental and spatial components.

Parent and colleagues are finding that different nodes in the phylogeny of these land snails are significantly associated with different combinations of spatial and environmental effects (Fig. 9.4).

The results show a complex mix of spatial and environmental effects on the distribution of these species. If we did not know anything about the biogeographic context and history of these distributions, the biogeographic effects identified in this analysis would show up as spatial ([S]) effects in the variation partitioning methods we described in Chapter 4. We would infer that these spatial effects reflect dispersal limitation, but we might mistakenly think that this spatial structure was due to NT or PD. While our inference that dispersal limitation is important would be correct, the purely historical effects that this analysis identifies are quite distinct from the ones assumed in NT and PD. Parent and Crespi (2006, 2009) also found that differentiation into different habitats within an island was more frequent for the more isolated islands, whereas multiple colonizations without habitat shifts from adjacent islands were more common in less isolated islands.

Sometimes taxa that are widespread and likely to be more influenced by SS processes might interact with taxa that are more narrowly distributed and constrained by biogeographic features. Leibold et al. (2010) illustrated this possibility in the case of crustacean zooplankton in lakes of the northeastern United States. They applied their method separately to two clades that co-occur in these lakes and compete for the same resources of phytoplankton algae but differ in their historical biogeographic constraints (Fig. 9.5). One group, daphniids, are cladocerans that are more rapid colonists than the other group, calanoid copepods. This disparity occurs largely because calanoids are obligately sexual and subject to substantial Allee effects (Sarnelle and Knapp 2004, Thum 2007), whereas daphniids can undergo numerous generations of asexual reproduction that likely reduce the Allee effects. The distribution of calanoids was strongly predicted by a previously published historical biogeographic hypothesis (Stemberger 1995) and was only weakly related to any of the measured environmental variables (mostly water quality and lake morphology). In contrast, daphniids showed no evidence that their distribution was constrained by the same biogeographic factors; instead they showed strong evidence for environmental filtering.

The community phylogeny approach and its extension to metacommunity and biogeographic processes is clearly still in the early stages of its development. Nevertheless, the findings to date are highly suggestive of a possible synthesis, or at least a reconciliation, that would more directly address Ricklefs's (2008) critique. Such methods are also likely to reveal numerous other possible interactions between metacommunity and biogeographic processes. Although the diagram we used to initially explore the possible domains of the four archetypes for metacommunity ecology is grossly oversimplified (Fig. 2.5), we can try to give these possible interactions between biogeography and metacommunity ecology some context by expanding that figure as shown in Figure 9.6. Just as we identified dispersal events (associated with different degrees of connectivity and spatial scale) in relation to demographic events and PD events, we can also identify the prevalence of biogeographic effects by relating dispersal to the driving events of biogeography (e.g., vicariance or island formation).

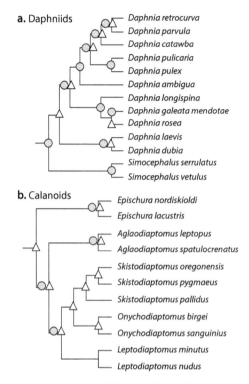

FIGURE 9.5. Contrasting relative effects of biogeography and environmental factors on daphniids (a) and calanoid copepods (b). In each part, solid dots indicate significant association of branching with environmental factors, and open triangles indicate significant association with a priori spatial biogeographic patterns. In daphniids, the large majority of effects are correlated with environmental factors, whereas historical biogeographic effects predominate in calanoids. Modified from Leibold et al. 2010.

9.3 SYNTHESIS: A RESEARCH AGENDA FOR INTEGRATING EVOLUTIONARY AND ECOLOGICAL PROCESSES THAT AFFECT BIODIVERSITY

One of the key features of metacommunity ecology is that it straddles the spatial and temporal scales of biogeography and community ecology (Fig. 9.2). To the degree that this is true, there can be feedback between ecological and evolutionary/biogeographic processes (Mittelbach and Schemske 2015). If so, Ricklefs's (2008) relegation of ecological processes to secondary status may be unwarranted. We have identified some of the ways this feedback could happen and how it could alter our expectations about metacommunity patterns and inform our understanding of biogeographic patterns.

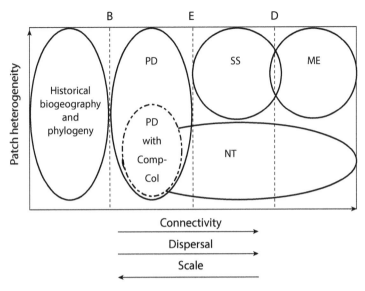

FIGURE 9.6. Modified scheme for metacommunity-historical biogeography links (compare with Fig. 2.5). Here we highlight that phylogeny and historical effects will occur when the connectivity is low compared with the rate of biogeographic events (vicariance events, island formation, etc.). We assume that the rate of biogeographic events (B) is much lower than the background local population extinction rate (E), though this need not always be the case.

The ideas presented here have important antecedents in some of the classic studies of adaptive radiation, especially those occurring on archipelagos, such as Darwin's finches in the Galapagos (Grant and Grant 2014), anoles in the Caribbean (Losos 2009), and several of the radiations in the Hawaiian Islands (Baldwin and Sanderson 1998, Gillespie 2016) and other archipelagos. These studies indicate that niche relations and competition within clades can strongly shape the nature of radiations. Past ideas about the way competition works in these models was, however, probably oversimplified because it was based primarily on assumed mechanisms that focused on the local scale (e.g., if competitors are present, they inhibit a potentially new species arising either from colonization or from in situ cladogenesis). Metacommunity ecology suggests a variety of other, perhaps more appropriate and certainly more comprehensive, mechanisms, such as PD or those involving microevolution in the metacommunity (Chap. 8). The insights that come from metacommunity ecology tell us that it is often going to be the landscape context that matters, including factors such as connectivity, isolation, disturbance, and habitat heterogeneity but perhaps also more subtle or complex processes such as multiple stable equilibria and endpoint assembly cycles (see Chaps. 5 and 11).

Indeed, current work seems to show patterns that correspond well to these predictions. Gillespie has perhaps done more than anyone else to compare and contrast radiations of different clades within a common landscape framework (the Hawaiian archipelago with its defining historical imprint of island creation, tectonic movement, and erosion; see, e.g., Gillespie 2016). Her work shows that the radiation of different clades can range widely and can roughly be categorized into four types: (1) Early-divergence radiation, in which an initial colonist to the islands (some of which don't exist anymore) subsequently radiated into various ecological niches in which descendent taxa retained their ancestors' niches throughout the rest of the history of the radiation (Fig. 9.7b). (2) Nondivergence radiation, in which initial colonists to the islands were distinguished by few ecological differences and remained so as they repeatedly colonized new islands (Fig. 9.7a, c). Although these appear to easily colonize new islands, they do not appear to coexist at local scales within islands. (3) More-complex radiations that incorporate some of each of the two patterns above (Fig. 9.7c). There is a general tendency toward having more recent species on newer islands, but even this isn't completely so. This mixture of patterns is more reminiscent of the pattern for Galapagos land snails shown in Figure 9.4. (4) No radiation, in which a single species is present throughout the archipelago even though it may show interesting patterns of ecotypic polymorphism throughout (not shown). Variation in the shape of these clades involving spiders as well as other taxa seems to be largely influenced by their apparent dispersal abilities but probably also involve other aspects of their biology.

Although one can interpret some of the simpler cases without necessarily taking an explicit metacommunity framework, this may not always be the case, and at some point the argument becomes, at least implicitly, a metacommunity process.

FIGURE 9.7. Biogeography and phylogeny in Hawaiian spiders. In each case the phylogeny is shown on the left and biogeography on the right. The biogeography identifies which island is inhabited by that species; the five major islands of Hawaii are shown in five columns in decreasing age of the island. The niches of the species are not identified in this figure but are discussed in Gillespie (2016). (a) *Orsonwelles* sheet web spiders: each island was apparently colonized by a single ancestor with subsequent radiation within the island. (b) *Mecaphesa* crab spiders: ecological niche radiation happened early on in the phylogeny (probably even before any of the extant islands were formed), and the formation of each island was followed by colonization by all the species with little or no ecological speciation. (c) *Ariamnes* stick spiders: As with *Orsonwelles*, each island appears to have been colonized by a single ancestor with subsequent niche radiation within islands. (d) *Tetragnatha* long-jawed spiders: the pattern is more complex; the early part of the clade is similar to *Orsonwelles* and *Ariamnes*, but the more recent phylogeny has features that resemble *Mecaphesa* but also some others; for example, there appears to have been at least one successful phylogenetic event with a species (*T. kukuiki*) from an older island with an apparent ancestor in a newer island. Modified from Gillespie 2016.

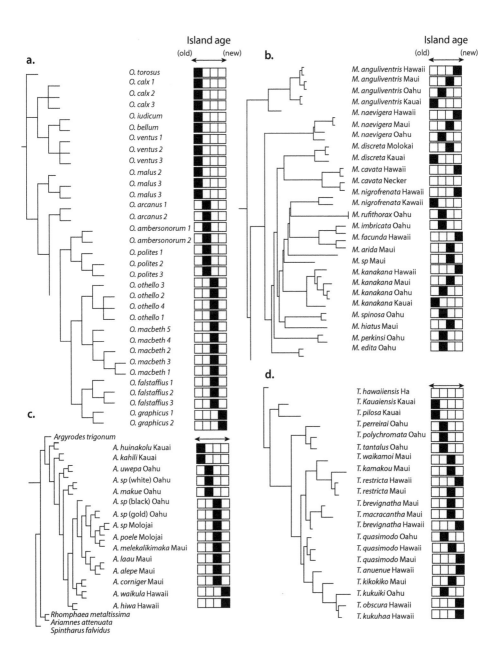

a.

Island age
(old) → (new)

O. torosus
O. calx 1
O. calx 2
O. calx 3
O. iudicum
O. bellum
O. ventus 1
O. ventus 2
O. ventus 3
O. malus 2
O. malus 3
O. malus 3
O. arcanus 1
O. arcanus 2
O. ambersonorum 1
O. ambersonorum 2
O. polites 1
O. polites 2
O. polites 3
O. othello 3
O. othello 2
O. othello 4
O. othello 1
O. macbeth 5
O. macbeth 4
O. macbeth 2
O. macbeth 3
O. macbeth 1
O. falstaffius 1
O. falstaffius 2
O. falstaffius 3
O. graphicus 1
O. graphicus 2

b.

Island age
(old) → (new)

M. anguliventris Hawaii
M. anguliventris Maui
M. anguliventris Oahu
M. anguliventris Kauai
M. naevigera Hawaii
M. naevigera Maui
M. naevigera Oahu
M. discreta Molokai
M. discreta Kauai
M. cavata Hawaii
M. cavata Necker
M. nigrofrenata Hawaii
M. nigrofrenata Kawaii
M. rufithorax Oahu
M. imbricata Oahu
M. facunda Hawaii
M. arida Maui
M. sp Maui
M. kanakana Hawaii
M. kanakana Maui
M. kanakana Oahu
M. kanakana Kauai
M. spinosa Oahu
M. hiatus Maui
M. perkinsi Oahu
M. edita Oahu

c.

Argyrodes trigonum

A. huinakolu Kauai
A. kahili Kauai
A. uwepa Oahu
A. sp (white) Oahu
A. makue Oahu
A. sp (black) Oahu
A. sp (gold) Oahu
A. sp Molojai
A. poele Molojai
A. melekalikimaka Maui
A. laau Maui
A. alepe Maui
A. corniger Maui
A. waikula Hawaii
A. hiwa Hawaii

Rhomphaea metaltissima
Ariamnes attenuata
Spintharus falvidus

d.

T. hawaiiensis Ha
T. Kauaiensis Kauai
T. pilosa Kauai
T. perreirai Oahu
T. polychromata Oahu
T. tantalus Oahu
T. waikamoi Maui
T. kamakou Maui
T. restricta Hawaii
T. restricta Maui
T. brevignatha Maui
T. macracantha Maui
T. brevignatha Hawaii
T. quasimodo Oahu
T. quasimodo Hawaii
T. quasimodo Maui
T. anuenue Hawaii
T. kikokiko Maui
T. kukuiki Oahu
T. obscura Hawaii
T. kukuhaa Hawaii

For example, Ricklefs (2015) argued that the biogeographic distribution of organisms (which in his case was largely focused on birds) can be seen as the equivalent of the Janzen-Connell hypothesis (see Chap. 5) extended to archipelago or continental scales involving, especially, host-pathogen coevolution. His argument is based on eight premises (slightly modified below), most of which converge remarkably with our exposition of metacommunity ecology: (1) Species richness is a property of the region. (2) Species tend to have broadly overlapping ecological (abiotic and resource) requirements. (3) Species can invade suitable environments independently of the presence of potential resource competitors. (4) The distributions and abundances of species are evolutionarily labile. (5) Variations in population size reflect coevolutionary outcomes with pathogens or other specialized antagonists (see Chap. 11). (6) The timescale of population change is long, reflecting the acquisition of rare mutations. (7) Population expansion and contraction drive diversification. (8) Host-pathogen coevolution can affect whole clades, causing clade collapse (okay this one is novel we admit).

We would argue that Ricklefs is thinking like a metacommunity ecologist even if he may not want to admit it!

9.4 CONCLUSIONS

We obviously think metacommunity thinking can bring a lot to the growing synthesis of ecology and evolution/biogeography. First, it provides a scale-appropriate way to think about the role of ecology. Previous and much current work is predicated on rather vague ideas about how community assembly might affect the dispersion of organisms (through changes in ranges, for example) and probably relies excessively on formulations based on local ecological models such as Lotka-Volterra models and Gause's axiom, formulations that are not as axiomatic within a broad metacommunity perspective. Second, metacommunity ecology provides a more scale-appropriate set of ideas, and it highlights novel features of landscapes that can affect the feedback between the biogeographic/evolutionary and ecological drivers of biodiversity and diversification. Critically, metacommunity ecology incorporates the various landscape features that might strongly mediate how these processes, such as isolation, disturbance, and environmental heterogeneity, interact. Third, metacommunity ecology can provide novel tools to facilitate this synthesis, for example, the method of Leibold et al. (2010) and related methods. Finally, metacommunity ecology provides ways of understand the ecological consequences of this interaction. How do such processes affect other matters we might care about, such as the functioning of ecosystems or the resilience of biotas to environmental change at different scales? We discuss some of these possibilities in Chapters 12 and 13.

CHAPTER TEN

The Macroecology of Metacommunities

Prospectus

1. Metacommunity ecology and macroecology share a lot of features because both explore how biodiversity is regulated at different spatial scales. Nevertheless, they differ in that metacommunity ecology is more firmly anchored in population biology and species interactions, whereas macroecology is more strongly based on statistical theory and biophysical constraints.

2. Similar to metacommunity ecology, macroecology also has numerous theories that predict commonly observed patterns, including many of those also addressed by metacommunity ecology. These include the SAD, the SAR, and the distance-decay relationship of β-diversity.

3. One of the most important ways that macroecology merges with metacommunity ecology comes from the former's focus on spatial scale. Instead of starting from the typical two-scale approach of classic metacommunity ecology, macroecology assumes that patterns will be scale dependent in a (semi-) continuous way. This approach provides one of the most powerful ways to make the transition from metacommunity sensu stricto to metacommunity sensu lato that we argued for in Chapter 1.

4. We can merge metacommunity thinking with macroecological approaches in the context of the nested SAR, which can help us to understand the role of metacommunity-level processes in the scaling relationships of island and habitat fragmentation studies.

5. This implicit scaling approach at the interface of macroecology and metacommunity ecology can also help us to understand the patterns and processes of how biodiversity varies along important biogeographic and environmental gradients, such as the species-energy relationship (i.e., how species richness varies with the productivity of ecosystems) and other natural and anthropogenic drivers.

Metacommunity ecology was born out of community ecology. The recognition of a need for a larger spatial perspective, including habitat heterogeneity, dispersal, and scale, led to a number of parallel approaches that explored different aspects of what a metacommunity could look like (including the SS, PD, ME, and NT paradigms) and what sorts of patterns of co-occurrence, relative abundance, and species richness they could predict. However, around the same time as the advent and explosion of modern perspectives on metacommunities, another perspective arose that tried to describe a number of ecological phenomena, often at much larger spatial scales—macroecology (Brown and Maurer 1989, Brown 1995).

As the name might suggest, macroecology deals with patterns of species distributions, abundances, and diversities, as well as other parameters (e.g., functional traits such as body size), typically across large spatial scales. Although the term "macroecology" was coined by Brown and Maurer (1989) and the field defined in subsequent volumes (Brown 1995, Gaston and Blackburn 2000), it has a much older history. For example, some of the questions central to macroecology, such as how species richness varies across productivity and latitudinal gradients, were present in the writings of Darwin (1859) and Wallace (1876), among others. Likewise, explorations of patterns of the scaling of species richness with area (SAR) and the relative commonness and rarity of species (SAD) go back 50–100 years (Arrhenius 1921, Gleason 1926, Fisher et al. 1943, MacArthur 1957, Preston 1960), as do questions of species energetics and body-size distributions (Lotka 1925, Hutchinson 1959), among others.

The modern versions of both metacommunity ecology and macroecology were born from similar concerns that motivated metacommunity ecology: an overemphasis on small-scale processes and limited numbers of species and interactions that defined the experimental revolution in community ecology from the 1970s to the 1990s. And metacommunity ecology and macroecology have common roots. For example, MacArthur and Wilson's (1963, 1967) equilibrium theory of island biogeography forms the basis for modern perspectives of metacommunities that involve mechanisms similar to NT and PD perspectives on metacommunities and also forms the basis for some of macroecology's most fundamental patterns, such as the SAR. Likewise, in "Homage to Santa Rosalia," Hutchinson (1959) used ideas of energetics and interspecific interactions grounded in coexistence theory that forms the basis of the SS perspective of metacommunities, but he used these to describe large-scale patterns of body-size distributions and limits to biodiversity that are more the realm of modern-day macroecological biogeography.

The fields of metacommunity ecology and macroecology have somewhat different foci. Metacommunity ecology has historically been more concentrated on small- to intermediate-scale patterns and the population-level mechanisms that create them, whereas macroecology has focused more on statistical properties at larger regional to global patterns, often combined with physiological principles, and has less of an emphasis on mechanisms (or at least mechanisms as metacommunity ecologists tend to think about them; McGill and Nekola 2010). Nevertheless, there seems to be no standard agreed-upon boundary at which metacommunity ecology stops and macroecology starts. Nor is it clear where other subfields that deal with similar questions (diversity, abundance, distribution) at similarly large scales, such as landscape ecology and biogeography, fit into this schema. And maybe this is OK.

Furthermore, metacommunity approaches have progressively moved toward trying to understand macroecological patterns of species compositional variation and biodiversity (Chisholm and Pacala 2010, Henriques-Silva et al. 2013, Meynard et al. 2013), and macroecological approaches have progressively moved toward the use of more refined data sets that allow a deeper exploration of mechanisms that emerge at the metacommunity level (Gotelli et al. 2009, Passy 2012, Maurer et al. 2013, D'Amen et al. 2015). Thus, the two fields may be merging in many respects. Figure 10.1 presents an analysis of Google Scholar searches for "macroecology" and "metacommunity" for each year from 1992, which was not long after both words were coined in the literature. Three things are evident in the figure. First, both fields rapidly gained interest in the literature, with leaps of hundreds of studies each year. Second, although the concept of macroecology rose faster in popularity, the concept of metacommunity rapidly caught up and was used in a nearly equivalent number of studies by 2016. Third, the numbers of studies that use both "macroecology" and "metacommunity ecology" also rose, from near zero to nearly 20% of studies using either "metacommunity" or "macroecology."

In this chapter, we discuss the patterns and processes that are typically the focus of macroecology, pointing to places of convergence and cross-fertilization with the metacommunity perspective that we have emphasized. Much of this will borrow from discussions and concepts from previous chapters but will be viewed through a different lens (a "macroscope," sensu Brown and Maurer 1989). We first define the history and contemporary macroecological perspective, pointing at overlap with other areas of this book. Next, we show how the metacommunity perspective provides a bridge from the local scale to the regional/global one. Here, it is also useful to emphasize that the link between macroecology and metacommunity ecology is probably much more easily envisioned using the sensu lato perspective of metacommunities that we described in Chapter 1, in which spatial scale is considered to be more of a gradient from smaller to larger, rather than the sensu stricto perspective, in which local and regional scales are thought to be more discretely defined. We identify several areas where macroecological perspectives can enhance the predictability of metacommunity ecology and areas where metacommunity perspectives can enhance the mechanistic understanding of macroecology.

10.1 WHAT IS MACROECOLOGY?

Brown and Maurer (1989) defined macroecology more as a "perspective" on large-scale patterns and processes, rather than as a truly separate subdiscipline. They were seeking generality and synthesis in patterns of species abundances, traits (e.g.,

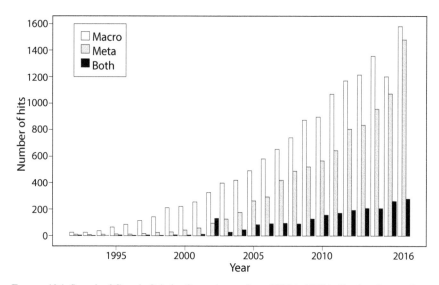

FIGURE 10.1. Search of Google Scholar for each year from 1992 to 2016 indicating the numbers of article hits for the terms "macroecology," "metacommunity," or both "macroecology" and "metacommunity."

body size), and range sizes, as well as the factors that emerged from these patterns (e.g., measures of biodiversity), and they used a basic underlying principle (i.e., mechanism) relevant to all organisms—energetic requirements—to understand those patterns. Arguably then, macroecology contrasts with metacommunity ecology (and other forms of community ecology) by being a "pattern-first" approach rather than a "process-first" approach. It is possible to disagree with this focus (Roughgarden 2009, Vellend 2010, 2016), but maybe reconciliation is a better option.

Brown (1995) extended the perspective to state that macroecology studied the "statistical patterns of abundance, distribution and diversity" (10), typically with underlying mechanisms that included fundamental principles of physiology (e.g., thermodynamics) and statistics (e.g., null expectations). Other important early treatises on macroecology (Maurer 1999, Gaston and Blackburn 2000) emphasized the multivariate nature of macroecology, suggesting it was at the "intersection of several other fields of biology, including ecology, biogeography and macroevolution" (Gaston and Blackburn 2000, 16). The macroecological perspective has also been applied to understand patterns of diversity and abundance from one or a few regionally confined plots (Green et al. 2003, Volkov et al. 2003, Supp et al. 2012, May et al. 2016) to global patterns (Gaston and Blackburn 2000, Currie et al. 2004, Kreft et al. 2008). Given its synthetic nature and its birth from among the intersection of different fields, it is perhaps not surprising that today, where macroecology starts and stops relative to related subdisciplines is difficult to pinpoint.

For convenience and ease of mapping onto our metacommunity construct, we here divide macroecological approaches into four conceptual realms, or "flavors" (albeit with some overlap), which are illustrated in the cartoon in Figure 10.2. These flavors include *biodiversity macroecology*, *contemporary biogeography*, *historical biogeography*, and *metabolic and functional macroecology*, each of which we discuss in more detail next.

1. *Biodiversity macroecology.* In this subfield, the focus is on questions about relative abundances (commonness and rarity of species) and spatial distributions of species (their range sizes and inter- and intraspecific clumping relationships). This subfield includes basic diversity partitioning (α-, β-, γ-diversity), as well as statistical patterns of diversity scaling and relative abundances that we will describe below. In Chapters 4 and 5, we noted that despite many attempts to use these sorts of macroecological patterns as a diagnostic for underlying metacommunity processes, such as neutral versus niche theory (Hubbell 2001, McGill 2003, Volkov et al. 2003, Chisholm and Pacala 2010), these patterns in and of themselves are not particularly diagnostic of which metacommunity mechanisms were operating (e.g., SS vs. NT) (Chave et al. 2002, Wilson et al. 2003, Chase et al. 2005). Nevertheless, regularities and variability in the shapes of these patterns, especially when combined with other information (e.g., environmental variability), can provide important information regarding the underlying processes creating them.

2. *Contemporary biogeography.* The focus of this subfield is on variation in the abundance and distribution of species from place to place. One aspect of this subfield is the patterns of the co-distribution of species across space, such as the checkerboard distributions, gradient replacements, and nestedness patterns we discussed in Chapter 4 and thus will largely ignore here. The other aspect considers how patterns of species composition and diversity vary from place to place due to some ecological/environmental variation along those places. It also includes patterns along ecological gradients, such as the island species-area relationship (ISAR, which is different from the nested SAR, discussed below) and the relationship between habitat isolation and diversity/abundance (Triantis et al. 2012), which emerge from the equilibrium theory of island biogeography (MacArthur and Wilson 1963, 1967) and other metacommunity models (Hanski and Gyllenberg 1997, Mouquet and Loreau 2003). It also includes

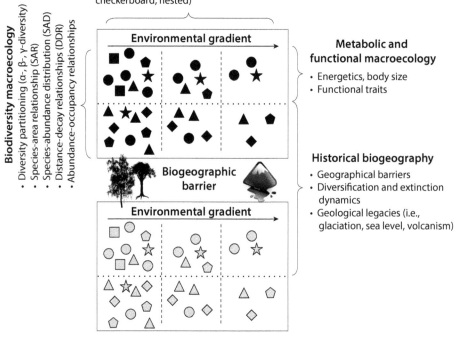

Contemporary biogeography
- Island biogeography (colonization, extinction)
- Diversity gradients (productivity, disturbance)
- Species associations and distributions (e.g., checkerboard, nested)

Biodiversity macroecology
- Diversity partitioning (α-, β-, γ-diversity)
- Species-area relationship (SAR)
- Species-abundance distribution (SAD)
- Distance-decay relationships (DDR)
- Abundance-occupancy relationships

Environmental gradient

Biogeographic barrier

Environmental gradient

Metabolic and functional macroecology
- Energetics, body size
- Functional traits

Historical biogeography
- Geographical barriers
- Diversification and extinction dynamics
- Geological legacies (i.e., glaciation, sea level, volcanism)

FIGURE 10.2. The many different "flavors" of macroecology that are prevalent in the literature mapped onto metacommunity-related concepts and their place in this book. In each locality (bordered by dashed lines) there are different numbers and types of species (denoted by different shapes). Within a locality or across localities, different kinds of biodiversity macroecology can be estimated, such as the species area relationship, diversity partitioning, or the species abundance distribution, among others. There are also more detailed aspects of the species, such as metabolic rates, body sizes, or functional traits—metabolic and functional macroecology. From left to right, under "contemporary biogeography," an environmental gradient is depicted (e.g., differences in energy), in which differences in the abundances, diversities, co-occurrences, and other biodiversity macroecology variables can be compared along the gradient. Finally, "historical biogeography" encompasses work at larger spatial scales, such as those where biogeographic boundaries (e.g., islands or even continents) are crossed, and the processes of dispersal limitation, geological history, and macroevolutionary processes play a larger role.

patterns along environmental gradients, such as the important but controversial relationships between energy (or productivity) and species diversity
(Wright 1983, Currie 1991, Rosenzweig and Abramsky 1993, Mittelbach
et al. 2001, Evans et al. 2005, Adler et al. 2011, Fraser et al. 2015), disturbance and species diversity (Connell 1978, Mackey and Currie 2001,
Svensson et al. 2012), and heterogeneity and species diversity (Stein
et al. 2014). Importantly, several patterns of interest probably result from a
combination of contemporary and historical factors, such as the latitudinal
biodiversity gradient (Willig et al. 2003, Hillebrand 2004, Mittelbach et al.
2007).

3. *Historical biogeography.* This subfield includes considerations of the
influence of a number of historical factors and how they influence
modern-day patterns of species abundance and distribution and the resulting patterns of biodiversity (Ricklefs 2004, 2015). For example, historical
factors such as sea-level change, plate tectonics, and volcanism, as well as
variation in speciation rates, have a strong influence on the diversity of
many islands independent of their current size and isolation (MacArthur
and Wilson 1963, Ricklefs and Lovette 1999, Losos and Schluter 2000,
Lomolino and Weiser 2001, Whittaker et al. 2007). Likewise, past
climates and variation in diversification rates can strongly influence
patterns of species diversity in habitats independent of their environmental
conditions (Wiens and Donoghue 2004, Jetz and Fine 2012, Harmon and
Harrison 2015). These historical factors can create patterns sometimes
known as *diversity anomalies*, whereby regions that may be otherwise
similar in environmental conditions are nevertheless greatly divergent in
the richness of their species pools, such as the differences between
temperate-tree richness in Asia (highest), North America (intermediate),
and Europe (lowest) (Ricklefs et al. 1999, Ricklefs and He 2016, Swenson
et al. 2016), or the highly diverse corals and coral reef fishes in the "coral
triangle" compared with areas outside of this region (Karlson et al. 2004,
Parravicini et al. 2013).

4. *Metabolic and functional macroecology.* This subfield includes patterns
and hypothesized processes that shape (1) distributions of traits within and
among local communities (e.g., shapes of body-size distributions, co-
occurrences of functional traits) and (2) distributions of traits along
environmental gradients (e.g., patterns of body size or leaf traits along
broad climatic gradients). Functional macroecology has tended to focus on
how species' functional traits (rather than their Latin binomials) might
influence patterns of species co-occurrences and distributions along
environmental gradients (McGill et al. 2006b) and is often conceptually

linked with phylogenetic analyses, assuming phylogenetic similarity is proportional to functional similarity (Webb et al. 2002, Cavender-Bares et al. 2009, Mouquet et al. 2012). We discussed this approach in Chapter 6. Here we focus more on the metabolic theory of macroecology (West et al. 1997, Brown et al. 2004), which is more predictive and uses fundamental physiological mechanisms to unify not only ecological features of systems (e.g., body size and abundance distributions, diversity relationships) but also a number of other patterns in biology—rates of molecular evolution (Gillooly et al. 2005), cell size and number (Savage et al. 2007), and the processes of ecosystems (Schramski et al. 2015). In terms of predictions regarding biodiversity, this is somewhat of a hybrid approach, whereby rates of molecular evolution and diversification (historical biogeography) are predicted to create gradients of variation in biodiversity, but these are strongly influenced by external environmental conditions (i.e., temperature) that modulate metabolic rates (contemporary biogeography) (Allen et al. 2002, Stegen et al. 2009, Brown 2014).

In this chapter, we will primarily focus on the first two flavors of macroecology—biodiversity and contemporary biogeography. For historical biogeography, Chapter 9 has already provided an overview of a number of ways that the metacommunity perspective could enhance our understanding of some of these historical processes, as well as some patterns in the distributions of species. We will, however, briefly discuss ways in which the metacommunity perspective can help us understand some of the historical biogeographic patterns that emerge. Likewise, we addressed many of the topics related to metabolic and functional macroecology in Chapter 6, although we will discuss a few other aspects of how such traits relate to a macroecological perspective.

10.2 SYNTHESIZING BIODIVERSITY MACROECOLOGY PROCESSES AND PATTERNS

The crux of the metacommunity perspective is the traditional α-, β-, and γ- partitioning of diversity, where α-diversity represents the numbers of species at the smaller scale (locality), γ-diversity represents the numbers of species at the larger scale (region), and β-diversity is an index of the site-to-site turnover of species linking the local to the regional. The subfield of macroecology that is focused on biodiversity also examines this partitioning, but not surprisingly, with a stronger emphasis on β- and γ-diversity. Importantly, however, "local" and "regional" scales are often defined quite differently, and sometimes rather arbitrarily, both within and among metacommunity and macroecology studies.

α-diversity in trees, for example, has been defined at anywhere from less than 0.5 ha (Currie et al. 2004, Kraft et al. 2011, Simova et al. 2011) to 1 ha (Condit et al. 2002, Ter Steege et al. 2003) to 20–50 ha (Ricklefs 2015, Ricklefs and He 2016) or to even entire states and provinces in North America (from 250,000 ha to several million ha; Qian and Ricklefs 2007). It is not surprising, then, that different definitions of just what α- or γ-diversity (and thus β-diversity) are can lead to dramatically different conclusions about the importance of different processes. For example, with trees, studies that treat α-diversity on smaller scales tend to find a greater influence of climate on patterns of biodiversity (Currie et al. 2004), whereas studies that treat α-diversity at larger scales often find a bigger role for historical and biogeographic processes (Ricklefs 2015, Ricklefs and He 2016). A more explicit consideration of these scaling relationships with an explicit metacommunity-scale perspective will go a long way toward synthesizing the relative importance of biogeographic processes with macroecological processes (Chase and Myers 2011).

In Chapter 4 we pointed out that initial empirical tests of some of the key metacommunity archetypes—specifically NT and its contrast with various forms of niche theory—were based on what are largely considered to be macroecological patterns, such as the shape of the SAD and SAR (Hubbell 2001, McGill 2003, Volkov et al. 2003). We concluded that these macroecological patterns were not particularly informative for distinguishing the relative importance of the metacommunity assembly processes, at least without further information (Hubbell 2001, Chave et al. 2002, Wilson et al. 2003). On an ever more general level, McGill (2010) recently synthesized a number of theories intended to make unified predictions of a number of macroecological ecological patterns and found that they each share a common core set of principles that can lead to a number of patterns observed in natural communities.

Figure 10.3 summarizes McGill's (2010) arguments by noting that each of six proposed "unified" theories of macroecology share three fundamental assumptions regarding the properties of biodiversity in communities, and that when these three properties are all at play, they can in turn produce most of the macroecological patterns of interest. Three of the unified theories reviewed by McGill (2010) are based more on statistical mechanics (e.g., maximum entropy, generalized fractal, clustered Poisson), which can be considered mechanisms in the sense that they create the patterns observed (McGill and Nekola 2010) but are not necessarily the types of mechanisms typically considered at the metacommunity scale, which are based more on ecological (rather than statistical) principles. The other three theories are much more closely aligned (or identical) to three of the four metacommunity archetypes on which we have focused in this book. Specifically, McGill (2010) described (1) Hubbell's (2001) unified NT; (2) Hanski and Gyllenberg's (1997) metapopulation model, which is a form of PD with no (or weak) species interactions; and (3) Gauch and Whittaker's (1972) formulation of *continuum*

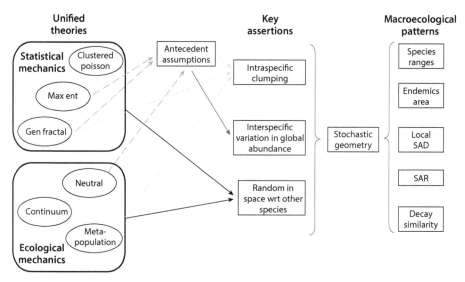

FIGURE 10.3. Conceptual overview and synthesis of theories that have purported to "unify" biodiversity macroecological variables. The left-hand column represents a list of six different types of theories—three based on statistical mechanics (clustered Poisson, maximum entropy, and generalized fractal), and three that are more process based and more closely aligned with metacommunity models: The continuum model, similar to species sorting; the metapopulation model, similar to patch dynamics; and the neutral model, identical to neutral theory. In each case, the models make three key assertions (though sometimes for different reasons)—that species are intraspecifically clumped, that there is variation among species in their abundances, and that species are distributed randomly with respect to each other at large scales. While some of the models make antecedent assumptions that lead to these assertions, others just make the assertions. Finally, each of these assertions generates the five biodiversity macroecology patterns through stochastic geometry (SAD = species abundance distribution; SAR = species area relationship). Modified from McGill 2010.

theory, which is essentially a SS model based on Gleason (1926) and others' arguments that species distributions along gradients are essentially independent of one another.

McGill (2010) argued that each of the unified theories either assumes—or predicts—three fundamental properties of biodiversity in communities. First, there is variation in the abundance (both locally and globally) of species, leading to a "hollow-shaped" SAD (see details below). Second, species are intraspecifically clumped; that is, individuals of any given species are more likely to be found nearer to one another than to individuals of other species on average, which is a commonly expected outcome when there is dispersal limitation or SS into favored habitats. This clumping leads to distance-decay relationships in community composition (communities that are closer together are more similar to one another in species composition than those that are further apart; e.g., Nekola and White 1999, Soininen et al. 2007). Third, species distributions are random with respect to other species. This assumption is perhaps the most difficult to reconcile with the mechanisms of metacommunity assembly (other than NT) but may still be true at larger spatial scales because the effects of interspecific interactions on species distributions may often be much less important than habitat preferences/tolerances and dispersal limitations (McGill 2010). Indeed, this is the one assumption that may not always hold; for example, May et al. (2016) showed that at the relatively small spatial scales of the 50-ha BCI forest plot, species did not appear to be random with respect to their distributions relative to other species (because of distinct habitat preferences and interspecific interactions).

In addition to identifying the core assumptions that seem to be at the heart of most macroecological theories of biodiversity, McGill (2010, 2011) also noted that these assumptions and emergent abundance and distribution of species can lead to the majority of biodiversity macroecology patterns of interest. There are at least fifteen such general biodiversity macroecology patterns, which can be roughly categorized into patterns at the α-scale (e.g., the SAD, rarefaction curve), the γ-scale (e.g., abundance-occupancy relationships), and the β-scale (e.g., distance-decay relationships). Here, we overview four of the most commonly studied macroecological biodiversity relationships and their connections to metacommunity ecology.

1. *The species-abundance distribution (SAD).* A "hollow-shaped" SAD is a fundamental property of communities: a few species are very common, and many species are rare, in any given community (Fisher et al. 1943, Preston 1960, McGill et al. 2007). As we discussed in Chapter 4, the SAD shape itself is not particularly diagnostic of the mechanisms of metacommunity assembly (Chave et al. 2002, Wilson et al. 2003, May et al. 2015). Nevertheless, the SAD, and its distribution, is a critical macroecological parameter, the curvature of which indicates how even (or uneven) the species in a given community are (albeit unlikely, a completely even community would have a flat SAD) and can be strongly influenced by mechanisms that allow coexistence in the metacommunity.

2. *The species-area relationship (SAR).* Here, we will generally ignore issues about the whether there is a "best" equation for describing the SAR and discuss only the two main types of patterns—*nested SARs*, which we discuss here and in Section 10.6, and *island SARs* (ISARs), which we discuss in Section 10.7. The nested SAR, which is often (but not always) depicted as a power law (e.g., $S = cA^z$, where S is species richness, A is

area, c is the intercept, and z is the scaling parameter), is one of the most universal patterns in ecology (Lomolino 2000, Scheiner et al. 2011).[1] The nested SAR is also strongly influenced by metacommunity assembly processes, including those that determine the coexistence of species locally (by influencing the intercept, c, and thus is related to α-diversity) and those that determine the variation in species composition from site to site (influencing the slope, z, of the SAR and thus is related to β-diversity).

3. *The endemics-area relationship (EAR).* Although not as old (or well-known) as the SAR, we discuss the EAR next because it is conceptually the inverse of the SAR. The EAR depicts the numbers of species that are endemic to a given sampling area (Harte and Kinzig 1997). At the largest extent examined, all of the species are endemic to that area, and as the area is reduced, there are fewer and fewer species endemic to the next smallest area. One useful application of the EAR is to predict how many species might be lost if a certain amount of habitat is lost (Kinzig and Harte 2000, He and Hubbell 2011). An important observation is that because species are typically aggregated or clumped in space (e.g., due to differential SS or dispersal limitation), the expected extinctions based on the EAR is not the same as one might predict using the inverse of the SAR, which was often previously used to predict species extinctions (Pimm et al. 1995). There are, however, many caveats to the specific use of the EAR to predict species extinctions, including the role of extinction debt and dynamic processes (Halley et al. 2013, Rybicki and Hanski 2013, Kitzes and Harte 2014, Matias et al. 2014) and the geometry of habitat loss (Pereira et al. 2012, Keil et al. 2015). Nevertheless, the EAR is a useful tool to describe and compare biodiversity patterns and varies as a result of metacommunity assembly processes in a manner similar to that for SARs described above.

4. *The distance-decay relationship (DDR).* The DDR is a measure of site-to-site variation in species composition (i.e., β-diversity); communities more distant from one another spatially are expected to be more distinct from one another compositionally. When examined at relatively large spatial scales, the DDR is typically strong, so communities that are further from one another in geographic distance are also further from one another in species composition (Nekola and White 1999, Morlon et al. 2008). This occurs because there is more dispersal limitation with

---

[1] We do not make the distinction between sampling curves (sometimes called collector's curves or species accumulation curves) and species-area curves; even though they can differ, this is more of a question of sampling and estimation rather than a conceptual one.

increasing distance, as well as the likelihood that sites more distant
geographically will also be more distant environmentally. At smaller
scales, the specific shape of the DDR can be more variable and will
depend both on the degree of habitat heterogeneity, as well as on the
relative importance of dispersal limitation versus SS experienced by
the metacommunity (Soininen et al. 2007).

Importantly, most of these biodiversity macroecology patterns are mathemati-
cally linked to one another (as well as the standard α-, β-, γ-diversity partition of
metacommunity ecology; Crist and Veech 2006) and can be derived from auto-
correlated sampling (McGill 2010, 2011). For example, the shape of the SAR is
directly determined by the shape of the SAD (i.e., how even or uneven the species
are in their relative abundances), as well as the degree to which species are spatially clumped (i.e., an indicator of β-diversity) (Preston 1962, May 1975, He and
Legendre 2002, Azaele et al. 2015). Figure 10.4 illustrates how some of the most
important different biodiversity macroecology patterns—the SAR, EAR, SAD,
and DDR—can be derived from the same landscape.

10.3 DISSECTING BIODIVERSITY MACROECOLOGY PATTERNS

The main biodiversity macroecology patterns described above and elsewhere (e.g.,
McGill 2011) are largely determined by four fundamental components underlying
the diversity and composition within any metacommunity—the total number of
species in the metacommunity (S), the numbers of individuals of all species in a
given area (N), the relative abundance of individuals in the entire metacommunity
(SAD), and the spatial aggregation (clumping) of individuals across the metacommunity (He and Legendre 2002, McGill 2010, 2011, Chase and Knight 2013). The
metacommunity processes that we have described throughout this book, such as
habitat sorting and dispersal limitation, are critical for determining these properties
and thus lie at the heart of understanding biodiversity macroecological patterns.

If any metacommunity-level process influences one or more of these underlying components of species richness—for example, if the relative commonness and
rarity of species changes as a result of changing amounts of habitat heterogeneity
or dispersal—then such influence will have a cascading effect on all of the
biodiversity macroecology patterns (e.g., SAR, EAR, SAD, DDR, etc.). Figure 10.5
illustrates how changing any one of the four underlying components (S, N, SAD,
aggregation) will alter the shape of the SAR (though we could also show how it in
turn influences the other biodiversity macroecology variables accordingly (Chase
and Knight 2013).

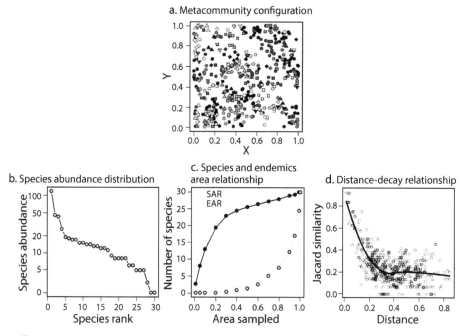

FIGURE 10.4. (a) A hypothetical group of individuals in which each point is an individual (e.g., in a forest plot) and shapes/shades indicate different species (e.g., individuals of the same shape/shade are the same species; 30 species in total); individuals are distributed in a spatially explicit landscape where the x and y coordinates are arbitrary spatial dimensions. (b) The species abundance distribution (SAD) depicting the ranking of species and their abundances. (c) The species-area relationship (SAR) shows the numbers of species as the sampled area increases, while the endemics-area relationship (EAR) shows the numbers of species lost as the sampled area decreases. (d) The distance-decay relationship (DDR) shows the similarity (measured as Jaccard's index) between two communities (within grids in this case) and how it declines with increasing distance among the grids.

10.3.1 The Influence of the Total Number of Species in the Metacommunity

The total number of species (S) that can persist in a metacommunity (i.e., γ-diversity) has many determinants. First, of course, are the macroevolutionary, biogeographic, and historical features of the regional species pool (see Chap. 9). Second, the number of species that can coexist regionally in a metacommunity results from environmental features of the metacommunity that influence species coexistence (e.g., the number of limiting factors and its spatial heterogeneity, as expected from SS) as well as spatial and species dispersal within that metacommunity (i.e., dispersal and colonization-extinction dynamics, as expected from PD and NT).

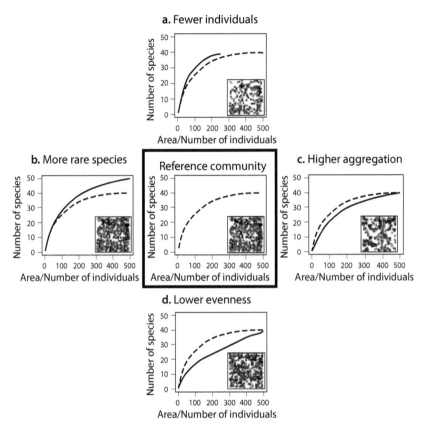

FIGURE 10.5. How changes to the four fundamental features of biodiversity influence the species-area relationship (SAR). The center panel illustrates a typical reference SAR that could be calculated from the inset landscape (similar to that shown in Fig. 10.4). This scenario has a species pool of 40, with 500 total individuals, a coefficient of variation of 1.0, and a mean cluster size (aggregation) of 0.1. Each of the outer four panels represents the case in which one of the four factors was changed, holding the others largely constant. (a) The total number of individuals (but not their relative abundance) was reduced by half. (b) The numbers of species was increased (by increasing the numbers of rare species with minimal effect on the SAD, by randomly replacing 10 individuals with rare species, so that the number of species is 50. (c) Aggregation was changed by increasing clustering size to 0.2. (d) The evenness was reduced by increasing the variation in abundance among species to 3.0.

From a simple SS perspective, a more heterogeneous metacommunity provides more possible opportunities for species to persist than a less heterogeneous metacommunity. And as discussed in Chapter 3, there is ample evidence that metacommunity heterogeneity has a strong influence on the numbers of species that can live there (Tews et al. 2004, Stein et al. 2014). However, it is also possible that when the area of a given metacommunity is finite (i.e., an island or patch), increased heterogeneity can eventually lead to fewer species (especially habitat specialists) because there is too little of each habitat available (Kadmon and Allouche 2007, Allouche et al. 2012).

In addition to habitat heterogeneity, a number of other spatial processes can influence the total number of species that can live in a metacommunity. These include dispersal limitation and ecological drift (as in NT), colonization-competition trade-offs (as in PD), and spatially mediated frequency dependence (such as Janzen-Connell effects or positive frequency dependence; see Chaps. 2 and 3). Importantly, just as with habitat heterogeneity, it is important to recognize that these spatial processes do not always have a unidirectional influence on the numbers of species in the metacommunity. For example, some moderate amount of dispersal is necessary in order for a number of these spatial processes to be able to enhance the numbers of species persisting in the metacommunity. However, if dispersal rates become too high, the whole metacommunity can become "homogenized," reducing the number of species that can persist at the metacommunity level (Mouquet and Loreau 2003).

10.3.2 The Influence of the Total Number of Individuals

The total numbers of individuals (N) in a metacommunity is determined by the spatial extent of that metacommunity as well as by the balance of species birth and death rates, and N will change from site to site when processes that influence birth and death rates fluctuate (e.g., productivity, disturbance). For organisms whose body size is variable, changes in the size distribution can also influence the total number of individuals (particularly when sampling areas are small). For example, when fertilized, many individual plants will grow quite large (regardless of species identity), and diversity may decline simply because the numbers of individuals decline (Oksanen 1996). Likewise, in disturbed areas such as forest gaps, large increases in the density of small trees can lead to greater diversity (Hubbell et al. 1999), while self-thinning as individuals grow can lead to fewer individuals and less diversity (Denslow 1995).

Unfortunately, although changes in the numbers of individuals can have a large influence on macroecological patterns, the main metacommunity theories we have outlined in this book are largely agnostic about processes that influence the numbers of individuals in a given site, often assuming species are simply present or absent (as in PD models) or that there is a zero-sum constraint (as in NT) explicitly constraining variation in the numbers of individuals.[2] Nevertheless, variation

---

[2] Some extensions of the NT, however, allow birth and death rates to vary with ecological drivers and are not "zero sum," but rather are neutral in the sense that all species' relative fitnesses are the same (Passy 2012).

in the numbers of individuals among localities within a metacommunity can have a dramatic effect on the shapes of the SAR, EAR, and DDR, even if there is no change in the shape of the SAD (i.e., if all species differ in exactly the same proportion). This is why it is often useful to use rarefaction (controlling for differences in the numbers of individuals) in order to determine whether any variation in biodiversity parameters among localities results from differences in the total numbers of individuals (in which case rarefied richness should not differ) or from the relative numbers of individuals (in which case rarefied richness should differ; Gotelli and Colwell 2001).

Although there are probably few drivers that change only the total abundances of individuals, with no disproportional effect altering the shape of the SAD (i.e., species respond neutrally to the driver), examples could include varying the amounts of energy and frequency of disturbances, which could hypothetically explain some of the patterns observed in nature. For example, the species-energy theory (Wright 1983) predicts that regions with higher energy inputs will have more species because energy will increase the total numbers of individuals that can be supported in that area and thus increase the number of species. This explanation has been termed the *more individuals hypothesis* (MIH; Srivastava and Lawton 1998). There has been some support for the MIH in describing some of the variation in species richness with energy inputs and scale (Evans et al. 2006, 2008; Yee and Juliano 2007; Beck et al. 2011; Schuler et al. 2015), but there is much more evidence indicating that often species respond quite disproportionately to changing resource availability (i.e., changing the shape of the SAD) (Srivastava and Lawton 1998, Currie et al. 2004, Hurlbert 2004, Hurlbert and Jetz 2010, Simova et al. 2011).

Ideas related to the MIH have also been examined with regard to how disturbances may influence patterns of diversity. For example, Hubbell et al. (1999) suggested that the responses of species in tree-fall gaps at BCI were largely neutral. Specifically, while they found higher richness in tree-fall gaps, they also found many more (small-stemmed) individuals in recently formed gaps than in areas with older (large-stemmed) trees, such that there was no influence of disturbance on species richness beyond its influence on the numbers of individuals (Hubbell et al. 1999). Importantly, this neutrality of the response of forest trees to disturbance does not mean the entire community of plants responds neutrally to disturbances. Schnitzer and Carson (2001) confirmed Hubbell et al.'s (1999) result that there were no differences in rarefied tree-species richness in tree-fall gaps and non-gaps on BCI, but that there was a large difference for the diversity of lianas, which are typically unmeasured in these forest plots but make up nearly 30% of the woody species at BCI. And of course, there are dozens, if not hundreds, of examples in which species respond disproportionately to disturbances due to trade-offs, leading to a shift in the SAD (e.g., Svensson et al. 2012), suggesting that there is typically more to the influence of disturbance on species diversity than simply a change in numbers of individuals.

10.3.3 The Influence of the Relative Abundance of Individuals

The SAD is an empirical pattern that has a universal property—a hollow-shaped curve indicating that all communities consist of a few common and many rare species. Nevertheless variation in the shape of the curve depends on features of the metacommunity that influence species dominance and evenness.

When metacommunities differ in important environmental or spatial properties, different species will be disproportionately influenced by these changes, and such disproportions often alter the shape of the SAD among metacommunities (i.e., more common or more rare species, changing evenness). Changes in the shape of the SAD will in turn influence the shapes of the SAR, EAR, and DDR among metacommunities. The vast majority of mechanisms that are examined under the guise of coexistence theory (e.g., Chesson 2000) are expected to influence the relative abundances of species; for example, when there are few processes that allow species to coexist, we would expect one or a few species to dominate, whereas when coexistence is possible through a variety of mechanisms, we would expect species to be more evenly distributed throughout the SAD. Thus, if some ecological driver, like disturbance or energy, has a disproportional influence on some species compared with others (i.e., they are nonneutral in their response), this can change the shape of the SAD and thus the SAR. There are, of course, multitudes of examples in metacommunities that would lead to these disproportionate effects (far too many to discuss in depth here), including reductions in habitat heterogeneity (Hortal et al. 2009, Stein et al. 2014), reductions in dispersal rates or habitat connectance (Cadotte 2006, Haddad et al. 2015), decreases or increases in disturbance rates (Svensson et al. 2012), shifts toward highly skewed resource ratios (Harpole and Tilman 2007, De Schrijver et al. 2011), and changes to food-web structure (e.g., loss or gain of herbivores or predators; Proulx and Mazumder 1998, Hillebrand et al. 2007).

It is possible, however, for metacommunities to differ in composition with no concomitant change in the shape of the SAD (or other macroecological patterns). For example, Supp et al. (2012) examined how the long-term removal of rodent seed predators altered a variety of macroecological patterns of the annual plant community in in the Chihuahuan Desert near Portal, Arizona. Although rodent removals caused a large shift in the composition of the plant communities, the shape of the SAD and other macroecological patterns within a given replicate of the treatment was largely unchanged in most of their analyses (except in a few circumstances in which the total number of individuals or species was also strongly influenced).

10.3.4 The Influence of the Intraspecific Aggregation (Clumping) of Species

The last component that influences the shape of the SAR is the degree to which species are intraspecifically aggregated (clumped) within the metacommunity. Most of the metacommunity processes that we have discussed in this book have an influence on the aggregation of species. Intraspecific aggregation is a key component that creates β-diversity in a metacommunity. For example, ecological drift and dispersal limitation will create landscapes with nonrandom distributions (aggregation) of species (Condit et al. 2002, Seidler and Plotkin 2006, May et al. 2015), and spatial frequency-dependent processes or priority effects can lead to similar patterns by favoring local clumping (Molofsky and Bever 2002, Chase 2007, 2010). Likewise, species sorting into heterogeneous habitats will create clumping of species in the metacommunity, where species are more prevalent in areas where environmental conditions favor their dominance (Cottenie 2005, Soininen 2014). In contrast, high rates of dispersal that produce source-sink relations, as in ME, can reduce the degree of intraspecific clumping through spatial homogenization of the metacommunity (Mouquet and Loreau 2003, Grainger and Gilbert 2016, Soininen 2016).

10.3.5 Interactions between the Components

While total abundance, relative abundance, and spatial aggregations all have predictable influences on the shape of the SAR, it is likely that more than one of these factors will differ whenever we compare two or more metacommunities, making the comparisons a bit more challenging (Chase and Knight 2013). For example, changes in the supply of resources, rates of disturbances, or prevalence of predators will often affect the total number of individuals in a given metacommunity as well as act disproportionately on some species relative to others (unless they are completely neutral), and thus also change the shape of the SAD. Luckily, when data are available on the abundances of individuals (rather than just on their presence or absence), it is relatively straightforward to control for differences in the shapes of the SARs that result from changes in total abundances by examining the accumulation of species as the numbers of individuals sampled increases, rather than as the total area sampled increases. When there is no aggregation, the resulting data plot is sometimes called the *individual-based rarefaction curve*, but the shape will differ if there is aggregation in the distribution of species with area, the plot of which is sometimes called the *sample-based rarefaction curve* (Gotelli and Colwell 2001). Comparing these rarefaction curves, rather than the full SAR, can allow us to determine whether there are differences in the shape of the SAR that go beyond simply changes in the number of individuals (i.e., changes in the relative abundance or aggregation of the system).

Disentangling variation among metacommunities due to differences in the shape of the SAD and the degree of aggregation in space is a bit more difficult. In Figure 10.5, we altered the SAD without changing aggregation and also altered aggregation without changing the SAD; but these are quite contrived scenarios. First, some degree of aggregation is a ubiquitous feature of metacommunities (Martin and Goldenfeld 2006, McGill 2010, He and Hubbell 2011). Second, the degree to which it is possible for species to aggregate spatially will depend on the overall shape of the SAD. Finally, when some process influences the ability of species to coexist in a metacommunity, such as the degree of habitat heterogeneity or dispersal limitation, it will influence both their spatial distributions (and intraspecific aggregations) as well as their SAD in the metacommunity. Thus, aggregation and the shape of the SAD are entangled and will covary among metacommunities. Nevertheless, overall patterns of diversity and its scaling in a metacommunity (i.e., the SAR) can be evaluated even when the relative influence of SAD and aggregation cannot be disentangled.

10.4 THE ROLE OF METACOMMUNITY ASSEMBLY PROCESSES IN BIODIVERSITY MACROECOLOGY PATTERNS

Many types of macroecology from Figure 10.1 are seemingly agnostic about mechanism and focus instead on using statistical approaches to describe larger-scale patterns. However, as macroecology has matured and diversified, the search for mechanisms has become more prominent (Smith et al. 2008, McGill and Nekola 2010, Beck et al. 2012). Some of these mechanisms may seem largely outside of the realm of biology, instead emphasizing statistical expectations and constraints, such as sampling theory (Etienne and Alonso 2005, Green and Plotkin 2007, Borda-de-Agua et al. 2012), the central limit theorem (Šizling et al. 2009), and maximum entropy (Harte and Newman 2014). Likewise, in so far as we can call the NT a metacommunity assembly process, its simplifying assumptions (i.e., all species demographically equivalent, no spatial heterogeneity) allow a number of macroecological patterns to emerge simply as a result of dispersal limitation and ecological drift, primarily due to statistical effects.

When these macroecological expectations are combined with metacommunity assembly processes, a great deal of insight can be gained. For example, statistical theories such as maximum entropy (MaxEnt) often do well when predicting one or a few patterns from natural communities (Harte et al. 2008, 2009). However, when multiple predictions are tested simultaneously with nature observations, one or more patterns often deviate from the MaxEnt predictions (Newman et al. 2014, McGlinn et al. 2015, Xiao et al. 2015). For example, McGlinn et al. (2015) showed how patterns from a number of systems have a negative power-law DDR, as predicted by MaxEnt, but very different slopes and intercepts. Here, MaxEnt was unable to predict the spatial signals of strong intraspecific aggregations in the distributions of species observed in natural metacommunities.

Likewise, May et al. (2015) sought to determine just how well the NT did when predicting multiple macroecology biodiversity patterns at a single time using the BCI forest-plot data. As we have discussed previously, early support for the NT came from macroecological patterns (e.g., the shape of the SAD and SAR) that appeared to conform to the structure of the BCI forest plot. May et al. developed an individual-based neutral-model simulation parameterized with BCI data and found a similar mismatch in the predicted patterns—the neutral model did quite well in predicting a number of individual macroecological patterns in the BCI forest. The same was true when the model was constrained to predict two of these patterns at a time. However, Figure 10.6 presents the results when May et al. constrained the model so that three or more of the macroecological patterns (e.g., the SAR, SAD, and DDR, among others) had to be predicted simultaneously. Here, the neutral model did less well; there was a good fit between the model predictions and data for the SAD (Fig. 10.6a), a moderate fit between predictions and the SAR (Fig. 10.6b), and a poor fit between predictions and the DDR (Fig. 10.6c). Again, a likely reason for the lack of conformity between the simple model predictions and observed patterns is that there is more intraspecific aggregation in the natural system than expected from the model.

In both of these examples, deviations from expectations can point toward processes of metacommunity assembly (other than neutrality) playing an important role. Specifically, statistical models such as MaxEnt (in cases of McGlinn et al. 2015 and Xiao et al. 2015) and the NT (in case of May et al. 2015) were unable to capture the amount of intraspecific aggregation that was observed in the natural communities. It is these very patterns of aggregations that are at the heart of predictions from metacommunity ecology. For example, in addition to resulting from dispersal limitation, strong intraspecific aggregations are expected to emerge as

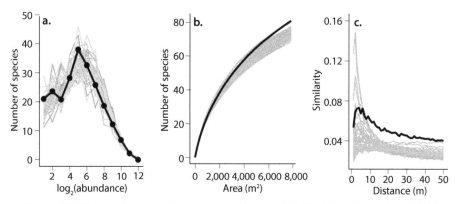

FIGURE 10.6. Expected macroecological patterns from an individual-based neutral model (thin gray lines representing many model iterations) compared with the actual data from the BCI forest dynamics plot (thick black line)). (a). The species abundance distribution (SAD). (b). The species-area relationship (SAR). (c). The distance-decay relationship (DDR); the *y*-axis was calculated as the probability that two randomly chosen individuals belong to the same species. Modified from May et al. 2015.

a result of nonneutral metacommunity assembly processes in heterogeneous landscapes (e.g., SS or ME) or trade-offs between dispersal and interspecific interactions (e.g., PD). By comparing the predictions of simplified macroecological models like MaxEnt or NT to the observed data across multiple predictions, one can begin to piece together the relative importance of these complexities in the spatial distributions of the species and of the mechanisms that create them.

10.5 DIVERSITY PARTITIONING AND THE SAR

The most obvious connection between metacommunity-level and macroecological studies on biodiversity is in the partitioning of local and regional diversity (i.e. α-, β-, and γ-diversity) and the SAC. We consider the classic SAR equation, $S = cA^z$ (see Sec. 10.2; Crist and Veech 2006, Rosenzweig et al. 2011). If a metacommunity assembly process, such as dispersal limitation or heterogeneity, influences β-diversity, we would also expect it to influence the slope (z) of the SAR. Likewise, the number of species that can coexist in a given locality (α-diversity) is influenced by local (e.g., resource ratios, resource availability) and spatial processes (e.g., dispersal), and these effects in turn determine the intercept (c) of the SAR.

Although the links between S and α-diversity and between z and β-diversity provide a useful heuristic, this relationship is complicated by a number of issues. First, although we have used the concept of SAR in a rather generic way, it is clear that operationalizing the SAR depends critically on how it is sampled and estimated, even when comparisons are made from within a continuous land-scape rather than among island (patch) isolates (i.e., the ISAR; Scheiner 2003, 2004; Gray et al. 2004). Second, while β-diversity is critically important as a concept in metacommunity ecology, illustrating how species composition shifts among localities within a larger region, the reality is that it is much uglier to operationalize, estimate, and compare patterns of β-diversity (see, e.g., Jost 2007; Baselga 2010; Tuomisto 2010a, 2010b; Veech and Crist 2010; Anderson et al. 2011; Legendre and De Caceres 2013).

Issues with calculations of β-diversity and their meaning is particularly criti-cal, but complex, when metacommunities are conceptualized sensu lato (i.e., con-tinuous landscapes), rather than sensu stricto (see Chap. 1). Here, it will often be the case that the choice of sampling grain and extent (Barton et al. 2013), as well as the size of the regional species pool relative to local diversity (Chase et al. 2011, Kraft et al. 2011), will critically influence estimates of β-diversity. As a result, below we avoid explicit comparisons of β-diversity patterns in a macroecological context and instead simply focus on the inherent scale dependence of the observed patterns, as well as its implications.

10.6 METACOMMUNITY ASSEMBLY AND THE NESTED SAR

That the number of species increases at a declining rate as the sampling window is enlarged—the nested SAR—is one of the few "laws" in ecology (Lawton 1999, Lomolino 2000). The shape of the nested SAR is given primarily by three under-lying mechanisms—sampling effects, habitat heterogeneity, and area per se (Con-nor and McCoy 1979). These three mechanisms are analogous to the statistical components underlying the SAR described above (N, SAD, S, aggregation).

Sampling effects are always operating as one moves up and down spatial scales (and N) with a given S, SAD, and aggregation. For any given metacommunity, if we know the shape of the SAD and the amount of aggregation, we can predict the numbers of species in the metacommunity (Colwell et al. 2012, Chao et al. 2014) over possibly very large spatial extents (Harte et al. 2009, Azaele et al. 2015).

Habitat heterogeneity and its differential influence on species distributions and coexistence is the signature process of the SS and the ME metacommunity ar-chetypes and is the second major process that leads to the SAR. In this case, it is assumed that (1) the number of habitats increases with increasing area (as a sam-pling process), and (2) species differentially favor different habitat types such that the inclusion of multiple habitat types allows higher biodiversity in larger sam-pled areas. As we described in Chapter 3, habitat heterogeneity, and thus SS/ME

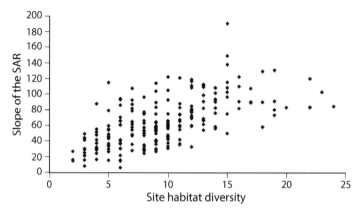

FIGURE 10.7. Estimate of the slope (z) of the species-area relationship (SAR) of plant diversity from 237 sites from Greece that were part of the Natura 2000 network plotted against the estimated habitat heterogeneity in each of those sites. Habitat heterogeneity was quantified by estimating the numbers of different habitat types present in a given region. Modified from Kallimanis et al. 2008.

processes, typically have a clear and strong influence on patterns of species coexistence and biodiversity in metacommunities (see also Hortal et al. 2009 and Stein et al. 2014). Likewise, variation in habitat heterogeneity can have a strong influence on the shape of the SAR of a number of different organism groups (Boecklen 1986, Drakare et al. 2006, Kallimanis et al. 2008, Shen et al. 2009). Figure 10.7 illustrates the influence of habitat heterogeneity on the slope (z) of the SAR from a number of sites in Greece.

The last mechanism proposed to drive the shape of the SAR is area per se (see, e.g., Connor and McCoy 1979), by which the influence of amount of space itself is a mechanism that influences species diversity. This mechanism is more often considered in the context of the ISAR, which we discuss in more detail below. However, there are a number of spatial processes that allow species to coexist simply because of area per se (i.e., in the absence of heterogeneity). For example, spatial extent affects the colonization-competition trade-offs in PD models, dispersal limitation in NT models, and a number of spatial frequency-dependent mechanisms (discussed in Chaps. 2 and 5). Consequently there are important effects of area in island archipelagos and habitat islands (due, e.g., to fragmentation), where other factors come into play that we discuss in the next two sections. And in Section 10.9 we expand this topic to describe cases in which variation in other ecological drivers (e.g., energy, latitude) change the SAR, leading to scale-dependent biodiversity responses.

10.7 METACOMMUNITY ASSEMBLY AND THE ISAR

The difference between the nested SAR and the ISAR is critical even though they are often confused with each other. Although they involve similar variables, they are not the same at all (Scheiner 2003); for example, the nested SAR must always have a positive slope given its simple sampling properties (see Fig. 9.3) while the ISAR can be more variable. There are two reasons why the ISAR is more variable and differs from the nested SAR; metacommunity thinking can help to understand this variability and clear up some of this confusion.

First, the ISAR can show variable relations between species richness and area, including no influence of area on species and even sometimes negative relationships. This is not surprising if other factors also vary with area. For example, simple island biogeography theory (sensu MacArthur and Wilson 1967) tells us that island size will interact with island isolation to determine the numbers of species on a given island; if larger islands are also more isolated, we would expect deviations from the simple positive ISAR expectation. Likewise, on most islands, both island area and habitat heterogeneity vary. Often area and heterogeneity covary in a positive way (i.e., larger islands are more heterogeneous), which would allow all three SAR mechanisms (sampling, heterogeneity, and area per se) to operate simultaneously and in the same direction, leading to a strong ISAR. However, island area and heterogeneity do not always covary positively, and there are many cases in which larger islands are less heterogeneous than smaller islands. By explicitly recognizing the interplay between habitat area and heterogeneity, one can understand cases in which a positive ISAR might not be expected to emerge. Just as important, one can explore the relative importance of heterogeneity versus area per se in driving ISAR patterns (Ricklefs and Lovette 1999, Triantis et al. 2003, Allouche et al. 2012).

Second, there is an important sampling issue that can lead to considerable variation and confusion regarding the true shape of the ISAR. Specifically, the ISAR and theories developed to predict it were originally intended to describe the *total* numbers of species on islands that vary in size (MacArthur and Wilson 1963, 1967). Indeed, this is what a large number of studies measure, both on "true" islands (Triantis et al. 2012) and on "habitat" islands (Matthews et al. 2014, 2016). Unfortunately, a large number of studies that purport to be measuring parameters of the ISAR (e.g., its slope, z) do not actually measure the total number of species on the island, but instead measure the numbers of species in a fixed area within an island or within a habitat patch. As shown in Figure 10.8, comparing the influence of area on the numbers of species in an entire island versus the numbers in a fixed area on an island are clearly not the same thing, and inferences from measured slopes using the two different methods can mean very different things.

Also illustrated in Figure 10.8 is the insight that can be gained if one were to compare the numbers of species in fixed quadrats on islands varying in size with the numbers of species on the entire island. We can compare a scenario in which the ISAR is driven completely by sampling (Fig. 10.8b, d) to one in which the ISAR is driven by heterogeneity or area per se (Fig. 10.8c, e). Here, although both scenarios lead to the same shape of the ISAR when measured as total richness on the island, they give different answers when species richness is measured in a fixed-sized quadrat regardless of island size. When the ISAR is driven by sampling, we expect no difference in species richness when measured with a fixed grain, whereas when the ISAR is driven by heterogeneity or area per se, we expect there to be on average more species per unit area (i.e., in the fixed grained quadrat). This latter result emerges because of metacommunity-level processes (e.g., heterogeneity or dispersal).

A quantitative review of the influence of sampling versus metacommunity effects on the ISAR is beyond the scope of what we can accomplish here. The reason is that sampling designs in both true and habitat island studies (i.e., natural and anthropogenic habitat fragmentation) are highly variable and often inconsistent, and abundance data and multiple scales are rarely collected or presented. Nevertheless, a qualitative review and a few case studies illustrate some of the possibilities.

For the ISAR in true islands (i.e., terrestrial islands surrounded by water; aquatic islands—lakes/ponds—surrounded by land) or natural habitat islands (i.e., not anthropogenic), few data are available that can definitively test for sampling effects versus metacommunity effects (i.e., heterogeneity and area per se). Of those studies in which multiple scales were collected within islands (i.e., fixed-grain vs. total island), the vast majority seem to show ISARs with slopes that are steeper than would occur with sampling effects alone; that is, there was a significant increase in species richness with total island area even when the sampling grain was fixed (more species per unit area in larger islands). Figure 10.9 shows some examples from disparate island types and organisms, both observational and experimental. We could not locate any study that found a significant effect of area involving fixed-grain sampling in islands, although it is possible that other methods might identify cases where sampling effects alone create the ISAR in some systems. Of course, while this situation allows us to conclude that sampling effects alone are not likely the cause of these ISARs and that spatial (i.e., metacommunity) effects are generally at play, there are in fact several mechanisms that could lead to such spatial effects, including increased habitat heterogeneity, edge effects, and a number of direct area-dependent mechanisms (colonization-extinction dynamics, spatiotemporal frequency dependence, etc.).

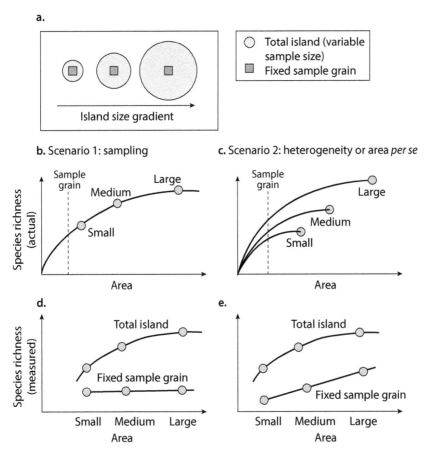

FIGURE 10.8. The influence of sampling total area versus fixed sample grain size on the island species area relationship (ISAR). (a) A schematic of the two different sampling approaches on three hypothetical islands that very from small, medium, and large. (b) The accumulation of species when the ISAR is driven just by sampling effects. (c) The accumulation of species on each island if heterogeneity or area per se plays a role in the ISAR. Note that each island has its own accumulation of species, with larger islands having more species than would be expected from simple sampling smaller islands. (d) and (e) The results for the "measured" numbers of species depending on whether sample grain is fixed, or the richness of the entire island is measured. Note that when only sampling effects are operating, as in (d), the ISAR is positive, but there is no influence of island size on fixed grain sampling. When heterogeneity or area per se is operating, as in (e), both total and fixed grain sampling show an increase with increasing island area, but the slope differs.

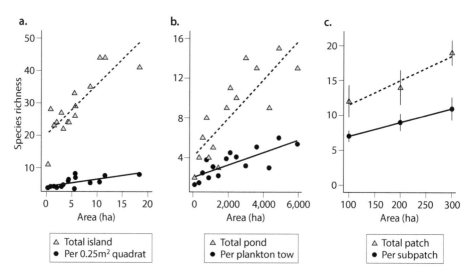

FIGURE 10.9. Measures of island species area curves (ISARs) in which richness was reported at both the entire-island scale, and a fixed-grain scale. (a) Data on herbaceous plants on small oceanic islands off of Shetland, UK; data from Kohn and Walsh 1994. (b) Data on crustacean zooplankton from fishless ponds in Missouri; data from Ryberg and Chase 2007. (c) Data on mollusks settling on artificial patches in the intertidal zone in New South Wales, Australia; units were 100 m² and treatments varied the numbers of units (1,2, or 3) in a given island patch; data from Matias et al. 2010.

There is also the possibility that the ISAR can be a bit more complicated when metacommunity dynamics are explicitly considered. For example, Kadmon and Allouche (2007) developed a model that combined island biogeography theory with niche theory (i.e., SS dynamics) and showed that when island size and habitat heterogeneity are considered together, they can interact in a nonlinear way. Specifically, even if smaller and larger islands have the same amount of habitat heterogeneity (say, discrete habitat types), the smaller island will have less of each habitat type than the larger island, so species may be unable to persist (go extinct) in a given habitat type on the smaller island because the small population size may enhance extinctions. This situation could result in small islands with high habitat heterogeneity actually having fewer species than small islands with lower habitat heterogeneity. Although controversial (Hortal et al. 2009, 2013), there is some evidence for this interaction of habitat heterogeneity and island size (Tamme et al. 2010, Allouche et al. 2012), particularly from a study on zooplankton diversity in freshwater mesocosms in which habitat area and heterogeneity were varied orthogonally (Schuler et al. 2017).

10.8 BIODIVERSITY IN THE ANTHROPOCENE

One species—humans—is having a huge impact on the world around us. So much so, that we have defined a new geological era—the Anthropocene, which is characterized by the great impact of humans on the geological and ecological properties of the world. And one of the key signatures of the Anthropocene is a rather dramatic rearrangement of the Earth's biota (Steffen et al. 2011). In what follows, we illustrate a few ways in which an explicit scale-dependent perspective of biodiversity can help us to understand how biodiversity is (or is not) changing in the Anthropocene and how metacommunity-level processes mediate these changes.

10.8.1 The ISAR, Habitat Loss and Fragmentation, and the Conservation of Biodiversity

We described the ISAR and metacommunities above in the context of natural islands. The other major context in which the ISAR has been examined is in the context of habitat loss or fragmentation; that is, where human activities make habitats that were once relatively continuous into smaller and patchier island-like systems. Habitat loss is universally cited as the number one cause of species endangerment and biodiversity decline (Pimm et al. 1995, Wilcove et al. 1998, Dirzo and Raven 2003, Pereira et al. 2012). This is perhaps not surprising given the universality of the SAR; if species richness increases with habitat area, decreases in habitat area should certainly lead to reductions in richness (Pimm and Askins 1995, Halley et al. 2013).

Upon development of their theory, MacArthur and Levins (1967) immediately noticed the application of basic island biogeography theory to understanding the influence of patterns of habitat loss or fragmentation on patterns of biodiversity, and the field quickly took off, though not without controversy. For some time, a major question was whether a single large or several small (SLOSS) habitat reserves would preserve more biodiversity (Diamond 1975b, Diamond and May 1976, Gilpin and Diamond 1980, Higgs and Usher 1980, Simberloff and Abele 1984, Simberloff 1988). Interestingly, and perhaps not coincidentally, at the center of this debate were Diamond and Simberloff, who were also embroiled in the question about whether structure due to competitive interactions could be detected from patterns of species co-occurrences using null models (see Chap. 4).

The SLOSS debate was never formally resolved but simply faded away with the recognition that the answer to whether a single large or several small reserves of the same total area would preserve more species depended greatly on context, such as which organisms and which systems might be involved. It is, however, possible to view this context dependence as being related to metacommunity-level processes. For example, a single large preserve could protect more diversity if small reserves experience sufficiently high rates of local extinctions (without recolonizations) or edge effects (i.e., negative influence of adjacent matrix habitats). However, several small reserves could preserve more biodiversity if they encompass higher amounts of habitat heterogeneity in total, if they allow spatial processes such as dispersal limitation and colonization-extinction dynamics to emerge and maintain regional diversity, or if they experience fewer mass effects from adjacent populations.

As a result of this uncertainty, the influence of habitat fragmentation on patterns of biodiversity has been an area of intense research during the past thirty-plus years and has included theoretical, observation, and experimental (both small- and large-scale) approaches. While a frequent conclusion is that habitat fragmentation can have dramatic negative effects on biodiversity and a number of other community attributes (Haddad et al. 2015, 2017), a set of controversies has emerged that has the same flavor as those from the SLOSS discussions of the past. At the core of this controversy is the same fundamental question that lies at the heart of the SAR and ISAR we described above: what is the relative importance of sampling, heterogeneity, and area per se?

The first controversy stems from He and Hubbell's (2011) use of the EAR (described above) to predict extinction rates with habitat loss (see also Kinzig and Harte 2000). They suggested that many estimates of species extinction rates due to habitat loss, which were typically derived by taking the inverse of the SAR, were in fact overestimates and that the situation was not as dire when estimates were taken using the EAR. Although a number of nuances certainly exist in the use of the EAR for predicting species extinction rates, including the geometry of habitat loss (e.g., Pereira et al. 2012, Keil et al. 2015), the basic premise of using the EAR for projecting extinction rates with habitat loss is that area influences species richness only through a sampling effect, effectively ignoring the influence of habitat heterogeneity and area per se.

However, as a number of authors have pointed out, if metacommunity dynamics are important, the mechanisms that allow species to coexist in the first place can be altered by habitat loss and fragmentation, and those changes can greatly influence our expectations of how biodiversity loss should reverberate from habitat loss. For example, Hanski et al. (2013) used metapopulation models of a large number of species (i.e., a simple PD approach) to show that simply changing the spatial configuration of area and isolation (among equally divided habitat fragments) leads to many fewer species than would have been expected from a simple sampling scenario (Fig. 10.10; see also Rybicki and Hanski 2013). That is, area per se matters! Likewise, Matias et al. (2014) showed that, among other things, the influence of habitat loss on the expected numbers of species loss depended critically on which metacommunity archetype was operating. Again, species loss was expected to be much greater with habitat loss under SS (i.e., habitat heterogeneity) and PD (i.e., area per se) dynamics, rather than when metacommunity assembly was governed by NT dynamics.

The second controversy stems from concepts recently described as the *habitat amount hypothesis* (Fahrig 2013), which suggests that habitat fragmentation per se has little influence on patterns of biodiversity and that what really matters is the total amount of habitat available (Fahrig 2001, 2013). Again, this hypothesis explicitly assumes that the primary mechanism underlying the influence of habitat area on diversity is the sampling effect and that area per se in the context of metacommunities has little effect (habitat heterogeneity is presumably assumed to be equally represented regardless of the configuration of remaining habitat). This hypothesis has also proved to be quite controversial, especially among ecologists who favor the perspective that area per se can often play an important role, for example, through dispersal limitation, colonization-extinction dynamics, or habitat alterations (e.g., edge effects; Haddad et al. 2015, 2017; Hanski 2015).

As we discussed above, disagreements about the relative importance of area per se (and to a lesser extent, heterogeneity) relative to sampling effects are at the center of these controversies; that is, just how important are the metacommunity processes that create area per se effects on biodiversity under habitat loss/fragmentation scenarios? To answer this, we turn to the empirical literature. Although a quantitative review is beyond the scope of what we can accomplish here, we can provide a qualitative overview.

There have been hundreds of studies that have purported to examine the influence of habitat fragmentation on biodiversity, both observational and experimental (reviewed in Debinski and Holt 2000, Ewers and Didham 2006, Haddad et al. 2015). And a multitude of results have emerged, including negative, positive, and neutral effects on biodiversity. Some of this variation arises because fragmentation is often not random with respect to habitat. For example, in a study on insect diversity in nature reserves of different sizes, Baldi (2008) found no effect of reserve area on diversity but noted that larger fragments had lower habitat heterogeneity. Much of this variation probably comes from inconsistent sampling designs and analyses. For example, some studies report effects on total richness in fragments of different sizes, while others report effects on fixed-grain samples in fragments of different sizes.

However, there also seems to be a good deal of variation in the influence of sampling, as opposed to metacommunity-level processes, indicating that heterogeneity and area per se are important. Many of these differences may have to do with the fact that some studies on fragmentation measure total richness whereas others measure fixed-grain species richness (and even more so than for the ISAR described above). Figure 10.11a shows an example of hoverflies in French woodlands (Ouin et al. 2006) in which total species richness increased with total fragment area but did not change at all when richness was measured on a fixed-grain scale (i.e., sampling effects appeared to explain the entire pattern). Figure 10.11b shows an example from trees in Ghanaian forest fragments (Hill et al. 1994) in which both total and fixed-grain species richness increased with fragment area, but the latter was much weaker (i.e., sampling effects played a large role, even though heterogeneity and/or area per se did as well). Finally, Figure 10.11c shows an example from birds in grassland fragments in Illinois in which both total and fixed-grained species richness increase with fragment area (Herkert 1994); even though the slope of the fixed-grain relationship was expectedly weaker than for total richness, this mismatch suggests a much stronger role for heterogeneity and/or area per se effects.

The variability in responses, and the relative importance of sampling versus metacommunity-level effects, can even emerge among different taxa in the same system. Ziv and colleagues (1995) conducted a series of studies in fragmented scrub habitat in Israel using similar sampling approaches that allowed them to dissect the influence of area per se. Yaacobi et al. (2007) found that diversity of beetles in fixed-grain sampling was uninfluenced by fragmentation; plant diversity in fixed-grain sampling decreased with fragmentation (Giladi et al. 2011); and the diversity of spiders in fixed-grain sampling actually increased with fragmentation (Gavish et al. 2012).

Despite the observed variability in responses, several authors have suggested that the negative influence of habitat fragmentation on species richness goes beyond sampling effects (i.e., heterogeneity and/or area per se) and is the most common response (Haddad et al. 2015, 2017; Hanski 2015). Yet, there are probably too few data, too many alternative explanations, and enough exceptions (e.g., Fahrig 2003, 2013; Yaacobi et al. 2007) to thus warrant deeper scrutiny into this relationship and the relative importance of sampling versus metacommunity-level effects in causing biodiversity loss in response to habitat loss and fragmentation.

10.8.2 Habitat Loss and Biodiversity Loss

Next, we discuss how habitat fragmentation and loss influence biodiversity in the entire landscape (not just the remaining habitat and the ISAR, as discussed above). In the first figure of their book, MacArthur and Levins (1967) used an illustration taken from Curtis (1956) showing how anthropogenic activities turned once

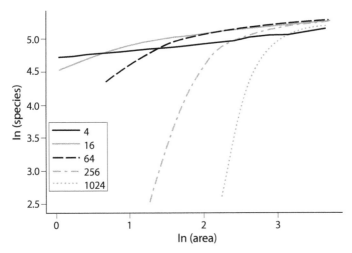

FIGURE 10.10. Results from the simulations of Hanski et al. (2013) showing how the species-area relationship changes dramatically among landscapes that have the same total amount of area, but the degree to which habitats are fragmented varies. Here, the same total area is divided into 4, 8, 16 . . . 1024 equal-sized fragments placed randomly within a landscape. This model assumes that simple metapopulation processes influence each species, but that species dynamics are independent from one another.

continuous habitat (an intact forested landscape) into a series of small isolated habitat fragments (habitat islands). We reproduce part of the illustration in Figure 10.12a and b to illustrate what was most striking to Curtis, MacArthur, and Wilson, and many others—that we as humans are making the world into a series of small islands, and as such, island biogeography theory can help us to understand the biodiversity consequences of these actions.

This perspective on habitat fragmentation as the creation of habitat islands has been useful and important in conservation biology. Indeed, it has been repeated many thousands of times that habitat destruction is the number-one cause of biodiversity loss on the planet (e.g., MacArthur and Wilson 1967, Pimm and Raven 2000, Millennium Ecosystem Assessment 2005, Pereira et al. 2012). This is undoubtedly true. However, by using the concept of "habitat destruction," we implicitly make the assumption that the total amount of area available in which biodiversity can thrive has been reduced and that the destroyed habitat is now nothing more than a matrix between habitats that is completely inhospitable to the species of interest. This is only true for the strictest of specialist species; usually the total amount of area has not been reduced per se but, rather, habitats have been altered. Clearly habitat alteration will disfavor a number of species (i.e., those that live in forests in the case of the Curtis maps), although other types of species will be favored by the creation of different kinds of habitat. And dispersal rates will depend on the isolation of different habitat types, as well as the degree to which species are able to utilize the different types of habitats.

Figure 10.12c presents a GIS image of the same location as the Curtis maps (Cardiz Township, Wisconsin) but differs in a few key aspects. Although other analyses have shown that the forested landscape has changed little since 1950 (e.g., Brown et al. 2013), Curtis's maps presented only intact forested islands that were 4 ha or greater. Our depiction illustrates a much more dynamic mosaic of forest patches, several of which are connected (or nearly so) by corridors and small "stepping stones." Additionally, our illustration emphasizes that while the forested landscape has certainly been lost, there are other kinds of landscape that have replaced it. In order to understand how species composition, diversity, and interactions have been altered in this landscape, it may thus be more useful to think about the landscape as a mosaic metacommunity that has altered its habitat heterogeneity and dispersal rather than as simply a series of islands of different sizes and isolation in an inhospitable matrix.

The perspective in which habitats are lost in an island biogeography context rather than simply changed in a metacommunity context has a profound influence on how we estimate the influence of anthropogenic activities on biodiversity loss. For example, a standard approach for estimating how habitat loss will lead to global species extinctions is to use the SAR (or as would be more appropriate, the EAR) to project how many species should be lost following a given amount of habitat loss (Pimm and Askins 1995, Pimm et al. 1995, Millennium Ecosystem Assessment 2005, He and Hubbell 2011, Halley et al. 2013). However, these types of estimates assume that area is simply gone following habitat destruction, which is not necessarily always, or even often, the case.

One solution to the problem that habitats are changed, not lost per se, is to "correct" predictions of species extinctions based on total habitat loss by taking into account that individuals of many species can often survive (and even thrive) in the area where habitat alteration took place (i.e., the matrix). Basically, this correction allows one to incorporate a bit more of the metacommunity perspective into habitat loss-species loss estimates (this is sometimes referred to as the "countryside" or "matrix calibrated" SAR to emphasize the role of the habitat mosaic in mediating extinction risk; Daily et al. 2003, Pereira and Daily 2006, Koh and Ghazoul 2010). Figure 10.13 presents an example of this approach from Koh and Ghazoul (2010), who compared the observed numbers of extinct bird species with those predicted from habitat loss when using traditional species-area estimates based on island biogeography theory as well as estimates that were corrected to include matrix-level (metacommunity) effects (see also Pereira et al. 2014 for a slightly modified approach, but with qualitatively similar result).

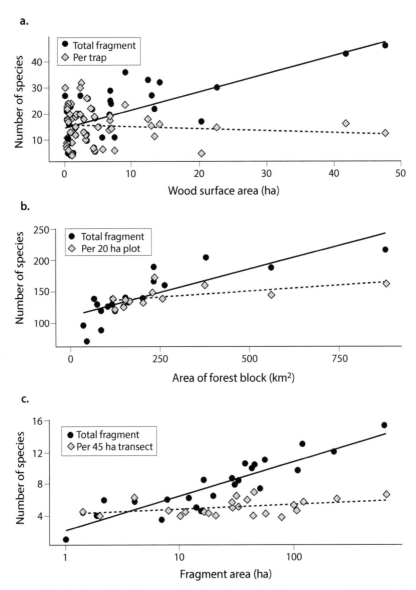

FIGURE 10.11. Total species richness and fixed-grain species richness measured in habitat fragments of different areas. (a) Hoverflies in forest fragments in France show a strong influence of fragment area on total richness, but no influence of fragment area on fixed-grain richness measured as the average richness per trap; from Ouin et al. 2006. (b) Trees from forest fragments in Ghana show a strong influence of fragment area on total richness but a weaker influence on fixed-grain richness measured as 20-ha plots; from Hill et al 1994. (c) Birds from grassland fragments in Illinois show a strong influence of fragment area on total richness and a strong (albeit slightly less so) influence on fixed-grain richness (measured on 300-m × 75-m transects); from Herkert 1994.

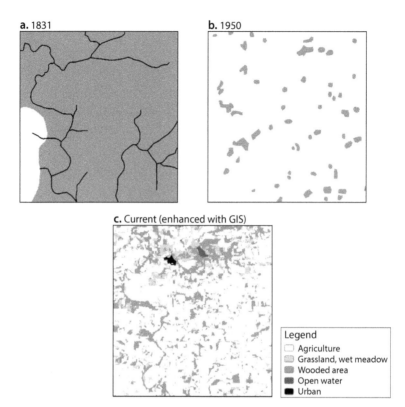

FIGURE 10.12. (a) The forested (gray-shaded) and non-forested (white) area of Cadiz Township, Wisconsin, in 1831. (b). The same area in 1950, illustrating a large loss in forested habitat in total, and fragmentation of the remaining habitats into small isolated islands. Both (a) and (b) modified from Curtis 1956. (c) A GIS image of the same land area today but including a number of other features not captured in Curtis's initial images. While Curtis's fragments can be seen, there are many smaller areas of forest and forested corridors that can influence dispersal within the landscape mosaic. In addition, many species utilize multiple parts of this mosaic habitat, suggesting that a metacommunity perspective considering habitat heterogeneity and complex dispersal dynamics may be a more useful way to conceptualize this landscape rather than simply as a series of fragmented islands surrounded by a sea of "nothing."

Taking this a step further, if we consider habitat mosaics that include various levels of anthropogenic modification as a metacommunity, we may be able to enhance our understanding of the ability to conserve species diversity over the entire mosaic. Rosenzweig (2003) rightly pointed out that if we simply view biodiversity conservation in the context of the habitats that are spared in isolated reserves (i.e., nature) and habitats that are used by humans (i.e., not nature), we will

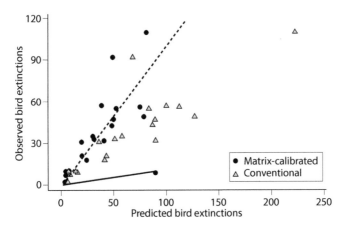

FIGURE 10.13. A comparison of the traditional island biogeography approach to predicting species extinctions due to habitat loss (labeled "conventional") relative to predictions assuming a
more "metacommunity-like" structure in which the matrix mediates expected extinctions
(labelled "matrix-calibrated"). Data are taken from 16 biodiversity hotspots across the world.
Modified from Koh and Ghazoul 2010.

continue to lose a great amount of biodiversity simply because of the species-area
relationship (i.e., the amount of land reserved for nature can only hold a fraction
of the species that could live in the entire landscape). Instead, Rosenzweig argues
that a "win-win" perspective is necessary whereby biodiversity is maintained as
part of a metacommunity consisting of a heterogeneous mosaic of habitats, some
of which are less used by people and some of which are more used by them.

    Finally, such a metacommunity perspective can greatly enhance our ability to
develop more biodiversity-sustainable policies regarding land use. For example,
one long-standing debate concerns whether biodiversity and other ecosystem services are best preserved under a "land-sparing" scenario, whereby high-intensity
agriculture and other land uses are relegated to certain areas, while other areas
are "spared" as reserves to preserve biodiversity (see, e.g., Fischer et al. 2014, Kremen 2015). Or, instead, whether a "land-sharing" scenario, whereby lower-intensity
land use that is more "biodiversity friendly" (e.g., shade-grown coffee, organic
farms) but requires more total area provides better biodiversity outcomes. Unfortunately, we cannot fully evaluate this debate because most studies have lacked
a metacommunity perspective; the vast majority of studies that have compared
the biodiversity outcomes of different approaches with agriculture and land use have
only measured biodiversity at a local scale (i.e., α-diversity), providing a highly
incomplete picture (von Wehrden et al. 2014, Kremen 2015).

*10.8.3 Invasions, Extinctions, and Diversity Change
in the Anthropocene*

In addition to habitat destruction, a second major anthropogenic modification of
the environment that is often considered detrimental to biodiversity is the macroecological pattern that seems to be strongly influenced by human activities: the
DDR. By moving species beyond their historical biogeographic boundaries, humans have broken down patterns of dispersal limitation, at least at the regional or
continental scale. This often leads to a reduction in (at least some metrics of)
β-diversity, which is known as *biotic homogenization* (McKinney and Lockwood
1999, Olden and Rooney 2006). As an example of this effect, Capinha et al. (2015)
compared the DDR of terrestrial snail assemblages between historical and modern times and found much steeper relationships historically compared with contemporary patterns (Fig. 10.14). Similar results were observed for plants by van
Kleunen et al. (2015).

From a metacommunity perspective, increases in dispersal rates leading to
biotic homogenization are often thought to be associated with reductions in regional-
scale (γ-) diversity. The idea is that if β-diversity declines, γ-diversity should also
decline. Indeed, this is the expectation we get from some metacommunity models,
such as those associated with ME (Mouquet and Loreau 2003). This need not be
the case, however, for many of the ways that we actually measure β-diversity (i.e.,
pairwise dissimilarity metrics such as those in Fig. 10.14). Especially when measured at biogeographic scales, biotic homogenization need not lead to declines in
diversity at the smaller or larger scales, unless movement of species causes localized
extinctions, which is often not the case (Sax and Gaines 2003). We next illustrate
the fundamental role of spatial scale in mediating our understanding of biodiversity
change in the face of localized extinctions and regional colonizations.

10.8.4 Scale-Dependent Biodiversity Change

Clearly, species are being lost at the global scale (i.e., extinctions), and the rate at
which they are doing so appears to be accelerated due to human activities (Pimm
et al. 2014, Ceballos et al. 2015). However, as we discussed in Chapter 5, the fact
that species are being lost at larger spatial scales (i.e., globally or regionally) does
not necessarily (or perhaps even likely) imply that diversity will decrease locally.
In fact, a number of recent analyses have suggested that, on average, local diversity
does not seem to be systematically declining over time (Vellend et al. 2013, Dornelas
et al. 2014, Elahi et al. 2015).

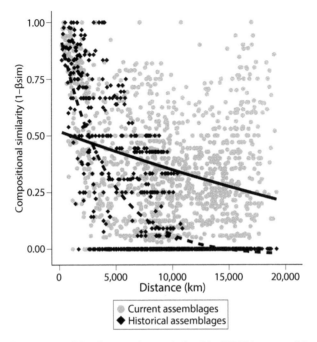

FIGURE 10.14. Comparison of the distance-decay relationship (DDR) in terrestrial snail assemblages native to given regions (and thus, the likely historical relationships) (black diamonds, dashed line) with those same assemblages including alien species that have been moved by humans over the past several hundred years (gray squares, solid line). The DDR was much steeper prior to human-mediated dispersal. Data from Capinha et al. 2015.

When we recognize the importance of scale dependence in a metacommunity, it becomes clear that there is not necessarily a one-to-one relationship between diversity gain and loss at different scales. It is possible that the same processes that cause species to go extinct from a larger regional extent will also make rare species rarer, reducing the numbers of species observed locally. But it is equally (or even more) plausible that within a metacommunity, some species will be lost from a given locality or region, whereas other species will gain in abundance and distribution; and this disparity in their responses will certainly be exacerbated with human-mediated dispersal of species (i.e., exotic species). Such change is evident in a number of larger-scale analyses of biodiversity change, which often show increases, rather than decreases, as a result of exotic species invasions (Sax and Gaines 2003, Essl et al. 2013). In fact, a number of versions of metacommunity theory (see Chap. 2) can predict decreases in species diversity at larger scales while simultaneously predicting increases at smaller scales, or vice versa. Consequently, we have to be very careful when discussing and analyzing how humans are (or are not) influencing biodiversity, and spatial scale must be a mitigating consideration (Cassey et al. 2006, Vellend et al. 2017).

To illustrate the importance of scale considerations for understanding patterns of biodiversity loss, Mary O'Connor and coworkers (unpublished data) developed a highly simplistic (noninteractive) metacommunity model. They modified the two processes known to be influenced by anthropogenic activities—local extinctions of species and regional dispersal of species. When only local extinction rates were increased, small-scale species richness declined, and when only dispersal rates were increased, small-scale richness increased. However, the magnitude of both of these effects decreased with increasing scale. Likewise, when the most likely anthropogenic scenario was implemented—increasing local extinction rates and regional dispersal rates—there was a hump-shaped relationship in which species richness declined at the smallest scales, was higher in the presence of human activities at intermediate scales, and then turned negative (but very slightly) at the largest scales (see also Cassey et al. 2006).

Thus, the general picture that emerges regarding biodiversity change is that at smaller spatial scales, communities have a lot of variation in their change, sometimes showing substantial losses, sometimes increases, and sometimes no significant changes. At intermediate scales, most biodiversity change is actually seen as increases (owing to the movement of exotic species). It is only at the largest scales that declines are seen due to global extinctions of species, but these declines are quite small compared with the total numbers of species, so they are difficult to detect. This scale-dependent effect may thus be a simple metacommunity process expressed at the global scale, when colonizations (exotic species) exceed localized extinctions. Of course, this does not mean that there is not a "biodiversity crisis," and clearly, exotic species might not provide the sorts of "desirable" biodiversity many conservation biologists strive to conserve (nor might they serve the same ecosystem function; see Chap. 12). It does mean, however, that the processes that maintain biodiversity from local to global scales in the face of anthropogenic change can be better understood by taking a scale-explicit metacommunity perspective.

10.9 METACOMMUNITIES AND CONTEMPORARY BIOGEOGRAPHY: SCALE-DEPENDENT PATTERNS OF SPECIES DIVERSITY ALONG ECOLOGICAL GRADIENTS

In addition to describing patterns of biodiversity itself, such as the SAD, SAR, and DDR, another major research area under the macroecology umbrella is variation in biodiversity (usually species richness) across biogeographic gradients. We referred above to this as contemporary biogeography, to contrast it with historical biogeography (see Chap. 9). Although there are many such gradients, we focus here on the species-energy relationship (i.e., how species richness varies with energy), because it is arguably one of the most important contemporary environmental gradients other than perhaps human land-use and impact (Rosenzweig 1995, Gaston 2000, Hawkins et al. 2003, Evans et al. 2005). In addition, other frequently studied large-scale biodiversity gradients, such as latitudinal and elevational gradients, may have a historical biogeographic component but are also likely to be strongly influenced by energy (Willig et al. 2003, Hillebrand 2004, Brown 2014).

Again our overview of species-energy relationships (often called the productivity-diversity relationship[3]) is not intended to be comprehensive, but rather to highlight the role that metacommunity-level processes may play in driving large-scale biodiversity patterns (for a more comprehensive overviews, see Mittelbach et al. 2001, Currie et al. 2004, Evans et al. 2005, Whittaker 2010, Cusens et al. 2012). The core idea is that energy entering an ecosystem is the key limiting resource for life; more energy provides faster growth and reproductive rates, allows more individuals to be able to persist in the same area, and provides more opportunity for niche differentiation (e.g., resource partitioning, spatial partitioning, higher trophic levels, etc.) and thus leads to greater biodiversity. If so, we would predict that there should be a monotonic relationship between species richness and productivity.

Despite the conceptual appeal of the energy-richness relationship, empirical studies show much more variable patterns, including flat, humped, negative, and U-shaped relationships. The data is confusing enough that even supposed "syntheses" of energy-richness relationships come to divergent overall conclusions (Mittelbach et al. 2001, Evans et al. 2005, Whittaker 2010, Cusens et al. 2012). Some of this variation most likely emerges because "energy" in an ecosystem is usually measured via proxies that are thought to influence energy inputs in a given ecosystem (e.g., water availability, temperature, nutrients), proxies for energy production per se (e.g., productivity, evapotranspiration), or proxies of the outcome of energy productivity via photosynthesis (e.g., standing biomass). These differences in energy measurements (or its proxies) certainly accounts for some of the variation observed in among studies in the shape and direction of the

---

[3] In this relationship, productivity is the independent variable, and diversity is the dependent variable. It is important to note that this is not at all the same field of study as the exploration of the relationship between biodiversity (independent variable) on productivity (known as the biodiversity-ecosystem function [BEF] relationship, even though they are often confused in the literature. Of course, biodiversity and productivity can be both cause and effect (Allouche and Kadmon 2009), but for now, we consider only productivity as a driver of diversity, not vice versa (see Chap. 12 for a discussion of the BEF relationship in the context of metacommunities).

species-energy relationship. However, these differences may also reflect the roles of other processes, including metacommunity dynamics.

One of the primary reasons for differences in the observed species-energy relationship is the spatial scale at which both independent (energy and proxies) and dependent (species diversity) variables are measured (Mittelbach et al. 2001, Chase and Leibold 2002, Evans et al. 2005, McBride et al. 2014; but see Cusens et al. 2012 for a dissenting opinion). Specifically, the relationship between measures of energy and diversity are often quite variable at small spatial scales, even to the point where different conclusions and debates emerge about what are essentially the same data (Adler et al. 2011, Fraser et al. 2015). At larger scales, however, the relationship is most often positive (though this depends a bit on how "scale" is defined) both in terrestrial (Currie 1991, Hawkins et al. 2003, Storch et al. 2005, Gardezi and Gonzalez 2008) and in aquatic (Tittensor et al. 2010) environments.

Given the strong effect of energy on diversity at larger scales, but its weaker effect at smaller scales, Chase and Leibold (2002) reasoned that a metacommunity perspective could synthesize these seemingly dissenting results. Specifically, the only mathematical way for there to be stronger (or even opposing) correlations between energy and γ-diversity than α-diversity would be if energy was also correlated with β-diversity. More specifically, if steeper slopes between richness and productivity occur at higher productivity, β-diversity must also increase with productivity. While useful, the hypothesis that β-diversity should increase with energy inputs has received mixed support (see Chalcraft et al. 2008). Instead, and given the arguments we presented above, it might be that β-diversity is not the best way to test the scale-dependence of the species-energy relationship. Are there other ways to look at this problem?

If energy-richness gradients are more strongly positive at larger scales, we would expect that the intercept or slope of the SAR should increase with energy; as a corollary, we would expect that the difference in species richness between smaller and larger scales would be greater in high-, relative to low-, energy systems. Figure 10.15 gives three examples across different extents and with different organisms and ecosystem types that illustrate this pattern. Figure 10.15a shows results with a global extent; the meta-analysis of SARs presented by Drakare et al. (2006) reveals that the slope of the SAR increased as latitude decreased (and likely energy inputs increased). Figure 10.15b shows results from a continental extent; the analysis by Wang et al. (2009b) illustrates that the relationship between temperature and tree-species richness across North America is strongly scale dependent, with stronger effects of scale at higher temperatures. Finally, Figure 10.15c shows the results from a regional scale survey of fish species richness in Canadian lakes (i.e., the province of Ontario); again, the influence of energy (measured by potential evapotranspiration) is stronger at smaller, relative to larger, spatial scales (Gardezi and Gonzalez 2008).

By thinking about species richness in a scale-explicit way (i.e., metacommunity thinking), we can gain important insights regarding the potential processes driving this relationship. Below, we provide two general types of mechanisms: one based on averaging in the landscape, and the other based on how energy moderates the influence of heterogeneity or area per se in driving the SAR in a metacommunity context.

First, scale-dependence can emerge if the independent variable used appropriately "averages" important variation better at larger, than at smaller, spatial scales. At large spatial scales, it is a reasonable generalization to say that there is a strong gradient in energy inputs as one moves from temperate to tropical terrestrial habitats, even when energy is measured in a number of different ways (e.g., temperature, rainfall, productivity, evapotranspiration) (Gillman et al. 2015); this is likely because the most important drivers of energy inputs are probably correlated at larger scales (i.e., warmer areas tend to be wetter). At smaller spatial scales, these factors can decouple, leading to much more variation. For example, in low-latitude deserts, temperatures are high, but precipitation is low, limiting the photosynthetic capacity (and thus energy) of the system. Likewise, even if precipitation or temperature is high, energy production will be low if the nutrient availability (e.g., nitrogen, phosphorus, etc.) necessary for photosynthetic capacity and growth are low.

Second, scale-dependence can emerge from metacommunity-level processes if energy inputs alter the shape of the SAR by altering S, N, SAD, or aggregation as we described above. Interestingly, the scale dependence most commonly observed in nature (i.e., a steeper relationship at larger scales) is not predicted if energy only alters N, as assumed by one of the original species-energy theories (Wright 1983); in this case, we would predict that the difference between low- and high-energy sites would be smaller at larger spatial scales (Schuler et al. 2015). Instead, scale dependence in which the relationship is stronger at larger scales can emerge only if energy changes more than just the number of individuals (N) (see also Hurlbert and Jetz 2010); that is, by also changing the total numbers of species that can persist in a system (S), as well as possibly the SAD and aggregation levels—that is, the relative importance of heterogeneity or the influence of area per se in the SAR seems to change with energy inputs.

The relative importance of habitat heterogeneity can increase with increasing energy if, for example, regions with higher average energy inputs also have more heterogeneity in those inputs (or the species in those areas perceive more heterogeneity). The relative importance of area per se can increase with increasing energy if,

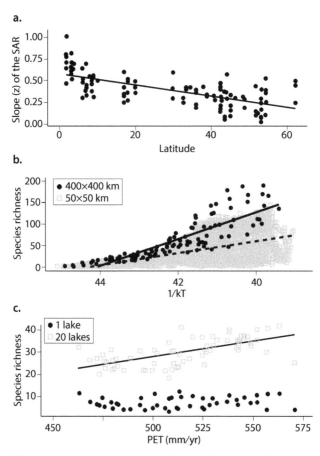

FIGURE 10.15. Putative relationships between estimates of energy and species richness from global to local scales. (a) Meta-analysis of nested species-area relationship (SAR) measurements, showing a strong decline in the slope (z) (and thus weaker scale-dependence) with increasing latitude (and thus decreasing energy); from Drakare et al. 2006. (b) Analysis of the relationship between tree species richness and the reciprocal of absolute temperature ($1/kT$, where k is Boltzmann's constant, 8.62×10^{-5} eV K^{-1}, and T is absolute temperature) in North America at two spatial grains (50×50 km and 400×400 km), showing a steeper slope at the larger grain; data from Wang et al. 2009b. (c) Analysis of the response of fish species richness in Ontario (Canada) lakes at the grain of 1 and 20 lakes per watershed to changes in energy inputs, measured as potential evapotranspiration (PET); data from Gardezi and Gonzalez 2008.

for example, regions with higher energy inputs experience greater degrees of spatially mediated coexistence, such as frequency-dependent interactions in the landscape (see Chap. 5). For example, some evidence suggests that more energy-rich sites can experience higher rates of asynchronous temporal turnover (e.g., assembly cycles), enhancing the likelihood of coexistence at the regional levels relative to local ones (Steiner and Leibold 2004, Steiner 2014). Likewise, some evidence suggests that more-energy-rich sites can experience higher levels of priority effects and multiple stable equilibria, also allowing higher regional coexistence (Chase and Leibold 2003, Chase 2010). We discuss these mechanisms in a bit more depth in Chapter 11 in the context of food-web interactions, but for our purposes here, either of these mechanisms can create a stronger influence of area per se at higher energy levels, leading to the observed scale-dependent energy-richness relationship.

It is important to note that the sort of scale-dependence in energy-richness relationships we discussed above is by no means universal. For example, a number of studies have found the opposite pattern: the intercept of the SAR is higher, but the slope lower, in high-energy regions relative to low-energy regions (Pastor et al. 1996, Weiher 1999, Storch et al. 2005, Chiarucci et al. 2006). However, this does not mean we cannot understand and synthesize the variation in scale dependence. As we mentioned above, this pattern would be predicted, for example, if there did not exist a strong difference in S between low- and high-energy systems, and the influence of energy was mostly by altering N.

Although we often imagine a positive relationship between energy inputs into an ecosystem and species diversity, there are good reasons why some factors that increase energy inputs into a system will lead to a decrease in species richness, especially at small scales. For example, inputs of some types of resources can increase energy to a degree but cause a stoichiometric imbalance that can result in the dominance of only a few species, leading to declines in species richness (i.e., the *resource-ratio hypothesis*; Harpole and Tilman 2007, Cardinale et al. 2009). Here, nutrient inputs can lead to scale-dependent declines in species richness, the opposite of what was discussed above; the SAR has a lower intercept and slope with higher, rather than lower, energy. Some evidence may be evident in systems where nutrient levels are already high, and increased inputs caused by humans (e.g., nitrogen deposition) increases the imbalance (Chalcraft et al. 2008, Donohue et al. 2009). Figure 10.16 shows the influence of the sampling spatial grain on the impact of nitrogen deposition on species richness in an experiment in Mongolian grasslands (data from Lan et al. 2015). Although there was a significantly negative relationship between the level of nitrogen addition and species richness at both spatial grains, the slope of the relationship was more than twice as steep at the larger (25 m^2) grain (slope $=-0.31$) relative to the smaller (1 m^2) grain (slope $=-0.13$).

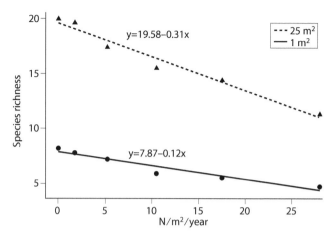

FIGURE 10.16. Effects of nitrogen (N) addition to grasslands (six treatments) on species richness measured at two spatial grains: 1 m² (circles, solid line) and 25 m² (triangles, dashed line). Regression equations for each spatial grain are given, showing the steeper slope observed at the larger spatial grain. Data from Lan et al. 2015.

Here, we have used scale dependence in species-energy relationships to highlight the role that explicit metacommunity thinking can play in understanding large-scale biodiversity gradients. As described above, any driver that alters the numbers of individuals (N), the relative abundance of individuals (SAD), or the aggregation of individuals will also alter the shape of the scaling relationship in that metacommunity, leading to scale dependence in the comparisons between metacommunities that differ in their environmental inputs. Most contemporary biogeography gradients, including energy, elevation, disturbance, habitat isolation, etc., are considered in a scale-agnostic way, and explicit consideration of scale would provide much more insight.

10.10 OTHER MACROECOLOGICAL PATTERNS

As we discussed at the beginning of this chapter, the study of macroecology is an interdisciplinary field that encompasses a number of patterns. In this chapter we have focused primarily on the subsections of biodiversity macroecology and contemporary biogeography, as well as on some other issues that are often included under the macroecology umbrella: biogeography (Chap. 9) and traits and phylogeny (Chaps. 6 and 7). Unfortunately, we have been unable to explicitly deal with many other patterns of interest to many macroecologists—patterns of energetics and metabolic rates and body-size distributions (except in the context of trait distributions), range size (except in so far as it influences occupancies), and abundance relationships (except relative to other species in the context of SADs).

Some progress is being made, however, in linking other macroecological patterns through metacommunity-level processes. Abundance, body size, and range size, for example, are strongly linked statistically, and the combination of them plays an important role in determining patterns of species richness across spatial scales (Damuth 1981, Brown 1995, White et al. 2007). Although these patterns are often described using a statistical mechanics framework (McGill and Nekola 2010), there is some progress in linking metacommunity-level processes. For example, Passy (2012) used a hierarchical metacommunity framework to link power-law patterns of niche breadth and abundance with body size and distributions and used this framework to predict patterns ranging from microbes to fish in stream metacommunities. Nevertheless, there is still much to do in linking the mechanisms that influence species traits, abundance, and distribution in metacommunities with the patterns at macroecological scales.

10.11 CONCLUSIONS

The fields of macroecology (at least the flavors of macroecology that stress biodiversity) and metacommunity ecology are converging; however, they still have different foci. While macroecology tends to concentrate on statistical patterns with a bit less of an emphasis on clear biological mechanisms, metacommunity ecology is more focused on mechanisms. However, metacommunity ecology is often "stuck" with a two-scale perspective of local (α-) and regional (γ-) diversity, and their scalar (β-diversity), whereas macroecology takes a more continuous perspective (i.e., the SAR) that can greatly enhance our understanding of the patterns that emerge and which metacommunity processes underlie.

This difference in emphasis between macroecology and metacommunity ecology—patterns versus mechanisms—might actually be an advantage when the fields are considered together. For example, deviations from patterns that are expected from simple power-law relationships in macroecology can be understood by examining the underlying metacommunity-level processes. Likewise, metacommunity-level studies can at times seem to get mired in a series of special cases regarding the importance of difference processes (dispersal, heterogeneity, etc.), whereas focusing on the macroecological patterns can help find generality in the patterns, as well as connect patterns that might otherwise seem quite disparate.

Finally, as we have emphasized in this chapter, the connection between metacommunity ecology and the kinds of macroecology that concentrate on biodiversity can provide important insights into the context of biodiversity conservation.

Human-mediated changes to the relative abundances of species and their spatial distributions—for example, through the movements of exotic species or alteration of the spatial context (heterogeneity, isolation, area) in which species interact and co-occur—directly affect metacommunity-level processes and macroecological patterns. A tighter connection between these fields will also help us to understand how human modifications to species and the landscapes in which they are embedded will ultimately influence patterns of species abundances and biodiversity.

CHAPTER ELEVEN

Food Webs in Metacommunities

Prospectus

1. Explicitly incorporating food webs (and other multitrophic interactions) into metacommunity ecology is one of the most pressing issues for moving metacommunity ecology forward.

2. Trophic interactions can affect metacommunity processes by several means, including by modifying dispersal, by modifying the ways habitat variables influence distributions, and by modifying the importance of stochastic effects. These changes can lead to strong differences in the ways in which metacommunity structure is partitioned (e.g., by environmental and spatial factors).

3. Trophic interactions can also alter the outcome of metacommunity-level processes, such as the shape of the species-area relationships.

4. Incorporating May's (1973) complexity-stability approach into a metacommunity framework shows that diversity is less constrained than previously thought and varies in direct proportion to the amount of environmental heterogeneity in the metacommunity.

5. Food-web interactions can generate complex feedback patterns that can generate spatial and stochastic distributions of species in a landscape. This is because food-web interactions are particularly prone to creating multiple stable equilibria and endpoint assembly cycles during metacommunity assembly.

6. One instance in which such effects may be particularly important is in the often-described (though controversial) unimodal relationship between productivity and diversity. Metacommunities with strong interactions in food webs suggests that this pattern may be more prevalent at smaller spatial scales and disappear at larger spatial scales, potentially resolving some of the controversy.

7. Combined with spatial linkages, the complex set of interactions that occur in food webs can dramatically increase the network of species interactions we may observe, as well as the spatial links that occur among patches in a metacommunity.

So far, we have almost completely focused on interactions within trophic levels (the theories of competitive metacommunities discussed in Chap. 2 and their

processes discussed in Chap. 3), or we have been agnostic about explicit interspecific interactions (the patterns of composition and biodiversity discussed in
Chaps. 4, 5, and 10). This emphasis reflects the bulk of the existing work on metacommunity theory and can often capture the core processes and patterns observed
in natural systems. However, there are many instances in which other types of
interactions—particularly those between enemies and victims, including predators,
herbivores, parasites, and pathogens embedded within metacommunities—might
fundamentally change the dynamics of these systems.

The inclusion of food-web interactions in the metacommunity perspective adds
complexity that can be viewed from two angles. First, how does the inclusion of
trophic interactions alter our expectations from competitive metacommunities? For
example, predators may respond differently to spatial processes (e.g., habitat heterogeneity, dispersal) than prey communities, and this difference can create expectations for patterns of coexistence and species competition that deviate from
those when only competitors are considered. Second, how does the inclusion of
spatial processes alter our expectations of food webs? For example, the addition
of space may alter the stability and strength of food-web interactions. Clearly, these
are two sides of the same coin (food webs modifying space and space modifying
food webs), but the distinction can help separate distinct ideas (and literatures)
about the interactions between spatial processes and food webs.

In what follows, we first consider how the addition of trophic interactions in
food webs modifies expectations of coexistence and community assembly across
scales relative to competitive metacommunities. Next, we examine how the addition of spatial factors modifies expectations of the strengths of food-web interactions and stability. Finally, we consider the complex feedbacks between trophic
interactions and spatial factors that can create metacommunity pattern and processes that can strongly alter the expectations from competitive metacommunities.

11.1 HOW DO SPATIAL PROCESSES AND TROPHIC INTERACTIONS COMBINE TO INFLUENCE COEXISTENCE IN SIMPLE METACOMMUNITIES?

Regardless of how species interact (e.g., competition, facilitation, trophic interactions), a core set of principles is necessary for stable local coexistence. If we ignore NT, species must experience a set of trade-offs such that each species responds
to, or impacts, the environment and other species slightly differently (e.g., stabilizing niche differences), but these trade-offs are balanced such that species cannot
be too different (equalizing fitness differences; Chesson 2000, Chase and Leibold
2003). Said another way, for coexistence, a species' intraspecific effects on itself

must be greater than its interspecific effects on other species (which may be through modification of resources or any other aspect of the environment, including enemies). A key feature about food webs is that the addition of trophic interactions provides a richer source for trade-offs that might (or might not) facilitate coexistence (Leibold 1996, Chase and Leibold 2003, Chesson and Kuang 2008), and this effect on coexistence can be reinforced by the inclusion of strong intraspecific density-dependence (McPeek 2012). At the same time, the addition of trophic interactions can lead to destabilization of species dynamics and to the loss of one or more species (Noonburg and Abrams 2005). In short, coexistence within trophically structured metacommunities follows the same basic principles of competitive systems, while at the same time leading to a richer array of possibilities for local and regional coexistence by altering the propensity for trade-offs and stability.

The addition of spatial processes—notably dispersal and heterogeneity—modifies the parameters of coexistence of predators and prey in food webs in roughly the same way as space modifies competitive interactions (see Chap. 2; also Amarasekare 2008a); that is, spatial processes can enhance the likelihood of local and regional coexistence of food-web interactions that are unstable in nonspatial systems. For example, the simple addition of dispersal can allow prey to coexist with a predator that can otherwise drive prey extinct in nonspatial systems, as is well known from Huffaker's (1958) classic experiments (see also Luckinbill 1974 and Holyoak and Lawler 1996). In such cases, the instability of the predator-prey interaction leads to frequent local extinctions, but when these are balanced by colonization rates of prey species, the regional dynamics are stabilized and predators and prey can coexist regionally just as competitors can coexist in PD models. This effect can be enhanced when prey have higher rates of dispersal than predators (Hastings 1977, Holt 1984). Habitat heterogeneity can also stabilize an unstable predator-prey interaction; for example, if the prey have a spatial refuge where predators cannot persist (Huffaker 1958, Amezcua and Holyoak 2000). Dispersal and habitat heterogeneity can also increase the likelihood for coexistence in other models that include trophic interactions, such as models of apparent competition, in which two prey species share a common predator (Holt 2002), or keystone predation, in which species share a common resource and a common predator (Shurin and Allen 2001).

Despite the often-positive influence of spatial processes on coexistence in food webs, the realization of stability and coexistence will depend on the balance of stabilizing and destabilizing processes (Amarasekare 2008a). For example, the rates at which predators and prey disperse, the mechanisms by which they disperse (e.g., directed or random), the degree of prey selectivity, the degree to which environments fluctuate, and the interactions between dispersal and habitat heterogeneity will all influence the likelihood that species can coexist in a meta-community (Holt and Barfield 2003, Östman and Chase 2007, Amarasekare 2008b, Gouhier et al. 2010, Cooper et al. 2012, Gounand et al. 2014, Haegeman and Loreau 2014).

11.2 HOW DO TROPHIC INTERACTIONS INFLUENCE METACOMMUNITY PROCESSES?

One of the most straightforward ways to explore the influence of predators on meta-community processes is to examine how metacommunity processes, such as the effects of dispersal, heterogeneity, or drift, are altered when predators are present or absent from a metacommunity of prey species. While in some cases the presence of top predators is simply manipulated by the experimenter, there are a number of cases in which predators vary in their presence, or at least show dramatic differences in their abundance, among metacommunities. For example, predators are often absent or reduced on small islands or ponds (Schoener 1989b, Wellborn et al. 1996, Martinson and Fagan 2014) or as a result of human activities that reduce top predators (e.g., harvesting and predator control). Additionally, there are instances in which herbivory is high (e.g., when top predators are absent, grazing is enhanced), and other instances in which it is lower.

The presence and distribution of these consumers can consequently often (though not always) influence the role of metacommunity-level processes via a number of ways.

11.2.1 Predation Modifies the Effects of Dispersal

A number of studies have examined how top predators alter the influence of dispersal (system openness) on prey metacommunities. A majority of these studies have shown that the presence of predators reduced the influence of dispersal on prey diversity. For example, predators minimized the effects of dispersal on the diversity of protists and zooplankton in experimental aquatic microcosms and mesocosms (Kneitel and Miller 2003, Cadotte et al. 2006). Similarly, vertebrate herbivores reduced the effect of seed additions on plant diversity (MacDougall and Wilson 2007, Maron et al. 2014a). However, sometimes the effects are reversed. Shurin (2001) found that predatory fish (*Lepomis macrochirus*), and to a lesser degree, insects (*Notonecta undulata*), enhanced the influence of dispersal on zooplankton diversity. Chase et al. (2010) found a qualitatively similar result when control ponds with predators present showed a positive relation between connectivity in a metacommunity and herbivore diversity, although the experimental removal of predators eliminated this effect.

The difference between studies showing that predators either reduce or enhance the influence of dispersal on species diversity likely arises because predators can have two types of effects that may differ in relative importance due to metacommunity context (Proulx and Mazumder 1998, Chase et al. 2002, Katano et al. 2015). Predators can increase species diversity of their prey when they preferentially consume competitively dominant species (i.e., a keystone predator; Paine 1966, Leibold 1996, Terborgh 2015), or they can decrease prey diversity by reducing overall abundances of prey, including particularly rare or vulnerable species, and thus reduce net dispersal rates (Spiller and Schoener 1998, Heinlein et al. 2010, Begley-Miller et al. 2014).

It is frequently observed that predation can diminish the way dispersal can enhance diversity, possibly because dispersal tends to allow species that are rarer or not particularly well suited to a given habitat to persist. These are the same types of species that might be particularly vulnerable to predation impacts, even if predators are not necessarily selective on these species. Alternatively, when predators reduce the more competitively dominant species, they can increase the opportunities for competitively subordinate species, which may persist due to competition-colonization trade-offs, to persist when dispersal rates are high (Shurin and Allen 2001).

11.2.2 Predation Modifies the Effects of Habitat Heterogeneity

Predators can also influence how prey communities respond to habitat heterogeneity. Kneitel and Chase (2004) found that protist community structure was strongly influenced by habitat heterogeneity in resource abundance, but that these resource effects on protist community composition were largely negated when larval mosquito predators were present. Likewise, Suzuki et al. (2013) found that the influence of habitat heterogeneity on plant community composition declined as the density of deer herbivores increased. In these cases, it seems that effective generalist predators can homogenize prey communities that would otherwise be likely to finely partition habitat heterogeneities.

The presence of predators can also increase the influence of habitat heterogeneity on coexisting prey species. For example, in marine systems, habitat heterogeneity in the form of tidal height (i.e., time exposed to air during low tide) creates a strong SS gradient, known as *vertical zonation*. Zonation creates bands that tend to be dominated by species that are more desiccation resistant at higher tidal heights and by species that are more predator resistant at the lower tidal heights (because predators tend to be more susceptible to desiccation). The removal of top predators from these systems, such as in Connell's (1961a) and Paine's (1966) classic experiments, allows species that are more susceptible to predators to venture deeper into lower tidal zones, suggesting that predators enhance the influence of habitat heterogeneity in these cases. These contrasting effects may depend on whether predators also respond to habitat heterogeneity; when they do, they may enhance the influence of heterogeneity (as in the case of the tidal marine systems), and when they do not, they may reduce the influence of habitat heterogeneity (as in the case of the protist system described above).

11.2.3 Predation Modifies the Effects of Drift

Because predation influences the role of dispersal and habitat heterogeneity, as well as local diversity and the size of the realized species pool, it follows that predation can influence the relative importance of stochastic drift. However, the influence of predators on metacommunity assembly will depend on the intensity of predation and the degree of specialization among the predators (Ryberg et al. 2012).

If predation is intense and predators are generalists, we expect that predators should reduce local community size (overall density or number of individuals per patch) but have little influence on the regional pool of species. Because of the reduced community size, the likelihood of ecological drift will be higher (Orrock and Fletcher 2005, Orrock and Watling 2010), and this will in turn result in higher among-site variation in community composition (higher β-diversity) even in the absence of habitat heterogeneity (Ryberg et al. 2012). Alternatively, if predators are more selective, or certain prey are more sensitive to predation, predators act as a strong filter that can potentially inhibit species' local communities; that is, they can reduce the "realized" pool of species in these local communities. This can in turn more strongly constrain variation in community composition and thus reduce stochasticity (Ryberg et al. 2012).

To date, only a handful of empirical studies have examined the influence of predators on metacommunity assembly and the relative importance of ecological drift or determinism. Nevertheless, it is clear that predators can often alter patterns in β-diversity, but the direction by which they do so varies. On the one hand, Stier et al. (2013, 2014a) examined the influence of two relatively generalist predatory fish (*Paracirrhites arcatus, Cephalopholis argus*) on community assembly of coral reef fish. After a settlement period, Stier et al. found that β-diversity (measured as null deviations from expectations) was higher on reefs with fish relative to reefs without fish, which they attributed to higher demographic stochasticity when local community size was small (as a result of intense predation). On the other hand, Chase et al. (2009) found that predatory fish in small ponds led to reduce β-diversity among invertebrates and amphibians compared with fishless ponds, a result they attributed to the fact that fish preferentially consume some species relative to others, restricting the realized species pool. This core result, in which predators reduced β-diversity, has also been found in a study of rodent seed predation on plants (Germain et al. 2013), *Daphnia* grazing on microbes (Berga et al. 2015), and salamander predation on zooplankton (Zokan and Drake 2015).

These results are consistent with the theoretical predictions of Ryberg et al. (2012), who suggested that the influence of predators on metacommunity assembly will be at least partially due to their level of specialization (generalists should increase, while specialists should decrease, β-diversity relative to controls). Johnston et al. (2016) more directly tested this idea in laboratory microcosms of protist prey, comparing the effect of a specialist (*Euplotes aediculatus*) with that of a generalist predator (*Stentor coeruleus*), and found partial support: the generalist increased β-diversity relative to the controls, whereas the specialist had no influence on β-diversity.

As we described in depth in Chapter 4, the simplest expectations of metacommunity assembly theory are that environmental characteristics [E] should explain a large proportion of the variation in species composition among sites when SS processes are important, but that spatial [S] and random [Resid] factors will explain more variation when PD or NT predominate (with ME being somewhat intermediate). However, as we discussed above, trophic interactions can either increase or decrease the response of prey communities to habitat heterogeneity, and thus the influence of [E]. In addition, as we described briefly in Chapter 5 and will do more in depth below, trophic interactions are also more likely to create spatial [higher [S]) or unexplained variation (higher [Resid]) as a result of frequency dependence (e.g., Connell-Janzen like effects), MSE, and EACs. Thus, the presence of strong trophic interactions can either enhance or reduce the structuring of metacommunities by environment and space, as well as the possible conclusions we might reach about the processes underlying their structure.

To date, the only clear examples that have examined the influence of predation on metacommunity structure are those in which predators are not strongly influenced by habitat heterogeneity, and we might thus expect that predators would reduce the influence of environmental heterogeneity (lower [E]) and enhance [S] and [Resid]. For example, revisiting the surveys of freshwater ponds like those described in Chapter 4, a number of ponds have top predatory fish, whereas others that are otherwise similar in environmental conditions do not (Chase et al. 2009). Although not presented in the original study, we analyzed the data from those surveys to compare the variation explained by [E], [S], and [Resid] (Table 11.1). What is immediately clear is that, similar to the results presented in Chapter 4, [S] explains little of the variation regardless of fish presence. However, the relative importance of [E] in structuring the metacommunity is much higher in the

TABLE 11.1. Comparison of Variation Partitioning Results from Ponds
with and without Top Predatory Fish

Treatment	Unexplained [Resid]	Total Space [S]	Total Environment [E]
Fish absent	0.59	0.02	0.39
Fish present	0.78	0.05	0.17

ponds without fish (~40%) compared with that in ponds where fish are present (less than 20%). Although the effect was much smaller, Ohashi and Hoshino (2014) suggested a similar predator-induced reduction in [E] when comparing plant metacommunity structure before and after the dramatic increase in the abundance of Sika deer (*Cervus nippon*) in a national park in Japan.

Another approach to examining the influence of predators on metacommunity assembly is to explicitly include them as explanatory variables in a variation partition. Livingston et al. (2017) used this approach in an experimental metacommunity of microcosms with three protist species—two prey (*Colpidium* and *Paramecium*) and a predator (*Didinium*) that primarily eats one (*Paramecium*) prey but not the other—along with environmental heterogeneity (autochthonous patches, where the protists are supported by algal production only, vs. allochthonous patches, where most of the production comes from the bacterial decomposition of a wheat seed). As above, because of the presence of the predator, the amount of variation in the abundances of the species that was explained by environmental conditions was low when the distribution of predators was ignored; much lower than in a previous experiment with these two prey species but no predator (Fukumori et al. 2015). However, when the abundance of the predatory *Didinium* was used as an explanatory factor, a much higher proportion of the variation was explained. Similarly, Ruppert et al. (2013) found that habitat [E] explained approximately 20% of the variation in coral-reef community structure among atolls off the northeastern coast of Australia, but the inclusion of spatial variation in fishing pressure on sharks (which released smaller predators, causing a trophic cascade) explained another approximately 13% of variation in community structure.

11.3 TOWARD A THEORY FOR TROPHICALLY STRUCTURED METACOMMUNITIES

MacArthur and Wilson's (1967) equilibrium theory of island biogeography (ETIB) is a highly simplistic view of metacommunities that has nevertheless proved to be an extremely valuable baseline for understanding a number of problems in ecology and evolution (Losos and Ricklefs 2009, Warren et al. 2015). We can use this as a starting point to think about the dynamic properties of trophically structured metacommunities using the same core assumptions as the ETIB (i.e., local communities structured as a sample of the larger regional pool, dynamics described by colonization-extinction dynamics, minimal differences among species, homogeneous environmental conditions). Simple modifications of the ETIB to include trophic interactions, while clearly far too simplistic, can provide useful insights into patterns expected when trophic structure is explicitly considered.

For example, based on the frequent observation that predators increase the likelihood of prey extinction (Huffaker 1958, Holyoak and Lawler 1996, Schoener et al. 2001b), Ryberg and Chase (2007) modified the ETIB with the simple assumption that the presence of a generalist predator on an island (or patch) should increase the extinction rates of prey species on those islands (Fig. 11.1a; see also Holt 2009). From this, they predicted that the increase in species richness with increasing island size should be smaller in the presence of predators than in their absence, leading to a lower slope (z) of the island species-area relationship (ISAR; see Chap. 10; Fig. 11.1b). Figure 11.2 shows support for this expectation from two different ecosystems (from Ryberg and Chase 2007): grasshoppers in rocky outcrops with and without lizard predators (Fig. 11.2a) and zooplankton in small freshwater ponds with and without fish predators (Fig. 11.2b).

The simple assumption of how predators can alter the ISAR presented by Ryberg and Chase (2007) is only one of a number of ways that predators can alter the colonization and extinction rates of prey (see Holt 2009). Taking a complementary viewpoint, Resetarits and Binckley (2013) argue that prey species can often choose habitats in the face of predation risk such that prey colonization rates, rather than extinction rates, might be reduced in the presence of predators. In this case, the expected outcome of shallower-sloped ISARs would be similar, but the mechanism (colonization vs. extinction) would be different.

Other, more complex, scenarios are also possible. For example, some types of consumers, like herbivores, may simultaneously increase the extinction rates of plants and also their colonization rates, leading to more complex outcomes (Olff and Ritchie 1998). De Bello et al. (2007) found that grazer effects on plants appeared to vary with climatic regime. In wetter areas, grazers reduced the slope of the nested SAR in a similar way as found in the ISAR patterns discussed above, whereas in more arid conditions, grazers increased the slope of the nested SAR, possibly as a result of altered colonization rates. Likewise, in another arid region, Supp et al. (2012) found a negligible influence of the long-term exclusion of seed-predatory rodents on nested SAR patterns in plants, even though plant species composition was rather distinct in plots with and without seed predators. In this

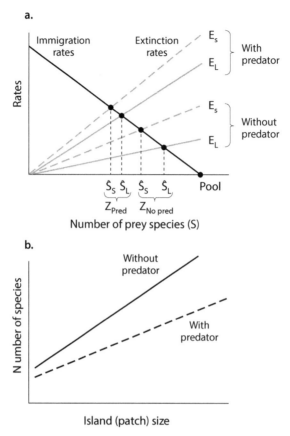

a.

b.

FIGURE 11.1. (a) Extension of MacArthur and Wilson's equilibrium theory of island biogeography by adding the simple assumption that predators increase the extinction rate of prey populations (i.e., by reducing their numbers and increasing their chances of stochastic extinction). There is only one immigration rate (I) assumed, while there are multiple extinction rates. E_s and E_L refer to extinction rates in small and large islands, respectively, and these can occur with or without predators. Expected equilibria values of species ($\hat{S}$) are given for small ($\hat{S}_s$) and large ($\hat{S}_L$) islands, either with or without predators. The slope of the island species-area relationship (ISAR) (z) is approximated by the difference between $\hat{S}_s$ and $\hat{S}_L$, which shows that z is expected to be higher (steeper) with no predators present. Modified from Ryberg and Chase 2007. (b) Extension of this model across multiple island sizes to indicate the expected ISARs on islands or patches without predators (steeper slope) compared with those with predators.

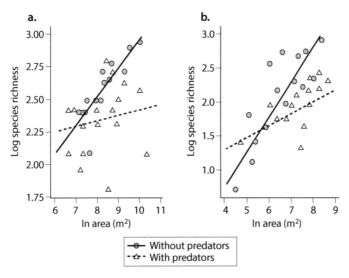

FIGURE 11.2. How predators alter the island species-area relationship of prey. (a) Species richness of grasshoppers (Acrididae and Tettigonidae) on rocky outcrops (glades) in Missouri with (gray triangles, dashed lines) and without (black circles, solid lines) predatory lizards (*Crotophytus collaris*); modified from Ryberg and Chase 2007. (b) Species richness of zooplankton (Cladocera and Copepoda) in small ponds with and without predatory fish (*Lepomis* spp.); modified from Ryberg and Chase 2007.

case, when seed predators were excluded, the plant composition shifted toward species that are susceptible to seed predators (i.e., those with large seeds). This differential influence of predators on some prey species rather than others, although probably quite common, clearly violates the assumptions of the simple ETIB model with predators added, which assumes that predators influence the extinction (or colonization) rates of all species equally.

Despite its simple (and perhaps often too simplistic) assumptions, modifications of the ETIB have also proved useful for exploring conditions under which both predators and prey have explicit colonization-extinction dynamics. For example, using a number of colonization-extinction-based models and arguments, Holt has suggested that predators and prey should differ in the shape of their SARs (Holt 1993, 1996, 2002, 2009; Holt et al. 1999). Because predators are often larger and have higher energy requirements than their prey, Holt assumed that predators also have lower population densities or higher area requirements, or both, and so are also more likely to go locally extinct if their prey are extinct. As a result, (1) predators should on average have higher rates of extinction than their prey species, and (2) predator extinction rates should be more sensitive to variation in habitat

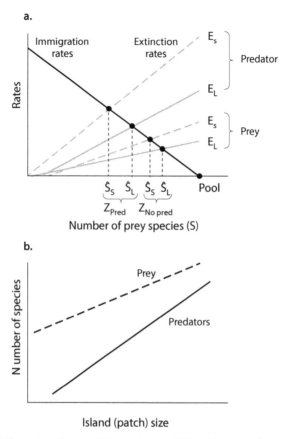

FIGURE 11.3. (a) Extension of the equilibrium theory of island biogeography similar to Fig. 11.1 but where extinction rates of prey and predators are compared. Here, the slope (z) of the island species-area relationship (ISAR) is expected to be higher (steeper) for predators relative to prey; modified from Holt 2009. (b) Illustration of the expected ISAR for predators and prey given the predictions of (a).

size than prey. With these two linked assumptions, one can manipulate the ETIB (Fig. 11.3a) to predict that predators should have a lower intercept but steeper-sloped SAR than that of prey (Fig. 11.3b).

There are several case studies that qualitatively support the prediction that predators should have lower intercepts but steeper slopes in their SARs than their prey. These include Spencer et al.'s (1999) comparison of predatory and prey macroinvertebrates in freshwater ponds, Kruess and Tscharntke's (2000) analysis of herbivores and parasitoids on plants in meadows of different sizes, Hoyle and

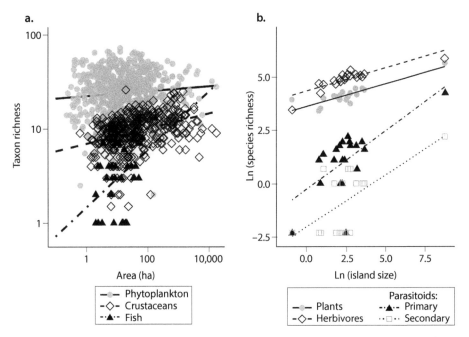

FIGURE 11.4. (a) Island species-area relationship (ISAR) patterns in North American lakes (expressed as log-log scaling plots) for planktonic algae (producers), planktonic zooplankton (mostly grazers), and fishes (mostly planktonic carnivores). Data are derived from single time-point surveys to avoid the possible effects of turnover and/or seasonality. Data for algae were collated from the EPA-National Eutrophication Lake Survey (Taylor et al. 1979), data for zoo-plankton were collated from the Eastern Lakes Survey (Tessier and Horwitz 1990), and data for fishes were collected from the Adirondacks Lake Acidification survey (Baker et al. 1990; lakes with pH < 5.0 were not included in this analysis to isolate the effects of lake area from effects of lake acidification). (b) ISAR patterns of plants, lepidopteran herbivores, primary parasitoids, and secondary parasitoids from Baltic Sea islands off of the coast of Finland; modi-fied from Roslin et al. 2014.

Gilbert's (2004) study of predatory and prey mites in experimental moss patches of different sizes, and van Noordwijk et al.'s (2015) analysis of beetles in calcare-ous grasslands across Europe. Figure 11.4 presents two cases that illustrate this generalized relationship. Figure 11.4a comes from our own collation of phytoplank-ton, zooplankton, and fish data from lakes across North America that vary in sur-face area. Figure 11.4b comes from Roslin et al. (2014), who studied the ISARs of plants, lepidopteran herbivores, and primary and secondary parasitoids on Baltic Sea islands.

One class of interactions that may be an exception to this expectation of lower intercepts and steeper slopes of predator SARs relative to those of prey is that comprising parasitic enemies (including insect herbivores, which act as parasites of plants). For example, a single species of plant can harbor tens to hundreds of specialized herbivorous insects (Price et al. 1995, Novotny et al. 2006), not to mention many hundreds more parasitic, commensal, and mutualistic microorganisms. Likewise, there are hundreds to thousands of parasitic, commensal, and mutualistic organisms that can live within a single species of animal (Poulin 1997, Ley et al. 2008). Here, the intercept of the SAR of these parasites should be higher than the intercept of their hosts (there are many more species of parasites than hosts). However, we would still expect the slope of the SAR to be steeper for parasites than for hosts, because each new species of host will bring along with it its own complement of parasites.

Because the ETIB considers habitat isolation and habitat area to have opposite effects on diversity patterns, we also have a theoretical expectation that predator diversity should be more strongly influenced by habitat isolation than prey diversity (Ryall and Fahrig 2006, Gravel et al. 2011a). This prediction is also borne out in several case studies. For example, Shulman and Chase (2007) placed small mesocosms at different distances from natural ponds and found that the richness of predator species declined much faster than that of prey species (Fig. 10.5; see also Chase et al. 2010). Likewise, Kruess and Tscharntke (1994) found steeper decreases in parasitoid species richness compared with herbivore richness in experimental "islands" of red clover with different levels of isolation, and Belmaker et al. (2005) found steeper declines of predator relative to prey richness on artificial coral reefs placed near and far from a large intact reef.

Not surprisingly, there are a number of reasons why the simple predictions of predator and prey SARs are not always expected or observed. Extinction rates and habitat sensitivity are assumed to be higher for predators than for prey in Holt's (2009) models, as well as its extensions (Gravel et al. 2011b). This might not be the case if predators are generalists that can use matrix, as well as island, habitats, or if predator and prey colonization rates differ. For example, Stier et al. (2014a) found no differences in the slopes of coral reef fish predator and prey species richness with habitat size (reef volume), and no differences in the effects of the presence or absence of generalist predatory groupers on prey fish richness with reef volume. Likewise, Stier et al. (2014b) suggested that predatory coral reef fish might have longer dispersal distances, and thus higher colonization rates at more distant patches, leading to predictions that matched their observation of higher predator-prey richness ratios in the more isolated reef systems, rather than the lower richness ratios that are expected when predators have lower colonization rates (Shulman and Chase 2007, Gravel et al. 2011b).

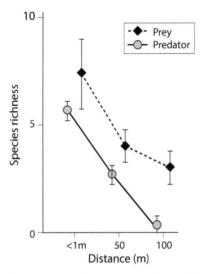

FIGURE 11.5. Effect of placing experimental mesocosms at three different distances from natural freshwater ponds. Predator species richness (e.g., dragonflies, beetles) declined more steeply with isolation than did prey species richness (e.g., flies, zooplankton). From Shulman and Chase 2007.

While the simple modifications of the ETIB to include differential effects on predators and prey (Ryberg and Chase 2007, Holt 2009) provided useful predictions of more general patterns, there are many more interesting observations that can be made about a trophically structured metacommunity than simply patterns of species richness and composition. One obvious point is that consumers are dependent on the presence of resources (i.e., they require adequate resources to be able to occupy a patch for more than a generation), whereas resources are not generally dependent on the presence of the predator (and typically suffer greater extinction when the predator is present). This is especially so for specialist consumers. Figure 11.6 shows a simple cartoon of this scenario by increasing the scale of observation from one to a few patches where only simple interactions are observed, to many patches where the combination of all interactions create a more complex food-web structure that has many of the attributes (such as link connectance) that are seen in real food webs but are often difficult to reproduce in theoretical models without space (Pillai et al. 2009, 2011).

When the differences between predators and prey are explicitly considered in a metacommunity model, we can also explore how trophic traits, such as diet breadth and interaction network connectance, might be influenced by spatial processes. For example, by assuming that consumer species differ in their diet breadth and associated persistence probabilities, Gravel et al. (2011b) tested predictions of

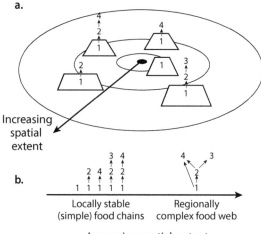

FIGURE 11.6. How a more complex food-web structure at larger spatial extents can emerge from a simple food-web structure at smaller spatial extents. (a) Each local patch (depicted by squares) has a simple one- (Species 1 alone), two- (Species 1 and Species 2, 3, or 4), or three-species food chain (Species 1, 2, and 3, or 1, 2, and 4). (b) When spatial extent is increased and the food webs are aggregated more complex food webs including more link types, and more species, can emerge. Modified from Pillai et al. 2011.

the trophic ETIB with an analysis of food webs (e.g., fish, zooplankton) among 50 lakes in the Adirondack Mountains of New York based on the data from Havens (1993). In accordance with their predictions, Gravel et al. found that the regional occupancy of more specialized species was lower than that of more generalized species and that the number of links in the food web scaled more rapidly with the numbers of species than expected if interactions were random.

Using a similar approach, Gravel et al. (2011a) argued that connectance (number of trophic links per consumer), diversity, and colonization rate can all enhance the persistence of complex food webs at the regional scale even when local-scale dynamics might be very dynamic and not persistent. Thus, while diversity and connectance constrain local stability, they enhance the persistence of much more complex webs at the regional scale. Importantly, these models typically assume that patches are homogeneous; the inclusion of environmental heterogeneity would presumably result in even greater diversity and persistence at the regional scale.

11.4 SPATIAL PROCESSES AND MAY'S
DIVERSITY-STABILITY THEORY

As we described above, natural food webs are often much more complex than what simple food-web theory can predict, and it is well known that adding a spatial (metacommunity) perspective can reconcile some of this discrepancy (Holt 1993, 1996, 2009; Polis 1999; Gravel et al. 2011a, 2011b; Pillai et al. 2011). One particularly pernicious prediction from simple food-web theory that remains a continual conundrum to both theoreticians and empiricists alike is May's (1972, 1973) prediction that more complex/diverse food webs should in fact be less stable than simpler food webs—a prediction that flew in the face of earlier conjectures by Elton (1958) and MacArthur (1955), who argued that diverse communities should be more stable and less easily perturbed.

May (1972, 1973) examined the consequences of random interaction matrices with different numbers of species (S), different proportions of linkages between pairs (c=connectivity), and different mean interaction strengths (σ) in comparison with intraspecific interaction strength (we will use the term m, but May standardized his equations so that his formulation of the problem is based on $m = 1$). Using the method of Gershgoring circles (see McCann 2011, for a particularly useful exposition), he showed that the dominant eigenvalue of the Jacobian of the interaction matrix (the matrix that summarizes all pairwise interactions in the community) was likely to be greater than zero, and thus that the equilibrium point of the system was unstable if

$$\sigma\sqrt{c(S-1)} < m$$

(following Allesina and Tang 2012, we express this result a bit differently and more precisely than May did).

This straightforward result of May's indicates that complex (high c), speciose (high S), and strongly interacting (high σ) food webs would be unlikely because they would violate this inequality unless intraspecific population regulation was also particularly strong (high m). In practice it seems difficult to create food webs with more than a dozen or so species and still ensure stability of the equilibrium point (see, e.g., Case 2000). Many ecologists were somewhat perplexed by May's (1973) predictions, and they did not believe the predictions to be true in natural communities, but it was hard to deny the fact that the predictions were reasonably robust unless a variety of features were added to the food web (e.g., nonrandom interactions, strong and weak interactions, spatial coupling; reviewed in McCann 2000, 2011; Rooney and McCann 2012, Loreau and de Mazancourt 2013). Neutel et al. (2007) called this tension between the theory (predicting very simple communities) and data (showing highly complex food webs) "May's paradox of diversity."

Even though a number of studies have suggested that spatial processes might be one mechanism that could stabilize what might otherwise be unstable interactions (McCann et al. 2005, Gravel et al. 2011a), these theoretical studies were not directly linked to May's analysis of the stability criterion. Gravel et al. (2016) tackled this problem by making a direct expansion of May's approach to incorporate metacommunity dynamics (we call this the *meta-May hypothesis*). To do so, they followed up on the suggestion by Svirezhev and Logofet (1983) that one could model a metacommunity by having distinct equations for each species in each patch of the metacommunity. The resulting matrix (Fig. 11.7) describing these interactions would consist of one set of submatrices that describe the interactions in each patch (i.e., matrices like May's) and can be arranged along the diagonal of the meta-May matrix. The off-diagonal matrices describe interactions between patches and thus, dispersal. It is worth noting that these off-diagonal matrices will generally have many zeros since species retain their identity when they disperse. The resulting meta-matrix would thus have low connectance (non-zero off-diagonal elements).

More formally we can define the meta-May matrix as a Jacobian matrix:

$$\mathbf{J} = \mathbf{M} + \mathbf{D} + \mathbf{A}$$

where

- $\mathbf{M}$ is the diagonal matrix with value $-m_{i=j}$ on the diagonal and 0 in the rest of the matrix, which describes the strength of intraspecific population regulation by the ith species in the jth patch (in May 1973, this effect is always standardized to value of -1);
- $\mathbf{D}$ is the matrix representing dispersal among patches;
- $\mathbf{A}$ is the collection of local Jacobian matrices, arranged as diagonal blocks that describe feedbacks due to species interactions in isolated communities (except for diagonal intraspecific regulation terms that are contained in the elements of $\mathbf{M}$).

May's matrices thus consist of combining $\mathbf{M}$ and $\mathbf{A}$ (we keep them separate for reasons we describe below), and the meta-May matrix $\mathbf{J}$ includes them in combination with $\mathbf{D}$. The $\mathbf{J}$ matrix is thus of size $n \times S$, where n is the number of patches in the metacommunity. Note that with this formulation, because there is a continuous supply of immigrants from other patches, all species are found in every patch, even those where they might have extremely low fitness.

The details of the analytical approach are too complex to go into detail here (and unnecessary for our main points), but see Gravel et al. 2016 for details. However, a few points are worth noting. First, assuming that S and n are both large

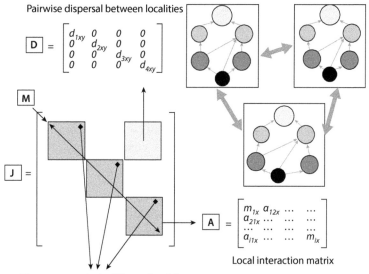

FIGURE 11.7. Schematic of the "meta-May" matrix of Gravel et al. (2016). The rows and columns are ordered by species nested within patches, and there is an equation (row) for each species of the metacommunity in each patch, forming a square submatrix (dark shading) that describes the local interactions within each patch. Each of these is similar to a May-matrix except for the additional role of emigration by each species that modifies the strength of intraspecific regulation (along the main diagonal). The remainder of the matrix consists of submatrices that describe dispersal among patches. In the figure this is shown for a single pair of patches with light shading. The elements of this matrix describe the rate of dispersal from one patch (identified by the column) into another (the row). It will have only diagonal elements since species don't change their identity when they disperse! It is possible to model a diverse array of situations with this framework, manipulating for example the dispersal network and accounting for asymmetrical dispersal although this remains to be done.

and that dispersal rates (d) are identical for all species across all patches and also large, we obtain the criterion

$$\sigma\sqrt{c(S-1)\left(\frac{1}{n}+\frac{n-1}{n}\rho\right)} < m,$$

where ρ is the among-patch correlation among interaction coefficients occupying the same place in the different submatrices in $\mathbf{A}$. If the correlation is zero, then the stability criterion is

$$\sigma\sqrt{c(S-1)/n} < m$$

This inequality indicates that the spatial effects stabilize the dynamics in proportion to the number of patches (n) in the metacommunity when compared with the criterion for isolated communities and thus quantifies the stabilizing effect under these conditions. To the degree that different local communities have similar conditions (ρ increases), the spatial effect is weakened, and if all the local communities are identical, it disappears. We can thus think of n as the number of ecologically "independent" local communities (for which the **A** matrices for the different subcommunities are uncorrelated with each other).

If d is small, the effect is different. Here the criterion is

$$\sigma\sqrt{c(S-1)} < m + d.$$

Thus, here it is only the dispersal rate (d) that increases stability, and it does so additively in comparison with the criterion for isolated communities.

To determine what happens if d is intermediate or if the metacommunity is not so large, Gravel et al. (2016) used numerical methods. They studied a food web with 15 species in a metacommunity that consisted of 10 patches (much less than the assumption of large S and n above; Fig. 11.8). They found that stability first increased with increasing dispersal (lower dominant eigenvalues) and was highest (lowest eigenvalue) at intermediate levels of dispersal. They attributed these findings to a generally stabilizing effect of dispersal among heterogeneous patches; however, at higher levels of dispersal, stability declined and became insensitive to further changes in dispersal. The decrease can be attributed to a homogenization effect on species densities in patches and a tendency to synchronize them. Nevertheless, this homogenization-synchronizing effect is never enough to bring the eigenvalue back to that of isolated communities because of the heterogeneous pattern of species interactions that stabilizes them, as in the analytical model.

These results thus indicate that metacommunity structure can be a powerful stabilizing feature and may help to resolve May's paradox. Except when dispersal is very low, it suggests that maximum permissible diversity (as well as complexity and interaction strength) may well be regulated more by the ecological size of the metacommunity than by the nature of species interactions. Explaining food webs that contain hundreds, or more, of species (even at the local scale) may require only understanding how large the metacommunity is. There is yet insufficient current knowledge about the variation in species interactions at the community level to really know how well this hypothesis may work to explain discrepancies between May's original model and data from highly diverse ecosystems.

Another intriguing consideration is how dispersal effects may differ in food-web models compared with competition models, such as the ME model by Mouquet and Loreau (2003). In that model, very high dispersal led to homogenization and

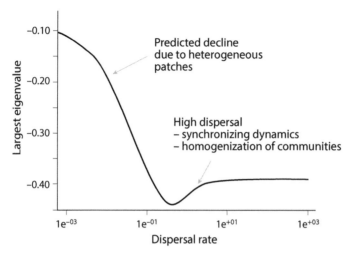

FIGURE 11.8. Effect of dispersal on the stability of the metacommunity. The stability is measured by the value of the largest eigenvalue (higher values imply less stability). The dispersal
rate is plotted on a log-transformed axis. At low values of dispersal the metacommunity is least
stable. As dispersal increases stability increases to a maximum value (here at values of approximately $2e^{-1}$) and then decreases at high dispersal rates (where the metacommunity becomes
increasingly homogenized). Modified from Gravel et al. 2016.

reduced regional diversity (and consequently also local diversity); this did not occur
in the metacommunity food-web model of Gravel et al. (2016). However, the two
models are not directly comparable because there is only one niche per patch in
the competition model, whereas there are many in the food-web model, but it does
indicate that food-web interactions might modify metacommunity dynamics and
their predictions in important ways.

Of course, there probably is still an important role for other mechanisms in
moderating the diversity-stability relationship (e.g., frequency-dependent habitat
switching and coupled weak-strong pathways). And the influence of dispersal can
be more complex than that assumed in the simple meta-May model of Gravel et al.
(2016). For example, Amarasekare (2008a) showed that variation in dispersal by
edible prey in a diamond-shaped food-web module (a predator with edible and
inedible prey) could affect coexistence in a metacommunity, whereas variation in
dispersal by the resistant prey did not. She coined the term *keystone dispersers* to
describe this general possibility. Although Gravel et al. (2016) ignore these sub-
tleties, their results do show that there is a broad general context for them in highly
diverse food webs in large metacommunities.

11.5 FREQUENCY DEPENDENCE AND FEEDBACKS BETWEEN
TROPHIC INTERACTIONS AND SPATIAL PROCESSES

A number of complexities of spatiotemporal organization of metacommunities are likely to emerge when trophic interactions are explicitly considered (see Chap. 5). For example, in the simplest case described above (Sec. 11.1), locally unstable dynamics, such as that between a predator and prey, are stabilized at larger scales when space is explicit and dispersal is localized (Huffaker 1958, Holyoak and Lawler 1996). In fact, this situation is likely to be a reasonably general phenomenon. The addition of spatial factors, including limited dispersal, heterogeneity, and simply averaging across scales, will generally lead to more diversity and more stability at the larger spatial scales. The interplay between time and space will create a mosaic of spatiotemporal patterns of species' relative abundances and coexistence.

These spatiotemporal patterns mediated by trophic interactions can, for example, weaken observed relationships between species compositional variation and environmental variation. However, a closer look at these spatiotemporal patterns, as mediated by trophic interactions, can also provide deeper insight into the mechanisms at play. For example, when trophically mediated spatiotemporal processes are important, patterns of compositional variation in a metacommunity might appear dispersal limited or even random, suggesting support for NT or PD, when in fact the process might actually be much more deterministic. Furthermore, these processes can create metacommunity-level patterns of co-occurrences and diversity relationships that differ significantly from those that might be expected under competitive scenarios.

A classic example of how spatiotemporal patterns mediated by trophic interactions can alter expected patterns of coexistence and spatial distributions of species in a metacommunity is the Janzen-Connell mechanism of coexistence (Janzen 1970, Connell 1971). Here, when enemies are sufficiently specialized and recruit to areas where their preferred prey are locally more common (giving locally rarer species a local advantage), there is sufficient space and aggregated dispersal of predators and prey, and this mechanism alone can allow the coexistence of many species. Importantly, coexistence is, among other things, contingent on the degree of specialization of the enemies (Sedio and Ostling 2013) and their scales of movement (Adler and Muller-Landau 2005, Stump and Chesson 2015). Support for this hypothesis, at least in terms of spatially differential damage and mortality mediated by local densities, has been reasonably forthcoming, at least in the forested systems for which the hypothesis was originally developed (Hyatt et al. 2003, Comita et al. 2010), as well as some evidence in other systems (Petermann et al. 2008). Likewise, the emerging field of plant-soil feedbacks, though more complex, leads to similar patterns of spatially mediated coexistence due to frequency-dependent interactions with both soil enemies and soil mutualists (reviewed in Bever 2003, Kulmatiski et al. 2008, and van der Putten et al. 2013).

While the Connell-Janzen hypothesis typically focuses on the role of enemies in mediating spatial composition and diversity via variation in local densities of predator and prey (or pathogen and host), recent work has also suggested that similar processes can create larger spatiotemporal variation when there are coevolutionary processes between enemies and victims. For example, Benitez et al. (2013) expanded the Janzen-Connell hypothesis to include local adaptation in the host-pathogen interaction, and Loeuille and Leibold (2014) similarly showed that local adaptation could drive distributions of ecotypes. Here, it is possible that a more generalized pathogen can maintain diversity by differentially adapting to the defenses of the most common host species. Likewise, Ricklefs (2011, 2012, 2013, 2015) argued that localized coevolutionary dynamics between parasites and hosts can allow the coexistence of a diversity of species with otherwise similar traits, but that it will create spatiotemporal dynamics in relative abundances (and pathogen resistance) of species (see also Bever et al. 2015).

A more complex spatiotemporally dynamic process can occur at the metacommunity scale when colonization is slow relative to the internal dynamics of the community. One useful theoretical tool to study this is the idea of "permanence" developed by Law and Morton (1993), in which the assembly process is an iterative sequence of alternative colonization events (by a single species at a time) and population dynamics to steady-state behavior. In this scenario, three long-term general outcomes are possible: (1) a single stable equilibrium can be achieved regardless of which species entered first; (2) MSEs contingent on initial conditions; and (3) EACs, in which assemblages are continually replacing each other, such as the classic "rock-paper-scissors" scenario of nontransitive interactions discussed in Chapter 5. Importantly, the sort of history/frequency dependence that lead to both MSEs and EACs are much more likely to occur when there are trophic interactions.

Holt (2009) gave a simple example of how the slow assembly process can lead to EACs (which he called "community churning") in islands or patches even when the full configuration of species would otherwise persist stably (Fig. 11.9). Although the full system of species (two predators, two prey) could potentially coexist (Fig. 11.9a), this situation is not realized when dispersal is relatively slow compared with local dynamics, such as might occur in patches or islands of a metacommunity (Fig. 11.9b). For example, if both prey colonize an empty patch before either predator does, the addition of one of the predators will lead to the overexploitation of the prey species on which it is most efficient and instead it will persist on the prey species that it is less effective at consuming. However, when

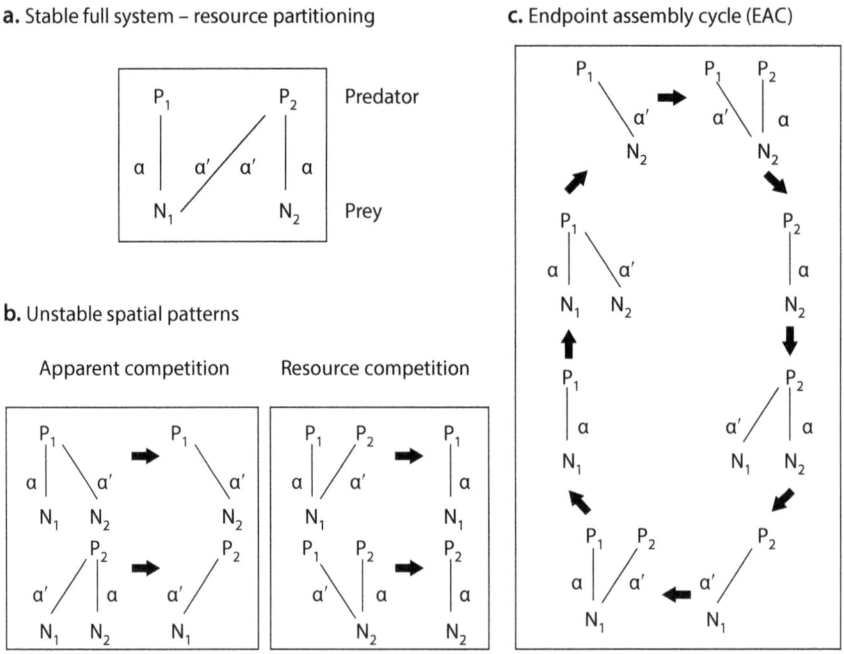

a. Stable full system – resource partitioning

b. Unstable spatial patterns

Apparent competition   Resource competition

c. Endpoint assembly cycle (EAC)

FIGURE 11.9. "Community churning" in a four-species community of two predators (P₁ and P₂) and two prey (N₁ and N₂). (a) All four species can stably coexist under a variety of parameter values as long as the two predators have a resource partitioning trade-off (P_1 is a better consumer of N_1, denoted with α, and a worse consumer of N_2, denoted with α', whereas the opposite is true of P_2) and the two prey do not compete with one another. (b) If we assume that local dynamics are faster than dispersal, it is possible that if both predators are present but only one of the prey species is present, the predator which is a less effective predator on that prey will go extinct, leaving only two-species predator-prey pairs (left side). Alternatively, if both prey species are present with one of the predator species, the prey species that is the superior apparent competitor (lower predation rate) will persist alone with the predator (right side). (c) The endpoint assembly cycle emerges during community assembly because each three-species combination is unstable, but each two-species combination is invasible. Modified from Holt 2009.

the second predator colonizes, it can then outcompete the first predator on that prey species. The system next becomes open to invasion by the other prey species, which is less consumed by that predator, and the first prey species is supplanted. The system is then open to the first predator to invade and the cycle begins anew (Fig. 11.9c).

The EAC presented in Figure 11.9, while interesting and (hopefully) relatively easy to understand, is just a beginning for understanding patterns of community assembly when many species and many environmental conditions are considered.

A number of studies have extended this framework to investigate the influence of variation in parameters of the species pool (e.g., number and types of species) and parameters of the environment (Law and Morton 1993, 1996; Morton and Law 1997; Fukami 2004b; Steiner and Leibold 2004; Law and Leibold 2005; Capitán et al. 2009, 2011). In the following sections, we discuss one issue in particular—the influence of variation in the productivity of the environment—and how the propensity of MSEs and EACs is expected to change. We illustrate how such effects might help explain the variation in diversity patterns related to ecosystem productivity we discussed in Chapter 10, but we imagine that they may equally work to explain a number of other possible cases.

11.6 FOOD-WEB METACOMMUNITY ASSEMBLY
AND THE SCALE-DEPENDENT
PRODUCTIVITY-DIVERSITY RELATIONSHIP

As we described in Chapter 10, a common macroecological observation is that the relationship between energy inputs into an ecosystem (productivity) and species richness is scale dependent (Mittelbach et al. 2001, Chase and Leibold 2002, Storch et al. 2005, Chalcraft et al. 2008, Gardezi and Gonzalez 2008, McBride et al. 2014). For example, Chase and Leibold (2002) surveyed natural ponds in Michigan to illustrate this relationship with benthic invertebrates by examining the relationship at two spatial scales—within ponds and among three ponds within a metacommunity (watershed); they showed a hump-shaped productivity-diversity relationship at the smaller scale and a positive relationship when the same data were organized at the regional scale (Fig. 11.10a). This scale dependence resulted because ponds in high-productivity watersheds were more variable in species composition (i.e., higher β-diversity) than ponds in low-productivity watersheds (Fig. 11.10b). See also Chase 2010 for experimental confirmation of some of these results.

Chase and Leibold (2002) hypothesized three possible mechanisms that could have led to the observed higher β-diversity in high-productivity watersheds: (1) high-productivity watersheds are more heterogeneous pond to pond; (2) high-productivity watersheds have a greater propensity for initial assembly conditions leading to MSEs; (3) high-productivity watersheds have a higher propensity for locally dynamic assembly due to EACs, which would lead to higher temporal variation in species composition, which in turn should create more spatial variation (i.e., β-diversity; see Chap. 5). Of course, each of these mechanisms is quite possible, but Chase and Leibold suggested that the first mechanism (higher heterogeneity) was unlikely in this case, leaving higher MSEs or EACs as plausible mechanisms in this system.

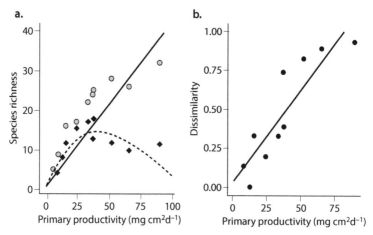

FIGURE 11.10. (a) The relationship between productivity measured within each pond (local) or the average of the productivity among three similar ponds (regional) and the species richness per pond (open circles) or three ponds (closed circles). (b) The relationship between regional productivity and the average dissimilarity (1 − Jaccard's index) among the three ponds in each watershed as an index of β-diversity. Modified from Chase and Leibold 2002.

In subsequent theoretical work, we have shown that both MSEs and EACs are more likely to emerge when strong trophic interactions are present in a system. On the one hand, Chase and Leibold (2003) used a simple keystone predator model to show how the likelihood of MSEs increases with increasing productivity when there are many species in the species pool because of the trade-offs between resource competitive abilities and predator defense. This is less likely to emerge in situations without higher trophic levels.

On the other hand, Steiner and Leibold (2004) modeled food webs under varying levels of productivity using the permanence criteria and found a greater propensity for EACs at higher levels of productivity (see also Capitán et al. 2011). Again, however, this complexity, leading to more β-diversity and scale-dependent productivity-diversity relationships, did not emerge when there were only two trophic levels present (Fig. 11.11a); it only did so when there were more trophic levels (Fig. 11.11b). Here, with increasing productivity, we expect a shift from predator-susceptible to predator-defended prey species, but when there are temporal-scale mismatches, this shift makes the new food chain vulnerable to invasion, much like the "community churning" scenario of Holt (2009) described above.

While both mechanisms—MSEs and EACs—are more likely to emerge when food-web interactions are considered, the empirical results are mixed. Chase (2010) performed an eight-year experiment in a seminatural metacommunity (artificial

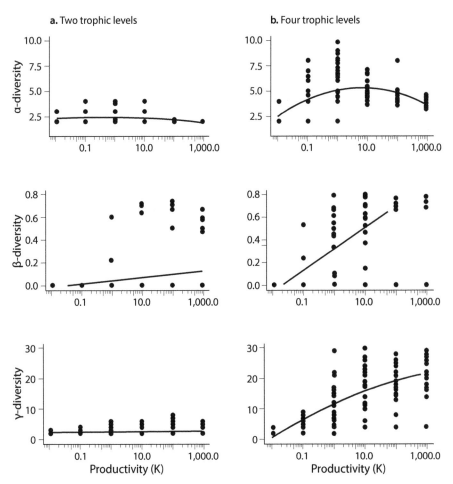

FIGURE 11.11. Results from the simulation models examining the resulting species diversity at local (α-diversity) and regional (γ-diversity) scales, as well as their β-diversity. (a) Only two trophic levels present. (b) Four trophic levels present (results with three trophic levels were qualitatively similar). Modified from Steiner and Leibold 2004.

ponds open to natural colonization-extinction dynamics). After the initial assembly period, the high-productivity ponds were much more divergent from one another spatially (greater β-diversity) than were the low-productivity ponds, but there was minimal temporal turnover in this system. Alternatively, Steiner (2014) found more temporal turnover with higher productivity in laboratory microcosms of zooplankton. Both of these studies, however, were a bit simplistic in terms of their food-web complexity and ability to detect change through time and space.

A more direct way to determine whether food-web-induced MSEs or EACs were behind the sorts of scale-dependent productivity-diversity relationships observed would be to make observational data available in natural systems that include both spatial and temporal variability (see Chap. 5 for some examples). Mathew Leibold, Bernadette Pinel-Alloul, and Ginette Methot (unpublished data) have been comparing patterns of zooplankton composition among lakes in watersheds that differ in their level of productivity (eutrophication) in different parts of Canada. They have found patterns remarkably consistent with the EAC predictions of Steiner and Leibold (2004; Fig. 11.11). At the local scale, the pattern was strongly unimodal, resembling patterns that are often observed in other lake and pond systems (Leibold 1999, Chase and Leibold 2002, Gardezi and Gonzalez 2008; Fig. 10.12a). At the regional scale, spatial β-diversity (Fig. 11.12b) increases, which is, again, consistent with the observations made by Chase and Leibold (2002) and Gardezi and Gonzalez (2008). What makes this data set intriguing is that year-to-year observations of these lake systems show a strong relationship between productivity and temporal turnover (Fig. 11.12c), as well as between spatial β-diversity and temporal turnover (Fig. 11.12d). We cannot discern whether these patterns are actually driven by the processes of EACs envisioned by the "churning" food chains predicted in theory because there are no data on phytoplankton turnover and we do not really know the underlying processes. However, the correspondence of these patterns with models of EACs is suggestive (Steiner and Leibold 2004, Capitán et al. 2011).

Future work examining the role of metacommunity assembly and trophic interactions in driving scale-dependent productivity-diversity relationships could further dissect these patterns and processes in a number of ways; for example, by comparing the diets of the consumers and resources as they cycle; by comparing patterns among metacommunities that vary in their dispersal rates; or by comparing among metacommunities with different types of trophic interactions.

Despite the utility of trophically mediated community assembly scenarios for understanding patterns and processes in metacommunities, some caveats are necessary. To begin with, the existence of EACs requires that each member of the species pool has a probability of colonizing each locality. This means, there either must be a "mainland" or other part of the metacommunity where species are present (Holt 2009) and that the dynamics of the EACs are asynchronous with respect to which species are in which localities (Law and Leibold 2005). Likewise, community assembly toward a single endpoint, MSE, or EAC often requires the presence of species not present in the final assemblage, a condition known as the "humpty-dumpty" effect (Law and Morton 1996; i.e., a community cannot be put "back together" with only the component species that are present in the final

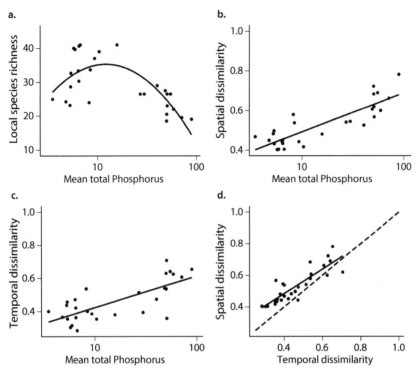

FIGURE 11.12. Diversity patterns in Canadian lake zooplankton. (a) Species richness of zoo-plankton (as the total number of species seen in three samples taken over a summer) is uni-modally related to productivity (using the proxy of log total phosphorous concentration) in 28 lakes. (b) Spatial turnover (as mean pairwise Jaccard distance with other lakes in the same watershed) increases with productivity. (c) Year-to-year turnover within a lake (also expressed as Jaccard distance) increases with productivity. (d) Year-to-year turnover is strongly correlated with spatial turnover, and almost equal to it; more limited data indicates that a two-year lag in temporal turnover is equivalent to spatial turnover in these lakes (not shown).

community). If we think of community assembly as being like the process of eco-logical succession (Young et al. 2001), the humpty-dumpty effect is the same as saying that the climax (endpoint) community cannot reassemble itself without some of the pioneer species. Thus, pioneer species must have some presence in the re-gional pool in order to evade the humpty-dumpty effect. Finally, if extinction is possible in a system with MSEs, one species will eventually win and exist alone (i.e., the system is neutral or may even have positive feedback) unless the species have some stabilizing factor, such as habitat heterogeneity, that allows them to per-sist at the regional scale (Shurin et al. 2004).

11.7 THE INFLUENCE OF METACOMMUNITY PROCESSES ON FOOD-WEB STRUCTURE AND INDIRECT INTERACTIONS

By altering species abundance, composition, and diversity, the fundamental components of metacommunities (e.g., dispersal, heterogeneity) can indirectly change food-web structure and the nature and strengths of species interactions. A general tenet of predator-prey theory is that predators typically require more energy than prey in order to maintain a viable population (Elton 1927, Hatton et al. 2015), leading to the expectation that the population dynamics of predators are more likely to be influenced by larger spatial scales than that of their prey (Holt 1996, Holt and Hoopes 2005, Pillai et al. 2009). A straightforward expectation from this tenet is that predators should be less likely to present, abundant, or diverse in smaller and/or more isolated habitats, which can in turn influence the structure of interaction webs.

Schoener's (1989b) *productive space hypothesis* suggests that the length of a food chain is constrained by the productivity of a habitat as well as by the total area of that habitat; that is, a small patch such as a pond or island can support only a small number of trophic levels, regardless of its productivity, whereas larger habitat areas can support more trophic levels. Evidence for this hypothesis has often been found in a number of different ecosystem types (Post et al. 2000, Takimoto et al. 2008, Takimoto and Post 2013, Petermann et al. 2015), although there are exceptions (e.g., Young et al. 2013). In addition, the number of links (connectance) within food webs typically increases with increasing habitat area (Brose et al. 2014). Likewise, more isolated and/or fragmented habitats typically have shorter food chains and less complex food-web networks (Tscharntke et al. 2002, Tylianakis et al. 2007, Calcagno et al. 2011, Gonzalez et al. 2011, Hagen et al. 2012, Valladares et al. 2012, LeCraw et al. 2014).

As a rather natural consequence of reductions in predator abundances and simplification of food webs, spatial processes can also indirectly affect the abundances and diversities of prey species. A dramatic example comes from the comparison of islands that were created after the construction of a reservoir in Venezuela with areas still connected to the mainland (reducing both habitat area and isolation; Terborgh et al. 2001, 2006). While the mainland areas and large islands were frequented by large mammalian predators as well as by a full complement of smaller vertebrate predators (e.g., birds, lizards), smaller and medium-sized islands were devoid of predators, allowing some types of herbivores (notably howler monkeys and leafcutter ants) to flourish and devastate plant growth (i.e., a habitat-loss-induced trophic cascade). Other examples of the phenomenon of isolation-mediated trophic cascades include studies in habitat fragmentation due to human land use (Crooks and Soule 1999, Tscharntke et al. 2012, Martinson and Fagan 2014) and freshwater ponds and mesocosms that vary in their degree of isolation (Chase et al. 2010, Fahimipour and Anderson 2015).

Food-web structure can also be strongly influenced by habitat heterogeneity. Much like the mass effects that can occur in competitive systems (e.g., see Chap. 2), flows of resources from the bottom up and predators from the top down can strongly influence food-web structure and the strengths of direct and indirect effects among adjacent habitats, particularly when habitats differ in the spatial flows of organisms and biomass (Leroux and Loreau 2008). Classic examples of resource subsidies among adjacent habitats include seaweed washing up on the beach (Polis and Hurd 1996) and insects emerging from aquatic habitats (Sabo and Power 2002) to subsidize consumers in terrestrial habitats, and leaves from trees and terrestrial insects that fall into streams or ponds, subsidizing aquatic consumers (e.g. Nakano et al. 1999). We will discuss implications of the flow of energy and materials among patches in a landscape in Chapter 13.

Finally, by moving among heterogeneous habitats, predators can exert strong direct and indirect effects on adjacent habitats—often referred to as *spillover predation* (Oksanen 1990, Oksanen et al. 1992, McCoy et al. 2009). For example, cod populations in the Baltic Sea can spillover into an adjacent "sink" area (the Gulf of Riga) during times of high population abundances. This spillover, in turn, can dramatically alter the abundances of planktivorous fish, an effect that cascades through the food web (Casini et al. 2012). Likewise, Knight et al. (2005) showed how fish predation in Florida ponds reduced the abundances of larval dragonflies in those ponds, which in turn reduced the abundance of adult dragonflies flying near ponds with fish relative to ponds without fish. Because adult dragonflies act as voracious predators of flying insects, including pollinating flies and bees, the rates of pollination on plants surrounding fish ponds were considerably higher than for plants near ponds that had no fish predators.

11.8 CONCLUSIONS

Much of what we have discussed prior to this chapter has explicitly ignored trophic interactions (i.e., the competitive metacommunity assembly archetypes and their tests). The good news is that, despite this omission, we were able to make several general conclusions about metacommunity assembly. The less good news is that by implicitly or explicitly ignoring trophic interactions, we have probably missed a lot of important processes that may even change some of our conclusions. For example, there are many scenarios in which temporal patterns of metacommunity structure can be altered by trophic interactions, leading us to draw different conclusions about the relative importance of different metacommunity assembly processes. A common example is the negative density and frequency dependence that can emerge from spatially explicit interactions of predators (or pathogens). Trophic interactions can also alter the likelihood and distribution of MSEs and EACs, altering the stability of complex food webs. Indeed, food-web dynamics may be one of the more important sources of unexplained variation encountered in observational studies. Looking forward to a more complete metacommunity ecology will certainly require explicit consideration of trophic interactions.

CHAPTER TWELVE

Community Assembly and the Functioning of Ecosystems in Metacommunities

Prospectus

1. The interaction of community-level processes and ecosystem dynamics is important for both scientific and applied approaches to ecology. This question has been primarily addressed by studying the effects of biodiversity on ecosystem functioning at local scales.
2. It is important to expand this focus to address all the mechanisms that influence community structure and not just its local biodiversity. We show how metacommunity assembly within a broader spatial context and the resulting composition and diversity of species influences ecosystem attributes.
3. A simple resource-competition model illustrates some of the possibilities, including (1) dispersal sufficiency, in which selection effects predominate and maximize ecosystem function but tend to result in low local diversity, (2) dispersal surplus, in which dispersal leads to enhanced local diversity while simultaneously reducing ecosystem functioning, and (3) dispersal limitation, in which incomplete community assembly can lead to enhanced complementarity and biodiversity but still result in lower ecosystem functioning than would occur with dispersal sufficiency.
4. A broader perspective on community assembly and its relation to ecosystem functioning can be obtained by thinking of ecosystems as complex adaptive systems (CASs) that are composed of locally distinct entities that respond to inputs of variation in their elements in ways that favors some features of the ecosystem. Dispersal and connectivity can be key modulating factors of these features because they alter the relative importance of internal versus external inputs.
5. If CASs favor some features of ecosystems, these should be apparent as emergent properties, such as trophic structure, stoichiometric balance, and size abundance scaling.
6. Nevertheless ecosystems may also be susceptible to disruption by "tipping points." We argue that dispersal within a metacommunity may make such tipping points less likely and that they may thus be more prevalent in isolated ecosystems.

An important motivation for many ecologists is to address the effects of anthropogenic change on the structure and functioning of natural ecosystems. We have described some of these effects, mostly in terms of changes in species abundance, composition, and biodiversity, elsewhere in this book (especially Chap. 10). However, simply examining how anthropogenic effects change patterns of species abundance and biodiversity does not capture the complex influence that these changes may have on ecosystems (i.e., the combination of biotic and abiotic components in a given area). Taking a larger-scale view—that of metacommunities—can provide important insights into a number of patterns and processes of interest in ecology. In this chapter and Chapter 13, we extend this larger-scale spatial perspective to also include considerations of ecosystem-level patterns and processes. In this chapter, we discuss how metacommunity-level processes can strongly influence the role that species diversity and composition has on the functioning of local ecosystems.[1] In Chapter 13, we discuss progress in our understanding and the consequences of the movement and heterogeneity not only of species but also of the materials themselves in the context of a *metaecosystem* (sensu Loreau et al. 2003b). Although both chapters are related to ecosystems, we keep them separate because the conceptual and theoretical approaches are so distinct. Here we draw on metacommunity principles that we have already discussed, whereas metaecosystems uses a whole new set of principles related to the movement of materials per se.

Motivated by concern about anthropogenic effects on both biodiversity and the functioning of ecosystems, a new research field emerged in the early 1990s—examining the biodiversity (the attributes of species, as well as their diversity and composition) in the functioning of ecosystems, known colloquially as biodiversity-ecosystem functioning (BEF) studies (Schulze and Mooney 1994). A major thrust of this research area seeks to examine whether losses of biodiversity are connected to changes in the functioning of ecosystems, establishing a justification for preserving biodiversity beyond simply for its own sake.

The BEF research field was largely launched following the near-simultaneous publication of two high-profile papers. One, by Tilman and Downing (1994), showed that experimental prairie plant communities in Minnesota that had lower species richness (in this case, because of chronic nitrogen additions) were much slower in their ability to rebound following an intense drought than communities that had higher species richness (without nitrogen addition). The other study, by Naeem et al. (1994), experimentally manipulated the biodiversity of plants and animals at multiple trophic levels in experimental chambers and monitored a

---

[1] "Ecosystem functioning" refers to any number of processes, including the fixation of carbon (i.e., productivity), which is often measured simply as standing biomass; the cycling of nutrients; decomposition rates; etc.

number of ecosystem functions, finding much higher functioning in high-diversity communities.

Though inspirational for a large number of subsequent studies, these early experiments also engendered a great deal of criticism (Huston 1997, Wardle 1999, 2000). One of the biggest complaints was that the results from these studies were limited in their ability to make inferences about the effects on ecosystem functioning of changing biodiversity per se versus changes in other factors (e.g., nitrogen, species compositional effects) that were "hidden treatments" within these studies. As a result, a twenty-plus-year (and ongoing) effort has focused much more specifically on experimentally disentangling the influence of biodiversity (and composition, traits, etc.) on the functioning of ecosystems. This effort has lead to a general consensus that when experimentally controlled in experiments, there is an important influence of the numbers of species (and types) on the functioning of ecosystems; that is, biodiversity seems to matter for ecosystems (reviewed in Cardinale et al. 2012, Hooper et al. 2012, and Tilman et al. 2014).

Conceptually, the approach almost universally taken in BEF experiments is very straightforward and locally focused. It asks, what happens to productivity if we introduce (or inoculate) different numbers (and types) of species into a local ecosystem? This experimental approach allows direct tests for the influence of biodiversity per se, rather than for simultaneous changes in biodiversity and species composition that result from changing environments. It also has allowed us to move beyond early tests of the BEF relationship in which biodiversity and ecosystem functioning were confounded (Tilman and Downing 1994, see Wardle et al. 1997). Nevertheless, many recent studies still treat biodiversity as an independent variable and ecosystem function (or multifunction) as the dependent variable even when it is clear that the majority of changes (especially anthropogenic ones) jointly change environmental factors, biodiversity, and ecosystem functioning. Figure 12.1 (from Loreau et al. 2001) illustrates the subtle, though important, differences between one common form of BEF relationship—the influence of diversity on productivity—and a commonly studied contemporary biogeography relationship—the influence of productivity on diversity (see Chap. 10). Merely correlating the two and putting one on the x-axis and one on the y-axis gives us little information about the actual BEF relationship (especially if the influence of productivity on diversity is complex, such as the hump-shaped relationship depicted in Fig. 12.1a,c) (Grace et al. 2016).

From local BEF experiments (Fig. 12.1b), the pattern is quite clear: that local ecosystem function declines with species loss, at least in some parts of the curve (Cardinale et al. 2012, Hooper et al. 2012, Tilman et al. 2012). There are two main factors responsible for this effect (Loreau and Hector 2001): selection, by which the likelihood that a highly productive species is present is higher if the number

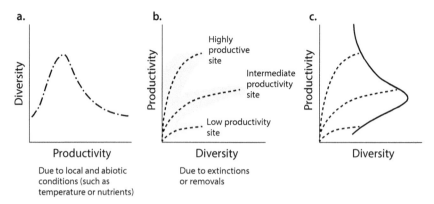

FIGURE 12.1. Reconciliation of two diversity-productivity relations. (a) One pattern seen in many natural systems is that species richness often shows a unimodal relationship with productivity. The main driver of productivity is an external factor such as soil nutrients or climate. (b) A second pattern (from experimental biodiversity-ecosystem functioning studies) shows that productivity is a monotonic function of imposed species richness under a common environmental setting (little or no variation in external factors); three such curves are shown. The two patterns differ as to which is the predictor variable or the response variable. (c) Combination of the data in the two graphs show that they are not contradictory. Note that (a) has been rotated 90° to match the axes of (b). For a given environment, the first relationship sets the maximum species richness and productivity (dashed line), but changes in species richness (by experimental manipulation for example) reduce productivity according to the second relationship (solid line). Modified from Loreau et al. 2001.

of species in the introduced species pool is higher; and complementarity, by which some trait differences among species allows them to exploit different resources and increase overall ecosystem function.

Despite many successes of the BEF research program, the focus on local richness effects as an independent variable (with experiments typically consisting of randomly assembled communities) ignores the metacommunity context that determines species richness (and composition). This omission has led some to question the relevance of the BEF research program amid the huge complexity by which both biodiversity and ecosystem function co-vary in nature (Srivastava and Vellend 2005, Wardle 2016). Adding to this skepticism is the observation that local richness is not always (or maybe even often) declining despite ongoing anthropogenic habitat conversion and global species extinctions (Vellend et al. 2013, Dornelas et al. 2014; but see Eisenhauer et al. 2016 and Gonzalez et al. 2016 for counterarguments).

In this chapter, we ask what the influence of biodiversity on ecosystem functioning is if we take a closer look at how patterns of species richness and composition form in a local community—through community assembly—in the first place. We begin by providing an overview of some of the various scenarios in which community assembly might be influenced by metacommunity processes in a simple resource competition model. We also look at the evidence that might support these ideas. Recognizing that the issue is likely to be much more complicated than these models, we next consider how the principles behind complex adaptive systems (CAS) might apply in metacommunities, focusing especially on community assembly. One important feature of CASs is that they are likely to have some important emergent properties that might correspond to numerous documented patterns in ecosystems. Finally, we discuss how the prevalence of "tipping points" in ecosystems might be influenced by spatial processes and community assembly.

12.1 THE ROLE OF SPATIAL PROCESSES IN MEDIATING BEF RELATIONSHIPS

The BEF research program has provided great insights into the influence of sampling and complementarity effects when species are assembled at random from a regional species pool (Eisenhauer et al. 2016). However, we agree with Wardle (2016) that much more can be learned by examining the relationship between diversity and productivity in more complex, and realistic, scenarios in which both community-level (species richness, composition) and ecosystem-level variables covary (see also Grace et al. 2016). Figure 12.2 illustrates three different ways the relationship between community and ecosystem levels can be conceptualized. Figure 12.2a represents the simplistic view of the experimental BEF perspective that we discussed above (and is reviewed in Cardinale et al. 2012 and Tilman et al. 2014), in which species richness (and sometimes species composition) is treated as an independent variable (controlled by the experimenter).

Figure 12.2b takes the perspective one level higher by assuming that local species richness is strongly influenced by the size of the regional species pool from the top down (Ricklefs 1987, 2004, 2015). Here, increasing the size of the regional species pool, or increasing dispersal of species from the species pool into localities (i.e., breaking down dispersal limitation), can often increase species richness (see Chap. 3). By extension, several authors have suggested a slightly different approach to the BEF relationship by examining how changes to the regional context—through dispersal—may influence ecosystem functioning indirectly via its direct effect on local species richness (Zobel et al. 2006). Although there are only a handful of studies that have taken this approach (all in grassland communities), the general consensus seems to be that breaking down dispersal limitation through seed additions can increase both species richness and ecosystem functioning, albeit with some important caveats depending on the environmental context

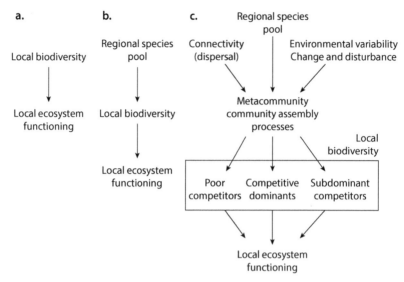

FigURE 12.2. Three increasingly complex conceptual views of how biodiversity affects ecosystem functioning. (a) The simplest view, in which there is a direct causal relation between local biodiversity and ecosystem functioning, but the effects of the regional species pool are ignored. (b) Recognition of the role of regional biodiversity as an ultimate driver of local biodiversity but with the assumption that this relationship is simple (e.g., monotonic). (c) The metacommunity view, in which multiple drivers (connectivity, and environmental variability and change as well as regional biodiversity) affect different components of local diversity; competitive dominants that drive other species extinct if present, subdominant competitors that can maintain local populations in the absence of better competitors, and poor competitors that exist as sink populations that then differentially affect ecosystem functioning. Modified from Leibold et al. 2017.

(e.g., site productivity, herbivory, etc.; Foster et al. 2004, Zeiter et al. 2006, Stein et al. 2008, Maron et al. 2014a). There is at least one important exception, however, where seed additions increased species richness but did not influence (in some cases) or decreased (in other cases) the productivity of the system (Wilsey and Polley 2003).

Finally, Figure 12.2c illustrates what we refer as the *community assembly and the functioning of ecosystems* (CAFÉ) perspective, in contrast to the simple BEF perspective. Under CAFÉ, species present in a given locality are there because of the complex interplay between dispersal, habitat heterogeneity, and regional species pools within a broader metacommunity. For example, species in a locality can be thought of as well adapted to a given locality (through SS dynamics), equally adapted to a given locality (NT), or poorly adapted to a given locality (ME), and the presence and numbers of these different types of species in a given locality will then conspire to influence the local functioning of that ecosystem.

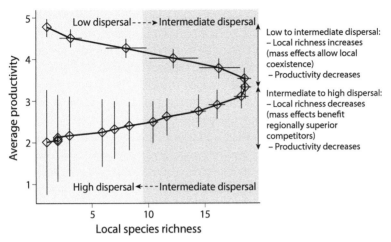

FIGURE 12.3. The relationship between local species richness and the average productivity of a given locality from the mass effects (ME) metacommunity model of Mouquet and Loreau (2003). Species are assumed to be differentially favored in different localities. Thus productivity is highest when low dispersal rates allow species that are most favored in a given locality to persist alone (local species richness is 1), as in the upper left part of the graph. As dispersal rates increase from the upper left to right, local species richness increases due to ME, but the average productivity of the locality decreases because species that are less well suited to a given site persist and utilize some of the space. At some dispersal rate, local species richness is maximized and increased levels of dispersal favor regionally competitive but locally unproductive species. Here, from right to lower left, species richness and productivity decline. Modified from Mouquet and Loreau 2003.

An early illustration of just how adding metacommunity-level spatial processes can alter BEF expectations showed that some mechanisms that lead to greater species richness in a given locality—such as ME—can create much more complex BEF relationships (Bond and Chase 2002, Mouquet et al. 2002, Mouquet and Loreau 2003). Specifically, in some ME scenarios, we expect that local species richness has a hump-shaped relationship with increasing dispersal rates in the metacommunity (Mouquet and Loreau 2003, Leibold and Miller 2004; see Chap. 2). However, because of the assumed traits of species (i.e., competitive abilities) that are favored at lower and higher dispersal rates, the productivity of the local communities can differ even if the numbers of species are the same.

Figure 12.3 illustrates this effect when species are differentially favored (competitively superior) in different localities and dispersal leading to ME alters both local richness and ecosystem functioning (Mouquet and Loreau 2003). Here, the highest productivity occurs when local diversity is the lowest (and dispersal is lowest) because each species is differentially competitively superior (ecologically "adapted") in each habitat locality. As dispersal rates increase, local diversity first increases due to ME, while productivity in a given patch declines (an inverse BEF relationship). However, after a threshold dispersal rate is passed, homogenization becomes important, and regional diversity decreases as local specialists are lost from the metacommunity. At this point, local diversity and local productivity decrease together as regionally competitive, but generally locally unproductive, species become favored everywhere. Of course, this scenario is only one way that adding a metacommunity perspective can alter expectations about BEF relationships, and different relationships can emerge, for example, if some degree of dispersal limitation means that all species are not in their favored habitats or if other spatial mechanisms (such as PD) drive local diversity and ecosystem functioning (Mouquet et al. 2002, Loreau et al. 2003b).

In the following sections we more explicitly explore how a CAFÉ perspective can enhance our understanding of the relationships between metacommunity-level patterns of diversity and species composition with ecosystem functioning. To do so, we combine the concepts discussed above that lead to both increases (sampling, complementarity) and decreases (mass effects) of ecosystem functioning with increasing species diversity into a more comprehensive framework that depends on the influence of dispersal, habitat heterogeneity, and local disturbances (Leibold et al. 2017).

12.2 A SIMPLE FRAMEWORK BASED ON RESOURCE COMPETITION IN A METACOMMUNITY CONTEXT

We start with the simple assumption that the metacommunity context (i.e., heterogeneity and dispersal) modulates ecosystem functioning by differentially favoring or disfavoring species with different traits (and consequent ecosystem functioning). Specifically, the metacommunity template on which we have focused, along heterogeneity-dispersal axes, can allow three qualitatively distinct "types" of species to persist in a given locality (note that these apply to particular locations; a given species can be of one type in one patch and another type elsewhere due to local environmental conditions):

1. *Competitively dominant species.* Species that are the superior competitor in a given habitat (i.e., have the lowest R* in an SS framework) are assumed to contribute the most to ecosystem functioning (e.g., Tilman et al. 1997, Gross and Cardinale 2007). The presence of such species in a local ecosystem contributes to ecosystem functioning through the selection mechanism, though they may or may not additionally contribute to complementarity.

2. *Poorly competitive species.* These are species that persist in a given locality because high dispersal rates allow them sufficient immigration to offset their poor competitive abilities and probable competitive exclusion in the absence of dispersal. The presence of these sorts of species in a locality can be associated with reduced ecosystem functioning, as in the model of Mouquet and Loreau (2003) described above. Such species would tend to reduce the strength of selection at the local scale in the same way that gene flow can prevent local adaptation in population genetics models; they may or may not also contribute to complementarity.

3. *Subdominant species.* These species are transient or weedy species, such as the colonizers in colonization-competition trade-off PD models (Mouquet et al. 2002). Here, however, we assume that such species need not be good colonizers per se but rather can be competitively dominant in other nearby habitats with different environmental conditions. These types of species will contribute to the selection effect moderately and may or may not contribute to complementarity.

Although highly simplistic, we assume that the relative prevalence of these three types of species varies depending on the metacommunity context including disturbance, dispersal and isolation, and environmental context, as well as the nature of the species pool (Fig. 12.2c; see also Ptacnik et al. 2010). In the following sections, we discuss several qualitative different metacommunity scenarios having variable levels of dispersal and environmental change that may help us to interpret the observations (experimental or correlative) that have been made about the ways biodiversity in general (at both local and regional scales) affects ecosystem functioning.

12.2.1 SS and Ecosystem Function

We start with an idealized SS case in which dispersal is assumed to be unimportant—that is, dispersal is high enough so that each species can find its favored habitat type, but not so high that ME emerge. Thus, in this scenario, we consider only species that are considered competitive dominants, not poorly competitive or subcompetitive species. Here, species from a regional pool differentially sort among localities according to how well they perform in different environmental conditions (Tilman 1982, Chase and Leibold 2003, Leibold et al. 2004). As we have described elsewhere, although there is variation, this process seems to play a strong role in the assembly of many, if not most, metacommunities (Cottenie 2005, Logue et al. 2011, Soininen 2014).

To model SS and ecosystem functioning, we start with the model of Tilman et al. (1997) and especially its elaboration by Gross and Cardinale (2007). We use graphical consumer-resource models that depict zero-net-growth isoclines (ZNGIs) and impact vectors (which include both consumption and possible recycling; Daufresne and Hedin 2005) for a simple model of resource competition for two essential abiotic resources across an environmental gradient in relative resource supply. We assume constraints on the shape of the ZNGIs so that the minimum requirements for one resource are negatively constrained by those on the other resource. We call the function that describes this constraint, R_{opt} (the bold dashed line in Fig. 12.4). For simplicity, we assume that all species have the "corner" of their ZNGIs on this function; this is the same as assuming that all species present have at least one set of environmental conditions under which they are "winners" and thus protected from regional extinction by competitive exclusion (Leibold 1998). To evaluate how variation in community composition can influence ecosystem function, we can compare how well a given species or pair of coexisting species (with resource requirements equal to R^*_{i}) actually suppresses resource levels (which we assume is strictly correlated with ecosystem function) compared with the levels expected if the resident species were the absolute best idealized competitor in that ecosystem (a species whose ZNGI included the corresponding value of R_{opt}). The difference between the actual resource levels and the predicted "best possible," which we call ΔR_{opt}, indicates the degree to which the functioning of an ecosystem with a given set of species deviates from its theoretical maximum (i.e., how much it deviates from maximum resource suppression and potentially maximum ecosystem function).

When two species compete for two resources (Fig. 12.4), intermediate resource ratios favor coexistence among the species, and ecosystem function is higher when both species are present than when either is alone (i.e., ΔR_{opt} is often lower for the two-species case than it is for either one-species case). The effect when two species are present is due to local-level resource complementarity as well as selection. Alternatively, when the supply of resource ratios is skewed toward one resource, the species that is able to use the less abundant resource more efficiently can outcompete the other and exist alone. Here, ecosystem function is maximized by only one species (i.e., ΔR_{opt} is lowest for the one-species case) as a result of selection effects.

Next, we expand this simple scenario slightly by showing how three competing species distribute themselves when resource ratios vary (Fig. 12.5). The results are a bit more complex because patterns among the coexistence of species pairs can vary. Despite this complexity, there is now a broad range in relative supply rates in which ecosystem functioning has been increased by the presence of all three species in the regional species pool. This effect over this range is a result

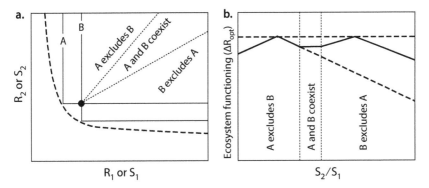

FIGURE 12.4. Hypothesized optimal traits that minimize R_i subject to physiological constraints. (a) Under perfect tracking, R_1^* and R_2^* are shown by the heavy dashed line, with a resulting R_{opt}. When two species (A and B) have zero-net-growth isoclines (solid lines) and impact vectors (lighter dotted lines) positioned so that they can possibly coexist, the final composition depends on the resource supply (S_1 and S_2). (b) The effects on ecosystem functioning quantified as the distance between the actual R_i and the nearest optimal R_i (ΔR_{opt}). ΔR_{opt} differs with varying nutrient supply ratios (S_2/S_1) if both species are present in the regional species pool (solid line) or if only species A is present (heavy dashed line). The fine dashed lines delimit how species composition varies if both species are present. Ecosystem functioning is higher with two species for all supply ratios where the solid line is above the dashed line. Modified from Leibold et al. 2017.

of SS and niche complementary. If one of the three species were absent in this metacommunity (regional extinction), then the remaining species in the metacommunity would expand their habitat distributions. As a result, the numbers of species that are present locally remains essentially unchanged (either one or two species, depending on resource ratios) even though regional diversity has declined by one-third. Importantly, however, despite the lack of change in local diversity, the average ecosystem function across the metacommunity will also decline as a result of the extinction. Again, selection and complementarity work in concert, but in this case complementarity plays a bigger role by operating across a larger range of environmental conditions than it did in Figure 12.4.

What happens in more diverse metacommunities? Although complicated by the details of which species go extinct and other factors, some insight can be gained by considering a limiting case. Specifically, we assume an infinite number of species with the elbow of their ZNGIs arrayed along the line showing the physiological constraint (R_{opt}, the heavy dashed lines in Fig. 12.5a, b). In a metacommunity with perfect species sorting along a relative resource supply gradient and an infinite species pool of species arrayed along the R_{opt} line, ecosystem functioning should be maximal, with a ΔR_{opt} of zero but with a local richness of one in every

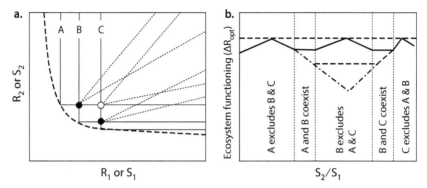

FIGURE 12.5. Effects of extinction on species sorting in a metacommunity. (a) Scenario in which species B goes extinct while A and C remain (the coexistence of A and C is shown by the open symbol, but all else as in Fig. 12.4). (b) The effects on ΔR_{opt} when (i) all species have access to local communities (solid line), (ii) species B is extinct but species A and C replace B where they can (heavy dashed line), (iii) species B is extinct and species A and C do not compensate (line with alternating dots and dashes). Here, ecosystem functioning has decreased over intermediate ranges in supply ratios when B is extinct even though local diversity has not necessarily decreased. A similar effect will happen in local communities if B is present in the metacommunity, but is dispersal limited enough that it has not arrived in suitable local communities. Modified from Leibold et al. 2017.

site (Tilman et al. 1997). This, together with the fact that each locality's ecosystem functioning is only truly at its optimum when there is only one species in a given locality, indicates that it is the high metacommunity diversity together with the process of SS, and not local diversity per se, that increases local ecosystem functioning.

The above scenarios highlight an important lesson for how SS in a metacommunity context can influence ecosystem function. Assuming a link between the levels of community resource use and ecosystem function, the match between the environment and the traits of species is critical. Therefore, having a diversity of traits in the regional species pool improves the probability that a species that maximizes ecosystem function can colonize; this is the *spatial insurance effect* of Loreau et al. (2003a). In our scenarios above, only one species is necessary to maximize function when resource ratios are skewed. However, if a metacommunity contains habitat types that vary in resource supply, then the identity of the species that maximizes ecosystem function will vary for each habitat. It is the SS process at the metacommunity scale that allows the species (and species combinations) that are most effective in a given environment to occur across the different habitat types. If a metacommunity contains habitats representing a variety of resource supply ratios, then a metacommunity containing only one species would have lower average functioning across this entire region than if multiple species were present and able to optimally sort themselves along environmental gradients.

Our simple model also shows that selection and complementarity are processes that work in tandem when SS in metacommunities is at work. Complementarity is important because it enhances ecosystem functioning when selection is imperfect, which happens if the competitively dominant species in a given habitat is absent from the species pool. If there is not a specific species in the pool that can maximize function at a site, then niche differences among species may allow several coexisting species to maintain ecosystem function closer to the theoretical optimum than any of them might if alone. The frequent observation that seed-addition experiments increase both species richness and ecosystem function (Foster et al. 2004, Maron et al. 2014a) suggests that species composition in many localities may be constrained away from optimal trait values because of the constraints of the regional species pool or because of other metacommunity processes, such as dispersal limitation (see below).

One other key prediction emerges. Under imperfect SS, there may be a negative relationship between the effects of selection and complementarity. If the patch is inhabited by the best set of competitors (selection effect) there is little opportunity for complementarity and thus a decrease in diversity. The strength of the selection effect should depend on the species pool and the distribution of ecological traits it encompasses.

In an intriguing analysis, Rychtecká et al. (2014) showed that despite the typically positive relationship between species richness and biomass in a long-term BEF experiment in Jena, Germany (solid line in Fig. 12.6), patterns within treatments where different pools of species were initially introduced and maintained seemed to show the opposite results. Even though species richness was initially controlled by the experimenters, many species went extinct from the local plots over time (especially the plots with the highest numbers of planted species). However, plots within a given initial richness treatment varied in exactly how many species went extinct over time, allowing exploration of how the "realized" species richness in plots was correlated with ecosystem functioning. The surprising result is that plots that had more extinctions within a given species richness treatment (lower realized diversity) had higher biomass production than plots with fewer extinctions (i.e., a negative BEF relationship within richness treatments, despite a positive BEF relationship among richness treatments; the dashed lines in Fig. 12.6). This result supports the idea that within a given plot, biomass will be high if competitively dominant species are present in the initial pool of species in a plot, but these competitive dominants also simultaneously reduce complementarity and extant diversity while enhancing biomass. However, the overall positive (albeit weak) effect of the size of the species pool on ecosystem function offsets this negative

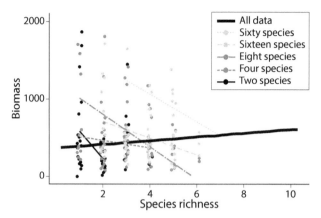

FIGURE 12.6. Analysis of diversity-biomass relations in the Jena experiment. The data are separated into the manipulations of species pool size (as shown in the inset). The ordinate shows the actual number of remaining species found in each plot. Correlations between biomass and realized diversity are all negative (three out of five are significantly so at $P < 0.05$) when the species pool diversity level is controlled for (lighter lines). The overall correlation across all diversity treatment levels however (bold solid line) is significantly positive. Data replotted from Rychtecká et al. 2014.

relation between biomass and extant species richness because it is associated with an enhanced likelihood of including locally dominant species (i.e., the sampling effect), even though one might otherwise assume the result is due to complementarity.

12.2.2 ME in Metacommunities under Dispersal Surplus

As we introduced above, high rates of dispersal can lead to species persisting in patches in which they are less well suited (i.e., poorly competitive species). In such cases, this can create ME, by which high dispersal rates can maintain populations of a species in localities where they could not otherwise persist (Loreau and Mouquet 1999, Mouquet and Loreau 2003); this effect in turn can lead to reduced ecosystem function (Bond and Chase 2002, Mouquet et al. 2002, Mouquet and Loreau 2003).

We illustrate this scenario in Figure 12.7 using the framework presented by Haegeman and Loreau (2015). This is a simple resource competition model closely related to the approach we describe above for resource competition, but it is more easily understood when incorporating ME (see Ryabov and Blasius 2011 for a more complex model that more closely relates to the model we have for SS here). As

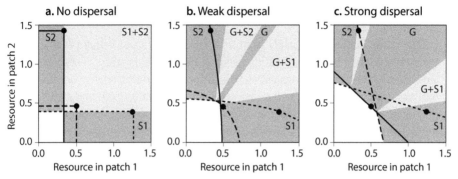

FIGURE 12.7. Effects of dispersal on resource levels and species composition in a mass effect model. Two habitat patches have independent resource concentrations in each patch, as shown in the two axes. There are three species in the species pool. In the absence of dispersal, S1 is a specialist competitor in Patch 1, S2 is a specialist in Patch 2, and G is a generalist with intermediate competitive ability in both patches as shown by zero-net-growth isoclines (ZNGIs) in Panel a. Resource concentrations at equilibrium are shown by the distance from the origin of the ZNGIs (the size of the white area in the lower left of the graph). In the absence of dispersal, G is eliminated, and S1 occupies Patch 1 and S2 occupies Patch 2 if the supply of resources in both patches is adequate to support either species. Resource concentrations are described by the point of intersection of the ZNGIs for S1 and S2 when the resource supply is balanced to allow coexistence. At intermediate dispersal (b), it is possible for G to exist in the metacommunity but only if the supply of resources is narrowly constrained to the zones labeled G+S2, G, or G+S1. Here the ZNGIs have been modified by the effects of dispersal, with the consequence that resource concentrations in the two patch types are higher (away from the origin of the graph). At very high dispersal (c), when species are homogenously mixed in the metacommunity, the ZNGIs are linear and the range of supply conditions where G excludes S1 and S2 is much larger. Resource concentrations are even higher than in (a) or (b). Modified from Haegeman and Loreau 2015.

dispersal increases, ecosystem functioning declines, for two reasons. First, because high dispersal implies high emigration, local populations are not just affected by the balance between birth and death but also by emigration and immigration. To maintain a population with high emigration (in addition to local death), the birth rate must be higher than in a closed population. Therefore, resource levels are also predicted to be higher at equilibrium. This pushes the ZNGIs away from R_{opt} relative to the SS case and thus reduces ecosystem functioning. Second, the exchanges among local communities create the potential for local "sink" populations of species that are poor competitors under local conditions, and local diversity is thus enhanced. In this situation, greater local diversity is associated with lower ecosystem functioning to the degree that poorly competitive species in sink populations occur and that species that are superior competitors have a reduced ability to deplete resources due to emigration (as in the model of Mouquet and Loreau 2003 in Fig. 12.3).

The second effect that results from high rates of dispersal is that populations are now regulated by a combination of local and regional effects. These effects are somewhat complex and depend on "spatial storage" effects, relative nonlinearities, and spatial structure as well as dispersal (Chesson 2000, 2012), but in general, they indicate that composition and diversity change from being strongly locally regulated when dispersal is low to being more regionally regulated when dispersal is high (Ryabov and Blasius 2011). Mouquet and Loreau (2003; see also Loreau and Mouquet 1999) show how this effect can influence the role of biodiversity on ecosystem function, as we described above (Fig. 12.3). The shift from local to regional control of local community composition means that the best regional competitors (those that do best overall across the range of environments in the metacommunity) will tend to drive extinct those that are more specialized in the less frequent or more distinct environments. Under extreme homogenization, the overall metacommunity and the resulting ecosystem functioning are less resistant to regional-level environmental change. Alternatively, if homogenization is less extreme, so that regional diversity remains high, dispersal can still enhance responses to environmental change because it facilitates the rate at which local composition can respond to change.

Despite the interesting potential role of ME in altering CAFÉ dynamics, there is little empirical evidence that these sorts of effects influence patterns of species composition and diversity (see Chap. 3), or for their mediation of the influence of diversity on ecosystem function. For example, Lindström and Östman (2011) manipulated immigration of bacteria from a regional biota into dialysis chambers in three different lakes. They found that it took at least 40% immigration per day to significantly alter the microbial composition, and even more (sometimes as much as 70% per day) to consistently reduce overall ecosystem production. Since even very small lakes rarely receive more than about 5% per day of inflow from upstream bodies of water (Logue and Lindström 2010), it is extremely unlikely that the high immigration rates needed to negatively affect production ever occur naturally in these lakes. In a conceptually similar study, Adams et al. (2014) found evidence for ME and reduced growth in a small pond during major storm events only when the inflow was about 100% per day! Similarly, Venail et al. (2008) showed that reduced production of microbial communities under controlled laboratory conditions also required extremely high dispersal (also when dispersal was 100% per day). Similar weak effects of high rates of dispersal on zooplankton species composition and abundance, as well as on various aspects of ecosystem functioning, have also been observed by Howeth and Leibold (2008, 2010a, 2010b). In all, these studies seem to indicate that reduced production due to ME is not likely to occur except under exceptional circumstances or at very small scales where dispersal exchanges are of similar magnitude to the population turnover rates of the organisms involved.

We might also speculate about interpreting the results from classic BEF experiments from this perspective. For example, in the classic BEF experiment in Cedar Creek, Minnesota, species richness treatments were established by initially seeding each plot with an equal amount of seed mass for each species (Tilman et al. 1996, 2001). It makes some sense to imagine that these seeding levels may thus be equivalent to immigrant sink populations for species if they are less well suited to local conditions. Assuming that the sampling scale of 2 m^2 roughly corresponds to the local scale in these communities, Tilman et al. (2001) showed that at richness levels of up to 8 species per plot, the observed number of species remaining at the end of the first year was not reduced from the number planted. At higher richness levels however (16 species planted), approximately 4 species on average went extinct in these sampling plots. In the subsequent four years, only about 1 additional species went extinct in these high-diversity plots, and the final richness was typical of the average richness in same-sized plots in the ambient community (Fig. 12.8a). Here, then, we might argue that SS (fitness differences) was high enough to prevent the establishment of poorly competitive species in this system when local richness was experimentally established at unnaturally high levels. Intriguingly, the BEF relations in this experiment showed that ecosystem functioning (biomass) increased (Fig.12.8a) even as richness decreased with time in these high-diversity plots (Fig. 12.8b). Ecosystem functioning also increased at lower diversity but did so proportionately less. These effects are consistent with the theory of ME dynamics on BEF relations (Mouquet and Loreau 2003), but they also suggest that to maintain this positive species richness-biomass relationship would require very high levels of immigration by poorly competitive species to occur (which was not possible with the experimental procedures).

Finally, it is important to note that most of the studies discussed above manipulated dispersal either directly or indirectly and still found relatively weak support for ME relationships. Under natural conditions, ME processes are perhaps even weaker because of various ways that organisms might bias their dispersal in response to local conditions (Leibold and Tessier 1998, Resetarits and Binkley 2013, Haegeman and Loreau 2015). If such biases are adaptive (e.g., adaptive habitat selection), ME relations and their consequent effects on ecosystem functioning may be even less likely. Clearly, more empirical assessment of the commonness of ME metacommunities needs to be conducted, but these studies indicate that they may not affect ecosystem functioning as much as we might have thought.

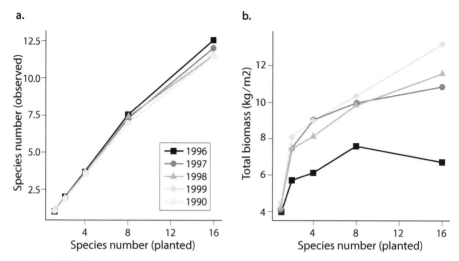

FIGURE 12.8. Relations between size of the species pool (species planted) and observed species richness (a) and total plant biomass (b). Modified from Tilman et al. 2001. Note that patterns involving diversity-biomass relations continued to change after 2000 in this experiment in complex ways that cannot be attributed to mass effects as we do here (Reich et al. 2012).

12.2.3 Dispersal-Limited Metacommunities

For SS to be maximized in a given metacommunity, species distributions have to respond rapidly to any spatial or temporal changes in environmental conditions so that the most dominant local competitors are able to establish in local communities, even after a local disturbance or environmental change. Thus, dispersal rates among localities within the metacommunity need to be high enough for such environmental tracking to occur but low enough to prevent ME effects from being important. What happens if dispersal is insufficient?

Dispersal limitation reduces the propensity for species to track their favored environmental conditions (Mouquet et al. 2002, Ozinga et al. 2005, Moore and Elmendorf 2006, Alexander et al. 2012). The degree to which dispersal limitation constrains the strength of the environment-species match, as expected from SS, depends on the size and degree of trait variation in the regional species pool as well as the amount of environmental heterogeneity, the prevalence of local or regional extinctions (as they may result from disturbances, environmental change, or demographic stochasticity), and the degree of dispersal (or connectivity) among local communities (Ptacnik et al. 2010).

We illustrate this point using the simple resource competition model we discussed above but with a more diverse species pool (Fig. 12.9). In this case, the

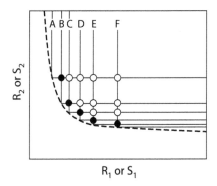

FIGURE 12.9. Local community assembly in a metacommunity with six species that differ in their position on the R_{opt} gradient (different preferred resource supply ratios) and that compete at local sites that can consist of pairs of species (with consequent resource levels described by dots) or as single species (with consequent resource levels along their respective zero-net-growth isoclines). Uninvadable local communities consist of pairs of species with adjacent positions along the R_{opt} gradient (with consequent resource levels shown as filled circles) or of single species closest to the R_{opt} constraint (dashed line). These are possible endpoints to community assembly. Other possible local communities (pairs of species with nonadjacent positions along the R_{opt} constraint or single species more distant from the R_{opt} line) are possible but are not end-points for community assembly. Any community inhabited by one of these non-endpoint communities has higher resources levels than the equivalent endpoint. Modified from Leibold et al. 2017.

endpoint community that will emerge in a given locality depends on the local environmental conditions (e.g., supply ratio of resources) and consists of either single species that have ZNGIs that are closest to the physiological constraint line (R_{opt}) for that supply ratio, or of pairs of coexisting species that are also closest to the physiological constraint. Community configurations consisting of such competitive dominant species for a given resource supply ratio are stable, but they can only be realized if those particular species can actually reach that location. Until those best-adapted species colonize, other species with ZNGIs that are more distant from R_{opt} (subdominant competitors) can inhabit this resource-state space (with consequent resource levels as shown by the open symbols in Fig. 12.9). In a meta-community with low dispersal or low connectivity among patches, it may take some time for community assembly to reach the most favored endpoint community that also maximizes ecosystem function. Thus, depending on colonization-extinction rates, it is possible that many local communities would be intermediate, rather than the final endpoint community predicted by SS. These intermediate communities can have the same number of species (i.e., richness) as the final one but will nevertheless have lower ecosystem functioning (be further from the R_{opt} line). Thus, we expect that species richness within localities might be considerably more robust than ecosystem functioning, even though ecosystem functioning across all the sites in the metacommunity still depends on the numbers of species in the regional species pool.

We describe two case studies that illustrate how SS and dispersal limitation within a metacommunity can influence local-scale ecosystem function. The first is a long-term study of a desert rodent assemblage in which a dominant group of kangaroo rats (a species in the genus *Dipodomys*) was experimentally removed from large, fenced plots (Ernest and Brown 2001, Ernest et al. 2008, Thibault et al. 2010) and the responses compared with controls where kangaroo rats were not removed. With the dominant kangaroo rats removed, other species compensated for these losses, so that the observed numbers of species in the perturbed rodent community was nearly the same as in the control plots (where kangaroo rats were present). However, total rodent energy use (a measure of ecosystem functioning) was substantially less (~86% less) in the removal plots relative to the control plots (Ernest and Brown 2001; Fig. 12.10), indicating that while subdominant competitors were able to compensate for kangaroo rat removal in terms of species richness, they were not able to compensate for them in terms of ecosystem functioning. Interestingly, after 18 years of kangaroo rat removal, *Chaetodipus baileyi* (Bailey's pocket mouse), a species that was present in the regional species pool (i.e., in sites no more than 5 km away) but not found until then in the experimental venue, colonized the plots (thus overcoming its dispersal limitation). With the colonization by *C. baileyi* in the kangaroo rat removal plots, the numbers of species actually decreased while at the same time energy use increased to nearly 80% of the values in the control plots. In sum, the removal of kangaroo rats led to strongly decreased ecosystem function with almost no change in diversity, whereas subsequent release of dispersal limitation by the colonization by *C. baileyi* led to almost complete recovery of ecosystem function (albeit with a reduction in diversity). These results illustrate how different aspects of community assembly other than those due to diversity (here dispersal limitation and selection) can affect ecosystem functioning.

A second case study (Petermann et al. 2010) comes from the same long-term BEF experiment in Jena, Germany, discussed above (Rychtecká et al. 2014). Here researchers manipulated the initial diversity of the species pool (as in the experiment by Tilman et al. 2001; see Fig. 12.8) but then allowed colonizations to occur in subplots of the experiment that they then compared with controls (no colonizations). These colonizations were either spontaneous (coming from seeds that were naturally dispersing in the area) or directly manipulated by the researchers (addition of an even mixture seeds from 60 specimens found in the area). Comparing these treatments, Petermann et al. (2010) showed that the traditional BEF relationship among plots that were closed to immigration continued to hold (Fig. 12.11).

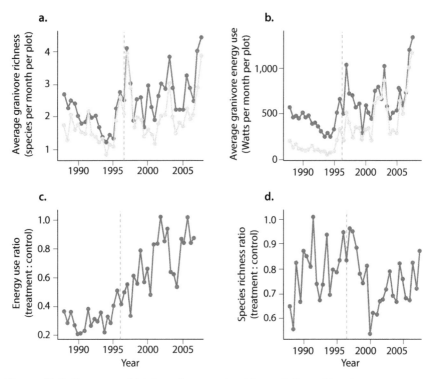

Figure 12.10. Responses of total granivore species richness (left panels) and energy use as a measure of ecosystem function (right panels) in experimental plots in which *Dipodomys* spp. were removed (gray symbols) compared with controls (black symbols) through time. The data is separated into two sections by the gray dashed line denoting the arrival of *Chaetodypus baileyi* to the site. The upper panels show the raw data, the lower panels express the same data as the ratio of the control to removal treatments (see Thibault et al. 2010 for details). Modified from Thibault et al. 2010.

However, plots subject to colonization showed a changing BEF relationship. Specifically, as soon as two years after opening these plots to the regional species pool through colonization or deliberate seed additions, there was a negative relationship between the realized species richness and ecosystem function (measured as community biomass; Fig. 12.11). This experiment thus showed that when communities are "open," the relationship between species richness and ecosystem function is not as straightforward as more controlled local BEF studies might have suggested and that processes involving other aspects of metacommunity assembly can be involved. Furthermore, plots where dispersal was assisted seemed to show this effect sooner than those where dispersal was natural, suggesting that this process is, at least to some degree, dispersal limited.

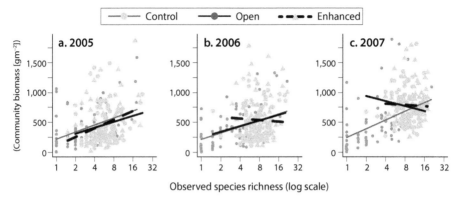

FIGURE 12.11. Regressions of total community biomass (y-axis) against observed species richness (x-axis) in the Jena experimental plots. Black dots and black solid line correspond to closed (labeled "control") plots, in which biodiversity was experimentally maintained via weeding. Gray dots and solid line correspond to plots that were initially the same as the control plots but in which weeding was stopped in 2005 (labeled "open"). Gray triangles and dashed lines correspond to plots identical to the open ones but in which an even mixture of seeds from 60 species were added in 2005 (labeled "enhanced dispersal"). Modified from Petermann et al. 2010.

In both of the case studies described above, dispersal limitation reduced the ability of species from colonizing suitable patches, with consequent impacts on ecosystem function. Additionally, both of these cases illustrate that dispersal-limited metacommunity assembly can alter our expectations of BEF relationships. While these case studies do not prove that dispersal-limited species sorting is a common process, they do suggest that the interaction of these processes can occur in nature and that when they do, they can have powerful influences on ecosystem function.

12.2.4 Combined Dispersal Limitation and Environmental Change

Because of concerns with climate change, dispersal limitation has been well studied in the context of species distributions (Sax et al. 2013). However, variable environmental conditions and dispersal limitation may also impact ecosystem functioning. We illustrate this point using the model framework introduced by Leibold and Norberg (2004) to consider how metacommunity responses to changing environmental conditions might be affected by dispersal limitation. Figure 12.12 shows how environmental change, by altering supply rates within a local community, is expected to alter diversity and ecosystem functioning with and without dispersal limitation from a given regional species pool. When there is local

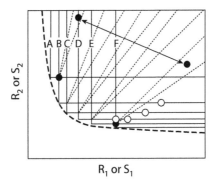

FIGURE 12.12. The influence of environmental change in resource supply at a single site (described by the arrow) within a metacommunity of six species (A–F), assuming that species sorting was complete prior to the change. Before the change, A and B coexist and resource availability is described by the upper-left gray point. After the change, species E and F are expected to coexist with resource levels described by the lower-right gray point so long as there is no dispersal limitation and species richness and ecosystem function is largely maintained (little to no change in R_{opt}). With dispersal limitation, it is possible that species B would remain, a secondarily adapted species (C or D) could establish, or only one of the final species (E or F), but not both, would establish (at the open symbols). None of these are final points to community assembly with reduced ecosystem functioning (further from R_{opt} line) than E together with F because all of them can be invaded by a missing species.

environmental change, dispersal limitation should reduce both local diversity and ecosystem functioning largely as a result of sampling effects—species best suited to the new local conditions are limited in their ability to disperse to newly suitable habitats.

An example that might indicate this sort of effect is illustrated by Baho et al.'s (2012) study of the response of a freshwater bacterioplankton community to perturbations in salinity. Experiments employed closed systems (chemostats), systems with a temporal refuge (chemostats with a biofilm that protected bacteria from washing out), and systems open to immigration from a regional species pool (via immigration at 10% per day from the undisturbed treatment). Even 18 days after the salinity perturbations (up to several hundred generations), the communities showed little signs of recovery. The highest degree of recovery occurred in the immigration treatments, but it was only partial. Thus, even in microbial systems thought to be highly responsive, lags in the responses of both composition and biomass are possible even over relatively long time periods (in comparison with the generation time of the organisms involved). The stronger recovery in the treatments open to immigration indicates that dispersal might play an important role in enhancing the recovery from perturbations and also enhance the ability of communities to track more permanent changes in the environment. Other studies have also shown that bacterial communities can exhibit low resistance and slow resilience to perturbations (Berga et al. 2012, Declerck et al. 2012, Andersson et al. 2014), although they can rapidly respond to environmental changes in other instances (Cleveland et al. 2007, DeAngelis et al. 2010) that may have something to do with differences in system openness. Microbial communities are not the only ones for which immigration appears to be important in moderating how ecosystem functioning responds to environmental change. Similar responses have been seen in plankton (Naeslund and Norberg 2006, Verreydt et al. 2012, Symons and Arnott 2013) and grasslands (Maron et al. 2014b).

Another way to think about the role of SS and possible dispersal limitation is to focus on the species themselves. Any single species may not always be particularly important to ecosystem functioning at all of the localities within a metacommunity, because they may be secondarily adapted species in some places. However, we might expect that there is typically at least one site in the metacommunity where each species is among the most adapted species present (Leibold 1998, Leibold and Norberg 2004). Isbell et al. (2011) examined how species from biodiversity experiments affect different components of ecosystem function, including contributions to function at different times and different locations, to different functions (i.e. biomass production, nitrogen accumulation, nutrient uptake, etc.), and in the face of different perturbations (e.g., drought, mowing). Although there are some biases inherent in their analyses (a substantial number of their analyses of spatial effects are from agricultural plots), Isbell et al. found that location was particularly important in affecting the proportion of species that contribute to ecosystem functioning. This observation suggests that many species may be present in a metacommunity because there are some locations where they strongly contribute to ecosystem functioning even if there are other locations where they are less important. Indeed, Isbell et al. found that a large proportion of species had negative effects on ecosystem functioning in at least some of the plots in which they were found.

The above studies and our proposed conceptual framework suggest that local richness and ecosystem function could respond differently to environmental change. If either the regional pool lacks species with suitable traits for the new environment or dispersal limitation prevents the most suitable species from colonizing, then communities undergoing environmental change may become dominated by secondarily adapted species (Fig. 12.1c). Because secondarily adapted species contribute less to ecosystem function, it is entirely possible for local diversity to be highly robust to disturbance and environmental change even though composition and ecosystem functioning are much less so. A meta-analysis of 114 studies of the influence of disturbance on animal communities by Supp and Ernest (2014) suggests that differences in the robustness of species richness, ecosystem function, and species composition can occur frequently. Specifically, they found that the influence of disturbance on individual species and community composition (measured as the median population level of components species) was quite strong, that species richness was much less influenced, and that total abundance (as a crude measure of ecosystem function) was intermediate.

12.3 ECOSYSTEMS WITHIN METACOMMUNITIES AS COMPLEX ADAPTIVE SYSTEMS?

So how do we put all of the metacommunity-level processes together with the ecosystem-level patterns that we discussed above into a single framework? The study of complex systems is one of the more exciting and challenging issues in science (not just ecology), and part of the excitement is that computational and other technical advances allow us to make incredible progress. At this point, complexity science is still nascent in terms of approaches and methods, so it is important to keep an open mind along with some skepticism. But it might be useful, at least for some questions, to consider metacommunities and the ecosystems in which they are embedded as complex systems.

One way to think about the complexity inherent at the intersection of metacommunity and ecosystems is through the concept of complex adaptive systems. CAS are one of those concepts that may at first sound a bit far-fetched and are perhaps associated with notions such as ecosystems as superorganisms that can be the product of selection—ideas that are often derided. Indeed, it is important to be as careful as we can in defining and applying the concept of CAS to metacommunities and ecosystems. CAS are systems consisting of networks of interacting components (in this case, species, nutrient pools, etc.) that show organization in their large-scale properties (examples are provided in Table 12.1). In CAS thinking, interest focuses on understanding the macroscopic features of a system in terms of how they are formed from interacting components (in this sense, CAS are similar to a number of macroecology approaches we described in Chap. 10). In other words, there is feedback between the macroscale and the microscale, so that the idea of CAS is bit more subtle than these criticisms imply. (For an excellent review of these issues, see Brownlee 2007.)

Levin (1998, 1999, 2005) has probably done the most to specify just how ecosystems might indeed qualify as CAS and how they differ from thinking that centers on ecosystem superorganisms and group selection. Specifically, he argues that the properties of CAS arise from internal interactions among components subject to at least the following three important criteria (but see Brownlee 2007 for contrasting views):

1. Maintenance of local variation, also described as "sustained diversity of individually interacting elements" (Levin 1998). Such interacting elements could consist of a diverse range of agents, but species populations within local communities might be one of the more important ones (multiple limiting nutrients probably also qualifies). Sustaining such diversity for species populations would likely also involve the processes that participate in metacommunity assembly. Some of these processes might be internal to the system (plasticity within genotypes, genetic change within local populations, changes in relative abundance of local populations; see Chap. 8), while others might involve external processes, particularly dispersal and movement of organisms or materials within metacommunities or metaecosystems (see Chap. 13).

2. Localized interactions among elements. Here the degree of localization determines the spatial extent of the ecosystem. We recognize that there is some fuzziness in this criterion (see Chap. 10), but there are typically scales that have a meaningful amount of such localized interaction. In some cases, these are obvious (e.g., lakes or habitat islands and other metacommunities sensu stricto) but in others less so (grasslands or oceans and other metacommunities sensu lato).

3. Interaction-dependent local selection, also described as an "autonomous process that favors some forms over others at the local scale" (Levin 1998). We view this question as one that involves the capacity of ecosystems to reorganize by following the rules and constraints of community assembly in response to local environmental conditions and to localized change in those conditions as described in this chapter. In well-mixed systems without localized interactions, such reorganization becomes impossible due to homogenization across progressively larger spatial scales even if local conditions vary.

Leibold and Norberg (2004, Norberg 2004) suggested that the metacommunity-level processes that we have focused on here can play an important role in moderating the CAS dynamics of ecosystems. Figure 12.13 illustrates the mathematical formulation of CAS dynamics (Norberg et al. 2001, Norberg 2004) in a simple cartoon. It shows how niche-trait variation (jointly determined by immigration and internal processes) could maintain local variation (criterion 1 above), how localized (within-patch) dynamics could alter niche-trait values in the form of community composition (criteria 2 and 3 above), and how the action of these three processes could then influence ecosystem function as a consequence of metacommunity assembly. An important feature of the approach is that the long-term outcome of metacommunity assembly is focused on an optimization principle

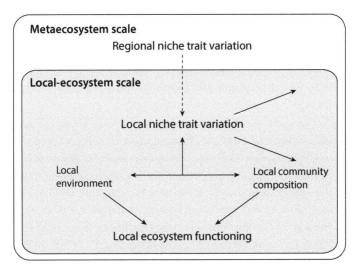

FIGURE 12.13. Local ecosystems as complex adaptive systems. Here a local ecosystem (shaded box) is embedded within a region (outer box, the metacommunity). The ability of the ecosystem to undergo change (self-organization and emergence) is driven by variation that comes from inputs from the metacommunity (dashed arrow) or is a result of the interaction between local community composition and local environment (black arrows). The interaction leads to changes in local community composition through the process of community assembly and local adaptive change in component species. A side effect of this is that species richness can also change. More importantly the joint effects of environment and composition determine the degree of ecosystem function. Modified from Norberg 2004.

that links mean trait values in the local community to environmental conditions (i.e., species sorting with no frequency dependence). Leibold and Norberg (2004) used more general dispersal-selection equations to argue that such optimization was reduced by isolation (insufficient maintenance of variation; criterion 1 above) or by excessive dispersal (loss of localized interaction and selection; criteria 2 and 3 above).

   In support of this contention, Leibold and Norberg (2004) examined the consequences of dispersal limitation and environmental change in deep-lake ecosystems that tend to be isolated from each other, in open-ocean systems that tend to be highly dispersive due to water movement, and in ponds that tend to be intermediate (inter-pond distances are substantially smaller than inter-lake distances in most landscapes). Leibold and Norberg reviewed anecdotal evidence that many lakes were dispersal limited, so that experimental or accidental introductions of species were often successful and often led to substantially altered ecosystem structures, suggesting that dispersal limitation prevented CAS selection from occurring very effectively. In ponds, dispersal limitation seemed to be substantially less important, and there is less evidence that ecosystem structure could be affected by invasions and thus show higher resilience as an emergent property. Finally, in oceans, dispersal surplus seemed more likely, because some important species showed very strong ME relations across large distances and were thus relatively insensitive to localized environmental change—with a consequent reduced effectiveness as a CAS.

These comparisons highlight two important points about the relationship between metacommunities and CAS. First, metacommunity processes contribute to the capacity of ecosystems to respond to environmental differences and environmental change in ways that conform to the idea of local ecosystems as CAS (similar to the spatial insurance hypothesis in Loreau et al. 2003a). Leibold and Norberg (2004) showed that this contribution can depend on dispersal in a unimodal way because of the way trait-input variance interacts with endogenous trait variation to alter the capacity of local ecosystems to act as CAS. In the absence of trait input from the metacommunity, other in situ processes that would do so include the following: (1) plasticity of resident species, (2) changes in the population structure of resident species, (3) local evolutionary processes by resident species, and (4) changes in the relative abundances of resident species. It is probable (but not necessary) that these other processes involve smaller trait differences than would differences involving the arrival of new species via colonization, suggesting that trait inputs via immigration from the metacommunity could be an important element contributing to emergent ecosystem attributes.

Second, the contribution of metacommunity processes to the adaptive response of the ecosystem is context dependent. For example, habitat isolation can be a major factor that determines the nature of the adaptive response. Ecosystems that are very isolated (an isolated oasis in the middle of the desert, for example) may not respond optimally, consistently, or as quickly as predicted by the idea of CAS. Similarly, local perturbations in a parcel of open ocean may also not respond as predictably to local change, due to the mixing and movement of water.

The bigger puzzle with the idea of CAS is, which features will be selected? The approach taken so far is to assume that there is an optimal value to the trait distribution of organisms for a particular environment (Leibold and Norberg 2004, Norberg 2004), an assumption similar to that above, in which we postulate an optimum relative resource use for any given environment (see Fig. 12.4). However, adaption in the system is not always so obvious, nor will adaptation always correspond to optimization. For example, in terrestrial plants that compete for light and nitrogen via allocation to different tissues, the winning strategy is not an optimum (the plant that would grow best in monoculture) but rather an evolutionarily stable strategy; that is, the species or species combination that can grow best relative to all other strategies (Dybzinski et al. 2011). The picture gets more complex

when trophic and other types of interactions (mutualisms, dependencies involving niche construction, etc.) are involved in trait optimization (Brown and Vincent
1992, Ripa et al. 2009). Here, in contrast to models describing resource competition (Sec. 12.2), it isn't clear which ecosystem-level property will be associated
with a given evolutionary stable strategy. This means that associating some property, such as productivity or biomass, with the endpoint of community assembly
(and evolution) may not be that reasonable. Although CAS dynamics of food webs
are difficult to study, more work needs to be done to identify which aspects of
ecosystem functioning are most likely.

12.4 DO ECOSYSTEMS HAVE REGULAR FEATURES?

So, if ecosystems can qualify (at least sometimes) as CAS, and if we think this
should result in some regular features of ecosystems (Levin 2005), what would these
be? In the absence of CAS, complex interactions within isolated communities lead
to highly contingent, context-dependent outcomes that hinge on the details of which
species are present and how they specifically interact (Schaffer 1981, Abrams 1993).
Do ecosystems that are nested in metacommunities show repeatable features characteristic of CAS, or are they idiosyncratic? There is a long history of thinking that
ecosystems are emergent, stable, and resilient entities (Forbes 1925, Margalef 1963,
Odum 1969, Holling 1992), having repeatable attributes such as the shape of trophic
pyramids (Elton 1927), food-web network structure (Pimm et al. 1991), and stoichiometric attributes (Redfield 1958). Table 12.1 lists some of these properties and
illustrates data that indicate their strength. Although the relationships often look
quite strong, it is important to note is that they are plotted on logarithmic scales;
they are much more variable when plotted on arithmetic scales, indicating that
whether you believe these regularities depends on whether you are more interested
in the broad patterns that are more likely to be driven by certain constraints, such
as those described above, or in the variation around those patterns.

   Arguably then, ecosystems may have some emergent properties that could be
linked to CAS dynamics. However, it is also likely that ecosystems may often be
vulnerable to environmental changes that could prevent these patterns from arising. This may even be true when changes in environmental conditions are small
(Scheffer et al. 2001). Here, ecosystems may show sudden tipping-points, or sharply
nonlinear change in response to small changes in driving variables (such as nutrient
inputs to aquatic systems or rainfall in terrestrial ones).

   Is the occurrence of these tipping points due purely to internal features of ecosystems, or are they also affected by spatial processes in the context of metacommunities? The answer seems to be yes to both. Perhaps the best-known example

TABLE 12.1. Examples of Proposed "Emergent" Patterns in Ecosystems

Name	Selected References	Illustrative Figure of Pattern in Nature	Brief Summary of Explanation
Trophic structure	McQueen et al. 1986, Leibold et al. 1997		Herbivore biomass scales with plant biomass with an exponent of 0.4 because excess herbivore mass favors inedible algae or predators and excess plant mass favors more efficient herbivores.
Predator:prey diversity ratios	Warren and Gaston 1992, Jeffries 2002, Donald and Anderson 2003		Predator:prey species richness are proportional; possible reasons include community assembly mechanisms.

Total nitrogen (TN) to total phosphorus (TP) relationships	Redfield 1958, Downing 1997, Sterner and Elser 2002		TN:TP converge to 16:1 because biological activities favor co-limitation by N and P.
Size-abundance scaling	Li 2002, Brown et al. 2004, Rossberg et al. 2008, Henderson and Magurran 2010		The abundance of different-sized organisms declines with and exponent of circa –3/4 either due to metabolic scaling or food-web assembly.

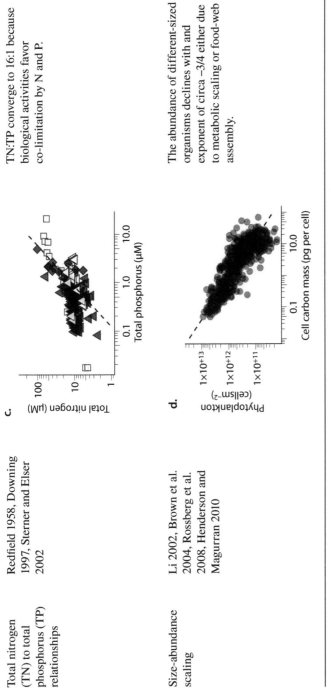

c.

Total nitrogen (μM)

Total phosphorus (μM)

d.

Phytoplankton (cells m^{-2})

Cell carbon mass (pg per cell)

Note: See references for details on how each of these is measured and quantified as well as for possible explanations.

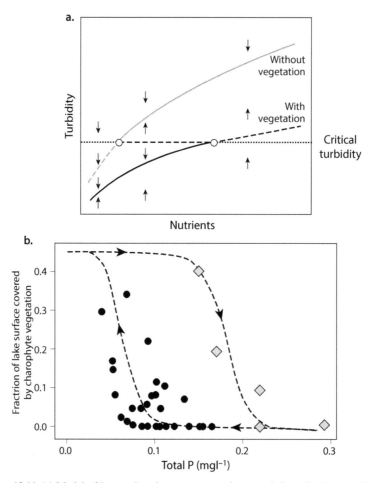

FIGURE 12.14. (a) Model of interactions between macrophytes and phytoplankton predicts bi-stability (associated with alternate stable states) over an intermediate range of nutrients (e.g., total phosphorous [P]) due to the effects of turbidity that determines how much light reaches the lake bottom where macrophytes germinate. At very low nutrient levels (lower line) and very high nutrient levels (upper line) there is only one stable turbidity level. At intermediate levels there are two stable turbidity levels, which are separated from each other (dashed line) in the range of turbidity levels that will converge on each. Across this range of nutrient levels, the system has two alternate stable states. If nutrient levels are reduced from initially high levels (from right to left along the upper line) turbidity stays high until nutrients are low enough to flip the system to the lower line (left end of the upper line where it crosses the dashed line). If nutrients increase from initially low levels (from left to right along the lower line), turbidity stays low until the right end of the line (crossing the dashed line) where it flips to the upper line. This feature is called hysteresis. The explanation is based on a simple (well-mixed) ecosystem model that takes into account the ways that emergent *adult* vegetation shades phytoplankton (due to floating and emergent leaves) but phytoplankton shades *germinating* vegetation (at the bottom of the water column. Modified from Scheffer 1990. (b) Apparent hysteresis in Lake Ve-luwe. Total P was thought to be low before the data were collected (upper left) and then increased due to cultural eutrophication (gray diamonds and rightward-pointing arrows). After eutrophication was reduced, total P decreased but vegetation did not recover until levels became very low (solid dots and left-pointing arrows). Modified from Scheffer et al. 2001.

of tipping points within an ecosystem involves the ways that nutrient inputs affect the turbidity of shallow lakes. It is well established that nutrient inputs (especially phosphorous) in lakes lead to high phytoplankton concentrations and consequent high turbidity (as well as other characteristics considered undesirable to humans, such as foul-tasting water). However, these systems may actually express a pattern of hysteresis (Fig. 12.14), whereby lakes that have intermediate levels of phosphorous can exist either in a relatively turbid (low macrophyte vegetation and high phytoplankton) or clear state (high macrophyte vegetation and low phytoplankton).

What effects do spatial processes have on such possible hysteresis? The case modeled by Scheffer et al. (2001) applies where there is complete mixing and homogenous conditions in the ecosystem. If there is environmental variation, this process can be altered. Here, there can be a continuous gradient in the environmental driver (e.g., total phosphorous in the lake scenario) that produces a relatively sharp discontinuity in the state of the environment (e.g., turbid vs. clear; van de Leemput et al. 2015). However, if the environmental heterogeneity is not along a continuous gradient but is distributed more randomly, then the domain of values that show hysteresis is smaller (van Nes and Scheffer 2003, Buenau et al. 2007).

Although hysteresis is quite interesting, it may not be as readily observed. In natural shallow lakes for example, Jeppesen et al. (2005, 2007) have done several careful studies of the dynamics of remediation in shallow lakes where phosphorous inputs were reduced, and they found important inconsistencies with the idea of tipping points in shallow lakes. First, they found that relationships between total phosphorous and turbidity (or chlorophyll concentrations, which drive turbidity) are no different between shallow lakes (where macrophytes can grow throughout the spatial extent of the whole lake) and deep lakes (where they are restricted to nearshore areas). Second, they also found no difference in the turbidity–total phosphorous trajectory of deep and shallow lakes that had similar initial total phosphorous levels. Third, they found that the apparent hysteresis between the trajectory of turbidity during the eutrophication and restoration phases can often be easily explained as lags (similarly present in both shallow and deep lakes), largely due to the time it takes to purge phosphorus from the lake after the reduction of inputs. Finally, they argued that apparent hysteresis may involve other processes with time lags, including altered top-down control and dispersal limitation of organisms. It isn't clear what having open metacommunities that connect community assembly to regional trait variation does to either real or apparent (lagged recovery) hysteresis. Obviously, a better understanding of CAS, especially in the context of limited dispersal, would help us grasp the complexity of ecosystem responses to changes either in the environment or in the metacommunity context (i.e., species pool, connectivity, environmental heterogeneity; see Fig. 12.2c).

12.5 CONCLUSIONS

Understanding the links between ecosystems and communities involves a long and contentious history (Naeem 2002, Loreau 2010). Regardless of how we interpret experiments and patterns that relate diversity, species composition, and ecosystem functioning, it is clear that it is the process of community assembly that establishes the link. The current state of research joining community structure to ecosystem functioning is very strongly focused on establishing ties between biodiversity (including both taxonomy- and trait-related components of diversity), but such research is much less conscious of the process of metacommunity assembly itself.

Here we have argued that variation in the process of community assembly is strongly regulated by metacommunity processes. We focused extensively on how this might work with simple models of resource competition, in which the processes are best understood, and found that this emphasis could improve our understanding of key results. Less obvious, however, is what happens under more complex scenarios such as might occur in food webs or other interaction webs (those involving mutualisms, facilitation, etc.). The study of communities as CAS provides a general set of insights into the behavior of these more complex possibilities, but much remains to be done. In particular, CAS have yet to adequately address the relative roles of external and internal factors that affect niche-trait variability, particularly in dispersal-limited systems. We anticipate that improving CAS and related approaches within a metacommunity framework could help us understand not only overall patterns in "emergent" properties of ecosystems, but also the residual variation present in these patterns (Table 12.1) and their dynamics; for example, how they may result in "tipping points" and other complexities.

CHAPTER THIRTEEN

From Metacommunities
to Metaecosystems

Prospectus

1. Metaecosystems ecology is an extension of metacommunity ecology that incorporates the movement of materials and the energy they contain.
2. Spatial processes, including the movement of materials and energy, are almost certainly important for almost all ecosystems because major biogeochemical processes require the movement of materials, and energy necessarily flows into and out of local ecosystems.
3. Burgeoning theory on metaecosystems shows that even very simple models can have complex outcomes that depend on the rates of movement of materials and energy, as well as on the movement of and interactions among organisms.
4. Variation in the rates of movement of different ecosystem components, such as inorganic nutrients, plants, herbivores, or detritus, often generate different outcomes.
5. We are only now beginning to understand the ways these movements affect the dynamics within metaecosystems, including biomass relations, productivity, and the stability of these dynamics under temporal and spatial variation in environmental conditions.

We have argued that metacommunity ecology can provide a synthetic framework that can integrate a number of findings from across ecology—ranging from competition to biodiversity to evolution and even the role of biodiversity in functioning of ecosystems. So why stop there? The most inclusive, highest-level construct for ecology is the ecosystem. Can we think about a metaecosystems ecology?

In Chapter 12, we examined the consequences of metacommunity dynamics for ecosystems. However, doing so captures only one aspect of ecosystem dynamics: that related to metacommunity assembly. When organisms move, especially among heterogeneous habitats, they also transport materials (in the form of their own biomass). Additionally, the materials themselves can move in space in inorganic or nonliving organic form by diffusion or transport by wind, water, or gravity. How does this movement influence ecosystems and community structure? Loreau et al. (2003b) were the first to propose that one could think about metaecosystems as a way to address such fundamental questions in ecology, but this approach echoes a number of related ones with a long history (reviewed in Massol et al. 2011). Switching from metacommunities to metaecosystems, however, is a big step. Much of what we discussed in previous chapters in this book, as well as other synthetic approaches to community ecology (e.g., Vellend 2010, 2016), focused on frequency- and density-dependent phenomena, which were mostly applicable to interspecific competition. In contrast, metaecosystems emphasize aspects of ecology related to the build-up of stocks of materials and energy into component compartments and the flows among those compartments. This makes the theory and practice of metaecosystems necessarily distinct.

Polis and collaborators provided some of the most compelling studies that demonstrated the importance of *ecological subsidies*: situations in which the input of materials from outside a well-delineated ecosystem could play a key role in the structure and dynamics within this ecosystem (Polis et al. 1997, 2004). However, the importance of spatial effects in ecosystems was recognized well before that time, including by a substantial history of work in landscape ecology (Troll 1939, Naveh and Lieberman 1984, Urban et al. 1987, Turner et al. 2001). Likewise, in river ecology, spatial movements of materials, such as that embodied in the *river continuum concept* (Vannote et al. 1980), has long been a key concept linking community and ecosystem processes to the movement of materials and organisms, as water flows from small headwater streams to larger rivers and eventually to the ocean. Similarly, Bormann and Likens (1967) proposed that terrestrial ecosystems could be delimited by watershed boundaries that constrained the downhill movement of materials through water flow and erosion.

In the context of metacommunity interactions, some very significant studies illustrate the potential importance of such subsidies. For example, the importance of upstream movement by anadromous fish, such as migrating salmon, can dramatically increase the phosphorus budget of headwater lakes and streams, which then cascades to terrestrial plants (Naiman et al. 2002) and consumers such as grizzly bears (Levi et al. 2012). On even finer spatial scales, spatial subsidies are sometimes surprisingly strong. For example, Nakano and Murakami (2001) found that almost half of the energy flux through stream-fish communities was terrestrial in origin (via the consumption of terrestrial insects) and that the same was reciprocally true for riparian-zone birds (via the consumption of emerging aquatic insects). In some cases, the movement of materials can create small-scale spatial patterns within an otherwise homogenous environment; for example, in the way termites concentrate nutrients and carbon toward their termite mounds, leaving interstices depauperate in soil organic matter (Pringle et al. 2010).

Alternatively, material fluxes can also have dramatic large-scale effects. For example, it is thought that iron-containing dust blown from northern Africa plays a key role in producing red tide algal blooms in the Gulf of Mexico (Walsh and Steidinger 2001)!

These and other approaches to spatial effects in ecosystems certainly justify the perspective of movement of materials along with that of organisms in a metaecosystem, as proposed by Loreau et al. (2003b). What made their suggestion important was that they argued that metaecosystem ecology could be developed within a similar framework as that for in metacommunity ecology. To make this idea operational, Massol et al. (2011) argued that it requires understanding the feedback between the movement of materials and organisms and how these are altered by species interactions.

13.1 WHY SPATIAL DYNAMICS ARE SO IMPORTANT IN ECOSYSTEMS

One can argue that most (and perhaps all) natural local ecosystems are in fact contained within metaecosystems, and thus connected to nearby localities in flows of materials and organisms. The reason is that key processes involving material conversions require distinct abiotic conditions that are not easily maintained except by separating them in space (Burke and Lauenroth 2011). For example, the nitrogen cycle has important components that require anoxic conditions (nitrogen fixation and denitrification), while other components are enhanced by or require aerobic conditions. It is possible for these processes to be separated in time (for example, if day-night fluctuations are associated with changes in oxygen concentrations), but this is not often the case. The exchanges between heterogeneous habitats (places with distinct abiotic conditions such as oxygen levels) are thus key to preventing the accumulation of materials in unusable or recalcitrant forms and redelivering them to habitats where they can be used. These exchanges can involve passive diffusion of materials between distinct habitats, where they are driven by concentration gradients, or active movement by organisms and materials via other processes (e.g., gravity, buoyancy, or water and air flow).

It is hard to think of ecosystems that do not have important spatial structure. Perhaps the most obvious candidates are closed artificial ecosystems (generally aquatic microcosms; Taub 1974), which can sometimes sustain life (mostly in the form of microbes, algae, and small invertebrate grazers) for several years. Some of these are even available commercially and you can have one for your very own![1] In nature,

---

[1] The Ecosphere (eco-sphere.com), which is advertised as the "world's first totally enclosed ecosystem." This ecosystem doesn't have any obvious spatial structure (except for the radiant light that enters from the Sun), but it generally exists for only a few years. Maybe there is a reason for that!

isolated cave systems sometimes prevent flows of organisms (such as the isolated caves in Australia that harbor the diving beetles shown in Fig. 8.13), but even here these communities must get material and energy subsidies from other habitats.

How do we incorporate an ecosystem perspective into *metaecology* thinking? Regardless of the scale involved, an ecosystem can be thought of as an entity that has inputs of materials and energy, then processes these inputs in various ways, and finally has outputs of materials and energy (Fig. 13.1).

The inputs are of two nonexclusive types. First, they involve materials, in the form of elements, and energy, in the form of bonds among them. The materials must follow mass balance, whereas the energy is necessarily lost during the processing stages (i.e., entropy), and thus ecosystems must gain energy from outside (e.g., radiant sunlight or other energy sources, or import from other localities within the metaecosystem). The second type of input involves the movement of organisms. Necessarily, these immigrants also bring in materials and energy, although this is often minor in comparison with inputs due to abiotic effects. Often, immigrants have a large effect because they alter community structure and can thus indirectly change the nature of species interactions (as discussed in Chap. 11, for example). These interactions in turn determine how materials and energy are processed within a local ecosystem and how effective the ecosystem is in capturing, converting, and recycling them. Local ecosystems also export individuals via emigration. Emigration, like immigration, also has the dual role of exporting materials and energy and thus altering community structure. Spatial movements can thus change ecosystems via two means: the direct transport of materials and energy, and indirect effects mediated by restructured community composition and species interactions.

The dynamics of metaecosystems are thus driven by a complex network of interactions that involve the movement of materials, energy, and individuals (Loreau et al. 2003b). Marleau et al. (2014) argued that these processes could be modeled using separate equations that tracked the movement of materials and energy in the form of organisms, dead organic matter, and inorganic forms in a network of patches depending on connectivity and diffusion and dispersal rates. This framework is very similar in spirit to the way Gravel et al. (2016) modeled metacommunity food webs (see Sec. 11.4), except that now there are additional equations for the movement of nonliving materials.

At present, metaecosystem ecology is still in the early stages of development, and it is difficult to make strong conclusions. Nevertheless, it is clear that it has the potential to alter how we understand ecosystem dynamics because it explicitly integrates the movement of organisms and species interactions of the metapopulation-metacommunity perspective with the more well-studied movement of materials and energy. Below, we illustrate some of the possibilities using selected examples from recent theoretical and empirical work.

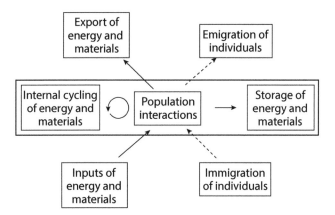

FIGURE 13.1. Schematic of a local ecosystem contained within a metaecosystem. Each local ecosystem has flows into it (bottom row of boxes) in the form of individuals that affect community assembly (right-hand box). These individuals can also lead to inputs of materials (and the energy they contain) that supplement nonliving inputs (in either organic or inorganic form; left-hand box). Similarly, each ecosystem can have outflows (upper boxes) in the form of individuals and materials (which may contain potential energy in chemical bonds). Within the ecosystem, there is internal cycling and modification of materials as well as storage (accumulation) of materials, and these are modified by the interactions of organisms.

13.2 ELEMENTS OF METAECOSYSTEMS ECOLOGY

Although the spatial dependencies of ecosystem processes have long been understood (e.g., Odum 1969), these dependencies have not been well integrated with metacommunity dynamics. Loreau et al. (2003b) were the first to suggest that this could be done by simultaneously considering the movement of living individuals (as in metacommunity ecology) as well as the movement of materials and energy (as in landscape ecosystem ecology; Pickett and Cadenasso 1995). Figure 13.2 shows how previous traditions in these two disciplines differed in the emphasis on the relative attention given to species interactions (dependent on their traits) versus landscape-dependent connections as they affect material exchanges (Massol et al. 2011). Metaecosystems ecology emphasizes that both types of exchanges are likely to work simultaneously and interactively.

To illustrate the potential importance of metaecosystem dynamics, Figure 13.3 shows a simple model from Loreau and Holt (2004). Specifically, they modeled nutrient-dependent plant growth in local ecosystems with different levels of openness and studied the influence of two possible inputs (immigration of plants [P] via propagule dispersal [I_P] and inorganic nutrient [N] via inflows or diffusion [I_N]) and outputs (export of plant material in any form [$e_p P$] and diffusive outputs of

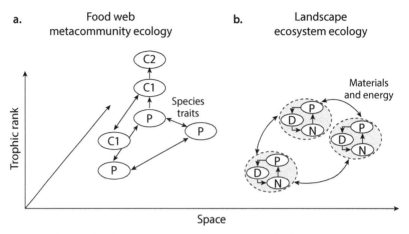

FIGURE 13.2. Contrasting food-web metacommunity ecology with landscape ecosystems ecology. Both approaches consider localities in space (x and z axes) but differ in several respects. (a) In food-web metacommunity ecology, it is the dispersal of distinct producers (P) and consumers (C1 and C2) that determine the resulting dynamics as they are mediated by niche relations (species traits) among them. (b) In landscape ecosystem ecology, there are local food webs in each locality that are based on material inputs such as nutrients (N) and material exports via either living organisms (here just considering producers, P) or via nonliving organic matter such as detritus (D). Modified from Massol et al. 2011.

inorganic nutrients [$e_N N$]). This simple formulation allowed them to contrast a closed ecosystem (I_i and e_i values equal to zero), with a chemostat model (values not equal to zero). When I_i and e_i values were nonzero and equal to each other, the behavior of the system was qualitatively similar to the case in which there was no movement. However, when I_i and e_i were not identical to each other, they found that model outcomes were substantially influenced by spatial movements, which are illustrated using classic isocline analysis in Figure 13.3b.

Though illustrative, this model is just for one local system with inputs and outputs. To inject a bit more complexity, Gravel et al. (2010) explored a model with two patch types between which there can be movement of detritus and/or herbivores, not just of plants and nutrients (see Fig. 13.4). Their results show that there can be important consequences depending on the type (whether movement is by nutrient, plants, detritus, or herbivores) and the strength of these exchanges. As an example, the scenario shown in Figure 13.4.a.ii illustrates a case in which nutrients diffuse between the two patches and detritus is also exchanged but plants and herbivores cannot disperse. The effect of the diffusion of the detritus on the biomass of plants is complex and nonlinear (Fig. 13.4b). When detritus is not exchanged (i.e., the systems are isolated from each other), one of the patches cannot

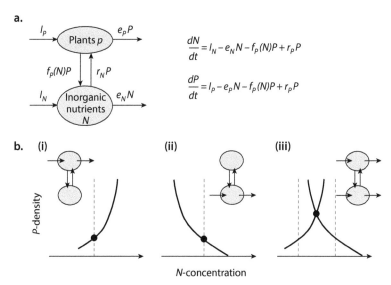

$$\frac{dN}{dt} = I_N - e_N N - f_p(N)P + r_p P$$

$$\frac{dP}{dt} = I_p - e_p N - f_p(N)P + r_p P$$

FIGURE 13.3. (a) A schematic description of a local ecosystem subject to inputs of either plants (via dispersal) and nutrients (via diffusion or transport) and exports of plants and nutrient via a diffusive process (export is proportional to standing levels of plants and nutrients). (b) Isocline analyses of the model under different scenarios. The first scenario (i) has only plant dispersal, the isocline for plants is sloped, the one for nutrients is vertical, and their intersection denotes the equilibrium point, which is achieved by damped oscillations. The second scenario (ii) has only nutrient diffusion, and the nutrient isocline is sloped, but the plant isocline is vertical. The third scenario (iii) has both diffusion and dispersal, and both isoclines are sloped. Different outcomes and dynamics are predicted in each case. Modified from Loreau and Holt 2004.

maintain any plants (because it is too poor in nutrient supply), but diffusion of detritus can allow the nutrient-poor system to support plants by subsidizing nutrient input in the form of detritus. However, this effect is stronger if the diffusion of available nutrients is low so that the sink habitat can retain this inflow in the form of detritus. Gravel et al. also consider other scenarios (Fig. 13.4a–d) and show that inter-patch movement of materials and organisms generally have similarly important indirect effects, and that these are dependent on the structure of the community (different consequences arise for the same exchange depending on the presence of herbivores, for example). An important follow-up study by Gounand et al. (2014) examined the effects of nutrient enrichment in metaecosystems with herbivores (Fig. 13.4a.iv) and found that consumer and herbivore movements were stabilizing (especially those of herbivores), whereas the flows of nutrients and detritus (especially nutrients) were destabilizing.

There are many possible outcomes for more-complex landscapes. For example, Marleau et al. (2014) showed that temporal fluctuations can be caused by the

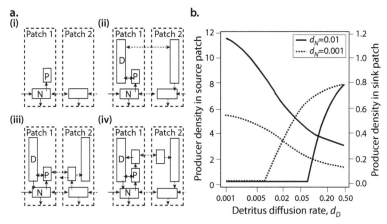

FIGURE 13.4. A two-patch metaecosystem model with nutrients (N), producers (P), herbivores (H), and detritus (D). (a) Gravel et al. (2010) considered four scenarios as illustrated in the four diagrams. In all cases, it is assumed that the plant can only be self-sustaining (absence of flows) in Patch 1. The first scenario (i) involves only plants and nutrients, with only nutrient flows. The second scenario (ii) involves flows of nutrients and detritus but not plants. The third scenario (iii) involves movement of plants and nutrients but not detritus. Finally the fourth scenario (iv) involves the movement of nutrients and herbivores but not plants or detritus. (b) Interactive effect of nutrient flows (dashed vs. solid line) and detritus flows (x-axis) on plant density in Patch 1 (line with decreasing slope) and Patch 2 (line with increasing slope) in the case of Scenario 2. There is a critical level of detritus flow that permits the establishment of plant populations in Patch 2 (otherwise unable to support a plant population), and the outflow of detritus means that overall nutrient levels are lower in Patch 1 because they are exported to Patch 2 and thus support lower plant biomass in Patch 1. Modified from Gravel et al. 2010.

movement of nutrients among patches depending on the specific ways by which patches are connected and the diffusion among the patches. While such a complex result is a bit foreboding (at least for empirical studies), as yet there are not any real-world examples that confirm these predictions.

The complexity and feedbacks that emerge from simple systems is perhaps reminiscent of early models of indirect effects that seemed similarly daunting to community ecologists in the late 1970s (Lawlor 1979, Schaffer 1981); perhaps we can take heart in the way that these have been resolved. For species interactions, the solution involved a combination of improved mechanistic understanding of their causes (Wootton 1994) and a better understanding about how they work gained from simplified models (Holt 1997), as well as emerging theoretical and analytical tools based on network ecology (Ings et al. 2009, Thebault and Fontaine 2010). For complexity in metaecosystems, we hope that similar insights will help to constrain what is possible.

The current message from metaecosytems ecology seems to be one of complexity. And although this can be an important point, it is also important to realize that the field is in its infancy and advances will likely come from simplifying and generalizing the approach. One way to do this is to take a generalized approach that explores nutrient flows and exchanges within model systems where we understand the behavior of closed systems, such as the resource competition scenarios overviewed in Chapter 12 (Tilman 1982, Chase and Leibold 2003). For example, Schade et al. (2005; see also Soranno et al. 2010), using the conventional chemostat model of Tilman (1982), considered a heuristic scenario in which nutrient inputs in a lake embedded within a watershed metaecosystem are determined by how nutrient levels in an upstream lake are depleted by producers; that is, they viewed each lake as if it were a chemostat linked to the output of the previous lake in a series. What happens in the first lake then has an indirect effect on nutrient competition in the next lake (but here the influences are unidirectional because water flows downhill). Schade et al. argued that this creates spatial indirect effects, whereby species interactions in an upstream location (perhaps driven by a factor such as the absence of piscivorous fish, for example) affect the flow of materials, and consequent species interactions, downstream.

13.3 AN EMERGING SET OF PRINCIPLES?

The models described above illustrate cases in which system "openness" can influence species interactions and coexistence, as well as material flows. In many cases, however, these effects seem to be highly context dependent and hard to predict in the absence of detailed knowledge of the processes involved. At the moment, we are far from a more general theory that is able to capture movements of organisms and their interactions, as well as materials and energy within metaecosystems.

We can, however, begin to summarize many of the specific findings that will help in the development of a more general framework. For example, Figure 13.5 illustrates a few generalities, which are summarized as follows (based on Massol et al. 2017): (1) the movement of organic materials (living or nonliving) is generally homogenizing (i.e., reduces environmental heterogeneity); (2) the movement of inorganic nutrients can amplify environmental heterogeneity because it is enhanced by low consumer abundance in low-productivity patches; (3) exchanges in materials among patches tend to dampen asynchronous temporal variation; (4) the movement of organisms often leads to source-sink relations in the presence of spatial environmental heterogeneity, which reduces overall productivity as a result of maladaptive mass effects (see Chap. 12); and (5) the movement of organisms can

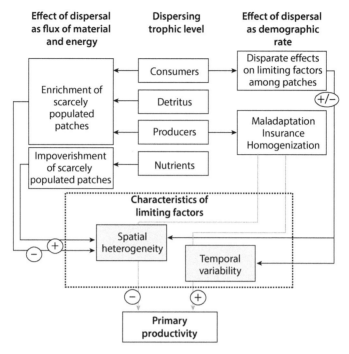

FIGURE 13.5. Schematic summary of metaecosystem models. The effects of movement among local ecosystems can be classified by the way they alter materials and energy (left side of the figure), or demographic effects that alter population dynamics of component species including species interactions (right side). The effects of these movements on ecosystems depend on whether ecosystems vary in space in some way (e.g., baseline nutrient levels) or whether they vary temporally (e.g., they are altered by temporally varying variables that are asynchronous among the patches). The flows of nutrients among patches tend to increase spatial heterogeneity, whereas the movement of materials via other components tends to reduce spatial heterogeneity. The movement of plants tends to lead to maladapted local populations (poorly competitive sink populations, in our terminology from Chap. 12) in purely spatially heterogeneous situations but increases the insurance effect under asynchronous temporally fluctuating conditions. The indirect effects of movement on regional productivity are generally negative in spatially heterogeneous systems but positive in temporally fluctuating ones. From Massol et al. 2017.

enhance the insurance effect, increasing production among asynchronously fluctuating ecosystems.

Metaecosystem theory seems to be full of surprises. Indeed, the complex nature of flows of materials and energy, rather than just of populations, can be responsible for numerous complex effects that are still poorly understood. Empirical work, both experimental and observational, is just beginning, and is sorely needed to evaluate these ideas.

13.4 CONCLUSIONS

Metaecosystems are perhaps the ultimate way to conceive of spatially distributed ecological dynamics. So far, the work indicates a wealth of possibilities that are difficult to synthesize into a well-structured understanding. This is not surprising since almost all of this work is recent and much less well connected to older work in ecology than has been the work we have described on metacommunities. One of the more salient features is the complex ways that feedbacks can be altered depending on which entities are moving (e.g., consumers or nutrients), especially the suggestion that the movement of abiotic nutrients can be particularly destabilizing and enhance spatial variation among ecosystems.

At this point it is hard to know how robust these conclusions might be, but a couple of issues are likely to be involved. First, the theory generally assumes that material flows are bidirectional, largely based on the idea that they diffuse from high- to low-concentration/density conditions. This bidirectional flow means that there can be feedback in the system. However, the movement of nonliving materials in ecosystems is perhaps more often directional (or at least strongly biased) by gravity and prevailing winds. This may reduce the opportunities for feedback via purely nonliving flows relative to that which has been studied in much metaecosystem theory. Of course, this does not diminish the possibility for feedback involving the movement of living organisms (with or without the movement of nonliving materials). For example, the strongly directional downstream movement of nutrients from stream headwaters to the ocean may be largely offset by the upstream movement of anadromous fishes. Second, the movement of organisms is often assumed to be passive, meaning that movement patterns are assumed to follow a density gradient rather than being adaptive. Yet "movement ecology" (Nathan et al. 2008) indicates that this is rarely so, and incorporating such directed movements will likely be a critical piece in metaecosystems ecology.

Ultimately, building and refining the metaecosystem concept may well be the key to an ultimate synthetic view of ecology. Naeem (2002) described the tensions between ecologists who focus on demographic and community-based approaches to ecology but ignore or subsume any interest in materials and energy, and ecologists who are more focused on materials and energy dynamics in ecosystems. Important progress has been made to better couple these two traditions, especially with the development of stoichiometric ecology, which emphasizes nutrients and materials (e.g., Sterner and Elser 2002), and metabolic theory, which emphasizes energy (e.g., Brown et al. 2004), as well as their interaction (e.g., Allen and Gillooly 2009). However, the link has not yet adequately taken spatial dynamics into account. It seems likely from the work described above that spatial effects may be critical in understanding this connection in nature.

CHAPTER FOURTEEN

A Coming Transition
in Metacommunity Ecology

Prospectus

1. Metacommunity ecology has come a long way since its start in the 1950s and 1960s with the development of island biogeography theory and meta-population theory, although it has much earlier roots (see Chap. 1); and it has even come a long way in the nearly fifteen years since the initial synthetic efforts by Leibold et al. (2004) and Holyoak et al. (2005).

2. The four archetypes of metacommunity structure that were envisioned and synthesized (SS, PD, ME, and NT) now seem a bit naïve and certainly inadequate, even though they were very important to the ontogeny of the field. Even though it was recognized early on that these archetypes are not mutually exclusive, much subsequent work focused more on them as alternative hypotheses rather than seeking integration among them.

3. By using scale as an explicit consideration, along with the fundamental processes related to habitat heterogeneity and dispersal, we suggest that metacommunity ecology can serve as a synthetic and conceptual hub for a broad view of ecology and its links to evolutionary biology and ecosystem processes.

4. Nevertheless, there are many important challenges in both theory and methodology that are needed to move the field forward. We are only now beginning to see how progress along these lines might proceed.

5. An important motivation for improving metacommunity ecology along these lines is to make it more relevant to applied ecology.

6. These challenges are not going to be easy to overcome, and solutions will certainly be more complex and pluralistic than some might like. But we hope to inspire others to join in this effort, which could lead to a new synthesis in ecology and allow it to address some of the biggest challenges confronting ecological systems.

When we both started moving from a predominantly "local scale" view of ecological pattern and process, which was predominant prior to 2000, toward a more explicit metacommunity perspective, neither of us had much of a sense of what we meant by the term, much less how far the idea would take us in our research and thinking. The work we did in the early 2000s as part of the Metacommunity Working Group at NCEAS, which lead to the syntheses by Leibold et al. (2004) and Holyoak et al. (2005), catalyzed our thinking in a way that was tremendously exciting. The metacommunity idea has been the driving force in both of our research programs ever since. We set out to write this book to figure out where this effort, together with the work of other ecologists (many of them colleagues and friends), has taken us and identify where it should go in the future.

Perhaps the most important insight from the NCEAS working group was the realization that there were numerous ways that ecologists had envisioned the dynamics of metacommunities, each of which made different core assumptions and predictions, but each of which also made a lot of sense, at least for some types of problems and systems. To organize this variability of approaches, we placed them into four categories that we called "paradigms" at the time (Leibold et al. 2004, Holyoak et al. 2005); in this book we settled on a more modest term of "archetypes" because they don't seem quite as critical today and it is clear that a more pluralistic and nuanced perspective is necessary (see also Chaps. 1–5; Brown et al. 2017). From its high and continued citation rate (more than 2500 citations when this book went to press, and more than 200 citations a year for the past five years), this categorization seems to have struck a nerve in the broader ecological community. We suspect it did so for several reasons.

First, it pointed out that there was not necessarily a single "correct" view of metacommunities. One of the more satisfying outcomes of our working group was that we felt we bypassed the polarizing debates that have characterized many important ideas in ecology: density dependence versus independence, Clements versus Gleason, assembly rules versus null models, competition versus predation, etc. Maybe this is because it is harder to polarize four points of view than two! In addition, it became obvious that our own initial biases and approaches were not likely to be fully adequate to address the full range of possible (and empirically obvious) dynamics.

Second, our categorization encouraged empiricists to look for ways to analyze their data and determine which of the divergent perspectives (i.e., "paradigms") might most closely correspond to their study system. One of the most influential of these is the variation partitioning approach that we have used as a reference point throughout much of this book, although there are several other approaches. Despite the limitations of each of them, we feel confident enough to say that species distributions are often strongly influenced by environmental conditions (i.e., SS dynamics), but that spatial and stochastic processes (i.e., PD and NT dynamics) can also often play an important role, and that the relative importance of these depends on scale, heterogeneity, and dispersal (there is, we feel, less clear evidence for strong ME dynamics in natural communities). It is also clear that we do not have the right tools that can fully deal with the complexity of metacommunity dynamics, at least not by themselves. Increasingly, studies are drawing on multiple analytical methods, as well as multiple observational and experimental approaches, to gain complementary insights about different aspects of metacommunity dynamics.

Third, the categorization generated testable predictions. However, metacommunity ecology poses huge challenges to experimental approaches because of the scale issue. We have to confront the fact that many of the more critical predictions address processes and phenomena that make large spatial-scale experiments logistically challenging (if not nearly impossible). That consequence has meant becoming increasingly interested in how to test predictions using statistical-observational approaches, making us think about metacommunities in more subtle ways and over broader conceptual domains than we might have done otherwise.

As we did the research for this book, we were increasingly impressed by the range of ideas that the concept had inspired and the progress it stimulated. Indeed, we found ourselves not mentioning numerous interesting components of metacommunity ecology that would have been great to include but would have lengthened the book greatly and not necessarily contributed to our main theses. And as we were working through the various chapters, we kept finding new publications that were so stimulating we found ourselves rewriting sections over and over again!

14.1 THE ACCOMPLISHMENTS OF METACOMMUNITY ECOLOGY VERSION 1.X

We view the paper by Leibold et al. (2004) and the edited volume by Holyoak et al. (2005) (both of which emerged from that initial NCEAS meeting discussed above) as the first organized attempts to develop metacommunity ecology into a broad synthetic approach. So perhaps we can call this first synthetic undertaking in metacommunity ecology "version 1.0." Throughout this book, we have reviewed and synthesized efforts to expand and modify v1.x (e.g., v1.1, 1.2, etc.); for example, by adding a temporal component in addition to space (Chap. 5), traits and phylogenies (Chaps. 6 and 7), evolutionary dynamics (Chaps. 8 and 9), food-web interactions (Chap. 11), etc. Nevertheless, most of this work is still closely tied to the four archetypes synthesized by Leibold et al. (2004), tempting us to label this book "version 1.9" with the idea that we will be heading toward "version 2.0" soon, but that it will need a rather significant overhaul from v1.x. Below, we will argue that the way toward v2.0 is to reframe the approach into a more integrated, subtle, and flexible one. This will involve a fundamental change in the theory, analysis, and interpretation of spatiotemporal ecological processes.

Despite looking forward to v2.0, we think it is worth looking at what v1.9 has accomplished beyond v1.0. Metacommunity ecology v1.x has led to important changes in the way the field understands and interprets ecological patterns and processes. We have described many of these in detail in previous chapters; however, it is worth giving a more macroscopic overview here.

We have shown that metacommunity ecology can be distilled down to the relative importance of three mechanisms—dispersal, trait-by-environment matching, and stochastic ecological drift—that are at the heart of the different metacommunity archetypes discussed throughout this book. However, these three mechanisms are strongly modified by three factors—scale, habitat heterogeneity, and connectivity— that ultimately influence the relative importance of the different mechanisms. Furthermore, these are modified by evolutionary processes (Chaps. 8 and 9), as well as by trophic interactions (Chap. 11), among other factors.

We contrast our synthesis of metacommunity ecology with another recent and very useful synthesis by Vellend (2010, 2016). Taking inspiration from evolutionary biology, Vellend suggested that community ecology can be viewed as the product of four basic high-level processes: *selection*, in which certain species are favored over others (often in a context-dependent way that may depend on environment or other factors); *drift*, in which species with similar ecological attributes change in relative abundance due to stochastic processes of demographic change; *dispersal*, in which the movement of organisms over larger spatial scales impacts the assembly processes that can occur at smaller scales; and *speciation*, in which the genesis of species within a larger spatial area (either by actual speciation within the area or by colonization from yet more remote locations) provides the basis for maintaining a diverse species pool. Clearly, there are strong analogies between Vellend's framework and ours; selection approximates our trait-by-environment matching (i.e., SS); drift approximates our drift (i.e., NT and PD); dispersal approximates our dispersal (i.e., NT and PD, as well as ME); and speciation approximates the species pool and macroevolutionary processes we have discussed throughout.

On the surface, Vellend's framework with four fundamental processes might seem a bit more elegant than our framework, which has three processes (dispersal, drift, trait-by-environment) modified by three factors (scale, heterogeneity, connectivity), which themselves are modified by a number of other influences (evolution, trophic interactions). However, we believe that our synthesis provides extra detail, allowing greater flexibility and the ability to explore patterns and

processes that are not captured as fully in Vellend's synthesis, such as macroecological patterns (Chap. 10), food webs (Chap. 11), and ecosystems (Chaps. 12 and 13).

Our approach to metacommunities, exploring the mechanisms of species interactions in a spatiotemporally heterogeneous matrix, also allows a more comprehensive link between evolutionary and ecological processes. This connection
resonates well with the emerging interest in eco-evolutionary feedbacks but places
them in a broader spatial context (see Chap. 8). This is important because both
community assembly and niche-trait evolution can be strongly influenced by dispersal and gene flow. What is even more exciting is the prospect that this link between ecology and evolution can also emerge at the macroevolutionary scale. We
argue that metacommunity assembly happens not just by colonization and extinctions, but also by the way the species pool evolves; the important insight is that
the species pool may also be strongly affected by the cumulative effect of species
interactions as they occur across landscape-regional spatial scales and not just as
an external process.

Finally, metacommunity ecology also provides a way to link community ecology with spatial and landscape ecology by providing more of a focus on ecosystem-
level processes, as well as by linking macroecology and biogeography, with an
emphasis on large-scale patterns of biodiversity. Although each of these areas
has a well-developed literature, there is little "cross talk" between them. By including individuals of species (and their traits) in a spatially explicitly perspective, metacommunity ecology provides a common currency for and translation
among the subfields that address similar problems from sometimes quite orthogonal directions.

14.2 SYNTHESIS THROUGH METACOMMUNITY ECOLOGY

In addition to the above successes, we argue that perhaps the most important thing
that version 1.x has done is to provide a framework for synthesis in ecology that
would have been difficult to obtain without the spatial context that comes from
metacommunity ecology. The basic structure of this book was meant to highlight
precisely this aspect of metacommunity ecology, as we worked to include evolutionary feedbacks, biogeography and phylogeny, macroecology, food webs, and
ecosystems as topics of interest within the broader metacommunity domain.

There are many examples for how metacommunity thinking, and in particular
the role of spatiotemporal scales, has helped us to achieve synthesis. For example,
as we described in depth in Chapter 1, one of the most enduring controversies in
community ecology has to do with "predictability." Starting at least with Gleason
versus Clements and probably before, there has been a century-long debate about

whether community structure in a given location can be predicted by the environmental conditions and species traits (akin to the modern-day SS perspective), or whether randomness reigns supreme and overrides the influence of environments and traits (akin to the modern NT). However, these disparate ideas need not be incompatible in a scale-explicit metacommunity framework, where smaller-scale patterns might look quite random, but at the same time, larger-scale patterns are more predictable (Wiens 1989, Levin 1992). Related synthetic perspectives that emerge from a spatially explicit metacommunity perspective include variation in the importance of metacommunity assembly mechanisms such as environment filtering, dispersal limitation, and drift with scale (Orrock and Watling 2010, Chase 2014), as well as scale dependence in the shape of biodiversity gradients (e.g., with productivity) (Chase and Leibold 2002, Gardezi and Gonzalez 2008).

Our hope is that this synthesis will continue to be a major feature of metacommunity ecology going forward. One could argue that in some cases, our discussions of these synthetic efforts have often been more speculative than conclusive. No doubt some of our speculations and hypotheses are simplistic and even perhaps a bit naïve. Nevertheless, our thesis is that the inherent scale dependence of ecological pattern and process, as encompassed by the metacommunity perspective, will allow once-seemingly disparate and incompatible observations and assumptions to be better integrated into an overarching synthetic perspective.

14.3 THE CURRENT STATUS AND LIMITATIONS OF METACOMMUNITY VERSION 1.9

So where are we now with version 1.9? What are the elements that characterize it and what are its shortcomings? In this book we have tried to outline the scope of metacommunity ecology as it went from its early synthesis (Leibold et al. 2004, Holyoak et al. 2005) to its current state. We have especially focused on exploring ideas that go beyond the four archetypes outlined by Leibold et al. (2004). These ideas include complex community assembly in space and time (Chaps. 4 and 5), local niche evolution (Chap. 8), historical biogeography (Chap. 9), and food-web interactions (Chap. 11). Nevertheless the key features that characterize variation in metacommunity ecology can still be broadly related to two key features: the amount of environmental heterogeneity and the connectance or dispersal in the metacommunity (Fig. 14.1).

Metacommunity ecology has generally emphasized emerging patterns at the scale of the entire metacommunity. For example, variation partitioning describes how variation across the entire metacommunity can be separated into environmental, spatial, and residual components. However, variation partitioning may not adequately

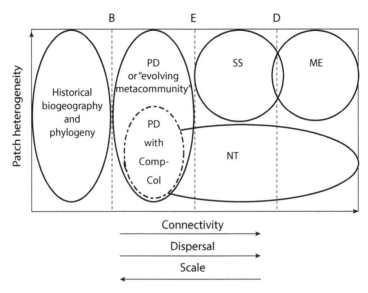

FIGURE 14.1. Synoptic for various archetypes of metacommunity structure and dynamics. They depend on how dispersal and patch heterogeneity interact. Here the rate of dispersal events (*x*-axis) is compared with the rate of per capita mortality (D), stochastic local population extinctions (E), and the rate of vicariance or other biogeographic events (B).

capture the details of processes within the metacommunity. As a thought experiment, we could ask what might happen if half of the metacommunity consists of very closely linked patches that are environmentally heterogeneous and the other half consists of more homogenous, highly dispersed patches. Here, variation partitioning would give us intermediate values for the environmental, spatial, and residual variation components. Although our analysis might be accurate, at the very broad scale it describes it will miss important components of the patterns that actually describe significant and, in this case, otherwise obvious processes.

Similarly, metacommunity ecology typically examines patterns across the entire set of species in the metacommunity. As with the possible differences in patch distributions in the landscape, differences among species could also be important. How do we separate patterns in which some species show associations with the environment (SS) and others show patterns of spatial limitation (PD or NT) from a pattern in which all species show intermediate spatial and environmental effects?

Metacommunity ecology also simplifies how spatial structure actually works. In its more extreme forms, where space is treated implicitly, there is no way to say anything about spatial effects per se. The work we have described often goes a bit further and has a spatially explicit form. Variation partitioning is the best example of this on the empirical side, but, increasingly, models have also expanded to be spatially explicit to address spatial effects (Economo and Keitt 2007, Leibold and Loeuille 2015, Fournier et al. 2017). That has been an important development but often still falls a bit short by assuming distinct spatial scales (the local patch and the metacommunity). What happens when species differ on the scale at which they perceive space, and what happens when we want to apply these ideas to meta-communities that do not have sharply delineated patches? How do we apply the insights we have developed for metacommunities sensu stricto to more continuous metacommunities sensu lato?

Finally, metacommunity ecology has mostly assumed that dispersal is a passive diffusion-like process. Yet we know that there are few organisms that disperse this way. We have suggested, for example, that the reason ME theory may be rather poorly supported is that most organisms have biased dispersal that depends on local conditions (Chap. 4). Resetarits and Silberbush (2016 and references therein) demonstrate that habitat selection behavior can have important effects on metacommunity patterns. Similarly, dispersal traits are subject to selection (Massol et al. 2017) and thus can influence metacommunities in ways that have not been adequately addressed.

This synopsis of the attributes and limitations of v1.9 are not meant to diminish what has been accomplished. Nevertheless, there is clearly much room for improvement.

14.4 GOING FROM VERSION 1.9 TO VERSION 2.0

So how can we make the transition to an approach to metacommunity ecology that deserves to be given a new version number (i.e., v2.0)? One way to think about it is to recognize that the predominant theoretical approach in v1.x was typically developed by thinking about species pairs (and not entire assemblages) interacting in simple ways (often direct competition for one or two resources) in simple landscapes (where all patches are roughly equivalent in how they contribute to pattern) and at discrete scales (local vs. regional). The limitations we keep running into often involve situations in which the variety of species and interactions, as well as the spatial features of landscapes, result in patterns and processes that cannot be adequately captured by this simple view because different patches, species, and mechanisms are interacting multidimensionally. This means that we need conceptual frameworks, statistical tools, and experimental methods that can help us understand these more complex patterns and processes. Thus, any upgrade to v2.0 needs to be framed within the following context:

1. It needs to be spatially explicit. While statistical methods such as variation partitioning are spatially explicit, we are only beginning to understand how this view alters pattern and process in metacommunities.

2. It needs to be multiscalar. One important insight from v1.9 is that pattern and process change with scale (Chaps. 4, 5, 10). The scaling needs to go from the domain of single individuals to macroecological and biogeographic scales. In addition to making metacommunity ecology multiscalar, this will also allow the transition from metacommunities sensu stricto to metacommunities sensu lato (Chap. 1) and illuminate the ways that the important ideas of metacommunity v1.x may apply more generally to ecosystems that are more continuous.

3. It needs to recognize multidimensional patterns of variation in ecological traits. This means allowing species to vary simultaneously in numerous traits that might affect their capacity to perform in different environmental condition, their capacity to interact with other species (including multitrophic and other interactions besides resource or space competition), and their dispersal strategies.

4. It needs to recognize that community assembly can be more complex than what is predicted by simple resource-competition models and include various forms of frequency dependence that can result in features such as priority effects and assembly cycles, often because they may be mediated by other types of interactors.

5. It needs to accommodate the dynamics of niche evolution.

This is a pretty tall order! Nevertheless, we are moving in the right direction. For example, the modeling of Fournier et al. (2017; see also Ai et al. 2013) discussed in Chapter 2 leaves behind the four archetypes and moves toward a more continuous perspective of metacommunities and their underlying processes. Here a landscape consists of a network of patches that can only be inhabited by a single individual, and the dynamics are modeled using a PD approach (sensu Levins and Culver 1971) in a "harlequin"-type landscape (Horn and MacArthur 1972, Leibold and Loeuille 2015). We view a framework that builds on this and expands the focal scale of analysis to understand process and generate predictions as an important step. For example, the spatially explicit "point pattern" analysis approach (Wiegand and Moloney 2014) provides a tool that can be useful in this context. A notable feature of this approach is that it can be used within a likelihood framework to test predictions.

From a statistical perspective, we need to find something that goes beyond current forms of variation partitioning, which have a number of limitations (many of which we have discussed throughout, especially in Chap. 4), To do so means that we must find more subtle ways to account for the interactions of the spatial and environmental features of landscapes, especially in ways that look at scale. We must also be able to study the individual contributions of different species (such as keystone dispersers sensu Amarasekare 2008b) and of different locations to the patterns of interest (i.e., keystone communities sensu Mouquet et al. 2013). Finally, we must account for how species combinations lead to patterns of coexistence (and not just joint responses to environment and space). One possibility builds from approaches based on latent variables and joint modeling (Warton et al. 2015). We described initial work on this idea in Chapter 4 (Sec. 4.7), but that work needs to be more fully expanded to address patterns across scales.

These are only some suggestions among a number of other features that will need to be included in an upgrade to metacommunity v2.0. Our point is not to make specific recommendations per se but rather to outline what types of approaches are needed and to highlight that ongoing work is already finding possible solutions.

14.5 FROM BASIC TO APPLIED METACOMMUNITY ECOLOGY

Much of our emphasis throughout this book has been on simply understanding patterns and processes in metacommunities and how they combine to influence the types, traits, and numbers of species (co-occurring) in time and across space. There is plenty of reason to do this just to obtain a basic understanding of how the world is put together. Increasingly, however, ecology has become the science of the Anthropocene (Schimel et al. 2013, Corlett 2015), in which it called upon to help understand and solve environmental issues related to human activities. Three of the main ways that humans have altered nature is by affecting landscape structure (including fragmentation and connectivity), dispersal (via the movement of species that have facilitated the existence of invasive species), and environmental conditions (including novel ecosystems and local as well as global effects). These are the very same factors we have emphasized as determining metacommunity dynamics. And indeed, we have discussed some of these applied problems in other contexts, such as alteration of the rates of dispersal and changing patterns of biodiversity and species composition due to habitat fragmentation or human-mediated dispersal in the context of restoration, assistant migration, and alien species (Chaps. 4 and 10).

Applied ecology is also increasingly incorporating a multiscalar perspective, in which processes at one location affects dynamics at another. This is certainly true for exotic species, conservation of species and habitats, and large-scale environmental change (e.g., climate change and nitrogen deposition), but it is probably also true for more subtle issues (e.g., pollution, urban and agro-ecosystems, etc.). Furthermore, methods and policy to address environmental issues are also increasingly called upon to do this at larger scales because it is thought that this is more likely to be effective (but see Lindenmayer et al. 2007 and Walker and Salt 2012). Finally, even when it does take spatial dynamics into consideration, conservation ecology often treats management as though species are independent of each other (e.g., metapopulations or groups of metapopulations) and does not typically adequately address possible effects involving multiple species and communities as a whole where these species interact. As such, metacommunity ecology has the potential to be important in applied ecology as well as for its own scientific interest (Bengtsson 2009, Siqueira et al. 2012a, Heino 2013b, Schiesari and coworkers, unpublished data).

Despite the clear potential for strong connections between metacommunity ecology and applied ecology, they have not yet been fully realized. Perhaps the reason is that until recently metacommunity theory was spatially implicit and largely conceptual (see Chap. 2), whereas most applied work was strongly based on spatially explicit problems and more specific biological context. In conservation ecology, for example, the most reliable data are in the form of range maps and some of the most important decisions concern prioritizing how conservation areas are to related topographic features (e.g., corridors, unique habitats). It is true that the description of metacommunity ecology has a longer record of incorporating spatial features (especially in the guise of variation partitioning and related methods; see Chap. 4), but this has focused on overall metrics rather than metrics that can be used to make spatially explicit decisions. Recent progress in spatially explicit metacommunity theory (e.g., Leibold and Loeuille 2015), combined with emerging methods that identify the role of individual patches (e.g., Fournier et al. 2017), could significantly increase the utility of metacommunity ecology for applied ecology. Metacommunity ecology should become even more useful as it continues to evolve from the simple two-scale perspective (metacommunities sensu stricto) to a more multiscale (continuous) perspective (metacommunities sensu lato).

In many instances, applied ecology is aimed at particular species (e.g., those of conservation interest) rather than at entire assemblages. To date, metacommunity ecology has paid less attention to the behavior of individual species than it has to overall community patterns. Again, however, theory and methods that identify the role of different species in a metacommunity, and whether rare species respond differently to spatial and environmental conditions (e.g., Siqueira et al. 2012b), can help develop a more integrative approach.

Reciprocally, applying metacommunity principles to issues in applied ecology will enrich and enhance metacommunity ecology itself. For example, although human-mediated dispersal of exotic species is a major environmental problem in many respects, it also provides an unintended experiment from which we can learn some important basic ecological principles (Sax et al. 2005), including metacommunity principles via altered dispersal rates. Likewise, assisted migration in the context of conservation actions during climate change (especially in fragmented habitats; McLachlan et al. 2007, Holyoak and Heath 2016) will also allow us to gain much knowledge about basic metacommunity principles. Finally, habitat fragmentation, as discussed briefly in Chapter 10, is probably one of humanity's greatest impacts on biodiversity and functioning natural ecosystems (Haddad et al. 2015, Isbell et al. 2015). By altering habitat heterogeneity, isolation, and size, habitat fragmentation possibly represents one of the best large-scale (albeit poorly controlled) experiments in metacommunity ecology to date.

14.6 CONCLUSIONS

Since the beginning of their discipline, ecologists have struggled to make generalizations. In his 1927 book, Elton said, "Many biologists view with despair the prospect of trying to learn anything about ecology, since the subject appears . . . as a mass of uncoordinated and indigestible facts. It is quite certain that some powerful digestive juice is required" (17). Elton contributed three important such "juices" in the form of four basic concepts: niche, food web, body-size relations, and trophic pyramids. Since then, other important concepts have been added, but we have not developed the capacity to completely break down the inherent complexity of ecological dynamics.

Our inability to fully understand the complexity of community ecology is manifest in the angst and repeated critiques of the field. Gleason (1927) was perhaps the first to argue that the distributions of species were almost totally idiosyncratic and rarely predictable. Andrewartha and Birch (1954) critiqued the use of simple models (e.g., Lotka-Volterra models) amid a complex and often unpredictable backdrop. In the 1980s, the null models controversy took many (if not most) of the important ideas about assembly rules and coexistence to task (Strong et al. 1984; see also Diamond and Case 1986) and argued that there was little evidence that interspecific competition played any significant role in community assembly. A decade later, Lawton (1999) reviewed what he saw as the rather sorry state of community ecology and encouraged scientists to leave community ecology behind and focus either more directly on the underlying populations or move toward a more mechanism-free macroecology focus. And a decade later still, Ricklefs (2008) argued that community ecology should be "disintegrated" because any effects of community-level mechanisms could not predict species distributions in the face of stochastic and biogeographical factors. Almost all of these criticisms invoke the claim that community perspectives are rarely able to explain more than a small fraction of the variation in community composition (see Chap. 4).

There are two general approaches for trying to "build a better community ecology" that can be more predictive and less subject to the criticisms it has endured since its inception. First, a number of authors have tried to develop elegant unified theories that are able to predict a number of patterns in nature with few assumptions. These include unified theories like Hubbell's (2001) neutral theory, Brown and colleagues' metabolic theory (West et al. 1997, Brown et al. 2004), maximum entropy theory (Shipley 2010, Harte 2011), and even unification of the unified theories (McGill 2010)! Although appealing, these unified theories often fail either to deal with natural complexities or to address multiple types of questions (e.g., food-web structure, ecosystem function), or they are phenomenological in their approach and are able to capture only natural patterns across large gradients and leave a considerable amount of variation unexplained.

The second approach is a necessarily more complex one that recognizes that there are multiple processes operating at any given time and develops a more generalized and synthetic framework for addressing problems, rather than being focused on any single "magic bullet." This is the strategy that we favor in this book (as also in Chase and Leibold 2003), where a more complex approach allows deeper insights into patterns and processes. In the case of metacommunities, we have shown that spatial scale can be critical for providing a synthetic understanding about the relative importance of the different metacommunity processes, and for identifying the role of these processes in driving patterns of biodiversity trait distributions, the influence of evolutionary processes, and the consequences for food-web structure and ecosystem function, perhaps even having some utility for understanding how biodiversity is changing during the Anthropocene.

The success of metacommunity ecology may even be reflected in an *actual* improvement in our ability to describe patterns in community variation, as we argued in Chapter 4. For example, Pedro Peres-Neto and Mathew Leibold (unpublished data) found that a typical data set (where environmental factors could predict only about 15% of community variance, as is typical of the studies used by the above critics) could be resolved much better by accounting for spatial effects and low-dimensional latent variables that may involve other deterministic factors, leading to the explanation of 76% of variance! Of course, there will always be variation in community data, so we are extremely unlikely to convert community ecology into a purely deterministic field. Nevertheless, this result (if it is general) is grounds for some optimism.

It is remarkable that the metacommunity concept has been able to produce such advances in a relatively short amount of time. We have highlighted the shortcomings of its current form and proposed developments that will hopefully lead to novel insights and utility (Sec. 13.4). We can't wait to see where the field will go once Metacommunity v.2.0 surpasses the beta version!

References

Abrams, P. A. 1987. Alternative models of character displacement and niche shift. 1. Adaptive shifts in resource use when there is competition for nutritionally nonsubstitutable resources. *Evolution* **41**:651–661.

Abrams, P. A. 1993. Effect of increased productivity on the abundances of trophic levels. *American Naturalist* **141**:351–371.

Abrams, P. A., and R. D. Holt. 2002. The impact of consumer–resource cycles on the coexistence of competing consumers. *Theoretical Population Biology* **62**:281–295.

Abramsky, Z., O. Ovadia, and M. L. Rosenzweig. 1994. The shape of a *Gerbillus pyramidum* (Rodentia, Gerbillinae) isocline: An experimental field-study. *Oikos* **69**:318–326.

Abramsky, Z., M. L. Rosenzweig, and B. Pinshow. 1991. The shape of a gerbil isocline measured using principles of optimal habitat selection. *Ecology* **72**:329–340.

Ackerly, D. D., and W. K. Cornwell. 2007. A trait-based approach to community assembly: Partitioning of species trait values into within- and among-community components. *Ecology Letters* **10**:135–145.

Adams, H. E., B. C. Crump, and G. W. Kling. 2014. Metacommunity dynamics of bacteria in an arctic lake: The impact of species sorting and mass effects on bacterial production and biogeography. *Frontiers in Microbiology* **5**:art. 82. doi:https://dx.doi.org/10.3389%2Ffmicb.2014.00082.

Adler, F. R., and H. C. Muller-Landau. 2005. When do localized natural enemies increase species richness? *Ecology Letters* **8**:438–447.

Adler, P. B. 2004. Neutral models fail to reproduce observed species-area and species-time relationships in Kansas grasslands. *Ecology* **85**:1265–1272.

Adler, P. B., A. Fajardo, A. R. Kleinhesselink, and N. J. Kraft. 2013. Trait-based tests of coexistence mechanisms. *Ecology Letters* **16**:1294–1306.

Adler, P. B., J. HilleRisLambers, and J. M. Levine. 2007. A niche for neutrality. *Ecology Letters* **10**:95–104.

Adler, P. B., and W. K. Lauenroth. 2003. The power of time: spatiotemporal scaling of species diversity. *Ecology Letters* **6**:749–756.

Adler, P. B., R. Salguero-Gomez, A. Compagnoni, J. S. Hsu, J. Ray-Mukherjee, C. Mbeau-Ache, and M. Franco. 2014. Functional traits explain variation in plant life history strategies. *Proceedings of the National Academy of Sciences USA* **111**:740–745.

Adler, P. B., E. W. Seabloom, E. T. Borer, H. Hillebrand, Y. Hautier, A. Hector, W. S. Harpole, et al. . 2011. Productivity is a poor predictor of plant species richness. *Science* **333**:1750–1753.

Adler, P. B., E. P. White, W. K. Lauenroth, D. M. Kaufman, A. Rassweiler, and J. A. Rusak. 2005. Evidence for a general species-time-area relationship. *Ecology* **86**:2032–2039.

Ai, D., D. Gravel, C. Chu, and G. Wang. 2013. Spatial structures of the environment and of dispersal impact species distribution in competitive metacommunities. *PLoS One* **8**:e68927.

Alexander, H. M., B. L. Foster, F. Ballantyne, C. D. Collins, J. Antonovics, and R. D. Holt. 2012. Metapopulations and metacommunities: Combining spatial and temporal perspectives in plant ecology. *Journal of Ecology* **100**:88–103.

Alexandrou, M. A., B. J. Cardinale, J. D. Hall, C. F. Delwiche, K. Fritschie, A. Narwani, P. A. Venail, et al. 2015. Evolutionary relatedness does not predict competition and co-occurrence in natural or experimental communities of green algae. *Proceedings of the Royal Society. Series B: Biological Sciences* **282**:20141745. doi: 10.1098/rspb.2014.1745.

Algar, A. C., J. T. Kerr, and D. J. Currie. 2011. Quantifying the importance of regional and local filters for community trait structure in tropical and temperate zones. *Ecology* **92**:903–914.

Allan, E., T. Jenkins, A.J.F. Fergus, C. Roscher, M. Fischer, J. Petermann, W. W. Weisser, and B. Schmid. 2013. Experimental plant communities develop phylogenetically overdispersed abundance distributions during assembly. *Ecology* **94**:465–477.

Allen, A. P., J. H. Brown, and J. F. Gillooly. 2002. Global biodiversity, biochemical kinetics, and the energetic-equivalence rule. *Science* **297**:1545–1548.

Allen, A. P., and J. F. Gillooly. 2009. Towards an integration of ecological stoichiometry and the metabolic theory of ecology to better understand nutrient cycling. *Ecology Letters* **12**:369–384.

Allesina, S., and J. M. Levine. 2011. A competitive network theory of species diversity. *Proceedings of the National Academy of Sciences USA* **108**:5638–5642.

Allesina, S., and S. Tang. 2012. Stability criteria for complex ecosystems. Nature **483**:205–208.

Allouche, O., and R. Kadmon. 2009. A general framework for neutral models of community dynamics. *Ecology Letters* **12**:1287–1297.

Allouche, O., M. Kalyuzhny, G. Moreno-Rueda, M. Pizarro, and R. Kadmon. 2012. Area-heterogeneity tradeoff and the diversity of ecological communities. *Proceedings of the National Academy of Sciences USA* **109**:17495–17500.

Alonso, D., R. Etienne, and A. McKane. 2006. The merits of neutral theory. *Trends in Ecology and Evolution* **21**:451–457.

Alonso, D., A. Pinyol-Gallemi, T. Alcoverro, and R. Arthur. 2015. Fish community reassembly after a coral mass mortality: Higher trophic groups are subject to increased rates of extinction. *Ecology Letters* **18**:451–461.

Amarasekare, P. 2000a. Coexistence of competing parasitoids on a patchily distributed host: Local vs. spatial mechanisms. *Ecology* **81**:1286–1296.

Amarasekare, P. 2000b. The geometry of coexistence. *Biological Journal of the Linnean Society* **71**:1–31.

Amarasekare, P. 2008a. Spatial dynamics of foodwebs. *Annual Review of Ecology, Evolution, and Systematics* **39**:479–500.

Amarasekare, P. 2008b. Spatial dynamics of keystone predation. *Journal of Animal Ecology* **77**:1306–1315.

Amarasekare, P., and R. M. Nisbet. 2001. Spatial heterogeneity, source-sink dynamics, and the local coexistence of competing species. *American Naturalist* **158**:572–584.

Amezcua, A. B., and M. Holyoak. 2000. Empirical evidence for predator- prey source-sink dynamics. *Ecology* **81**:3087–3098.

Anderson, M. J., T. O. Crist, J. M. Chase, M. Vellend, B. D. Inouye, A. L. Freestone, N. J. Sanders, et al. 2011. Navigating the multiple meanings of beta diversity: A roadmap for the practicing ecologist. *Ecology Letters* **14**:19–28.

Andersson, M. G., M. Berga, E. S. Lindström, and S. Langenheder. 2014. The spatial structure of bacterial communities is influenced by historical environmental conditions. *Ecology* **95**:1134–1140.

Andrewartha, H. G., and L. C. Birch. 1954. *The distribution and abundance of animals.* University of Chicago Press, Chicago.

Angert, A. L., T. E. Huxman, P. Chesson, and D. L. Venable. 2009. Functional tradeoffs determine species coexistence via the storage effect. *Proceedings of the National Academy of Sciences USA* **106**:11641–11645.

Araújo, M. B., and A. T. Peterson. 2012. Uses and misuses of bioclimatic envelope modeling. *Ecology* **93**:1527–1539.

Araya, Y. N., J. Silvertown, D. J. Gowing, K. J. McConway, H. P. Linder, and G. Midgley. 2011. A fundamental, eco-hydrological basis for niche segregation in plant communities. *New Phytologist* **189**:253–258.

Armstrong, R. A., and R. McGehee. 1976. Coexistence of species competing for shared resources. *Theoretical Population Biology* **9**:317–328.

Armstrong, R. A., and R. McGehee. 1980. Competitive-exclusion. *American Naturalist* **115**:151–170.

Arnan, X., X. Cerda, and J. Retana. 2012. Distinctive life traits and distribution along environmental gradients of dominant and subordinate Mediterranean ant species. *Oecologia* **170**:489–500.

Arrhenius, O. 1921. Species and area. *Journal of Ecology* **9**:95–99.

Azaele, S., A. Maritan, S. J. Cornell, S. Suweis, J. R. Banavar, D. Gabriel, and W. E. Kunin. 2015. Towards a unified descriptive theory for spatial ecology: Predicting biodiversity patterns across spatial scales. *Methods in Ecology and Evolution* **6**:324–332.

Baas-Becking, L.G.M. 1934. *Geobiologie of inleiding toto de milieukunde.* W. P. Van Stockum and Zoon, The Hague.

Badgley, C., and D. L. Fox. 2000. Ecological biogeography of North American mammals: Species density and ecological structure in relation to environmental gradients. *Journal of Biogeography* **27**:1437–1467.

Baho, D. L., H. Peter, and L. J. Tranvik. 2012. Resistance and resilience of microbial communities–Temporal and spatial insurance against perturbations. *Environmental Microbiology* **14**:2283–2292.

Bailey, J. K., J. A. Schweitzer, F. Ubeda, J. Koricheva, C. J. LeRoy, M. D. Madritch, B. J. Rehill, et al. 2009. From genes to ecosystems: A synthesis of the effects of plant genetic factors across levels of organization. *Philosophical Transactions of the Royal Society B–Biological Sciences* **364**:1607–1616.

Baiser, B., and J. L. Lockwood. 2011. The relationship between functional and taxonomic homogenization. *Global Ecology and Biogeography* **20**:134–144.

Baiser, B., J. D. Olden, S. Record, J. L. Lockwood, and M. L. McKinney. 2012. Pattern and process of biotic homogenization in the New Pangaea. *Proceedings of the Royal Society. Series B: Biological Sciences* **279**:4772–4777.

Baker, J. P., S. A. Gherini, R. K. Munson, S. W. Christensen, C. T. Driscoll, J. Gallagher, R. M. Newton, K. H. Reckbow, and C. L. Schofield. 1990. *Adirondack lakes survey: An interpretive analysis of fish communities and water chemistry 1984–1987.* Oak Ridge National Laboratory, Ray Brook, NY.

Baldeck, C. A., K. E. Harms, J. B. Yavitt, R. John, B. L. Turner, R. Valencia, H. Navarrete, et al. 2013a. Habitat filtering across tree life stages in tropical forest communities.

Proceedings of the Royal Society. Series B: Biological Sciences **280**:20130548. doi: 10.1098/rspb.2013.0548.

Baldeck, C. A., K. E. Harms, J. B. Yavitt, R. John, B. L. Turner, R. Valencia, H. Navarrete, et al. 2013b. Soil resources and topography shape local tree community structure in tropical forests. *Proceedings of the Royal Society. Series B: Biological Sciences* **280**:20122532. doi: 10.1098/rspb.2012.2532.

Baldi, A. 2008. Habitat heterogeneity overrides the species-area relationship. *Journal of Biogeography* **35**:675–681.

Baldwin, B. G., and M. J. Sanderson. 1998. Age and rate of diversification of the Hawaiian silversword alliance (Compositae). *Proceedings of the National Academy of Sciences USA* **95**:9402–9406.

Barabás, G., R. D'Andrea, R. Rael, G. Meszéna, and A. Ostling. 2013. Emergent neutrality or hidden niches? *Oikos* **122**:1565–1572.

Barbour, M. A., M. A. Fortuna, J. Bascompte, J. R. Nicholson, R. Julkunen-Tiitto, E. S. Jules, and G. M. Crutsinger. 2016. Genetic specificity of a plant-insect food web: Implications for linking genetic variation to network complexity. *Proceedings of the National Academy of Sciences USA* **113**:2128–2133.

Barone, J. A., J. Thomlinson, P. A. Cordero, and J. K. Zimmerman. 2008. Metacommunity structure of tropical forest along an elevation gradient in Puerto Rico. *Journal of Tropical Ecology* **24**:525–534.

Barton, P. S., S. A. Cunningham, A. D. Manning, H. Gibb, D. B. Lindenmayer, and R. K. Didham. 2013. The spatial scaling of beta diversity. *Global Ecology and Biogeography* **22**:639–647.

Baselga, A. 2010. Partitioning the turnover and nestedness components of beta diversity. *Global Ecology and Biogeography* **19**:134–143.

Beck, J., L. Ballesteros-Mejia, C. M. Buchmann, J. Dengler, S. A. Fritz, B. Gruber, C. Hof, et al. 2012. What's on the horizon for macroecology? *Ecography* **35**:673–683.

Beck, J., G. Brehm, and K. Fiedler. 2011. Links between the environment, abundance and diversity of Andean moths. *Biotropica* **43**:208–217.

Begley-Miller, D. R., A. L. Hipp, B. H. Brown, M. Hahn, and T. P. Rooney. 2014. White-tailed deer are a biotic filter during community assembly, reducing species and phylogenetic diversity. *AoB PLANTS* **6**:plu030. doi: https://doi.org/10.1093/aobpla/plu030.

Beisner, B. E., D. T. Haydon, and K. Cuddington. 2003. Alternative stable states in ecology. *Frontiers in Ecology and the Environment* **1**:376–382.

Bell, G. 2000. The distribution of abundance in neutral communities. *American Naturalist* **155**:606–617.

Bell, G. 2001. Neutral macroecology. *Science* **293**:2413–2418.

Bell, G., and A. Gonzalez. 2011. Adaptation and evolutionary rescue in metapopulations experiencing environmental deterioration. *Science* **332**:1327–1330.

Belmaker, J., and W. Jetz. 2011. Cross-scale variation in species richness-environment associations. *Global Ecology and Biogeography* **20**:464–474.

Belmaker, J., and W. Jetz. 2013. Spatial scaling of functional structure in bird and mammal assemblages. *American Naturalist* **181**:464–478.

Belmaker, J., N. Shashar, and Y. Ziv. 2005. Effects of small-scale isolation and predation on fish diversity on experimental reefs. *Marine Ecology Progress Series* **289**:273–283.

Bengtsson, J. 1989. Interspecific competition increases local extinction rate in a metapopulation system. *Nature* **340**:713–715.

Bengtsson, J. 1993. Interspecific competition and determinants of extinction in experimental populations of three rockpool *Daphnia* species. *Oikos* **67**:451–464.

Bengtsson J. 2009. Applied (meta)community ecology: Diversity and ecosystem services at the intersection of local and regional processes. Pages 115–130 *in* H. A. Verhoef and P. J. Morin, editors, *Community ecology: Processes, models, and applications.* Oxford University Press, Oxford.

Bengtsson, J. 2010. Applied (meta)community ecology: Diversity and ecosystem services at the intersection of local and regional processes. Pages 115–130 *in* H. A. Verhoef and P. J. Morin, editors, *Community ecology: Processes, models, and applications.* Oxford University Press, Oxford.

Benitez, M. S., M. H. Hersh, R. Vilgalys, and J. S. Clark. 2013. Pathogen regulation of plant diversity via effective specialization. *Trends in Ecology and Evolution* **28**:705–711.

Bennett, J. A., E. G. Lamb, J. C. Hall, W. M. Cardinal-McTeague, and J. F. Cahill, Jr. 2013. Increased competition does not lead to increased phylogenetic overdispersion in a native grassland. *Ecology Letters* **16**:1168–1176.

Bennett, J. R., and B. Gilbert. 2016. Contrasting beta diversity among regions: How do classical and multivariate approaches compare? *Global Ecology and Biogeography* **25**:368–377.

Berga, M., Ö. Östman, E. S. Lindström, and S. Langenheder. 2015. Combined effects of zooplankton grazing and dispersal on the diversity and assembly mechanisms of bacterial metacommunities: Grazing and dispersal effects on bacterial metacommunities. *Environmental Microbiology* **17**:2275–2287.

Berga, M., A. J. Székely, and S. Langenheder. 2012. Effects of disturbance intensity and frequency on bacterial community composition and function. *PLoS One* **7**:e36959.

Best, R. J., N. C. Caulk, and J. J. Stachowicz. 2013. Trait vs. phylogenetic diversity as predictors of competition and community composition in herbivorous marine amphipods. **16**:72–80.

Bever, J. 1999. Dynamics within mutualism and the maintenance of diversity: inference from a model of interguild frequency dependence. *Ecology Letters* **2**:52–61.

Bever, J. D. 2003. Soil community feedback and the coexistence of competitors: conceptual frameworks and empirical tests. *New Phytologist* **157**:465–473.

Bever, J. D., I. A. Dickie, E. Facelli, J. M. Facelli, J. Klironomos, M. Moora, M. C. Rillig, W. D. Stock, M. Tibbett, and M. Zobel. 2010. Rooting theories of plant community ecology in microbial interactions. *Trends in Ecology and Evolution* **25**:468–478.

Bever, J. D., S. A. Mangan, and H. M. Alexander. 2015. Maintenance of plant species diversity by pathogens. *Annual Review of Ecology, Evolution, and Systematics* **46**:305–325.

Blackburn, T. M., and B. A. Hawkins. 2004. Bergmann's rule and the mammal fauna of northern North America. *Ecography* **27**:715–724.

Blanchet, F. G., P. Legendre, and D. Borcard. 2008. Forward selection of explanatory variables. *Ecology* **89**:2623–2632.

Boecklen, W. J. 1986. Effects of habitat heterogeneity on the species area relationships of forest birds. *Journal of Biogeography* **13**:59–68.

Böhnke, M., N. Kreissig, W. Kröber, T. Fang, and H. Bruelheide. 2012. Wood trait-environment relationships in a secondary forest succession in South-East China. *Trees* **26**:641–651.

Bohonak, A. J., and D. G. Jenkins. 2003. Ecological and evolutionary significance of dispersal by freshwater invertebrates. *Ecology Letters* **6**:783–796.

Bolnick, D. I., P. Amarasekare, M. S. Araujo, R. Burger, J. M. Levine, M. Novak, V.H.W. Rudolf, S. J. Schreiber, M. C. Urban, and D. A. Vasseur. 2011. Why intraspecific trait variation matters in community ecology. *Trends in Ecology and Evolution* **26**:183–192.

Bolnick, D. I., R. Svanback, J. A. Fordyce, L. H. Yang, J. M. Davis, C. D. Hulsey, and M. L. Forister. 2003. The ecology of individuals: Incidence and implications of individual specialization. *American Naturalist* **161**:1–28.

Bond, E. M., and J. M. Chase. 2002. Biodiversity and ecosystem functioning at local and regional spatial scales. *Ecology Letters* **5**:467–470.

Borcard, D., P. Legendre, and P. Drapeau. 1992. Partialling out the spatial component of ecological variation. *Ecology* **73**:1045–1055.

Borda-de-Agua, L., P.A.V. Borges, S. P. Hubbell, and H. M. Pereira. 2012. Spatial scaling of species abundance distributions. *Ecography* **35**:549–556.

Bormann, F. H., and G. E. Likens. 1967. Nutrient cycling. *Science* **155**:424–429.

Borregaard, M. K., C. Rahbek, J. Fjeldså, J. L. Parra, R. J. Whittaker, and C. H. Graham. 2014. Node-based analysis of species distributions. *Methods in Ecology and Evolution* **5**:1225–1235.

Bossuyt, H., J. Six, and P. Hendrix. 2004. Rapid incorporation of carbon from fresh residues into newly formed stable microaggregates within earthworm casts. *European Journal of Soil Science* **55**:393–399.

Bourret, A., M. A. McPeek, and J. Turgeon. 2012. Regional divergence and mosaic spatial distribution of two closely related damselfly species (*Enallagma hageni* and *Enallagma ebrium*). *Journal of Evolutionary Biology* **25**:196–209.

Boyce, M. S., C. V. Haridas, C. T. Lee, and NCEAS Stochasetic Demography Working Group. 2006. Demography in an increasingly variable world. *Trends in Ecology and Evolution* **21**:141–148.

Brind'Amour, A., D. Boisclair, S. Dray, and P. Legendre. 2011. Relationships between species feeding traits and environmental conditions in fish communities: A three-matrix approach. *Ecological Applications* **21**:363–377.

British Ecological Society. 1944. Symposium: Ecology of closely allied species. *Journal of Animal Ecology* **13**:176–177.

Britton, N. 1989. Aggregation and the competitive exclusion principle. *Journal of Theoretical Biology* **136**:57–66.

Brose, U., A. Ostling, K. Harrison, and N. D. Martinez. 2004. Unified spatial scaling of species and their trophic interactions. *Nature* **428**:167–171.

Brown, A. M., D. I. Warton, N. R. Andrew, M. Binns, G. Cassis, and H. Gibb. 2014. The fourth-corner solution: Using predictive models to understand how species traits interact with the environment. *Methods in Ecology and Evolution* **5**:344–352.

Brown, B. L., E. R. Sokol, J. Skelton, and B. Tornwall. 2017. Making sense of metacommunities: dispelling the mythology of a metacommunity typology. *Oecologia* **183**:643–652.

Brown, C., D. Burslem, J. B. Illian, L. Bao, W. Brockelman, M. Cao, L. W. Chang, et al. 2013. Multispecies coexistence of trees in tropical forests: Spatial signals of topographic niche differentiation increase with environmental heterogeneity. *Proceedings of the Royal Society. Series B: Biological Sciences* **280**:20130502. doi:10.1098/rspb.2013.0502.

Brown, J. H. 1995. *Macroecology*. University of Chicago Press, Chicago.

Brown, J. H. 1998. The desert granivory experiments at Portal. Pages 71–95 *in* W. J. Restetarits and J. Bernardo, editors, *Experimental ecology: Issues and perspectives*. Oxford University Press, New York.

Brown, J. H. 2014. Why are there so many species in the tropics? *Journal of Biogeography* **41**:8–22.

Brown, J. H., S.K.M. Ernest, J. M. Parody, and J. P. Haskell. 2001. Regulation of diversity: Maintenance of species richness in changing environments. *Oecologia* **126**:321–332.

Brown, J. H., J. F. Gillooly, A. P. Allen, V. M. Savage, and G. B. West. 2004. Toward a metabolic theory of ecology. *Ecology* **85**:1771–1789.

Brown, J. H., and A. Kodric-Brown. 1977. Turnover rates in insular biogeography: Effect of immigration on extinction. *Ecology* **58**:445–449.

Brown, J. H., and B. A. Maurer. 1989. Macroecology: The division of food and space among species on continents. *Science* **243**:1145–1150.

Brown, J. S., and T. L. Vincent. 1992. Organization of predator-prey communities as an evolutionary game. *Evolution* **46**:1269–1283.

Brown, W. L., and E. O. Wilson. 1956. Character displacement. *Systematic Zoology* **5**:49–64.

Brownlee, J. 2007. Complex adaptive systems. Technical Report. Complex Intelligent Systems Laboratory, Centre for Information Technology Research, Faculty of Information Communication Technology, Swinburne University of Technology, Melbourne, Australia.

Buenau, K. E., A. Rassweiler, and R. M. Nisbet. 2007. The effects of landscape structure on space competition and alternative stable states. *Ecology* **88**:3022–3031.

Burke, I. C., and W. K. Lauenroth. 2011. Theory of ecosystem ecology. Pages 243–258 *in* S. Scheiner and M. Willig, editors, *Theory of ecology*. Oxford University Press, Oxford.

Burke, M. J., and J. Grime. 1996. An experimental study of plant community invasibility. *Ecology* **77**:776–790.

Burns, J. H., and S. Y. Strauss. 2011. More closely related species are more ecologically similar in an experimental test. *Proceedings of the National Academy of Sciences USA* **108**:5302–5307.

Burns, K. C., and C. J. Neufeld. 2009. Plant extinction dynamics in an insular metacommunity. *Oikos* **118**:191–198.

Buss, L. W., and J.B.C. Jackson. 1979. Competitive networks: Nontransitive competitive relationships in cryptic coral reef environments. *American Naturalist* **113**:223–234.

Cadotte, M. W. 2006. Dispersal and species diversity: A meta-analysis. *American Naturalist* **167**:913–924.

Cadotte, M. W., C. H. Albert, and S. C. Walker. 2013. The ecology of differences: assessing community assembly with trait and evolutionary distances. *Ecology Letters* **16**:1234–1244.

Cadotte, M. W., J. Cavender-Bares, D. Tilman, and T. H. Oakley. 2009. Using phyloge-netic, functional and trait diversity to understand patterns of plant community pro-ductivity. *PLoS One* **4**:e5695.

Cadotte, M. W., and T. J. Davies. 2016. *Ecophylogenetics: Using phylogenies in ecology.* Princeton University Press, Princeton NJ.

Cadotte, M. W., T. J. Davies, J. Regetz, S. W. Kembel, E. Cleland, and T. H. Oakley. 2010. Phylogenetic diversity metrics for ecological communities: integrating species rich-ness, abundance and evolutionary history. *Ecology Letters* **13**:96–105.

Cadotte, M. W., A. M. Fortner, and T. Fukami. 2006. The effects of resource enrichment, dispersal, and predation on local and metacommunity structure. *Oecologia* **149**: 150–157.

Cahill, J. F., Jr., S. W. Kembel, E. G. Lamb, and P. A. Keddy. 2008. Does phylogenetic relatedness influence the strength of competition among vascular plants? Perspec-tives in Plant Ecology Evolution and Systematics **10**:41–50.

Calcagno, V., F. Massol, N. Mouquet, P. Jarne, and P. David. 2011. Constraints on food chain length arising from regional metacommunity dynamics. *Proceedings of the Royal Society of London. Series B: Biological Sciences* **278**:3042–3049.

Calcagno, V., N. Mouquet, P. Jarne, and P. David. 2006. Coexistence in a metacommunity: The competition–colonization trade-off is not dead. *Ecology Letters* **9**:897–907.

Capinha, C., F. Essl, H. Seebens, D. Moser, and H. M. Pereira. 2015. The dispersal of alien species redefines biogeography in the Anthropocene. *Science* **348**:1248–1251.

Capitán, J. A., J. A. Cuesta, and J. Bascompte. 2009. Statistical mechanics of ecosystem assembly. *Physical Review Letters* **103**:168101.

Capitán, J. A., J. A. Cuesta, and J. Bascompte. 2011. Species assembly in model ecosys-tems, II: Results of the assembly process. *Journal of Theoretical Biology* **269**:344–355.

Cardinale, B. J., J. E. Duffy, A. Gonzalez, D. U. Hooper, C. Perrings, P. Venail, A. Nar-wani, G. M. Mace, D. Tilman, and D. A. Wardle. 2012. Biodiversity loss and its im-pact on humanity. *Nature* **486**:59–67.

Cardinale, B. J., H. Hillebrand, W. S. Harpole, K. Gross, and R. Ptacnik. 2009. Separating the influence of resource "availability" from resource "imbalance" on productivity-diversity relationships. *Ecology Letters* **12**:475–487.

Cardoso, P., F. Rigal, and J. C. Carvalho. 2015. BAT—Biodiversity Assessment Tools, an R package for the measurement and estimation of alpha and beta taxon, phyloge-netic and functional diversity. *Methods in Ecology and Evolution* **6**:232–236.

Carlson, A., and P. Edenhamn. 2000. Extinction dynamics and the regional persistence of a tree frog metapopulation. *Proceedings of the Royal Society. Series B: Biological Sciences* **267**:1311–1313.

Case, T. J. 2000. *An illustrated guide to theoretical ecology.* Oxford University Press, Oxford.

Casini, M., T. Blenckner, C. Mollmann, A. Gardmark, M. Lindegren, M. Llope, G. Ko-rnilovs, M. Plikshs, and N. C. Stenseth. 2012. Predator transitory spillover induces trophic cascades in ecological sinks. *Proceedings of the National Academy of Sci-ences USA* **109**:8185–8189.

Cassey, P., T. M. Blackburn, J. L. Lockwood, and D. F. Sax. 2006. A stochastic model for integrating changes in species richness and community similarity across spatial scales. *Oikos* **115**:207–218.

Caswell, H. 1976. Community structure: A neutral model analysis. *Ecological Monographs* **46**:327–354.

Caswell, H. 1978. A general formula for the sensitivity of population growth rate to changes in life history parameters. *Theoretical Population Biology* **14**:215–230.

Cavender-Bares, J., A. Keen, and B. Miles. 2006. Phylogenetic structure of Floridian plant communities depends on taxonomic and spatial scale. *Ecology* **87**:S109–S122.

Cavender-Bares, J., K. Kitajima, and F. A. Bazzaz. 2004. Multiple trait associations in relation to habitat differentiation among 17 Floridian oak species. *Ecological Monographs* **74**:635–662.

Cavender-Bares, J., K. H. Kozak, P.V.A. Fine, and S. W. Kembel. 2009. The merging of community ecology and phylogenetic biology. *Ecology Letters* **12**:693–715.

Ceballos, G., P. R. Ehrlich, A. D. Barnosky, A. Garcia, R. M. Pringle, and T. M. Palmer. 2015. Accelerated modern human-induced species losses: Entering the sixth mass extinction. *Science Advances* **1**:e1400253.

Chalcraft, D. R., S. B. Cox, C. Clark, E. E. Cleland, K. N. Suding, E. Weiher, and D. Pennington. 2008. Scale-dependent responses of plant biodiversity to nitrogen enrichment. *Ecology* **89**:2165–2171.

Chalmandrier, L., T. Munkemuller, S. Lavergne, and W. Thuiller. 2015. Effects of species' similarity and dominance on the functional and phylogenetic structure of a plant meta-community. *Ecology* **96**:143–153.

Chang, L. W., D. Zeleny, C. F. Li, S. T. Chiu, and C. F. Hsieh. 2013. Better environmental data may reverse conclusions about niche- and dispersal-based processes in community assembly. *Ecology* **94**:2145–2151.

Chao, A., C. H. Chiu, and L. Jost. 2014. Unifying species diversity, phylogenetic diversity, functional diversity and related similarity and differentiation measures through Hill numbers. *Annual Review of Ecology, Evolution, and Systematics* **45**:297–324.

Charles-Dominique, T., H. Beckett, G. F. Midgley, and W. J. Bond. 2015. Bud protection: A key trait for species sorting in a forest-savanna mosaic. *New Phytologist* **207**:1052–1060.

Chase, J. M. 2003a. Community assembly: When should history matter? *Oecologia* **136**:489–498.

Chase, J. M. 2003b. Strong and weak trophic cascades along a productivity gradient. *Oikos* **101**:187–195.

Chase, J. M. 2007. Drought mediates the importance of stochastic community assembly. *Proceedings of the National Academy of Sciences USA* **104**:17430–17434.

Chase, J. M. 2010. Stochastic community assembly causes higher biodiversity in more productive environments. *Science* **328**:1388–1391.

Chase, J. M. 2014. Spatial scale resolves the niche versus neutral theory debate. *Journal of Vegetation Science* **25**:319–322.

Chase, J. M., P. A. Abrams, J. P. Grover, S. Diehl, P. Chesson, R. D. Holt, S. A. Richards, R. M. Nisbet, and T. J. Case. 2002. The interaction between predation and competition: A review and synthesis. *Ecology Letters* **5**:302–315.

Chase, J. M., P. Amarasekare, K. Cottenie, A. Gonzalez, R. D. Holt, M. Holyoak, M. F. Hoopes, M. A. Leibold, M. Loreau, and N. Mouquet. 2005. Competing theories for competitive metacommunities. Pages 335–354 *in* M. A. Holyoak, M. Leibold, and R. D Holt, editors, *Metacommunities: Spatial dynamics and ecological communities*. University of Chicago Press, Chicago.

Chase, J. M., E. G. Biro, W. A. Ryberg, and K. G. Smith. 2009. Predators temper the relative importance of stochastic processes in the assembly of prey metacommunities. *Ecology Letters* **12**:1210–1218.

Chase, J. M., A. A. Burgett, and E. G. Biro. 2010. Habitat isolation moderates the strength of top-down control in experimental pond food webs. *Ecology* **91**:637–643.

Chase, J. M., and T. M. Knight. 2013. Scale-dependent effect sizes of ecological drivers on biodiversity: Why standardised sampling is not enough. *Ecology Letters* **16**:17–26.

Chase, J. M., N.J.B. Kraft, K. G. Smith, M. Vellend, and B. D. Inouye. 2011. Using null models to disentangle variation in community dissimilarity from variation in alpha-diversity. *Ecosphere* **2**:art. 24. doi: 10.1890/ES10-00117.1.

Chase, J. M., and M. A. Leibold. 2002. Spatial scale dictates the productivity-biodiversity relationship. *Nature* **416**:427–430.

Chase, J. M., and M. A. Leibold. 2003. *Ecological niches: Linking classical and contemporary approaches.* University of Chicago Press, Chicago.

Chase, J. M., and J. A. Myers. 2011. Disentangling the importance of ecological niches from stochastic processes across scales. *Philosophical transactions of the Royal Society B: Biological Sciences* **366**:2351–2363.

Chase, J. M., and W. A. Ryberg. 2004. Connectivity, scale-dependence, and the productivity-diversity relationship. *Ecology Letters* **7**:676–683.

Chase, J. M., W. G. Wilson, and S. A. Richards. 2001. Foraging trade-offs and resource patchiness: theory and experiments with a freshwater snail community. *Ecology Letters* **4**:304–312.

Chave, J. 2004. Neutral theory and community ecology. *Ecology Letters* **7**:241–253.

Chave, J. 2013. The problem of pattern and scale in ecology: What have we learned in 20 years? *Ecology Letters* **16**:4–16.

Chave, J., H. C. Muller-Landau, and S. A. Levin. 2002. Comparing classical community models: theoretical consequences for patterns of diversity. *American Naturalist* **159**:1–23.

Chesson, P. L. 1985. Coexistence of competitors in spatially and temporally varying environments: A look at the combined effects of different sorts of variability. *Theoretical Population Biology* **28**:263–287.

Chesson, P. L. 2000. Mechanisms of maintenance of species diversity. *Annual Review of Ecology and Systematics* **31**:343–366.

Chesson, P. L. 2012. Scale transition theory: Its aims, motivations and predictions. *Ecological Complexity* **10**:52–68.

Chesson, P. L., and N. Huntly. 1997. The roles of harsh and fluctuating conditions in the dynamics of ecological communities. *American Naturalist* **150**:519–553.

Chesson, P. L., and J. J. Kuang. 2008. The interaction between predation and competition. *Nature* **456**:235–238.

Chesson, P. L., and R. R. Warner. 1981. Environmental variability promotes coexistence in lottery competitive systems. *American Naturalist* **117**:923–943.

Chiarucci, A., D. Viciani, C. Winter, and M. Diekmann. 2006. Effects of productivity on species-area curves in herbaceous vegetation: Evidence from experimental and observational data. *Oikos* **115**:475–483.

Chisholm, R. A., and S. W. Pacala. 2010. Niche and neutral models predict asymptotically equivalent species abundance distributions in high-diversity ecological communities. *Proceedings of the National Academy of Sciences USA* **107**:15821–15825.

Chiu, C. H., L. Jost, and A. Chao. 2014. Phylogenetic beta diversity, similarity, and differentiation measures based on Hill numbers. *Ecological Monographs* **84**:21–44.

Chuyong, G. B., D. Kenfack, K. E. Harms, D. W. Thomas, R. Condit, and L. S. Comita. 2011. Habitat specificity and diversity of tree species in an African wet tropical forest. *Plant Ecology* **212**:1363–1374.

Cisneros, L. M., K. R. Burgio, L. M. Dreiss, B. T. Klingbeil, B. D. Patterson, S. J. Presley, and M. R. Willig. 2014. Multiple dimensions of bat biodiversity along an extensive tropical elevational gradient. *Journal of Animal Ecology* **83**:1124–1136.

Clark, C. M., E. E. Cleland, S. L. Collins, J. E. Fargione, L. Gough, K. L. Gross, S. C. Pennings, K. N. Suding, and J. B. Grace. 2007. Environmental and plant community determinants of species loss following nitrogen enrichment. *Ecology Letters* **10**:596–607.

Clark, J. S. 2009. Beyond neutral science. *Trends in Ecology and Evolution* **24**:8–15.

Clauss, M., M. T. Dittmann, D.W.H. Muller, C. Meloro, and D. Codron. 2013. Bergmann's rule in mammals: A cross-species interspecific pattern. *Oikos* **122**:1465–1472.

Clavel, J., R. Julliard, and V. Devictor. 2011. Worldwide decline of specialist species: Toward a global functional homogenization? *Frontiers in Ecology and the Environment* **9**:222–228.

Clements, F. E. 1916. *Plant succession: An analysis of the development of vegetation.* Carnegie Institution of Washington, Washington, DC.

Clements, F. E. 1936. Nature and structure of the climax. *Journal of Ecology* **24**:252–284.

Cleveland, C. C., D. R. Nemergut, S. K. Schmidt, and A. R. Townsend. 2007. Increases in soil respiration following labile carbon additions linked to rapid shifts in soil microbial community composition. *Biogeochemistry* **82**:229–240.

Cody, M. L., and J. M. Diamond, editors. 1975. *Ecology and evolution of communities.* Harvard University Press, Cambridge, MA.

Cohen, J. E., 1970. A Markov contingency-table model for replicated Lotka-Volterra systems near equilibrium. *American Naturalist* **104**:547–560.

Collinge, S. K. 2000. Effects of grassland fragmentation on insect species loss, colonization, and movement patterns. *Ecology* **81**:2211–2226.

Collins, M. D., D. Simberloff, and E. F. Connor. 2011. Binary matrices and checkerboard distributions of birds in the Bismarck Archipelago. *Journal of Biogeography* **38**:2373–2383.

Colwell, R. K., A. Chao, N. J. Gotelli, S. Y. Lin, C. X. Mao, R. L. Chazdon, and J. T. Longino. 2012. Models and estimators linking individual-based and sample-based rarefaction, extrapolation and comparison of assemblages. *Journal of Plant Ecology* **5**:3–21.

Comita, L. S., H. C. Muller-Landau, S. Aguilar, and S. P. Hubbell. 2010. Asymmetric density dependence shapes species abundances in a tropical tree community. *Science* **329**:330–332.

Condit, R., R. A. Chisholm, and S. P. Hubbell. 2012. Thirty years of forest census at Barro Colorado and the importance of immigration in maintaining diversity. *PLoS One* **7**:6.

Condit, R., N. Pitman, E. G. Leigh, J. Chave, J. Terborgh, R. B. Foster, P. Nunez, et al. 2002. Beta-diversity in tropical forest trees. *Science* **295**:666–669.

Connell, J. H. 1961a. Effects of competition, predation by *Thais lapillus*, and other factors on natural populations of the barnacle *Balanus balanoides*. *Ecological Monographs* **31**:61–104.

Connell, J. H. 1961b. The influence of interspecific competition and other factors on the distribution of the barnacle *Chthamalus stellatus*. *Ecology* **42**:710–723.

Connell, J. H. 1971. On the role of natural enemies in preventing competitive exclusion in some marine animals and in rain forest trees. Pages 298–310 *in* P. J. den Boer and G. R. Gradwell, editors, *Dynamics of populations*. Proceedings of the Advanced Study Institute on Dynamics of Numbers in Populations, Oosterbeek, 1970. Centre for Agricultural Publishing and Documentation, Wageningen.

Connell, J. H. 1978. Diversity in tropical rain forests and coral reefs. *Science* **199**: 1302–1310.

Connell, J. H. 1983. On the prevalence and relative importance of interspecific competition: Evidence from field experiments. *American Naturalist* **122**:661–696.

Connor, E. F., M. D. Collins, and D. Simberloff. 2013. The checkered history of checkerboard distributions. *Ecology* **94**:2403–2414.

Connor, E. F., M. D. Collins, and D. Simberloff. 2015. The checkered history of checkerboard distributions: reply. *Ecology* **96**:3388–3389.

Connor, E. F., and E. D. McCoy. 1979. Statistics and biology of the species-area relationship. *American Naturalist* **113**:791–833.

Connor, E. F., and D. Simberloff. 1978. Species number and compositional similarity of the Galapagos flora and avifauna. *Ecological Monographs* **48**:219–248.

Connor, E. F., and D. Simberloff. 1979. The assembly of species communities: Chance or competition? *Ecology* **60**:1132–1140.

Connor, E. F., and D. Simberloff. 1983. Interspecific competition and species co-occurrence patterns on islands: Null models and evaluation of evidence. *Oikos* **41**:455–465.

Cooper, J. K., J. Li, and D.J.S. Montagnes. 2012. Intermediate fragmentation per se provides stable predator-prey metapopulation dynamics. *Ecology Letters* **15**:856–863.

Cooper, N., J. Rodriguez, and A. Purvis. 2008. A common tendency for phylogenetic overdispersion in mammalian assemblages. *Proceedings of the Royal Society. Series B: Biological Sciences* **275**:2031–2037.

Corlett, R. T., 2015. The Anthropocene concept in ecology and conservation. *Trends in Ecology and Evolution* 3:36–41.

Cornell, H. V. 1985. Local and regional richness of cynipine gall wasps on California oaks. *Ecology* **66**:1247–1260.

Cornwell, W. K., and D. D. Ackerly. 2009. Community assembly and shifts in plant trait distributions across an environmental gradient in coastal California. *Ecological Monographs* **79**:109–126.

Cornwell, W. K., M. Westoby, D. S. Falster, R. G. FitzJohn, B. C. O'Meara, M. W. Pennell, D. J. McGlinn, et al. 2014. Functional distinctiveness of major plant lineages. *Journal of Ecology* **102**:345–356.

Cothran, R. D., K. A. Henderson, D. Schmidenberg, and R. A. Relyea. 2013. Phenotypically similar but ecologically distinct: Differences in competitive ability and predation risk among amphipods. *Oikos* **122**:1429–1440.

Cothran, R. D., P. Noyes, and R. A. Relyea. 2015. An empirical test of stable species coexistence in an amphipod species complex. *Oecologia* **178**:819–831.

Cottenie, K. 2005. Integrating environmental and spatial processes in ecological community dynamics. *Ecology Letters* **8**:1175–1182.

Cottenie, K., and L. De Meester. 2003. Connectivity and cladoceran species richness in a metacommunity of shallow lakes. *Freshwater Biology* **48**:823–832.

Cottenie, K., E. Michels, N. Nuytten, and L. De Meester. 2003. Zooplankton metacommunity structure: Regional vs. local processes in highly interconnected ponds. *Ecology* **84**:991–1000.

Craine, J. M., T. W. Ocheltree, J. B. Nippert, E. G. Towne, A. M. Skibbe, S. W. Kembel, and J. E. Fargione. 2013. Global diversity of drought tolerance and grassland climate-change resilience. *Nature Climate Change* **3**:63–67.

Crist, T. O., and J. A. Veech. 2006. Additive partitioning of rarefaction curves and species-area relationships: Unifying alpha-, beta- and gamma-diversity with sample size and habitat area. *Ecology Letters* **9**:923–932.

Crooks, K. R., and M. E. Soule. 1999. Mesopredator release and avifaunal extinctions in a fragmented system. *Nature* **400**:563–566.

Crutsinger, G. M., M. D. Collins, J. A. Fordyce, Z. Gompert, C. C. Nice, and N. J. Sanders. 2006. Plant genotypic diversity predicts community structure and governs an ecosystem process. *Science* **313**:966–968.

Crutsinger, G. M., L. Souza, and N. J. Sanders. 2008. Intraspecific diversity and dominant genotypes resist plant invasions. *Ecology Letters* **11**:16–23.

Currie, D. J. 1991. Energy and large-scale patterns and plant-species richness. *American Naturalist* **137**:27–49.

Currie, D. J., G. G. Mittelbach, H. V. Cornell, R. Field, J. F. Guegan, B. A. Hawkins, D. M. Kaufman, et al. 2004. Predictions and tests of climate-based hypotheses of broad-scale variation in taxonomic richness. *Ecology Letters* **7**:1121–1134.

Curtis, J. T. 1956. The modification of mid-latitude grasslands and forests by man. Pages 507–521 *in* W.L.J. Thomas, editor, *Man's role in changing the face of the Earth.* University Chicago Press, Chicago.

Cusens, J., S. D. Wright, P. D. McBride, and L. N. Gillman. 2012. What is the form of the productivity-animal-species-richness relationship? A critical review and meta-analysis. *Ecology* **93**:2241–2252.

D'Amen, M., C. Rahbek, N. E. Zimmermann, and A. Guisan. 2015. Spatial predictions at the community level: From current approaches to future frameworks. *Biological Reviews* **92**:169–187.

Daily, G. C., G. Ceballos, J. Pacheco, G. Suzan, and A. Sanchez-Azofeifa. 2003. Countryside biogeography of neotropical mammals: Conservation opportunities in agricultural landscapes of Costa Rica. *Conservation Biology* **17**:1814–1826.

Dallas, T., and S. J. Presley. 2014. Relative importance of host environment, transmission potential and host phylogeny to the structure of parasite metacommunities. *Oikos* **123**:866–874.

Dalling, J. W., H. C. Muller-Landau, S. J. Wright, and S. P. Hubbell. 2002. Role of dispersal in the recruitment limitation of neotropical pioneer species. *Journal of Ecology* **90**:714–727.

Damschen, E. I., N. M. Haddad, J. L. Orrock, J. J. Tewksbury, and D. J. Levey. 2006. Corridors increase plant species richness at large scales. *Science* **313**:1284–1286.

Damuth, J. 1981. Population-density and body size in mammals. *Nature* **290**:699–700.

Dangremond, E. M., E. A. Pardini, and T. M. Knight. 2010. Apparent competition with an invasive plant hastens the extinction of an endangered lupine. *Ecology* **91**:2261–2271.

Darwin, C. 1859. *On the origin of species by means of natural selection, or the preservation of favoured races in the struggle for life.* Cowles and Sons, London.

daSilva, P. G., and M.I.M. Hernández. 2015. Scale-dependence of processes structuring dung beetle metacommunities using functional diversity and community deconstruction approaches. *PLoS One* **10**:e0123030.

Daufresne, T., and L. O. Hedin. 2005. Plant coexistence depends on ecosystem nutrient cycles: Extension of the resource-ratio theory. *Proceedings of the National Academy of Sciences USA* **102**:9212–9217.

Davidson, D. W. 1998. Resource discovery versus resource domination in ants: A functional mechanism for breaking the trade-off. *Ecological Entomology* **23**:484–490.

Davies, K. F., J. Cavender-Bares, and N. Deacon. 2011a. Native communities determine the identity of exotic invaders even at scales at which communities are unsaturated. *Diversity and Distributions* **17**:35–42.

Davies, K. F., C. R. Margules, and K. F. Lawrence. 2000. Which traits of species predict population declines in experimental forest fragments? *Ecology* **81**:1450–1461.

Davies, K. F., B. A. Melbourne, and C. R. Margules. 2001. Effects of within- and between-patch processes on community dynamics in a fragmentation experiment. *Ecology* **82**:1830–1846.

Davies, S. J., S. Tan, J. V. LaFrankie, and M. D. Potts. 2005. Soil-related floristic variation in a hyperdiverse dipterocarp forest in Lambir Hills, Sarawak. Pages 22–34 *in* D. W. Roubik, S. Sakai, and A.A.H. Karim, editors, *Pollination ecology and rain forest diversity: Sarawak studies.* Springer-Verlag, New York.

Davies, T. J., A. P. Allen, L. Borda-de-Agua, J. Regetz, and C. J. Melian. 2011b. Neutral biodiversity theory can explain the imbalance of phylogenetic trees but not the tempo of their diversification. *Evolution* **65**:1841–1850.

Davis, M. B. 1981. Quaternary history and the stability of forest communities. Pages 132–153 *in* D. C. West, H. H. Shugart, and D. B. Botkin, editors, *Forest succession: Concepts and application.* Springer-Verlag, New York.

Dayan, T., and D. Simberloff. 2005. Ecological and community-wide character displacement: The next generation. *Ecology Letters* **8**:875–894.

de Bello, F., J. Lepš, and M.-T. Sebastià. 2007. Grazing effects on the species-area relationship: Variation along a climatic gradient in NE Spain. *Journal of Vegetation Science* **18**:25.

De Bie, T., L. De Meester, L. Brendonck, K. Martens, B. Goddeeris, D. Ercken, H. Hampel, et al. 2012. Body size and dispersal mode as key traits determining metacommunity structure of aquatic organisms. *Ecology Letters* **15**:740–747.

De Caceres, M., P. Legendre, R. Valencia, M. Cao, L.-W. Chang, G. Chuyong, R. Condit, et al. 2012. The variation of tree beta diversity across a global network of forest plots. *Global Ecology and Biogeography* **21**:1191–1202.

de Knegt, H. J., F. van Langevelde, M. B. Coughenour, A. K. Skidmore, W. F. de Boer, I.M.A. Heitkonig, N. M. Knox, R. Slotow, C. van der Waal, and H.H.T. Prins. 2010. Spatial autocorrelation and the scaling of species-environment relationships. *Ecology* **91**:2455–2465.

de la Sancha, N. U., C. L. Higgins, S. J. Presley, and R. E. Strauss. 2014. Metacommunity structure in a highly fragmented forest: Has deforestation in the Atlantic Forest altered historic biogeographic patterns? *Diversity and Distributions* **20**:1058–1070.

De Meester, L., A. Gomez, B. Okamura, and K. Schwenk. 2002. The Monopolization Hypothesis and the dispersal-gene flow paradox in aquatic organisms. *Acta Oecologica* **23**:121–135.

De Meester, L., G. Louette, C. Duvivier, C. Van Damme, and E. Michels. 2007. Genetic composition of resident populations influences establishment success of immigrant species. *Oecologia* **153**:431–440.

De Meester, L., J. Vanoverbeke, L. J. Kilsonk, and M. C. Urban. 2016. Evolving perspectives on monopolization and priority effects. *Trends in Ecology and Evolution* 31:136–146.

De Schrijver, A., P. De Frenne, E. Ampoorter, L. Van Nevel, A. Demey, K. Wuyts, and K. Verheyen. 2011. Cumulative nitrogen input drives species loss in terrestrial ecosystems. *Global Ecology and Biogeography* **20**:803–816.

DeAngelis, K. M., W. L. Silver, A. W. Thompson, and M. K. Firestone. 2010. Microbial communities acclimate to recurring changes in soil redox potential status. *Environmental Microbiology* **12**:3137–3149.

Debinski, D. M., and R. D. Holt. 2000. A survey and overview of habitat fragmentation experiments. *Conservation Biology* **14**:342–355.

Declerck, S. A., C. Winter, J. B. Shurin, C. A. Suttle, and B. Matthews. 2012. Effects of patch connectivity and heterogeneity on metacommunity structure of planktonic bacteria and viruses. *ISME Journal* **7**:533–542.

Denslow, J. S. 1995. Disturbance and diversity in tropical rain-forest: The density effect. *Ecological Applications* **5**:962–968.

Devictor, V., R. Julliard, J. Clavel, F. Jiguet, A. Lee, and D. Couvet. 2008. Functional biotic homogenization of bird communities in disturbed landscapes. *Global Ecology and Biogeography* **17**:252–261.

Diamond, J. M. 1969. Avifaunal equilibria and species turnover rates on the Channel Islands of California. *Proceedings of the National Academy of Sciences USA* **64**:57–63.

Diamond, J. M. 1975a. Assembly of species communities. Pages 342–444 *in* M. L. Cody and J. M. Diamond, editors, *Ecology and evolution of communities*. Harvard University Press, Cambridge, MA.

Diamond, J. M. 1975b. The island dilemma: Lessons of modern biogeographic studies for the design of natural reserves. *Biological Conservation* **7**:129–146.

Diamond, J. M., and T. J. Case. 1986. *Community ecology*. Harper & Row, New York.

Diamond, J. M., and M. E. Gilpin. 1982. Examination of the "null" model of Connor and Simberloff for species co-occurrences on islands. *Oecologia* **52**:64–74.

Diamond, J. M., and R. May. 1976. Island biogeography and the design of nature reserves. Pages 163–186 *in* R. M. May, editor, *Theoretical ecology: Principles and applications*. Blackwell, Oxford.

Diamond, J. M., S. L. Pimm, and J. G. Sanderson. 2015. The checkered history of checkerboard distributions: Comment. *Ecology* **96**:3386–3388.

Diaz, S., M. Cabido, and F. Casanoves. 1998. Plant functional traits and environmental filters at a regional scale. *Journal of Vegetation Science* **9**:113–122.

Diaz, S., J. Kattge, J.H.C. Cornelissen, I. J. Wright, S. Lavorel, S. Dray, B. Reu, et al. 2016. The global spectrum of plant form and function. *Nature* **529**:167–171.

Diez, J. M., J. J. Sullivan, P. E. Hulme, G. Edwards, and R. P. Duncan. 2008. Darwin's naturalization conundrum: dissecting taxonomic patterns of species invasions. *Ecology Letters* **11**:674–681.

Diniz, J.A.F., T. Siqueira, A. A. Padial, T. F. Rangel, V. L. Landeiro, and L. M. Bini. 2012. Spatial autocorrelation analysis allows disentangling the balance between neutral and niche processes in metacommunities. *Oikos* **121**:201–210.

Dionne, K. 2015. Coexistence d'un complexe d'espèces cryptiques à des échelles locales et régionales. Ph.D. diss., Université du Quebec á Rimouski, Rimouski, Quebec, Canada.

Dirzo, R., and P. H. Raven. 2003. Global state of biodiversity and loss. *Annual Review of Environment and Resources* **28**:137–167.

Doledec, S., D. Chessel, C.J.F. terBraak, and S. Champely. 1996. Matching species traits to environmental variables: A new three-table ordination method. *Environmental and Ecological Statistics* **3**:143–166.

Donald, D. B., and R. Stewart Anderson. 2003. Resistance of the prey-to-predator ratio to environmental gradients and to biomanipulations. *Ecology* **84**:2387–2394.

Donohue, I., A. L. Jackson, M. T. Pusch, and K. Irvine. 2009. Nutrient enrichment homogenizes lake benthic assemblages at local and regional scales. *Ecology* **90**:3470–3477.

Donoso, D. A. 2014. Assembly mechanisms shaping tropical litter ant communities. *Ecography* **37**:490–499.

Dornelas, M., S. R. Connolly, and T. P. Hughes. 2006. Coral reef diversity refutes the neutral theory of biodiversity. *Nature* **440**:80–82.

Dornelas, M., N. J. Gotelli, B. McGill, H. Shimadzu, F. Moyes, C. Sievers, and A. E. Magurran. 2014. Assemblage time series reveal biodiversity change but not systematic loss. *Science* **344**:296–299.

Dornelas, M., A. E. Magurran, S. T. Buckland, A. Chao, R. L. Chazdon, R. K. Colwell, T. Curtis, et al. 2013. Quantifying temporal change in biodiversity: Challenges and opportunities. *Proceedings of the Royal Society. Series B: Biological Sciences* **280**:20121931. doi:10.1098/rspb.2012.1931.

Douglas, W. Y., and H. B. Wilson. 2001. The competition-colonization trade-off is dead; Long live the competition-colonization trade-off. *American Naturalist* **158**:49–63.

Downing, J. A. 1997. Marine nitrogen: Phosphorus stoichiometry and the global N:P cycle. *Biogeochemistry* **37**:237–252.

Drakare, S., J. J. Lennon, and H. Hillebrand. 2006. The imprint of the geographical, evolutionary and ecological context on species-area relationships. *Ecology Letters* **9**:215–227.

Drake, J. A. 1991. Community-assembly mechanics and the structure of an experimental species ensemble. *American Naturalist* **137**:1–26.

Dray, S., P. Choler, S. Dolédec, P. R. Peres-Neto, W. Thuiller, S. Pavoine, and C.J.F. ter Braak. 2014. Combining the fourth-corner and the RLQ methods for assessing trait responses to environmental variation. *Ecology* **95**:14–21.

Dray, S., and P. Lebgendre. 2008. Testing the species traits-environment relationships: the fourth-corner problem revisited. *Ecology* **89**:3400–3412.

Dray, S., R. Pelissier, P. Couteron, M. J. Fortin, P. Legendre, P. R. Peres-Neto, E. Bellier, et al. 2012. Community ecology in the age of multivariate multiscale spatial analysis. *Ecological Monographs* **82**:257–275.

Duivenvoorden, J. F., J. C. Svenning, and S. J. Wright. 2002. Beta diversity in tropical forests. *Science* **295**:636–637.

Durrett, R., and S. Levin. 1998. Spatial aspects of interspecific competition. *Theoretical Population Biology* **53**:30–43.

Dybzinski, R., C. Farrior, A. Wolf, P. B. Reich, and S. W. Pacala. 2011. Evolutionarily stable strategy carbon allocation to foliage, wood, and fine roots in trees competing for light and nitrogen: An analytically tractable, individual-based model and quantitative comparisons to data. *American Naturalist* **177**:153–166.

Economo, E. P., and T. H. Keitt. 2007. Species diversity in neutral metacommunities: A network approach. *Ecology Letters* **11**:52–62.

Edwards, K. F., E. Litchman, and C. A. Klausmeier. 2013. Functional traits explain phytoplankton responses to environmental gradients across lakes of the United States. *Ecology* **94**:1626–1635.

Egler, F. E. 1954. Vegetation science concepts I. Initial floristic composition, a factor in old-field vegetation development. *Vegetatio* **4**:412–417.

Ehrenfeld, J. G., B. Ravit, and K. Elgersma. 2005. Feedback in the plant-soil system. *Annual Review of Environment and Resources* **30**:75–115.

Ehrlén, J., and O. Eriksson. 2000. Dispersal limitation and patch occupancy in forest herbs. *Ecology* **81**:1667–1674.

Ehrlén, J., Z. Münzbergova, M. Diekmann, and O. Eriksson. 2006. Long-term assessment of seed limitation in plants: Results from an 11-year experiment. *Journal of Ecology* **94**:1224–1232.

Eisenhauer, N., A. D. Barnes, S. Cesarz, D. Craven, O. Ferlian, F. Gottschall, J. Hines, et al. 2016. Biodiversity-ecosystem function experiments reveal the mechanisms underlying the consequences of biodiversity change in real world ecosystems. *Journal of Vegetation Science* **27**:1061–1070.

Ejrnæs, R., H. H. Bruun, and B. J. Graae. 2006. Community assembly in experimental grasslands: Suitable environment or timely arrival? *Ecology* **87**:1225–1233.

Elahi, R., Mary I. O'Connor, Jarrett E. K. Byrnes, J. Dunic, Britas K. Eriksson, Marc J. S. Hensel, and Patrick J. Kearns. 2015. Recent trends in local-scale marine biodiversity reflect community structure and human impacts. *Current Biology* **25**:1938–1943.

Ellis, A. M., L. P. Lounibos, and M. Holyoak. 2006. Evaluating the long-term metacommunity dynamics of tree hole mosquitoes. *Ecology* **87**:2582–2590.

Ellner, S. P. 2013. Rapid evolution: From genes to communities, and back again? *Functional Ecology* **27**:1087–1099.

Ellner, S. P., M. A. Geber, and N. G. Hairston. 2011. Does rapid evolution matter? Measuring the rate of contemporary evolution and its impacts on ecological dynamics. *Ecology Letters* **14**:603–614.

Elton, C. S. 1927. *Animal ecology.* University of Chicago Press, Chicago.

Elton, C. S. 1946. Competition and the structure of ecological communities. *Journal of Animal Ecology* **15**:54–68.

Elton, C. S. 1958. *Ecology of invasions by animals and plants.* Chapman and Hall, London.

Enquist, B. J., J. Norberg, S. P. Bonser, C. Violle, C. T. Webb, A. Henderson, L. L. Sloat, and V. M. Savage. 2015. Scaling from traits to ecosystems. *Advances in Ecological Research* **52**:249–318.

Ernest, S.K.M., and J. H. Brown. 2001. Homeostasis and compensation: The role of species and resources in ecosystem stability. *Ecology* **82**:2118–2132.

Ernest, S. M., J. H. Brown, K. M. Thibault, E. P. White, and J. R. Goheen. 2008. Zero sum, the niche, and metacommunities: Long-term dynamics of community assembly. *American Naturalist* **172**:E257–E269.

Eros, T., P. Saly, P. Takacs, C. L. Higgins, P. Biro, and D. Schmera. 2014. Quantifying temporal variability in the metacommunity structure of stream fishes: the influence of non-native species and environmental drivers. *Hydrobiologia* **722**:31–43.

Essl, F., D. Moser, T. Dirnbock, S. Dullinger, N. Milasowszky, M. Winter, and W. Rabitsch. 2013. Native, alien, endemic, threatened, and extinct species diversity in European countries. *Biological Conservation* **164**:90–97.

Etienne, R. S., and D. Alonso. 2005. A dispersal-limited sampling theory for species and alleles. *Ecology Letters* **8**:1147–1156.

Evans, K. L., N. A. James, and K. J. Gaston. 2006. Abundance, species richness and energy availability in the North American avifauna. *Global Ecology and Biogeography* **15**:372–385.

Evans, K. L., S. E. Newson, D. Storch, J.J.D. Greenwood, and K. J. Gaston. 2008. Spatial scale, abundance and the species–energy relationship in British birds. *Journal of Animal Ecology* **77**:395–405.

Evans, K. L., P. H. Warren, and K. J. Gaston. 2005. Species–energy relationships at the macroecological scale: a review of the mechanisms. *Biological Reviews* **80**:1–25.

Ewers, R. M., and R. K. Didham. 2006. Confounding factors in the detection of species responses to habitat fragmentation. *Biological Reviews* **81**:117–142.

Fahimipour, A. K., and K. E. Anderson. 2015. Colonisation rate and adaptive foraging control the emergence of trophic cascades. *Ecology Letters* **18**:826–833.

Fahrig, L. 2001. How much habitat is enough? *Biological Conservation* **100**:65–74.

Fahrig, L. 2003. Effects of habitat fragmentation on biodiversity. *Annual Review of Ecology, Evolution, and Systematics* **34**:487–515.

Fahrig, L. 2013. Rethinking patch size and isolation effects: The habitat amount hypothesis. *Journal of Biogeography* **40**:1649–1663.

Farjalla, V. F., D. S. Srivastava, N.A.C. Marino, F. D. Azevedo, V. Dib, P. M. Lopes, A. S. Rosado, R. L. Bozelli, and F. A. Esteves. 2012. Ecological determinism increases with organism size. *Ecology* **93**:1752–1759.

Farneda, F. Z., R. Rocha, A. López-Baucells, M. Groenenberg, I. Silva, J. M. Palmeirim, P.E.D. Bobrowiec, and C.F.J. Meyer. 2015. Trait-related responses to habitat fragmentation in Amazonian bats. *Journal of Applied Ecology* **52**:1381–1391.

Feeley, K. 2003. Analysis of avian communities in Lake Guri, Venezuela, using multiple assembly rule models. *Oecologia* **137**:104–113.

Fellers, J. H. 1987. Interference and exploitation in a guild of woodland ants. *Ecology* **68**:1466–1478.

Ferdy, J.-B., and J. Molofsky. 2002. Allee effect, spatial structure and species coexistence. *Journal of Theoretical Biology* **217**:413–424.

Fernandes, I. M., R. Henriques-Silva, J. Penha, J. Zuanon, and P. R. Peres-Neto. 2014. Spatiotemporal dynamics in a seasonal metacommunity structure is predictable: The case of floodplain-fish communities. *Ecography* **37**:464–475.

Fine, P.V.A., I. Mesones, and P. D. Coley. 2004. Herbivores promote habitat specialization by trees in Amazonian forests. *Science* **305**:663–665.

Fine, P.V.A., Z. J. Miller, I. Mesones, S. Irazuzta, H. M. Appel, M.H.H. Stevens, I. Saaksjarvi, L. C. Schultz, and P. D. Coley. 2006. The growth-defense trade-off and habitat specialization by plants in Amazonian forests. *Ecology* **87**:S150–S162.

Fischer, J., D. J. Abson, V. Butsic, M. J. Chappell, J. Ekroos, J. Hanspach, T. Kuemmerle, H. G. Smith, and H. von Wehrden. 2014. Land sparing versus land sharing: Moving forward. *Conservation Letters* **7**:149–157.

Fišer, Ž., F. Altermatt, V. Zakšek, T. Knapič, and C. Fišer. 2015. Morphologically cryptic amphipod species are "ecological clones" at regional but not at local scale: A case study of four *Niphargus* species. *PLoS One* **10**:e0134384.

Fisher, R. A., A. S. Corbet, and C. B. Williams. 1943. The relation between the number of species and the number of individuals in a random sample of an animal population. *Journal of Animal Ecology* **12**:42–58.

Fitzpatrick, C. R., A. A. Agrawal, N. Basiliko, A. P. Hastings, M. E. Isaac, M. Preston, and M.T.J. Johnson. 2015. The importance of plant genotype and contemporary evolution for terrestrial ecosystem processes. *Ecology* **96**:2632–2642.

Flinn, K. M., T. C. Gouhier, M. J. Lechowicz, and M. J. Waterway. 2010. The role of dispersal in shaping plant community composition of wetlands within an old-growth forest. *Journal of Ecology* **98**:1292–1299.

Forbes, A. E., and J. M. Chase. 2002. The role of habitat connectivity and landscape geometry in experimental zooplankton metacommunities. *Oikos* **96**:433–440.

Forbes, S. A. 1925. The lake as a microcosm. *Illinois Natural History Survey Bulletin* **15**:536–550.

Foster, B. L., T. L. Dickson, C. A. Murphy, I. S. Karel, and V. H. Smith. 2004. Propagule pools mediate community assembly and diversity-ecosystem regulation along a grassland productivity gradient. *Journal of Ecology* **92**:435–449.

Foster, B. L., E. J. Questad, C. D. Collins, C. A. Murphy, T. L. Dickson, and V. H. Smith. 2011. Seed availability constrains plant species sorting along a soil fertility gradient. *Journal of Ecology* **99**:473–481.

Fournier, B., N. Mouquet, M. A. Leibold, and D. Gravel. 2017. An integrative framework of coexistence mechanisms in competitive metacommunities. *Ecography* **40**: 630–641.

Fox, B. J. 1987. Species assembly and the evolution of community structure. *Evolutionary Ecology* **1**:201–213.

Fox, B. J., and J. H. Brown. 1993. Assembly rules for functional groups in North American desert rodent communities. *Oikos* **67**:358–370.

Fox, J. W., and D. C. Smith. 1997. Variable outcomes of protist-rotifer competition in laboratory microcosms. *Oikos* **79**:489–495.

Fox, J. W., and D. A. Vasseur. 2008. Character convergence under competition for nutritionally essential resources. *American Naturalist* **172**:667–680.

Fraser, L. H., J. Pither, A. Jentsch, M. Sternberg, M. Zobel, D. Askarizadeh, S. Bartha, et al. 2015. Worldwide evidence of a unimodal relationship between productivity and plant species richness. *Science* **349**:302–305.

Fritschie, K. J., B. J. Cardinale, M. A. Alexandrou, and T. H. Oakley. 2014. Evolutionary history and the strength of species interactions: Testing the phylogenetic limiting similarity hypothesis. *Ecology* **95**:1407–1417.

Fukami, T. 2004a. Assembly history interacts with ecosystem size to influence species diversity. *Ecology* **85**:3234–3242.

Fukami, T. 2004b. Community assembly along a species pool gradient: Implications for multiple-scale patterns of species diversity. *Population Ecology* 46:137–147.

Fukami, T. 2015. Historical contingency in community assembly: Integrating niches, species pools, and priority effects. *Annual Review of Ecology, Evolution, and Systematics* 46:1–23.

Fukami, T., H. J. Beaumont, X. X. Zhang, and P. B. Rainey. 2007. Immigration history controls diversification in experimental adaptive radiation. *Nature* **446**:436–439.

Fukami, T., T. M. Bezemer, S. R. Mortimer, and W. H. Putten. 2005. Species divergence and trait convergence in experimental plant community assembly. *Ecology Letters* **8**:1283–1290.

Fukami, T., I. A. Dickie, J. P. Wilkie, B. C. Paulus, D. Park, A. Roberts, P. K. Buchanan, and R. B. Allen. 2010. Assembly history dictates ecosystem functioning: evidence from wood decomposer communities. *Ecology Letters* **13**:675–684.

Fukumori, K., G. Livingston, and M. A. Leibold. 2015. Disturbance-mediated colonization-extinction dynamics in experimental protist metacommunities. *Ecology* **96**: 3234–3242.

Gardezi, T., and A. Gonzalez. 2008. Scale dependence of species-energy relationships: Evidence from fishes in thousands of lakes. *American Naturalist* **171**:800–815.

Garnier, E., M.-L. Navas, and K. Grigulis. 2016. *Plant functional diversity: Organism traits, community structure, and ecosystem properties.* Oxford University Press, Oxford.

Garzon-Lopez, C. X., P. A. Jansen, S. A. Bohlman, A. Ordonez, and H. Olff. 2014. Effects of sampling scale on patterns of habitat association in tropical trees. *Journal of Vegetation Science* **25**:349–362.

Gaston, K. J. 2000. Global patterns in biodiversity. *Nature* **405**:220–227.

Gaston, K. J., and T. M. Blackburn. 2000. *Pattern and process in macroecology.* Blackwell Science, Oxford.

Gauch, H. G., and R. H. Whittaker. 1972. Coenocline simulation. *Ecology* **53**:446–451.

Gause, G. F. 1934. Experimental analysis of Vito Volterra's mathematical theory of the struggle for existence. *Science* **79**:16–17.

Gause, G. F. 1936. The principles of biocoenology. *Quarterly Review of Biology* **11**:320–336.

Gavin, M. C., and N. Sibanda. 2012. The island biogeography of languages. *Global Ecology and Biogeography* **21**:958–967.

Gavish, Y., Y. Ziv, and M. L. Rosenzweig. 2012. Decoupling fragmentation from habitat loss for spiders in patchy agricultural landscapes. *Conservation Biology* **26**:150–159.

Gavrilets, S. 2000. Rapid evolution of reproductive barriers driven by sexual conflict. *Nature* **403**:886–889.

Gerhold, P., J. F. Cahill, M. Winter, I. V. Bartish, and A. Prinzing. 2015. Phylogenetic patterns are not proxies of community assembly mechanisms (they are far better). *Functional Ecology* **29**:600–614.

Gerla, D. J., and W. M. Mooij. 2014. Alternative stable states and alternative endstates of community assembly through intra- and interspecific positive and negative interactions. *Theoretical Population Biology* **96**:8–18.

Germain, R. M., L. Johnson, S. Schneider, K. Cottenie, E. A. Gillis, and A. S. MacDougall. 2013. Spatial variability in plant predation determines the strength of stochastic community assembly. *American Naturalist* **182**:169–179.

Ghilarov, A. M. 1984. The paradox of the plankton reconsidered; or, why do species coexist? *Oikos* **43**:46–52.

Gibson, L., A. J. Lynam, C.J.A. Bradshaw, F. He, D. P. Bickford, D. S. Woodruff, S. Bumrungsri, and W. F. Laurance. 2013. Near-complete extinction of native small mammal fauna 25 years after forest fragmentation. *Science* **341**:1508–1510.

Giladi, I., Y. Ziv, F. May, and F. Jeltsch. 2011. Scale-dependent determinants of plant species richness in a semi-arid fragmented agro-ecosystem. *Journal of Vegetation Science* **22**:983–996.

Gilbert, B., and J. R. Bennett. 2010. Partitioning variation in ecological communities: Do the numbers add up? *Journal of Applied Ecology* **47**:1071–1082.

Gilbert, B., and M. J. Lechowicz. 2004. Neutrality, niches, and dispersal in a temperate forest understory. *Proceedings of the National Academy of Sciences USA* **101**: 7651–7656.

Gilbert, F., A. Gonzalez, and I. Evans-Freke. 1998. Corridors maintain species richness in the fragmented landscapes of a microecosystem. Proceedings of the Royal Society. *Series* B: Biological Sciences **265**:577–582.

Gilbert-Norton, L., R. Wilson, J. R. Stevens, and K. H. Beard. 2010. A meta-analytic review of corridor effectiveness. *Conservation Biology* **24**:660–668.

Gillespie, R. G. 2016. Island time and the interplay between ecology and evolution in species diversification. *Evolutionary Applications* **9**:53–73.

Gillman, L. N., S. D. Wright, J. Cusens, P. D. McBride, Y. Malhi, and R. J. Whittaker. 2015. Latitude, productivity and species richness. *Global Ecology and Biogeography* **24**:107–117.

Gillooly, J. F., A. P. Allen, G. B. West, and J. H. Brown. 2005. The rate of DNA evolution: Effects of body size and temperature on the molecular clock. *Proceedings of the National Academy of Sciences USA* **102**:140–145.

Gilpin, M. E., and J. M. Diamond. 1980. Subdivision of nature reserves and the maintenance of species-diversity. *Nature* **285**:567–568.

Gleason, H. A. 1926. The individualistic concept of the plant association. *Bulletin of the Torrey Botanical Club* **53**:7–26.

Gleason, H. A. 1927. Further views on the succession-concept. *Ecology* **8**:299–326.

Godoy, O., N.J.B. Kraft, and J. M. Levine. 2014. Phylogenetic relatedness and the determinants of competitive outcomes. *Ecology Letters* **17**:836–844.

Goldberg, D. E., and K. Landa. 1991. Competitive effect and response: Hierarchies and correlated traits in the early stages of competition. *Journal of Ecology* **79**: 1013–1030.

Gomulkiewicz, R., and R. D. Holt. 1995. When does evolution by natural selection prevent extinction? *Evolution* **49**:201–207.

Goncalves-Souza, T., J.A.F. Diniz, and G. Q. Romero. 2014. Disentangling the phylogenetic and ecological components of spider phenotypic variation. *PLoS One* **9**:e89314.

Gonzalez, A., B. J. Cardinale, G.R.H. Allington, J. Byrnes, K. Arthur Endsley, D. G. Brown, D. U. Hooper, F. Isbell, M. I. O'Connor, and M. Loreau. 2016. Estimating local biodiversity change: a critique of papers claiming no net loss of local diversity. *Ecology* **97**:1949–1960.

Gonzalez, A., J. Lawton, F. Gilbert, T. Blackburn, and I. Evans-Freke. 1998. Metapopulation dynamics, abundance, and distribution in a microecosystem. *Science* **281**: 2045–2047.

Gonzalez, A., B. Rayfield, and Z. Lindo. 2011. The disentangled bank: How loss of habitat fragments and disassembles ecological networks. *American Journal of Botany* **98**:503–516.

Gonzalez-Caro, S., J. L. Parra, C. H. Graham, J. A. McGuire, and C. D. Cadena. 2012. Sensitivity of metrics of phylogenetic structure to scale, source of data and species pool of hummingbird assemblages along elevational gradients. *PLoS One* **7**:e35472.

Goodnight, C. J. 2000. Heritability at the ecosystem level. *Proceedings of the National Academy of Sciences USA* **97**:9365–9366.

Goodnight, C. J. 2011. Evolution in metacommunities. *Philosophical Transactions of the Royal Society B: Biological Sciences* **366**:1401–1409.

Gotelli, N. J., M. J. Anderson, H. T. Arita, A. Chao, R. K. Colwell, S. R. Connolly, D. J. Currie, et al.. 2009. Patterns and causes of species richness: a general simulation model for macroecology. *Ecology Letters* **12**:873–886.

Gotelli, N. J., and R. K. Colwell. 2001. Quantifying biodiversity: procedures and pitfalls in the measurement and comparison of species richness. *Ecology Letters* **4**:379–391.

Gotelli, N. J., and D. J. McCabe. 2002. Species co-occurrence: A meta-analysis of J. M. Diamond's assembly rules model. *Ecology* **83**:2091–2096.

Gotelli, N. J., and B. J. McGill. 2006. Null versus neutral models: What's the difference? *Ecography* **29**:793–800.

Götzenberger, L., F. de Bello, K. A. Brathen, J. Davison, A. Dubuis, A. Guisan, J. Leps, et al. 2012. Ecological assembly rules in plant communities—Approaches, patterns and prospects. *Biological Reviews* **87**:111–127.

Gouhier, T. C., F. Guichard, and A. Gonzalez. 2010. Synchrony and stability of food webs in metacommunities. *American Naturalist* **175**:E16–E34.

Gounand, I., N. Mouquet, E. Canard, F. Guichard, C. Hauzy, and D. Gravel. 2014. The paradox of enrichment in metaecosystems. *American Naturalist* **184**:752–763.

Grace, J. B., T. M. Anderson, E. W. Seabloom, E. T. Borer, P. B. Adler, W. S. Harpole, Y. Hautier, et al. 2016. Integrative modelling reveals mechanisms linking productivity and plant species richness. *Nature* **529**:390–393.

Graham, C. H., and P.V.A. Fine. 2008. Phylogenetic beta diversity: Linking ecological and evolutionary processes across space in time. *Ecology Letters* **11**:1265–1277.

Graham, C. H., J. L. Parra, B. A. Tinoco, F. G. Stiles, and J. A. McGuire. 2012. Untangling the influence of ecological and evolutionary factors on trait variation across hummingbird assemblages. *Ecology* **93**:S99–S111.

Grainger, T. N., and B. Gilbert. 2016. Dispersal and diversity in experimental metacommunities: Linking theory and practice. *Oikos* **125**:1213–1223.

Grant, P. R. 1972. Convergent and divergent character displacement. *Biological Journal of the Linnean Society* **4**:39–68.

Grant, P. R., and B. R. Grant. 2014. *40 years of evolution: Darwin's finches on Daphne Major Island*. Princeton University Press, Princeton, NJ.

Gravel, D., E. Canard, F. Guichard, and N. Mouquet. 2011a. Persistence increases with diversity and connectance in trophic metacommunities. *PLoS One* **6**:e19374.

Gravel, D., C. D. Canham, M. Beaudet, and C. Messier. 2006. Reconciling niche and neutrality: The continuum hypothesis. *Ecology Letters* **9**:399–409.

Gravel, D., F. Guichard, M. Loreau, and N. Mouquet. 2010. Source and sink dynamics in meta-ecosystems. *Ecology* **91**:2172–2184.

Gravel, D., F. Massol, E. Canard, D. Mouillot, and N. Mouquet. 2011b. Trophic theory of island biogeography. *Ecology Letters* **14**:1010–1016.

Gravel, D., F. Massol, and M. A. Leibold. 2016. Stability and complexity in model metaecosystems. *Nature Communications* **7**:12457.

Gray, J. S., K. I. Ugland, and J. Lambshead. 2004. Species accumulation and species area curves—A comment on Scheiner (2003). *Global Ecology and Biogeography* **13**:473–476.

Green, J. L., J. Harte, and A. Ostling. 2003. Species richness, endemism and abundance patterns: Tests of two fractal models in a serpentine grassland. *Ecology Letters* **6**:919–928.

Green, J. L., and J. B. Plotkin. 2007. A statistical theory for sampling species abundances. *Ecology Letters* **10**:1037–1045.

Grime, J. P. 1974. Vegetation classification by reference to strategies. *Nature* **250**:26–31.

Grime, J. P. 1977. Evidence for existence of three primary strategies in plants and its relevance to ecological and evolutionary theory. *American Naturalist* **111**: 1169–1194.

Grinnell, J. 1904. The origin and distribution of the chestnut-backed chickadee. *Auk* 21:375–377.

Grinnell, J. 1917. Field tests of theories concerning distributional control. *American Naturalist* **51**:115–128.

Groening, J., and A. Hochkirch. 2008. Reproductive interference between animal species. *Quarterly Review of Biology* **83**:257–282.

Gross, K. 2008. Fusing spatial resource heterogeneity with a competition–colonization trade-off in model communities. *Theoretical Ecology* **1**:65–75.

Gross, K., and B. J. Cardinale. 2007. Does species richness drive community production or vice versa? Reconciling historical and contemporary paradigms in competitive communities. *American Naturalist* **170**:207–220.

Grover, J. P. 1994. Assembly rules for communities of nutrient-limited plants and specialist herbivores. *American Naturalist* **143**:258–282.

Gueze, M., J. Paneque-Galvez, A. C. Luz, J. Pino, M. Orta-Martinez, V. Reyes-Garcia, and M. J. Macia. 2013. Determinants of tree species turnover in a southern Amazonian rain forest. *Journal of Vegetation Science* **24**:284–295.

Gunatilleke, C.V.S., I.A.U.N. Gunatilleke, S. Esufali, K. E. Harms, P.M.S. Ashton, D.F.R.P. Burslem, and P. S. Ashton. 2006. Species-habitat associations in a Sri Lankan dipterocarp forest. *Journal of Tropical Ecology* **22**:371–384.

Gurevitch, J., L. L. Morrow, A. Wallace, and J. S. Walsh. 1992. A meta-analysis of competition in field experiments. *American Naturalist* **140**:539–572.

Haddad, N. M., L. A. Brudvig, J. Clobert, K. F. Davies, A. Gonzalez, R. D. Holt, T. E. Lovejoy, et al. 2015. Habitat fragmentation and its lasting impact on Earth's ecosystems. *Science Advances* **1**:e1500052.

Haddad, N. M., A. Gonzalez, L. A. Brudvig, M. A. Burt, D. J. Levey, and E. I. Damschen. 2017. Experimental evidence does not support the Habitat Amount Hypothesis. *Ecography* **40**:48–55.

Haegeman, B., and M. Loreau. 2014. General relationships between consumer dispersal, resource dispersal and metacommunity diversity. *Ecology Letters* **17**:175–184.

Haegeman, B., and M. Loreau. 2015. A graphical-mechanistic approach to spatial resource competition. *American Naturalist* **185**:E1–E13.

Hagen, M., W. D. Kissling, C. Rasmussen, M.A.M. De Aguiar, L. E. Brown, D. W. Carstensen, I. Alves-Dos-Santos, et al. 2012. Biodiversity, species interactions and ecological networks in a fragmented world. *Advances in Ecological Research* **46**:189–210.

Hairston, N. G., S. P. Ellner, M. A. Geber, T. Yoshida, and J. A. Fox. 2005. Rapid evolution and the convergence of ecological and evolutionary time. *Ecology Letters* **8**:1114–1127.

Hajek, M., J. Rolecek, K. Cottenie, K. Kintrova, M. Horsak, A. Poulickova, P. Hajkova, M. Frankova, and D. Dite. 2011. Environmental and spatial controls of biotic assemblages in a discrete semi-terrestrial habitat: Comparison of organisms with different dispersal abilities sampled in the same plots. *Journal of Biogeography* **38**:1683–1693.

Halley, J. M., V. Sgardeli, and N. Monokrousos. 2013. Species-area relationships and extinction forecasts. *Annals of the New York Academy of Sciences* **1286**:50–61.

Haloin, J. R., and S. Y. Strauss. 2008. Interplay between ecological communities and evolution: Review of feedbacks from microevolutionary to macroevolutionary scales. *Annals of the New York Academy of Sciences* **1133**:87–125.

Hanski, I. 1982. Dynamics of regional distribution: The core and satellite species hypothesis. *Oikos* **38**:210–221.

Hanski, I. 1983. Coexistence of competitors in patchy environment. *Ecology* **64**:493–500.

Hanski, I. 1987. Carrion fly community dynamics: Patchiness, seasonality and coexistence. *Ecological Entomology* **12**:257–266.

Hanski, I., 1999. *Metapopulation ecology*. Oxford University Press, Oxford.

Hanski, I. 2012. Eco-evolutionary dynamics in a changing world. *Annals of the New York Academy of Sciences* **1249**:1–17.

Hanski, I. 2015. Habitat fragmentation and species richness. Journal of Biogeography **42**:989–993.

Hanski, I., and M. Gilpin. 1991. Metapopulation dynamics: Brief history and conceptual domain. *Biological Journal of the Linnean Society* **42**:3–16.

Hanski, I., and M. Gyllenberg. 1997. Uniting two general patterns in the distribution of species. *Science* **275**:397–400.

Hanski, I., T. Pakkala, M. Kuussaari, and G. C. Lei. 1995. Metapopulation persistence of an endangered butterfly in a fragmented landscape. *Oikos* **72**:21–28.

Hanski, I., and E. Ranta. 1983. Coexistence in a patchy environment: Three species of *Daphnia* in rock pools. *Journal of Animal Ecology* **52**:263–279.

Hanski, I., G. A. Zurita, M. I. Bellocq, and J. Rybicki. 2013. Species-fragmented area relationship. *Proceedings of the National Academy of Sciences USA* **110**:12715–12720.

Hanson, C. A., J. A. Fuhrman, M. C. Horner-Devine, and J. B. Martiny, 2012. Beyond biogeographic patterns: Processes shaping the microbial landscape. *Nature Reviews Microbiology* **10**:497–506.

Hardin, G. 1960. The competitive exclusion principle. *Science* **131**:1292–1297.

Hardy, O. J. 2008. Testing the spatial phylogenetic structure of local communities: Statistical performances of different null models and test statistics on a locally neutral community. *Journal of Ecology* **96**:914–926.

Harmon, L. J., and S. Harrison. 2015. Species diversity is dynamic and unbounded at local and continental scales. *American Naturalist* **185**:584–593.

Harms, K. E., R. Condit, S. P. Hubbell, and R. B. Foster. 2001. Habitat associations of trees and shrubs in a 50-ha neotropical forest plot. *Journal of Ecology* **89**:947–959.

Harpole, W. S., and D. Tilman. 2006. Non-neutral patterns of species abundance in grassland communities. *Ecology Letters* **9**:15–23.

Harpole, W. S., and D. Tilman. 2007. Grassland species loss resulting from reduced niche dimension. *Nature* **446**:791–793.

Harrison, R. D., S. Tan, J. B. Plotkin, F. Slik, M. Detto, T. Brenes, A. Itoh, and S. J. Davies. 2013. Consequences of defaunation for a tropical tree community. *Ecology Letters* **16**:687–694.

Harrison, S. 1991. Local extinction in a metapopulation context: An empirical evaluation. *Biological Journal of the Linnean Society* **42**:73–88.

Harrison, S., H. Cornell, and K. A. Moore. 2010. Spatial niches and coexistence: Testing theory with tarweeds. *Ecology* **91**:2141–2150.

Harrison, S., and Grace, J. B., 2007. Biogeographic affinity helps explain productivity-richness relationships at regional and local scales. *American Naturalist* **170**: S5–S15.

Harrison, S., D. D. Murphy, and P. R. Ehrlich. 1988. Distribution of the Bay checkerspot butterfly, *Euphydryas editha bayensis*: Evidence for a metapopulation model. *American Naturalist* **132**:360–382.

Harte, J. 2011. *Maximum entropy and ecology: A theory of abundance, distribution, and energetics.* Oxford University Press, Oxford.

Harte, J., and A. P. Kinzig. 1997. On the implications of species-area relationships for endemism, spatial turnover and food web patterns. *Oikos* **80**:417–427.

Harte, J., and E. A. Newman. 2014. Maximum information entropy: A foundation for ecological theory. *Trends in Ecology and Evolution* **29**:384–389.

Harte, J., A. B. Smith, and D. Storch. 2009. Biodiversity scales from plots to biomes with a universal species-area curve. *Ecology Letters* **12**:789–797.

Harte, J., T. Zillio, E. Conlisk, and A. B. Smith. 2008. Maximum entropy and the state-variable approach to macroecology. *Ecology* **89**:2700–2711.

Harvey, P. H., R. K. Colwell, J. W. Silvertown, and R. M. May. 1983. Null models in ecology. *Annual Review of Ecology and Systematics* **14**:189–211.

Hastings, A. 1977. Spatial heterogeneity and the stability of predator-prey systems. *Theoretical Population Biology* **12**:37–48.

Hastings, A. 1980. Disturbance, coexistence, history, and competition for space. *Theoretical Population Biology* **18**:363–373.

Hatosy, S. M., J. B. Martiny, R. Sachdeva, J. Steele, J. A. Fuhrman, and A. C. Martiny. 2013. Beta diversity of marine bacteria depends on temporal scale. *Ecology* **94**:1898–1904.

Hatton, I. A., K. S. McCann, J. M. Fryxell, T. J. Davies, M. Smerlak, A.R.E. Sinclair, and M. Loreau. 2015. The predator-prey power law: Biomass scaling across terrestrial and aquatic biomes. *Science* **349**:aac6284.

Havens, K. E. 1993. Pelagic food web structure in Adironkack Mountain, USA, lakes of varying acidity. *Canadian Journal of Fisheries and Aquatic Sciences* **50**:149–155.

Hawkins, B. A., R. Field, H. V. Cornell, D. J. Currie, J. F. Guegan, D. M. Kaufman, J. T. Kerr, et al. 2003. Energy, water, and broad-scale geographic patterns of species richness. *Ecology* **84**:3105–3117.

He, F. L., and S. P. Hubbell. 2011. Species-area relationships always overestimate extinction rates from habitat loss. *Nature* **473**:368–371.

He, F. L., and P. Legendre. 2002. Species diversity patterns derived from species-area models. *Ecology* **83**:1185–1198.

Heinlein, J. M., A. C. Stier, and M. A. Steele. 2010. Predators reduce abundance and species richness of coral reef fish recruits via non-selective predation. *Coral Reefs* **29**:527–532.

Heino, J. 2013a. Does dispersal ability affect the relative importance of environmental control and spatial structuring of littoral macroinvertebrate communities? *Oecologia* **171**:971–980.

Heino, J. 2013b. The importance of metacommunity ecology for environmental assessment research in the freshwater realm. *Biological Reviews* **88**:166–178.

Heino, J., T. Nokela, J. Soininen, M. Tolkkinen, L. Virtanen, and R. Virtanen. 2015. Elements of metacommunity structure and community-environment relationships in stream organisms. *Freshwater Biology* **60**:973–988.

Helmus, M. R., and A. R. Ives. 2012. Phylogenetic diversity-area curves. *Ecology* **93**:S31–S43.

Henderson, P. A., and A. E. Magurran. 2010. Linking species abundance distributions in numerical abundance and biomass through simple assumptions about community structure. *Proceedings of the Royal Society. Series B: Biological Sciences* **277**:1561–1570.

Hendry, A. P. 2016. *Eco-evolutionary dynamics*. Princeton University Press, Princeton, NJ.

Hendry, A. P., and M. T. Kinnison. 1999. Perspective: The pace of modern life: Measuring rates of contemporary microevolution. *Evolution* **53**:1637–1653.

Henriques-Silva, R., Z. Lindo, and P. R. Peres-Neto. 2013. A community of metacommunities: exploring patterns in species distributions across large geographical areas. *Ecology* **94**:627–639.

Herkert, J. R. 1994. The effects of habitat fragmentation on midwestern grassland bird communities. *Ecological Applications* **4**:461–471.

Hibbing, M. E., C. Fuqua, M. R. Parsek, and S. B. Peterson. 2010. Bacterial competition: Surviving and thriving in the microbial jungle. *Nature Reviews Microbiology* **8**:15–25.

Higgs, A. J., and M. B. Usher. 1980. Should nature reserves be large or small. *Nature* **285**:568–569.

Hill, J. L., P. J. Curran, and G. M. Foody. 1994. The effect of sampling on the species-area curve. *Global Ecology and Biogeography Letters* **4**:97–106.

Hillebrand, H. 2004. On the generality of the latitudinal diversity gradient. *American Naturalist* **163**:192–211.

Hillebrand, H., D. S. Gruner, E. T. Borer, M.E.S. Bracken, E. E. Cleland, J. J. Elser, W. S. Harpole, et al. 2007. Consumer versus resource control of producer diversity depends on ecosystem type and producer community structure. *Proceedings of the National Academy of Sciences USA* **104**:10904–10909.

HilleRisLambers, J., P. Adler, W. Harpole, J. Levine, and M. Mayfield. 2012. Rethinking community assembly through the lens of coexistence theory. *Annual Review of Ecology, Evolution, and Systematics* **43**:227–248.

Holling, C. S. 1992. Cross-scale morphology, geometry, and dynamics of ecosystems. *Ecological Monographs* **62**:447–502.

Holt, B. G., J.-P. Lessard, M. K. Borregaard, S. A. Fritz, M. B. Araujo, D. Dimitrov, P. H. Fabre, et al. 2013. An update of Wallace's zoogeographic regions of the world. *Science* **339**:74–78.

Holt, R. D. 1984. Spatial heterogeneity, indirect interactions, and the coexistence of prey species. *American Naturalist* **124**:377–406.

Holt, R. D. 1985. Population dynamics in two-patch environments: Some anomalous consequences of an optimal habitat distribution. *Theoretical Population Biology* **28**:181–208.

Holt, R. D. 1993. Ecology at the mesoscale: The influence of regional processes on local communities. Pages 77–88 *in* R. E. Ricklefs and D. Schluter, editors, *Species diversity in ecological communities: Historical and geographical perspectives.* University of Chicago Press, Chicago.

Holt, R. D. 1996. Food webs in space: An island biogeographic perspective. Pages 313–323 *in* G. A. Polis and K. O. Winemiller, editors, *Food webs.* Springer, New York.

Holt, R. D. 1997. From metapopulation dynamics to community structure: some consequences of spatial heterogeneity. Pages 149–164 *in* M. E. Gilpin, editor, *Metapopulation biology: Ecology, genetics and evolution.* Academic Press, San Diego, CA.

Holt, R. D. 2002. Food webs in space: On the interplay of dynamic instability and spatial processes. *Ecological Research* **17**:261–273.

Holt, R. D. 2009. Toward a trophic island biogeography. Reflections on the interface of island biogeography of island biogeography and food web ecology. Pages 143–185 *in* J. B. Losos and R. E. Ricklefs, editors, *The theory of island biogeography revisited.* Princeton University Press, Princeton, NJ.

Holt, R. D., and M. Barfield. 2003. Impacts of temporal variation on apparent competition and coexistence in open ecosystems. *Oikos* **101**:49–58.

Holt, R. D., J. Grover, and D. Tilman. 1994. Simple rules for interspecific dominance in systems with exploitative and apparent competition. *American Naturalist* **144**:741–771.

Holt, R. D., and M. F. Hoopes. 2005. Food web dynamics in a metacommunity context. Pages 68–93 *in* M. Holyoak, M. Leibold, and R. D. Holt, editors, *Metacommunities: Spatial dynamics and ecological communities.* University of Chicago Press, Chicago.

Holt, R. D., T. M. Knight, and M. Barfield. 2004. Allee effects, immigration, and the evolution of species' niches. *American Naturalist* 163:253–262.

Holt, R. D., J. H. Lawton, G. A. Polis, and N. D. Martinez. 1999. Trophic rank and the species-area relationship. *Ecology* **80**:1495–1504.

Holyoak, M. 2000. Habitat subdivision causes changes in food web structure. *Ecology Letters* **3**:509–515.

Holyoak, M., and S. K. Heath. 2016. The integration of climate change, spatial dynamics, and habitat fragmentation: A conceptual overview. *Integrative Zoology* **11**:40–59.

Holyoak, M., and S. P. Lawler. 1996. Persistence of an extinction-prone predator-prey interaction through metapopulation dynamics. *Ecology* **77**:1867–1879.

Holyoak, M., M. A. Leibold, and R. D. Holt. 2005. *Metacommunities: Spatial dynamics and ecological communities.* University of Chicago Press, Chicago.

Hooper, D. U., E. C. Adair, B. J. Cardinale, J.E.K. Byrnes, B. A. Hungate, K. L. Matulich, A. Gonzalez, J. E. Duffy, L. Gamfeldt, and M. I. O'Connor. 2012. A global synthesis reveals biodiversity loss as a major driver of ecosystem change. *Nature* **486**:105–108.

Horn, H. S., and R. MacArthur. 1972. Competition among fugitive species in a harlequin environment. *Ecology* **53**:749–752.

Hortal, J., L. M. Carrascal, K. A. Triantis, E. Thebaulte, S. Meiri, and S. Sfenthourakis. 2013. Species richness can decrease with altitude but not with habitat diversity. *Proceedings of the National Academy of Sciences USA* **110**:E2149–E2150.

Hortal, J., K. A. Triantis, S. Meiri, E. Thébault, and S. Sfenthourakis. 2009. Island species richness increases with habitat diversity. *American Naturalist* **174**:E205–E217.

Hoverman, J. T., C. J. Davis, E. E. Werner, D. K. Skelly, R. A. Relyea, and K. L. Yurewicz. 2011. Environmental gradients and the structure of freshwater snail communities. *Ecography* **34**:1049–1058.

Howeth, J. G., and M. A. Leibold. 2008. Planktonic dispersal dampens temporal trophic cascades in pond metacommunities. *Ecology Letters* **11**:245–257.

Howeth, J. G., and M. A. Leibold. 2010a. Prey dispersal rate affects prey species composition and trait diversity in response to multiple predators in metacommunities: Multiple predators in prey metacommunities. *Journal of Animal Ecology* **79**:1000–1011.

Howeth, J. G., and M. A. Leibold. 2010b. Species dispersal rates alter diversity and ecosystem stability in pond metacommunities. *Ecology* **91**:2727–2741.

Hoyle, M., and F. Gilbert. 2004. Species richness of moss landscapes unaffected by short-term fragmentation. *Oikos* **105**:359–367.

Hubbell, S. P. 1979. Tree dispersion, abundance, and diversity in a tropical dry forest. *Science* **203**:1299–1309.

Hubbell, S. P. 1997. A unified theory of biogeography and relative species abundance and its application to tropical rain forests and coral reefs. *Coral Reefs* **16**:S9–S21.

Hubbell, S. P. 2001. *The unified neutral theory of biodiversity and biogeography.* Princeton University Press, Princeton, NJ.

Hubbell, S. P. 2006. Neutral theory and the evolution of ecological equivalence. *Ecology* **87**:1387–1398.

Hubbell, S. P., and R. B. Foster. 1986. Biology, chance, and history and the structure of tropical rain forest tree communities. Pages 314–329 *in* J. Diamond and T. J. Case, editors, *Community ecology.* Harper and Row, New York.

Hubbell, S. P., R. B. Foster, S. T. O'Brien, K. E. Harms, R. Condit, B. Wechsler, S. J. Wright, and S. L. de Lao. 1999. Light-gap disturbances, recruitment limitation, and tree diversity in a neotropical forest. *Science* **283**:554–557.

Hubert, N., E. Paradis, H. Bruggemann, and S. Planes. 2011. Community assembly and diversification in Indo-Pacific coral reef fishes. *Ecology and Evolution* **1**:229–277.

Huffaker, C. 1958. Experimental studies on predation: Dispersion factors and predator-prey oscillations. *Hilgardia* **27**:343–383.

Hughes, A. R., B. D. Inouye, M.T.J. Johnson, N. Underwood, and M. Vellend. 2008. Ecological consequences of genetic diversity. *Ecology Letters* **11**:609–623.

Hugueny, B., H. V. Cornell, and S. Harrison. 2007. Metacommunity models predict the local-regional species richness relationship in a natural system. *Ecology* **88**:1696–1706.

Huisman, J., and F. J. Weissing. 1999. Biodiversity of plankton by species oscillations and chaos. *Nature* **402**:407–410.

Hurlbert, A. H. 2004. Species-energy relationships and habitat complexity in bird communities. *Ecology Letters* **7**:714–720.

Hurlbert, A. H., and W. Jetz. 2010. More than "more individuals": The nonequivalence of area and energy in the scaling of species richness. *American Naturalist* **176**:E50–E65.

Hurtt, G. C., and S. W. Pacala. 1995. The consequences of recruitment limitation: Reconciling chance, history and competitive differences between plants. *Journal of Theoretical Biology* **176**:1–12.

Huston, M. A. 1997. Hidden treatments in ecological experiments: Re-evaluating the ecosystem function of biodiversity. *Oecologia* **110**:449–460.

Hutchinson, G. E. 1951. Copepodology for the ornithologist. *Ecology* **32**:571–577.

Hutchinson, G. E. 1957. Concluding remarks. *Cold Spring Harbor Symposium on Quantitative Biology* **22**:415–421.

Hutchinson, G. E. 1959. Homage to Santa Rosalia, or why are there so many kinds of animals? *American Naturalist* **93**:145–159.

Hutchinson, G. E. 1961. The paradox of the plankton. *American Naturalist* **95**:137–145.

Hutchinson, G. E. 1965. *The ecological theater and the evolutionary play.* Yale University Press, New Haven, CT.

Hyatt, L. A., M. S. Rosenberg, T. G. Howard, G. Bole, W. Fang, J. Anastasia, K. Brown, et al. 2003. The distance dependence prediction of the Janzen-Connell hypothesis: a meta-analysis. *Oikos* **103**:590–602.

Ibanez, C., J. Belliard, R. M. Hughes, P. Irz, A. Kamdem-Toham, N. Lamouroux, P. A. Tedesco, and T. Oberdorff. 2009. Convergence of temperate and tropical stream fish assemblages. *Ecography* **32**:658–670.

Ibarra, J. T., and K. Martin. 2015. Biotic homogenization: Loss of avian functional richness and habitat specialists in disturbed Andean temperate forests. *Biological Conservation* **192**:418–427.

Ingram, T., and J. B. Shurin. 2009. Trait-based assembly and phylogenetic structure in northeast Pacific rockfish assemblages. *Ecology* **90**:2444–2453.

Ings, T. C., J. M. Montoya, J. Bascompte, N. Bluthgen, L. Brown, C. F. Dormann, F. Edwards, et al. 2009. Ecological networks—Beyond food webs. *Journal of Animal Ecology* **78**:253–269.

Isbell, F., V. Calcagno, A. Hector, J. Connolly, W. S. Harpole, P. B. Reich, M. Scherer-Lorenzen, B. Schmid, D. Tilman, and J. van Ruijven. 2011. High plant diversity is needed to maintain ecosystem services. *Nature* **477**:199–202.

Isbell, F., D. Tilman, S. Polasky, and M. Loreau. 2015. The biodiversity-dependent ecosystem service debt. *Ecology Letters* **18**:119–134.

Ishii, Y., and M. Shimada. 2012. Learning predator promotes coexistence of prey species in host–parasitoid systems. *Proceedings of the National Academy of Sciences USA* **109**:5116–5120.

Jabot, F., and J. Chave. 2009. Inferring the parameters of the neutral theory of biodiversity using phylogenetic information and implications for tropical forests. *Ecology Letters* **12**:239–248.

Jackrel, S. L., T. C. Morton, and J. T. Wootton. 2016. Intraspecific leaf chemistry drives locally accelerated ecosystem function in aquatic and terrestrial communities. *Ecology* **97**:2125–2135

Jamil, T., C. Kruk, and C.J.F. ter Braak. 2014. A unimodal species response model relating traits to environment with application to phytoplankton communities. *PLoS One* **9**:e97583.

Janzen, D. H. 1970. Herbivores and the number of tree species in tropical forests. *American Naturalist* **104**:501–528.

Jean, K., W. R. Burnside, L. Carlson, K. Smith, and J.-F. Guegan. 2016. An equilibrium theory signature in the island biogeography of human parasites and pathogens. *Global Ecology and Biogeography* **25**:107–116.

Jeffries, M. J. 2002. Evidence for individualistic species assembly creating convergent predator:prey ratios among pond invertebrate communities. *Journal of Animal Ecology* **71**:173–184.

Jenkins, D. G., and R. E. Ricklefs. 2011. Biogeography and ecology: Two views of one world. *Philosophical Transactions of the Royal Society B: Biological sciences* **366**:2331–2335.

Jeppesen, E., M. Sondergaard, J. P. Jensen, K. E. Havens, O. Anneville, L. Carvalho, M. F. Coveney, et al. 2005. Lake responses to reduced nutrient loading—An analysis of contemporary long-term data from 35 case studies. *Freshwater Biology* **50**: 1747–1771.

Jeppesen, E., M. Søndergaard, M. Meerhoff, T. L. Lauridsen, and J. P. Jensen. 2007. Shallow lake restoration by nutrient loading reduction—Some recent findings and challenges ahead. *Hydrobiologia* **584**:239–252.

Jetz, W., and P.V.A. Fine. 2012. Global gradients in vertebrate diversity predicted by historical area-productivity dynamics and contemporary environment. *PLoS Biology* **10**:11.

Jetz, W., C. H. Sekercioglu, and K. Bohning-Gaese. 2008. The worldwide variation in avian clutch size across species and space. *PLoS Biology* **6**:2650–2657.

Jiang, L., and S. N. Patel. 2008. Community assembly in the presence of disturbance: A microcosm experiment. *Ecology* **89**:1931–1940.

John, R., J. W. Dalling, K. E. Harms, J. B. Yavitt, R. F. Stallard, M. Mirabello, S. P. Hubbell, et al. 2007. Soil nutrients influence spatial distributions of tropical tree species. *Proceedings of the National Academy of Sciences USA* **104**:864–869.

Johnson, P.T.J., J. T. Hoverman, V. J. McKenzie, A. R. Blaustein, and K.L.D. Richgels. 2013. Urbanization and wetland communities: Applying metacommunity theory to understand the local and landscape effects. *Journal of Applied Ecology* **50**:34–42.

Johnston, N. K., Z. Pu, and L. Jiang. 2016. Predator identity influences metacommunity assembly. *Journal of Animal Ecology* **85**:161–170.

Jones, C. G., J. H. Lawton, and M. Shachak. 1994. Organisms as ecosystem engineers. *Oikos* 69:373–386.

Jones, M. M., S. Ferrier, R. Condit, G. Manion, S. Aguilar, and R. Pérez. 2013. Strong congruence in tree and fern community turnover in response to soils and climate in central Panama. *Journal of Ecology* **101**:506–516.

Jost, L. 2007. Partitioning diversity into independent alpha and beta components. *Ecology* **88**:2427–2439.

Kadmon, R., and O. Allouche. 2007. Integrating the effects of area, isolation, and habitat heterogeneity on species diversity: A unification of island biogeography and niche theory. *American Naturalist* **170**:443–454.

Kadmon, R., and K. Tielborger. 1999. Testing for source-sink population dynamics: An experimental approach exemplified with desert annuals. *Oikos* **86**:417–429.

Kallimanis, A. S., A. D. Mazaris, J. Tzanopoulos, J. M. Halley, J. D. Pantis, and S. P. Sgardelis. 2008. How does habitat diversity affect the species-area relationship? *Global Ecology and Biogeography* **17**:532–538.

Kalyuzhny, M., E. Seri, R. Chocron, C. H. Flather, R. Kadmon, and N. M. Shnerb. 2014. Niche versus neutrality: A dynamical analysis. *American Naturalist* **184**: 439–446.

Kardol, P., N. J. Cornips, M. M. van Kempen, J. T. Bakx-Schotman, and W. H. van der Putten. 2007. Microbe-mediated plant-soil feedback causes historical contingency effects in plant community assembly. *Ecological Monographs* **77**:147–162.

Karlson, R. H., H. V. Cornell, and T. P. Hughes. 2004. Coral communities are regionally enriched along an oceanic biodiversity gradient. *Nature* **429**:867–870.

Kaspari, M., B. S. Stevenson, J. Shik, and J. F. Kerekes. 2010. Scaling community structure: How bacteria, fungi, and ant taxocenes differentiate along a tropical forest floor. *Ecology* **91**:2221–2226.

Katano, I., H. Doi, B. K. Eriksson, and H. Hillebrand. 2015. A cross-system meta-analysis reveals coupled predation effects on prey biomass and diversity. *Oikos* **124**:1427–1435.

Keddy, P. A. 1992. Assembly and response rules: Two goals for predictive community ecology. *Journal of Vegetation Science* 3:157–164.

Keil, P., O. Schweiger, I. Kuhn, W. E. Kunin, M. Kuussaari, J. Settele, K. Henle, et al. 2012. Patterns of beta diversity in Europe: The role of climate, land cover and distance across scales. *Journal of Biogeography* **39**:1473–1486.

Keil, P., D. Storch, and W. Jetz. 2015. On the decline of biodiversity due to area loss. *Nature Communications* **6**:8837. doi:10.1038/ncomms9837.

Keith, S. A., A. C. Newton, M. D. Morecroft, D. J. Golicher, and J. M. Bullock. 2011. Plant metacommunity structure remains unchanged during biodiversity loss in English woodlands. *Oikos* **120**:302–310.

Keitt, T. H., M. A. Lewis, and R. D. Holt. 2001. Allee effects, invasion pinning, and species' borders. *American Naturalist* **157**:203–216.

Kelly, C. K., M. G. Bowler, O. Pybus, and P. H. Harvey. 2008. Phylogeny, niches, and relative abundance in natural communities. Ecology **89**:962–970.

Kembel, S. W. 2009. Disentangling niche and neutral influences on community assembly: Assessing the performance of community phylogenetic structure tests. *Ecology Letters* **12**:949–960.

Kembel, S. W., and S. P. Hubbell. 2006. The phylogenetic structure of a neotropical forest tree community. *Ecology* **87**:S86–S99.

Kerr, B., M. A. Riley, M. W. Feldman, and B. J. Bohannan. 2002. Local dispersal promotes biodiversity in a real-life game of rock–paper–scissors. *Nature* **418**:171–174.

Khalig, I., S. A. Fritz, R. Prinzinger, M. Pfenninger, K. Böhning-Gaese, and C. Hof. 2015. Global variation in thermal physiology of birds and mammals: evidence for phylogenetic niche conservatism only in the tropics. *Journal of Biogeography* **42**:2187–2196.

Kimura, M. 1968. Evolutionary rate at the molecular level. *Nature* **217**:624–626.

Kimura, M. 1983. *The neutral theory of molecular evolution.* Cambridge University Press, Cambridge.

Kinzig, A. P., and J. Harte. 2000. Implications of endemics-area relationships for estimates of species extinctions. *Ecology* **81**:3305–3311.

Kitzes, J., and J. Harte. 2014. Beyond the species-area relationship: Improving macroecological extinction estimates. *Methods in Ecology and Evolution* **5**:1–8.

Kleyer, M., S. Dray, F. Bello, J. Lepš, R. J. Pakeman, B. Strauss, W. Thuiller, and S. Lavorel. 2012. Assessing species and community functional responses to environmental gradients: Which multivariate methods? *Journal of Vegetation Science* **23**: 805–821.

Klironomos, J. N. 2002. Feedback with soil biota contributes to plant rarity and invasiveness in communities. *Nature* **417**:67–70.

Kneitel, J. M., and J. M. Chase. 2004. Disturbance, predator and resource interactions alter container community composition. *Ecology* **85**:2088–2093.

Kneitel, J. M., and T. E. Miller. 2003. Dispersal rates affect species composition in meta-communities of *Sarracenia purpurea* inquilines. *American Naturalist* **162**:165–171.

Knight, T. M., M. W. McCoy, J. M. Chase, K. A. McCoy, and R. D. Holt. 2005. Trophic cascades across ecosystems. *Nature* **437**:880–883.

Koh, L. P., and J. Ghazoul. 2010. A matrix-calibrated species-area model for predicting biodiversity losses due to land-use change. *Conservation Biology* **24**:994–1001.

Kohn, D. D., and D. M. Walsh. 1994. Plant-species richness—The effect of island size and habitat diversity. *Journal of Ecology* **82**:367–377.

Korhonen, J. J., J. Soininen, and H. Hillebrand. 2010. A quantitative analysis of temporal turnover in aquatic species assemblages across ecosystems. *Ecology* **91**:508–517.

Kraft, N.J.B., and D. D. Ackerly. 2010. Functional trait and phylogenetic tests of community assembly across spatial scales in an Amazonian forest. *Ecological Monographs* **80**:401–422.

Kraft, N.J.B., P. B. Adler, O. Godoy, E. C. James, S. Fuller, and J. M. Levine. 2015a. Community assembly, coexistence and the environmental filtering metaphor. *Functional Ecology* **29**:592–599.

Kraft, N.J.B., L. S. Comita, J. M. Chase, N. J. Sanders, N. G. Swenson, T. O. Crist, J. C. Stegen, et al. 2011. Disentangling the drivers of beta diversity along latitudinal and elevational gradients. *Science* **333**:1755–1758.

Kraft, N.J.B., G. M. Crutsinger, E. J. Forrestel, and N. C. Emery. 2014. Functional trait differences and the outcome of community assembly: an experimental test with vernal pool annual plants. *Oikos* **123**:1391–1399.

Kraft, N.J.B., O. Godoy, and J. M. Levine. 2015b. Plant functional traits and the multidimensional nature of species coexistence. *Proceedings of the National Academy of Sciences USA* **112**:797–802.

Kraft, N.J.B., R. Valencia, and D. D. Ackerly. 2008. Functional traits and niche-based tree community assembly in an Amazonian forest. *Science* **322**:580–582.

Kreft, H., W. Jetz, J. Mutke, G. Kier, and W. Barthlott. 2008. Global diversity of island floras from a macroecological perspective. *Ecology Letters* **11**:116–127.

Kremen, C. 2015. Reframing the land-sparing/land-sharing debate for biodiversity conservation. *Annals of the New York Academy of Sciences* **1355**:52–76.

Kruess, A., and T. Tscharntke. 1994. Habitat fragmentation, species loss, and biological control. *Science* **264**:1581–1584.

Kruess, A., and T. Tscharntke. 2000. Species richness and parasitism in a fragmented landscape: Experiments and field studies with insects on *Vicia sepium*. *Oecologia* **122**:129–137.

Kulmatiski, A., K. H. Beard, J. R. Stevens, and S. M. Cobbold. 2008. Plant-soil feedbacks: A meta-analytical review. *Ecology Letters* **11**:980–992.

Kunin, W. E. 1998. Biodiversity at the edge: A test, of the importance of spatial "mass effects" in the Rothamsted Park Grass experiments. *Proceedings of the National Academy of Sciences USA* **95**:207–212.

Kunstler, G., D. S. Falster, D. A. Coomes, F. Hui, R. M. Kooyman, D. C. Laughlin, L. Poorter, et al. 2016. Plant functional traits have globally consistent effect on competition. *Nature* **529**:204–207.

Kunstler, G., S. Lavergne, B. Courbaud, W. Thuiller, G. Vieilledent, N. E. Zimmermann, J. Kattge, and D. A. Coomes. 2012. Competitive interactions between forest trees are driven by species' trait hierarchy, not phylogenetic or functional similarity: Implications for forest community assembly. *Ecology Letters* **15**:831–840.

Lai, J., X. Mi, H. Ren, and K. Ma. 2009. Species-habitat associations change in a subtropical forest of China. *Journal of Vegetation Science* **20**:415–423.

Lamouroux, N., N. L. Poff, and P. L. Angermeier. 2002. Intercontinental convergence of stream fish community traits along geomorphic and hydraulic gradients. *Ecology* **83**:1792–1807.

Lan, Z., G. D. Jenerette, S. Zhan, W. Li, S. Zheng, and Y. Bai. 2015. Testing the scaling effects and mechanisms of N-induced biodiversity loss: Evidence from a decade-long grassland experiment. *Journal of Ecology* **103**:750–760.

Lankau, R. A. 2011. Resistance and recovery of soil microbial communities in the face of *Alliaria petiolata* invasions. *New Phytologist* **189**:536–548.

Lankau, R. A., and S. Y. Strauss. 2007. Mutual feedbacks maintain both genetic and species diversity in a plant community. *Science* **317**:1561–1563.

Lavorel, S., S. Díaz, J. H. Cornelissen, E. Garnier, S. P. Harrison, S. McIntyre, J. G. Pausas, N. Perez-Harguindeguy, C. Roumet, and C. Urcelay. 2007. Plant functional types: Are we getting any closer to the Holy Grail? Pages 149–164 *in* J. G. Canadell, D. E. Pataki, and L. F. Pitelka, editors, *Terrestrial ecosystems in a changing world*. Springer, New York.

Lavorel, S., and E. Garnier. 2002. Predicting changes in community composition and ecosystem functioning from plant traits: Revisiting the Holy Grail. *Functional Ecology* **16**:545–556.

Law, R., and M. Leibold. 2005. Assembly dynamics in metacommunities. Pages 263–278 *in* M. Holyoak, M. Leibold, and R. Holt, editors, *Metacommunities: Spatial dynamics and ecological communities*. University of Chicago Press, Chicago.

Law, R., and R. D. Morton. 1993. Alternative permanent states of ecological communities. *Ecology* **74**:1347–1361.

Law, R., and R. D. Morton. 1996. Permanence and the assembly of ecological communities. *Ecology* **77**:762–775.

Lawlor, L. R. 1979. Direct and indirect effects of n-species competition. *Oecologia* **43**:355–364.

Lawton, J. H. 1999. Are there general laws in ecology? *Oikos* **84**:177–192.

LeCraw, R. M., P. Kratina, and D. S. Srivastava. 2014. Food web complexity and stability across habitat connectivity gradients. *Oecologia* **176**:903–915.

Legendre, P., and M. De Caceres. 2013. Beta diversity as the variance of community data: Dissimilarity coefficients and partitioning. *Ecology Letters* **16**:951–963.

Legendre, P., R. Galzin, and M. L. HarmelinVivien. 1997. Relating behavior to habitat: Solutions to the fourth-corner problem. *Ecology* **78**:547–562.

Legendre, P., and L. F. Legendre. 2012. *Numerical ecology*. 3rd ed. Elsevier, Amsterdam.

Legendre, P., X. Mi, H. Ren, K. Ma, M. Yu, I.-F. Sun, and F. He. 2009. Partitioning beta diversity in a subtropical broad-leaved forest of China. *Ecology* **90**:663–674.

Lei, G. C., and I. Hanski. 1998. Spatial dynamics of two competing specialist parasitoids in a host metapopulation. *Journal of Animal Ecology* **67**:422–433.

Leibold, M. A. 1995. The niche concept revisited: mechanistic models and community context. *Ecology* **76**:1371–1382.

Leibold, M. A. 1996. A graphical model of keystone predators in food webs: Trophic regulation of abundance, incidence, and diversity patterns in communities. *American
Naturalist* **147**:784–812.

Leibold, M. A. 1998. Similarity and local co-existence of species in regional biotas. *Evolutionary Ecology* **12**:95–110.

Leibold, M. A. 1999. Biodiversity and nutrient enrichment in pond plankton communities.
Evolutionary Ecology Research **1**:73–95.

Leibold, M. A. 2011. The metacommunity concept and its theoretical underpinnings. Pages
163–183 *in* S. M. Scheiner and M. R. Willig, editors, *The theory of ecology.* University of Chicago Press, Chicago.

Leibold, M. A., J. M. Chase, and S.K.M. Ernest. 2017. Community assembly and the functioning of ecosystems: How metacommunity processes alter ecosystem attributes.
Ecology **98**:909–919.

Leibold, M. A., J. M. Chase, J. B. Shurin, and A. L. Downing. 1997. Species turnover and
the regulation of trophic structure. *Annual Review of Ecology and Systematics*
28:467–494.

Leibold, M. A., E. P. Economo, and P. Peres-Neto. 2010. Metacommunity phylogenetics:
Separating the roles of environmental filters and historical biogeography. *Ecology
Letters* **13**:1290–1299.

Leibold, M. A., M. Holyoak, N. Mouquet, P. Amarasekare, J. Chase, M. Hoopes, R. Holt,
J. Shurin, R. Law, and D. Tilman. 2004. The metacommunity concept: A framework for multi-scale community ecology. *Ecology Letters* **7**:601–613.

Leibold, M. A., and N. Loeuille. 2015. Species sorting and patch dynamics in harlequin
metacommunities affect the relative importance of environment and space. *Ecology* **96**:3227–3233.

Leibold, M. A., and M. A. McPeek. 2006. Coexistence of the niche and neutral perspectives in community ecology. *Ecology* **87**:1399–1410.

Leibold, M. A., and G. M. Mikkelson. 2002. Coherence, speces turnover, and boundary
clumping: Elements of meta-community structure. *Oikos* **97**:237–250.

Leibold, M. A., and T. E. Miller. 2004. From metapopulations to metacommunities. Pages
133–150 *in* I. Hanski and O. E. Gaggiotti, editors, *Ecology, genetics, and evolution
of metapopulations.* Academic Press, Cambridge, MA.

Leibold, M. A., and J. Norberg. 2004. Biodiversity in metacommunities: Plankton as complex adaptive systems? *Limnology and Oceanography* **49**:1278–1289.

Leibold, M. A., and A. J. Tessier. 1998. Experimental compromise and mechanistic approaches to the evolutionary ecology of interacting *Daphnia* species. Pages 96–112
in W. J. Resetarits and J. Bernardo, editors, *Perspectives in experimental ecology.*
Oxford University Press, Oxford.

Leroux, S. J., and M. Loreau. 2008. Subsidy hypothesis and strength of trophic cascades
across ecosystems. *Ecology Letters* **11**:1147–1156.

Leslie, P. H., T. Park, and D. B. Mertz. 1968. The effect of varying the initial numbers on the
outcome of competition between two *Tribolium* species. *Journal of Animal Ecology*
37:9–23.

Lessard, J. P., J. Belmaker, J. A. Myers, J. M. Chase, and C. Rahbek. 2012. Inferring local
ecological processes amid species pool influences. *Trends in Ecology and Evolution* **27**:600–607.

Lessard, J. P., R. R. Dunn, and N. J. Sanders. 2009. Temperature-mediated coexistence in temperate forest ant communities. *Insectes Sociaux* **56**:149–156.

Lessard, J. P., B. G. Weinstein, M. K. Borregaard, K. A. Marske, D. R. Martin, J. A. Mc-Guire, J. L. Parra, C. Rahbek, and C. H. Graham. 2016. Process-based species pools reveal the hidden signature of biotic interactions amid the influence of temperature. *American Naturalist* **187**:75–88.

Letten, A. D., and W. K. Cornwell. 2015. Trees, branches and (square) roots: why evolutionary relatedness is not linearly related to functional distance. *Methods in Ecology and Evolution* **6**:439–444.

Letten, A. D., P.-J. Ke, and T. Fukami. 2017. Linking modern coexistence theory and contemporary niche theory. *Ecological Monographs* **87**:161–177.

Levi, T., C. T. Darimont, M. MacDuffee, M. Mangel, P. Paquet, and C. C. Wilmers. 2012. Using grizzly bears to assess harvest-ecosystem tradeoffs in salmon fisheries. *PLoS Biology* **10**:e1001303.

Levin, S. A. 1970. Community equilibria and stability, and an extension of the competitive exclusion principle. *American Naturalist* **104**:413–423.

Levin, S. A. 1974. Dispersion and population interactions. *American Naturalist* **108**:207–228.

Levin, S. A. 1992. The problem of pattern and scale in ecology. *Ecology* **73**:1943–1967.

Levin, S. A. 1998. Ecosystems and the biosphere as complex adaptive systems. *Ecosystems* **1**:431–436.

Levin, S. A. 1999. *Fragile dominion: Complexity and the commons.* Perseus Publishing, Cambridge, MA.

Levin, S. A. 2005. Self-organization and the emergence of complexity in ecological systems. *BioScience* **55**:1075–1079.

Levin, S. A., and R. T. Paine. 1974. Disturbance, patch formation, and community structure. *Proceedings of the National Academy of Sciences USA* **71**:2744–2747.

Levine, J. M., and J. HilleRisLambers. 2009. The importance of niches for the maintenance of species diversity. *Nature* **461**:254–257.

Levine, J. M., and D. J. Murrell. 2003. The community-level consequences of seed dispersal patterns. *Annual Review of Ecology, Evolution, and Systematics* **34**:549–574.

Levins, R. 1969. Some demographic and genetic consequences of environmental heterogeneity for biological control. *Bulletin of the Entomological Society of America* **15**:237–240.

Levins, R., and D. Culver. 1971. Regional coexistence of species and competition between rare species. *Proceedings of the National Academy of Sciences USA* **68**:1246–1248.

Lewin, R. 1983. Santa Rosalia was a goat. *Science* **221**:636–639.

Ley, R. E., M. Hamady, C. Lozupone, P. J. Turnbaugh, R. R. Ramey, J. S. Bircher, M. L. Schlegel, et al. 2008. Evolution of mammals and their gut microbes. *Science* **320**:1647–1651.

Li, W.K.W. 2002. Macroecological patterns of phytoplankton in the northwestern North Atlantic Ocean. *Nature* **419**:154–157.

Lindenmayer, D., J. Fischer, A. Felton, R. Montague-Drake, A. D. Manning, D. Simberloff, K. Youngentob, et al. 2007. The complementarity of single-species and ecosystem-oriented research in conservation research. *Oikos* **116**:1220–1226.

Lindström, E. S., and Ö. Östman. 2011. The importance of dispersal for bacterial community composition and functioning. *PLoS One* **6**:e25883.

Litchman, E., and C. A. Klausmeier. 2008. Trait-based community ecology of phytoplankton. *Annual Review of Ecology Evolution and Systematics* **39**:615–639.

Liu, X., N. G. Swenson, J. Zhang, and K. Ma. 2013. The environment and space, not phylogeny, determine trait dispersion in a subtropical forest. *Functional Ecology* **27**:264–272.

Livingston, G., K. Fukumori, D. Provete, M. Kawachi, N. Takamura, and M. A. Leibold. 2017. Predators regulate prey species sorting and spatial distribution in microbial landscapes. *Journal of Animal Ecology* **86**:501–510.

Livingston, G., M. Matias, V. Calcagno, C. Barbera, M. Combe, M. A. Leibold, and N. Mouquet. 2012. Competition-colonization dynamics in experimental bacterial metacommunities. *Nature Communications* **3**:1243. doi:10.1038/ncomms2239.

Loeuille, N., and M. A. Leibold. 2008. Evolution in metacommunities: On the relative importance of species sorting and monopolization in structuring communities. *American Naturalist* **171**:788–799.

Loeuille, N., and M. A. Leibold. 2014. Effects of local negative feedbacks on the evolution of species within metacommunities. *Ecology Letters* **17**:563–573.

Logue, J. B., and E. S. Lindström. 2010. Species sorting affects bacterioplankton community composition as determined by 16S rDNA and 16S rRNA fingerprints. *ISME Journal* **4**:729–738.

Logue, J. B., N. Mouquet, H. Peter, and H. Hillebrand. 2011. Empirical approaches to metacommunities: a review and comparison with theory. *Trends in Ecology and Evolution* **26**:482–491.

Lomolino, M. V. 2000. Ecology's most general, yet protean pattern: The species-area relationship. *Journal of Biogeography* **27**:17–26.

Lomolino, M. V., and M. D. Weiser. 2001. Towards a more general species-area relationship: Diversity on all islands, great and small. *Journal of Biogeography* **28**:431–445.

López-González, C., S. J. Presley, A. Lozano, R. D. Stevens, and C. L. Higgins. 2012. Metacommunity analysis of Mexican bats: Environmentally mediated structure in an area of high geographic and environmental complexity. *Journal of Biogeography* **39**:177–192.

Loreau, M. 2000. Are communities saturated? On the relationship between α, β and γ diversity. *Ecology Letters* **3**:73–76.

Loreau, M. 2010. *From populations to ecosystems: Theoretical foundations for a new ecological synthesis.* Princeton University Press, Princeton, NJ.

Loreau, M., and C. de Mazancourt. 2013. Biodiversity and ecosystem stability: A synthesis of underlying mechanisms. *Ecology Letters* **16**:106–115.

Loreau, M., and A. Hector. 2001. Partitioning selection and complementarity in biodiversity experiments. *Nature* **412**:72–76.

Loreau, M., and R. D. Holt. 2004. Spatial flows and the regulation of ecosystems. *American Naturalist* **163**:606–615.

Loreau, M., and N. Mouquet. 1999. Immigration and the maintenance of local species diversity. *American Naturalist* **154**:427–440.

Loreau, M., N. Mouquet, and A. Gonzalez. 2003a. Biodiversity as spatial insurance in heterogeneous landscapes. *Proceedings of the National Academy of Sciences USA* **100**:12765–12770.

Loreau, M., N. Mouquet, and R. D. Holt. 2003b. Meta-ecosystems: A theoretical framework for a spatial ecosystem ecology. *Ecology Letters* **6**:673–679.

Loreau, M., S. Naeem, P. Inchausti, J. Bengtsson, J. P. Grime, A. Hector, D. U. Hooper, et al. 2001. Biodiversity and ecosystem functioning: current knowledge and future challenges. *Science* **294**:804–808.

Losos, J. B. 2008. Phylogenetic niche conservatism, phylogenetic signal and the relationship between phylogenetic relatedness and ecological similarity among species. *Ecology Letters* **11**:995–1003.

Losos, J. 2009. *Lizards in an evolutionary tree: Ecology and adaptive radiation of Anoles.* University of California Press, Los Angeles.

Losos, J. B., and R. E. Ricklefs. 2009. Adaptation and diversification on islands. *Nature* **457**:830–836.

Losos, J. B., and D. Schluter. 2000. Analysis of an evolutionary species-area relationship. *Nature* **408**:847–850.

Lotka, A. J. 1925. *Elements of physical biology.* Williams & Wilkins, Baltimore.

Luckinbill, L. S. 1974. The effects of space and enrichment on a predator-prey system. *Ecology* **55**:1142–1147.

Lukaszewski, Y., S. E. Arnott, and T. M. Frost. 1999. Regional versus local processes in determining zooplankton community composition of Little Rock Lake, Wisconsin, USA. *Journal of Plankton Research* **21**:991–1003.

Lutscher, F., and T. Iljon. 2013. Competition, facilitation and the Allee effect. *Oikos* **122**:621–631.

Luza, A. L., G. L. Gonçalves, and S. M. Hartz. 2015. Phylogenetic and morphological relationships between nonvolant small mammals reveal assembly processes at different spatial scales. *Ecology and Evolution* **5**:889–902.

Lynch, J. F., and N. K. Johnson. 1974. Turnover and equilibria in insular avifaunas, with special reference to California Channel Islands. Condor **76**:370–384.

Mabry, C. M. 2004. The number and size of seeds in common versus restricted woodland herbaceous species in central Iowa, USA. *Oikos* **107**:497–504.

MacArthur, R. 1955. Fluctuations of animal populations and a measure of community stability. *Ecology* **36**:533–536.

MacArthur, R. H. 1957. On the relative abundance of bird species. *Proceedings of the National Academy of Sciences USA* **43**:293–295.

MacArthur, R. H. 1958. Population ecology of some warblers of northeastern coniferous forests. *Ecology* **39**:599–619.

MacArthur, R. H. 1972. *Geographical ecology: patterns in the distribution of species.* Princeton University Press, Princeton, NJ.

MacArthur, R., and R. Levins. 1964. Competition, habitat selection, and character displacement in a patchy environment. *Proceedings of the National Academy of Sciences USA* **51**:1207–1210.

MacArthur, R., and R. Levins. 1967. The limiting similarity, convergence, and divergence of coexisting species. *American Naturalist* **110**:377–385.

MacArthur, R., and J. W. MacArthur. 1961. On bird species diversity. *Ecology* **42**:594–598.

MacArthur, R. H., and E. O. Wilson. 1963. An equilibrium theory of insular zoogeography. *Evolution* **17**:373–387.

MacArthur, R. H., and E. O. Wilson. 1967. *The theory of island biogeography.* Princeton University Press, Princeton, NJ.

MacDougall, A. S., and S. D. Wilson. 2007. Herbivory limits recruitment in an old-field seed addition experiment. *Ecology* **88**:1105–1111.

Mackey, R. L., and D. J. Currie. 2001. The diversity-disturbance relationship: Is it generally strong and peaked? *Ecology* **82**:3479–3492.

Maguire, B. 1963. The passive dispersal of small aquatic organisms and their colonization of isolated bodies of water. *Ecological Monographs* **33**:161–185.

Magurran, A. E., M. Dornelas, F. Moyes, N. J. Gotelli, and B. McGill. 2015. Rapid biotic homogenization of marine fish assemblages. *Nature Communications* **6**:8405. doi:10.1038/ncomms9405.

Malcom, J. W. 2011. Gene networks and metacommunities: Dispersal differences can override adaptive advantage. *PLoS One* **6**:e21541.

Mangan, S. A., S. A. Schnitzer, E. A. Herre, K.M.L. Mack, M. C. Valencia, E. I. Sanchez, and J. D. Bever. 2010. Negative plant-soil feedback predicts tree-species relative abundance in a tropical forest. *Nature* **466**:752–710.

Margalef, R. 1963. Certain unifying principles in ecology. *American Naturalist* **97**:357–374.

Margalef, R. 1978. Life-forms of phytoplankton as survival alternatives in an unstable environment. *Oceanologica Acta* **1**:493–509.

Marini, L., H. H. Bruun, R. K. Heikkinen, A. Helm, O. Honnay, J. Krauss, I. Kühn, R. Lindborg, M. Pärtel, and R. Bommarco. 2012. Traits related to species persistence and dispersal explain changes in plant communities subjected to habitat loss. *Diversity and Distributions* **18**:898–908.

Markl, J. S., M. Schleuning, P. M. Forget, P. Jordano, J. E. Lambert, A. Traveset, S. J. Wright, and K. Boehning-Gaese. 2012. Meta-analysis of the effects of human disturbance on seed dispersal by animals. *Conservation Biology* **26**:1072–1081.

Marleau, J. N., F. Guichard, and M. Loreau. 2014. Meta-ecosystem dynamics and functioning on finite spatial networks. *Proceeds of the Royal Society. Series B: Biological Sciences* **281**:20132094.

Maron, J. L., H. Auge, D. E. Pearson, L. Korell, I. Hensen, K. N. Suding, and C. Stein. 2014a. Staged invasions across disparate grasslands: Effects of seed provenance, consumers and disturbance on productivity and species richness. *Ecology Letters* **17**:499–507.

Maron, J. L., J. Klironomos, L. Waller, and R. M. Callaway. 2014b. Invasive plants escape from suppressive soil biota at regional scales. *Journal of Ecology* **102**:19–27.

Martin, H. G., and N. Goldenfeld. 2006. On the origin and robustness of power-law species-area relationships in ecology. *Proceedings of the National Academy of Sciences USA* **103**:10310–10315.

Martinson, H. M., and W. F. Fagan. 2014. Trophic disruption: A meta-analysis of how habitat fragmentation affects resource consumption in terrestrial arthropod systems. *Ecology Letters* **17**:1178–1189.

Martiny, J.B.H., B. J. Bohannan, J. H. Brown, R. K. Colwell, J. A. Fuhrman, J. L. Green,, M. C. Horner-Devine, et al. 2006. Microbial biogeography: putting microorganisms on the map. *Nature Reviews Microbiology* **4**:102–112.

Massol, F., F. Altermatt, I. Gounand, D. Gravel, M. A. Leibold, and N. Mouquet. 2017. How life-history traits affect ecosystem properties: Effects of dispersal in meta-ecosystems. *Oikos* **126**:532–546.

Massol, F., D. Gravel, N. Mouquet, M. W. Cadotte, T. Fukami, and M. A. Leibold. 2011. Linking community and ecosystem dynamics through spatial ecology: An integrative approach to spatial food webs. *Ecology Letters* **14**:313–323.

Matias, M. G., D. Gravel, F. Guilhaumon, P. Desjardins-Proulx, M. Loreau, T. Munkemuller, and N. Mouquet. 2014. Estimates of species extinctions from species-area relationships strongly depend on ecological context. *Ecography* **37**:431–442.

Matias, M. G., A. J. Underwood, D. F. Hochuli, and R. A. Coleman. 2010. Independent effects of patch size and structural complexity on diversity of benthic macroinvertebrates. *Ecology* **91**:1908–1915.

Matthews, T. J., F. Guilhaumon, K. A. Triantis, M. K. Borregaard, and R. H. Whittaker. 2016. On the form of species-area relationships in habitat islands and true islands. *Global Ecology and Biogeography* **25**:847–858.

Matthews, T. J., M. J. Steinbauer, E. Tzirkalli, K. A. Triantis, and R. J. Whittaker. 2014. Thresholds and the species-area relationship: A synthetic analysis of habitat island datasets. *Journal of Biogeography* **41**:1018–1028.

Maurer, B. A. 1999. *Untangling ecological complexity: The macroscopic perspective*. University of Chicago Press, Chicago.

Maurer, B. A., S. W. Kembel, A. J. Rominger, and B. J. McGill. 2013. Estimating metacommunity extent using data on species abundances, environmental variation, and phylogenetic relationships across geographic space. *Ecological Informatics* **13**:114–122.

May, F., A. Huth, and T. Wiegand. 2015. Moving beyond abundance distributions: Neutral theory and spatial patterns in a tropical forest. *Proceedings of the Royal Society. Series B: Biological Sciences* **282**:20141657. doi: 10.1098/rspb.2014.1657.

May, F., T. Wiegand, S. Lehmann, and A. Huth. 2016. Do abundance distributions and species aggregation correctly predict macroecological biodiversity patterns in tropical forests? *Global Ecology and Biogeography* **25**:575–585.

May, R. M., 1972. Will a large complex system be stable? *Nature* **238**:413–414.

May, R. M. 1973. *Stability and complexity in model ecosystems*. Princeton University Press, Princeton, NJ.

May, R. M. 1975. Patterns of species abundance and diversity. Pages 81–120 *in* M. L. Cody and J. M. Diamond, editors, *Ecology and evolution of communities*. Harvard University Press, Cambridge, MA.

Mayfield, M. M., and J. M. Levine. 2010. Opposing effects of competitive exclusion on the phylogenetic structure of communities. *Ecology Letters* **13**:1085–1093.

Mayr, E., and J. M. Diamond. 2001. *The birds of northern Melanesia*. Oxford University Press New York.

Mazel, F., J. Renaud, F. Guilhaumon, D. Mouillot, D. Gravel, and W. Thuiller. 2015. Mammalian phylogenetic diversity–area relationships at a continental scale. *Ecology* **96**:2814–2822.

McBride, P. D., J. Cusens, and L. N. Gillman. 2014. Revisiting spatial scale in the productivity-species richness relationship: fundamental issues and global change implications. *AoB Plants* **6**:plu057.

McCann, K. S. 2000. The diversity–stability debate. *Nature* **405**:228–233.

McCann, K. S. 2011. *Food webs*. Princeton University Press, Princeton, NJ.

McCann, K. S., J. B. Rasmussen, and J. Umbanhowar. 2005. The dynamics of spatially coupled food webs. *Ecology Letters* **8**:513–523.

McCauley, S. J. 2007. The role of local and regional processes in structuring larval dragonfly distributions across habitat gradients. *Oikos* **116**:121–133.

McCauley, S. J., C. J. Davis, R. A. Relyea, K. L. Yurewicz, D. K. Skelly, and E. E. Werner. 2008. Metacommunity patterns in larval odonates. *Oecologia* **158**:329–342.

McCoy, M. W., M. Barfield, and R. D. Holt. 2009. Predator shadows: Complex life histories as generators of spatially patterned indirect interactions across ecosystems. *Oikos* **118**:87–100.

McGill, B. J. 2003. A test of the unified neutral theory of biodiversity. *Nature* **422**: 881–885.

McGill, B. J. 2010. Towards a unification of unified theories of biodiversity. *Ecology Letters* **13**:627–642.

McGill, B. J. 2011. Linking biodiversity patterns by autocorrelated random sampling. *American Journal of Botany* **98**:481–502.

McGill, B. J., B. Enquist, E. Weiher, and M. Westoby. 2006a. Rebuilding community ecology from functional traits. *Trends in Ecology and Evolution* **21**:178–185.

McGill, B. J., R. S. Etienne, J. S. Gray, D. Alonso, M. J. Anderson, H. K. Benecha, M. Dornelas, et al. 2007. Species abundance distributions: Moving beyond single prediction theories to integration within an ecological framework. *Ecology Letters* **10**:995–1015.

McGill, B. J., B. A. Maurer, and M. D. Weiser. 2006b. Empirical evaluation of neutral theory. *Ecology* **87**:1411–1423.

McGill, B. J., and J. C. Nekola. 2010. Mechanisms in macroecology: AWOL or purloined letter? Towards a pragmatic view of mechanism. *Oikos* **119**:591–603.

McGlinn, D. J., X. Xiao, J. Kitzes, and E. P. White. 2015. Exploring the spatially explicit predictions of the Maximum Entropy Theory of Ecology. *Global Ecology and Biogeography* **24**:675–684.

McIntosh, R. P. 1986. *The background of ecology: Concept and theory.* Cambridge University Press, Cambridge.

McIntosh, R. P. 1998. The myth of community as organism. *Perspectives in Biology and Medicine* **41**:426–438.

McKinney, M. L., and J. L. Lockwood. 1999. Biotic homogenization: A few winners replacing many losers in the next mass extinction. *Trends in Ecology and Evolution* **14**:450–453.

McLachlan, J. S., J. J. Hellmann, and M. W. Schwartz. 2007. A framework for debate of assisted migration in an era of climate change. *Conservation Biology* **21**: 297–302.

McPeek, M. A. 1998. The consequences of changing the top predator in a food web: A comparative experimental approach. *Ecological Monographs* **68**:1–23.

McPeek, M. A. 2008. The ecological dynamics of clade diversification and community assembly. *American Naturalist* **172**:E270–E284.

McPeek, M. A. 2012. Intraspecific density dependence and a guild of consumers coexisting on one resource. *Ecology* **93**:2728–2735.

McPeek, M. A., and J. M. Brown. 2000. Building a regional species pool: Diversification of the Enallagma damselflies in eastern North America. *Ecology* **81**:904–920.

McPeek, M. A., and S. Gavrilets. 2006. The evolution of female mating preferences: Differentiation from species with promiscuous males can promote speciation. *Evolution* **60**:1967–1980.

McQueen, D. J., J. R. Post, and E. L. Mills. 1986. Trophic relationships in freshwater pelagic ecosystems. *Canadian Journal of Fisheries and Aquatic Sciences* **43**:1571–1581.

Menezes, S., D. J. Baird, and A.M.V.M. Soares. 2010. Beyond taxonomy: A review of macroinvertebrate trait-based community descriptors as tools for freshwater biomonitoring. *Journal of Applied Ecology* **47**:711–719.

Meynard, C. N., V. Devictor, D. Mouillot, W. Thuiller, F. Jiguet, and N. Mouquet. 2011. Beyond taxonomic diversity patterns: How do α, β and γ components of bird functional and phylogenetic diversity respond to environmental gradients across France? *Global Ecology and Biogeography* **20**:893–903.

Meynard, C. N., S. Lavergne, I. Boulangeat, L. Garraud, J. Van Es, N. Mouquet, and W. Thuiller. 2013. Disentangling the drivers of metacommunity structure across spatial scales. *Journal of Biogeography* **40**:1560–1571.

Michels, E., K. Cottenie, L. Neys, and L. De Meester. 2001. Zooplankton on the move: First results on the quantification of dispersal of zooplankton in a set of interconnected ponds. *Hydrobiologia* **442**:117–126.

Millennium Ecosystem Assessment. 2005. *Ecosystems and human well-being: Biodiversity synthesis*. World Resources Institute, Washington, DC.

Miller, C. B., D. R. Lynch, F. Carlotti, W. Gentleman, and C. V. Lewis. 1998. Coupling of an individual-based population dynamic model of *Calanus finmarchicus* to a circulation model for the Georges Bank region. *Fisheries Oceanography* **7**:219–234.

Miller, T. E., J. M. Kneitel, and J. H. Burns. 2002. Effect of community structure on invasion success and rate. *Ecology* **83**:898–905.

Miller, T. E., E. R. Moran, and C. P. terHorst. 2014. Rethinking niche evolution: Experiments with natural communities of protozoa in pitcher plants. *American Naturalist* **184**:277–283.

Miller, T. E., and C. P. terHorst. 2012. Testing successional hypotheses of stability, heterogeneity, and diversity in pitcher-plant inquiline communities. *Oecologia* **170**: 243–251.

Miller, T. J., P. F. Quintana-Ascencio, S. Maliakal-Witt, and E. S. Menges. 2012. Metacommunity dynamics over 16 Years in a pyrogenic shrubland. *Conservation Biology* **26**:357–366.

Mims, M. C., and J. D. Olden. 2012. Life history theory predicts fish assemblage response to hydrologic regimes. *Ecology* **93**:35–45.

Mittelbach, G. G. 2012. *Community ecology*. Sinauer Associates, Sunderland, MA.

Mittelbach, G. G., and D. W. Schemske. 2015. Ecological and evolutionary perspectives on community assembly. *Trends in Ecology and Evolution* **30**:241–247.

Mittelbach, G. G., D. W. Schemske, H. V. Cornell, A. P. Allen, J. M. Brown, M. B. Bush, S. P. Harrison, et al. 2007. Evolution and the latitudinal diversity gradient: Speciation, extinction and biogeography. *Ecology Letters* **10**:315–331.

Mittelbach, G. G., C. F. Steiner, S. M. Scheiner, K. L. Gross, H. L. Reynolds, R. B. Waide, M. R. Willig, S. I. Dodson, and L. Gough. 2001. What is the observed relationship between species richness and productivity? *Ecology* **82**:2381–2396.

Moles, A. T., and M. Westoby. 2002. Seed addition experiments are more likely to increase recruitment in larger-seeded species. *Oikos* **99**:241–248.

Molofsky, J., and J. D. Bever. 2002. A novel theory to explain species diversity in landscapes: Positive frequency dependence and habitat suitability. *Proceedings of the Royal Society. Series B: Biological Sciences* **269**:2389–2393.

Molofsky, J., J. D. Bever, and J. Antonovics. 2001. Coexistence under positive frequency dependence. *Proceedings of the Royal Society. Series B: Biological Sciences* **268**:273–277.

Molofsky, J., J. D. Bever, J. Antonovics, and T. J. Newman. 2002. Negative frequency dependence and the importance of spatial scale. *Ecology* **83**:21–27.

Molofsky, J., R. Durrett, J. Dushoff, D. Griffeath, and S. Levin. 1999. Local frequency dependence and global coexistence. *Theoretical Population Biology* **55**:270–282.

Moore, K. A., and S. C. Elmendorf. 2006. Propagule vs. niche limitation: Untangling the mechanisms behind plant species' distributions. *Ecology Letters* **9**:797–804.

Mordecai, E. A., A. G. Jaramillo, J. E. Jashford, R. F. Hechinger, and K. D. Lafferty. 2016. The role of competition-colonization tradeoffs and spatial heterogeneity in promoting trematode coexistence. *Ecology* **97**:1484–1149.

Morin, P. J. 2011. *Community ecology.* Wiley-Blackwell, Hoboken, NJ.

Morlon, H., G. Chuyong, R. Condit, S. Hubbell, D. Kenfack, D. Thomas, R. Valencia, and J. L. Green. 2008. A general framework for the distance-decay of similarity in ecological communities. *Ecology Letters* **11**:904–917.

Morlon, H., D. W. Schwilk, J. A. Bryant, P. A. Marquet, A. G. Rebelo, C. Tauss, B.J.M. Bohannan, and J. L. Green. 2011. Spatial patterns of phylogenetic diversity. *Ecology Letters* **14**:141–149.

Morrison, L. W. 2010. Long-term non-equilibrium dynamics of insular floras: A 17-year record. *Global Ecology and Biogeography* **19**:663–672.

Morrison, L. W., and D. A. Spiller. 2008. Patterns and processes in insular floras affected by hurricanes. *Journal of Biogeography* **35**:1701–1710.

Morton, R. D., and R. Law. 1997. Regional species pools and the assembly of local ecological communities. *Journal of Theoretical Biology* **187**:321–331.

Mouillot, D., D. R. Bellwood, C. Baraloto, J. Chave, R. Galzin, M. Harmelin-Vivien, M. Kulbicki, et al. 2013a. Rare species support vulnerable functions in high-diversity ecosystems. *PLoS Biology* **11**:e1001569.

Mouillot, D., N.A.J. Graham, S. Villeger, N.W.H. Mason, and D. R. Bellwood. 2013b. A functional approach reveals community responses to disturbances. *Trends in Ecology and Evolution* **28**:167–177.

Moulton, M. P., and S. L. Pimm. 1987. Morphological assortment in introduced Hawaiian passerines. *Evolutionary Ecology* **1**:113–124.

Mouquet, N., V. Devictor, C. N. Meynard, F. Munoz, L.-F. Bersier, J. Chave, P. Couteron, et al. 2012. Ecophylogenetics: Advances and perspectives. *Biological Reviews* **87**:769–785.

Mouquet, N., D. Gravel, F. Massol, and V. Calcagno. 2013. Extending the concept of keystone species to communities and ecosystems. *Ecology Letters* **16**:1–8.

Mouquet, N., and M. Loreau. 2003. Community patterns in source-sink metacommunities. *American Naturalist* **162**:544–557.

Mouquet, N., J. L. Moore, and M. Loreau. 2002. Plant species richness and community productivity: Why the mechanism that promotes coexistence matters. *Ecology Letters* **5**:56–65.

Munday, P. L. 2004. Competitive coexistence of coral-dwelling fishes: The lottery hypothesis revisited. *Ecology* **85**:623–628.

Munguia, P., and T. E. Miller. 2008. Habitat destruction and metacommunity size in pen shell communities. *Journal of Animal Ecology* **77**:1175–1182.

Münkemüller, T., F. de Bello, C. N. Meynard, D. Gravel, S. Lavergne, D. Mouillot, N. Mou-quet, and W. Thuiller. 2012. From diversity indices to community assembly processes: A test with simulated data. *Ecography* **35**:468–480.

Münkemüller, T., L. Gallien, S. Lavergne, J. Renaud, C. Roquet, S. Abdulhak, S. Dullinger, L. Garraud, A. Guisan, and J. Lenoir. 2014. Scale decisions can reverse conclusions on community assembly processes. Global Ecology and Biogeography **23**:620–632.

Murrell, D. J., and R. Law. 2003. Heteromyopia and the spatial coexistence of similar competitors. *Ecology Letters* **6**:48–59.

Muschick, M., A. Indermaur, and W. Salzburger. 2012. Convergent evolution within an adaptive radiation of cichlid fishes. *Current Biology* **22**:2362–2368.

Myers, J. A., J. M. Chase, I. Jimenez, P. M. Jorgensen, A. Araujo-Murakami, N. Paniagua-Zambrana, and R. Seidel. 2013. Beta-diversity in temperate and tropical forests reflects dissimilar mechanisms of community assembly. *Ecology Letters* **16**:151–157.

Myers, J. A., and K. E. Harms. 2009. Seed arrival, ecological filters, and plant species richness: A meta-analysis. *Ecology Letters* **12**:1250–1260.

Naeem, S. 2002. Ecosystem consequences of biodiversity loss: The evolution of a paradigm. *Ecology* **83**:1537–1552.

Naeem, S., L. J. Thompson, S. P. Lawler, J. H. Lawton, and R. M. Woodfin. 1994. Declining biodiversity can alter the performance of ecosystems. *Nature* **368**:734–737.

Naeslund, B., and J. Norberg. 2006. Ecosystem consequences of the regional species pool. *Oikos* **115**:504–512.

Nagelkerke, C. J., and S. B. Menken. 2013. Coexistence of habitat specialists and generalists in metapopulation models of multiple-habitat landscapes. *Acta Biotheoretica* **61**:467–480.

Naiman, R. J., R. E. Bilby, D. E. Schindler, and J. M. Helfield. 2002. Pacific salmon, nutrients, and the dynamics of freshwater and riparian ecosystems. *Ecosystems* **5**: 399–417.

Nakano, S., H. Miyasaka, and N. Kuhara. 1999. Terrestrial-aquatic linkages: Riparian arthropod inputs alter trophic cascades in a stream food web. *Ecology* **80**:2435–2441.

Nakano, S., and M. Murakami. 2001. Reciprocal subsidies: Dynamic interdependence between terrestrial and aquatic food webs. *Proceedings of the National Academy of Sciences USA* **98**:166–170.

Narwani, A., M. A. Alexandrou, J. Herrin, A. Vouaux, C. Zhou, T. H. Oakley, and B. J. Cardinale. 2015. Common ancestry is a poor predictor of competitive traits in freshwater green algae. *PLoS One* **10**:e0137085.

Narwani, A., M. A. Alexandrou, T. H. Oakley, I. T. Carroll, and B. J. Cardinale. 2013. Experimental evidence that evolutionary relatedness does not affect the ecological mechanisms of coexistence in freshwater green algae. *Ecology Letters* **16**: 1373–1381.

Nathan, R., W. M. Getz, E. Revilla, M. Holyoak, R. Kadmon, D. Saltz, and P. E. Smouse. 2008. A movement ecology paradigm for unifying organismal movement research. *Proceedings of the National Academy of Sciences USA* **105**:19052–19059.

Naughton, H. R., M. A. Alexandrou, T. H. Oakley, and B. J. Cardinale. 2015. Phylogenetic distance does not predict competition in green algal communities. *Ecosphere* **6**:1–19.

Naveh, Z., and A. S. Lieberman. 1984. *Landscape ecology: Theory and application.* Springer, New York.

Nekola, J. C., and P. S. White. 1999. The distance decay of similarity in biogeography and ecology. *Journal of Biogeography* **26**:867–878.

Neutel, A.-M., J.A.P. Heesterbeek, J. van de Koppel, G. Hoenderboom, A. Vos, C. Kaldeway, F. Berendse, and P. C. de Ruiter. 2007. Reconciling complexity with stability in naturally assembling food webs. *Nature* **449**:599–602.

Newbold, T., J.P.W. Scharlemann, S.H.M. Butchart, C. H. Sekercioglu, R. Alkemade, H. Booth, and D. W. Purves. 2013. Ecological traits affect the response of tropical forest bird species to land-use intensity. *Proceedings of the Royal Society. Series B: Biological Sciences* **280**:20122131. doi: 10.1098/rspb.2012.2131.

Newman, E. A., M. E. Harte, N. Lowell, M. Wilber, and J. Harte. 2014. Empirical tests of within- and across-species energetics in a diverse plant community. *Ecology* **95**:2815–2825.

Newmark, W. D. 1987. A land-bridge island perspective on mammalian extinctions in Western North American parks. *Nature* **325**:430–432.

Newton, A. C., R. M. Walls, D. Golicher, S. A. Keith, A. Diaz, and J. M. Bullock. 2012. Structure, composition and dynamics of a calcareous grassland metacommunity over a 70-year interval. *Journal of Ecology* **100**:196–209.

Nilsson, S. G., and I. N. Nilsson. 1983. Are estimated species turnover rates on islands largely sampling errors? *American Naturalist* **121**:595–597.

Noonburg, E. G., and P. A. Abrams. 2005. Transient dynamics limit the effectiveness of keystone predation in bringing about coexistence. *American Naturalist* **165**:322–335.

Norberg, J. 2004. Biodiversity and ecosystem functioning: A complex adaptive systems approach. *Limnology and Oceanography* **49**:1269–1277.

Norberg, J., D. P. Swaney, J. Dushoff, J. Lin, R. Casagrandi, and S. A. Levin. 2001. Phenotypic diversity and ecosystem functioning in changing environments: A theoretical framework. *Proceedings of the National Academy of Sciences USA* **98**: 11376–11381.

Norberg, J., M. C. Urban, M. Vellend, C. A. Klausmeier, and N. Loeuille, N., 2012. Eco-evolutionary responses of biodiversity to climate change. *Nature Climate Change* **2**:747–751.

Novotny, V., P. Drozd, S. E. Miller, M. Kulfan, M. Janda, Y. Basset, and G. D. Weiblen. 2006. Why are there so many species of herbivorous insects in tropical rainforests? *Science* **313**:1115–1118.

Odling-Smee, J., D. H. Erwin, E. P. Palkovacs, M. W. Feldman, and K. N. Laland. 2013. Niche construction theory: A practical guide for ecologists. *Quarterly Review of Biology* **88**:4–28.

Odling-Smee, F. J., K. N. Laland, and M. W. Feldman. 2003. *Niche construction: The neglected process in evolution.* Princeton University Press, Princeton, NJ.

Odum, E. P. 1969. Strategy of ecosystem development. *Science* **164**:262–270.

Ohashi, H., and J. Hoshino. 2014. Disturbance by large herbivores alters the relative importance of the ecological processes that influence the assembly pattern in heterogeneous meta-communities. *Ecology and Evolution* **4**:766–775.

Oksanen, J. 1996. Is the humped relationship between species richness and biomass an artefact due to plot size? *Journal of Ecology* **84**:293–295.

Oksanen, T. 1990. Exploitation ecosystems in heterogeneous habitat complexes. *Evolutionary Ecology* **4**:220–234.

Oksanen, T., L. Oksanen, and M. Gyllenberg. 1992. Exploitation ecosystems in heterogeneous habitat complexes. 2. Impact of small-scale heterogeneity on predator prey dynamics. *Evolutionary Ecology* **6**:383–398.

Olden, J. D., and T. P. Rooney. 2006. On defining and quantifying biotic homogenization. *Global Ecology and Biogeography* **15**:113–120.

Olff, H., and M. E. Ritchie. 1998. Effects of herbivores on grassland plant diversity. *Trends in Ecology and Evolution* **13**:261–265.

Olson, V. A., R. G. Davies, C.D.L. Orme, G. H. Thomas, S. Meiri, T. M. Blackburn, K. J. Gaston, I.P.F. Owens, and P. M. Bennett. 2009. Global biogeography and ecology of body size in birds. *Ecology Letters* **12**:249–259.

Orrock, J. L., and R. J. Fletcher. 2005. Changes in community size affect the outcome of competition. *American Naturalist* **166**:107–111.

Orrock, J. L., and J. I. Watling. 2010. Local community size mediates ecological drift and competition in metacommunities. *Proceedings of the Royal Society. Series B: Biological Sciences* **277**:2185–2191.

Orrock, J. L., M. S. Witter, and O. J. Reichman. 2008. Apparent competition with an exotic plant reduces native plant establishment. *Ecology* **89**:1168–1174.

Ossola, A., M. A. Nash, F. J. Christie, A. K. Hahs, and S. J. Livesley. 2015. Urban habitat complexity affects species richness but not environmental filtering of morphologically-diverse ants. *PeerJ* **3**:e1356.

Östman, Ö., and J. M. Chase. 2007. Predator selectivity alters the effect of dispersal on coexistence among apparent competitors. *Oikos* **116**:387–394.

Östman, Ö., J. M. Kneitel, and J. M. Chase. 2006. Disturbance alters habitat isolation's effect on biodiversity in aquatic microcosms. *Oikos* **114**:360–366.

Ouin, A., J. P. Sarthou, B. Bouyjou, M. Deconchat, J. P. Lacombe, and C. Monteil. 2006. The species-area relationship in the hoverfly (Diptera, Syrphidae) communities of forest fragments in southern France. *Ecography* **29**:183–190.

Ozinga, W. A., J. H. Schaminée, R. M. Bekker, S. Bonn, P. Poschlod, O. Tackenberg, J. Bakker, and J. M. v. Groenendael. 2005. Predictability of plant species composition from environmental conditions is constrained by dispersal limitation. *Oikos* **108**:555–561.

Pacala, S. W., and J. Roughgarden. 1982. Spatial heterogeneity and interspecific competition. *Theoretical Population Biology* **21**:92–113.

Paine, R. T. 1966. Food web complexity and species diversity. *American Naturalist* **100**:65–75.

Palmgren, A. 1926. Chance as an element in plant geography. Pages 591–602 *in* B. M. Duggar, editor, *Proceedings of the International Congress of Plant Sciences*. George Banta Publishing Company, Menasha, WI.

Pandit, S. N., and J. Kolasa. 2012. Opposite effects of environmental variability and species richness on temporal turnover of species in a complex habitat mosaic. *Hydrobiologia* **685**:145–154.

Panhuis, T. M., R. Butlin, M. Zuk, and T. Tregenza. 2001. Sexual selection and speciation. *Trends in Ecology and Evolution* **16**:364–371.

Pantel, J. H., C. Duvivier, and L. D. Meester. 2015. Rapid local adaptation mediates zooplankton community assembly in experimental mesocosms. *Ecology Letters* **18**:992–1000.

Pantel, J. H., M. A. Leibold, and T. E. Juenger. 2011. Population differentiation in *Daphnia* alters community assembly in experimental ponds. *American Naturalist* **177**: 314–322.

Parent, C. E., and B. J. Crespi. 2006. Sequential colonization and diversification of Galapagos endemic land snail genus *Bulimulus* (Gastropoda, Stylommatophora). *Evolution* **60**:2311–2328.

Parent, C. E., and B. J. Crespi. 2009. Ecological opportunity in adaptive radiation of Galapagos endemic land snails. *American Naturalist* **174**:898–905.

Park, T. 1957. Experimental studies of interspecies competition. III. Relation of initial species proportion to competitive outcome in populations of *Tribolium*. *Physiological Zoology* **30**:22–40.

Parmentier, I., M. Rejou-Mechain, J. Chave, J. Vleminckx, D. W. Thomas, D. Kenfack, G. B. Chuyong, and O. J. Hardy. 2014. Prevalence of phylogenetic clustering at multiple scales in an African rain forest tree community. *Journal of Ecology* **102**:1008–1016.

Parravicini, V., M. Kulbicki, D. R. Bellwood, A. M. Friedlander, J. E. Arias-Gonzalez, P. Chabanet, S. R. Floeter, et al. 2013. Global patterns and predictors of tropical reef fish species richness. *Ecography* **36**:1254–1262.

Pärtel, M., M. Zobel, K. Zobel, and E. van der Maarel. 1996. The species pool and its relation to species richness: evidence from Estonian plant communities. *Oikos* **74**:111–117.

Passy, S. I. 2012. A hierarchical theory of macroecology. *Ecology Letters* **15**:923–934.

Pastor, J., A. Downing, and H. E. Erickson. 1996. Species-area curves and diversity-productivity relationships in beaver meadows of Voyageurs National Park, Minnesota, USA. *Oikos* **77**:399–406.

Patterson, B. D., and W. Atmar. 1986. Nested subsets and the structure of insular mammalian faunas and archipelagos. *Biological Journal of the Linnean Society* **28**: 65–82.

Pavoine, S. 2016. A guide through a family of phylogenetic dissimilarity measures among sites. *Oikos* **125**:1719–1732.

Pavoine, S., M. Baguette, and M. B. Bonsall. 2010. Decomposition of trait diversity among the nodes of a phylogenetic tree. *Ecological Monographs* **80**:485–507.

Pavoine, S., M. Baguette, V. M. Stevens, M. A. Leibold, C. Turlure, and M. B. Bonsall. 2014. Life history traits, but not phylogeny, drive compositional patterns in a butterfly metacommunity. *Ecology* **95**:3304–3313.

Pavoine, S., and M. B. Bonsall. 2011. Measuring biodiversity to explain community assembly: A unified approach. *Biological Reviews* **86**:792–812.

Pearce-Duvet, J. C., and D. H. Feener, Jr. 2010. Resource discovery in ant communities: Do food type and quantity matter? *Ecological Entomology* **35**:549–556.

Peay, K. G., M. Belisle, and T. Fukami. 2012. Phylogenetic relatedness predicts priority effects in nectar yeast communities. *Proceedings of the Royal Society. Series B: Biological Sciences* **279**:749–758.

Peay, K. G., M. Garbelotto, and T. D. Bruns. 2010. Evidence of dispersal limitation in soil microorganisms: Isolation reduces species richness on mycorrhizal tree islands. *Ecology* **91**:3631–3640.

Pelletier, F., D. Garant, and A. P. Hendry. 2009. Eco-evolutionary dynamics. *Philosophical Transactions of the Royal Society. Series B: Biological Sciences* **364**:1483–1489.

Pereira, H. M., and G. C. Daily. 2006. Modeling biodiversity dynamics in countryside land-scapes. *Ecology* **87**:1877–1885.

Pereira, H. M., L. M. Navarro, and I. S. Martins. 2012. Global biodiversity change: The bad, the good, and the unknown. *Annual Review of Environment and Resources* **37**:25–50.

Pereira, H. M., G. Ziv, and M. Miranda. 2014. Countryside species-area relationship as a valid alternative to the matrix-calibrated species-area model. *Conservation Biology* **28**:874–876.

Peres-Neto, P. R., and P. Legendre. 2010. Estimating and controlling for spatial structure in the study of ecological communities. *Global Ecology and Biogeography* **19**:174–184.

Peres-Neto, P. R., P. Legendre, S. Dray, and D. Borcard. 2006. Variation partitioning of species data matrices: estimation and comparison of fractions. *Ecology* **87**: 2614–2625.

Peres-Neto, P. R., M. A. Leibold, and S. Dray. 2012. Assessing the effects of spatial contingency and environmental filtering on metacommunity phylogenetics. *Ecology* **93**:S14–S30.

Peres-Neto, P. R., J. D. Olden, and D. A. Jackson. 2001. Environmentally constrained null models: site suitability as occupancy criterion. *Oikos* **93**:110–120.

Perry, G.L.W., B. P. Miller, N. J. Enright, and B. B. Lamont. 2014. Stochastic geometry best explains spatial associations among species pairs and plant functional types in species-rich shrublands. *Oikos* **123**:99–110.

Petermann, J. S., A.J.F. Fergus, C. Roscher, L. A. Turnbull, A. Weigelt, and B. Schmid. 2010. Biology, chance, or history? The predictable reassembly of temperate grassland communities. *Ecology* **91**:408–421.

Petermann, J. S., A.J.F. Fergus, L. A. Turnbull, and B. Schmid. 2008. Janzen-Connell effects are widespread and strong enough to maintain diversity in grasslands. *Ecology* **89**:2399–2406.

Petermann, J. S., P. Kratina, N. A. Marino, A. A. MacDonald, and D. S. Srivastava. 2015. Resources alter the structure and increase stochasticity in bromeliad microfauna communities. *PLoS One* **10**:e0118952.

Pfennig, K. S., and D. W. Pfennig. 2009. Character displacement: Ecological and reproductive responses to a common evolutionary problem. *Quarterly Review of Biology* **84**:253–276.

Pickett, S.T.A., and F. A. Bazzaz. 1978. Organization of an assemblage of early successional species on a soil-moisture gradient. *Ecology* **59**:1248–1255.

Pickett, S.T.A., and M. L. Cadenasso. 1995. Landscape ecology: Spatial heterogeneity in ecological systems. *Science* **269**:331–334.

Pigot, A. L., and R. S. Etienne. 2015. A new dynamic null model for phylogenetic community structure. *Ecology Letters* **18**:153–163.

Pigot, A. L., C. H. Trisos, and J. A. Tobias. 2016. Functional traits reveal the expansion and packing of ecological niche space underlying an elevational diversity gradient in passerine birds. *Proceedings of the Royal Society. Series B: Biological Sciences* **283**:20152013.

Pilière A.F.H. 2015. Modelling environmental responses of freshwater fish and invertebrates. Ph.D. diss., Radboud University, Nijmegen, Netherlands.

Pillai, P., A. Gonzalez, and M. Loreau. 2011. Metacommunity theory explains the emergence of food web complexity. *Proceedings of the National Academy of Sciences USA* **108**:19293–19298.

Pillai, P., M. Loreau, and A. Gonzalez. 2009. A patch-dynamic framework for food web metacommunities. *Theoretical Ecology* **3**:223–237.

Pillar, V. D., and L.D.S. Duarte. 2010. A framework for metacommunity analysis of phylogenetic structure. *Ecology Letters* **13**:587–596.

Pimm, S. L., and R. A. Askins. 1995. Forest losses predict bird extinctions in eastern North America. *Proceedings of the National Academy of Sciences USA* **92**:9343–9347.

Pimm, S. L., C. N. Jenkins, R. Abell, T. M. Brooks, J. L. Gittleman, L. N. Joppa, P. H. Raven, C. M. Roberts, and J. O. Sexton. 2014. The biodiversity of species and their rates of extinction, distribution, and protection. *Science* **344**:1246752. doi: 10.1126/science.1246752.

Pimm, S. L., J. H. Lawton, and J. E. Cohen. 1991. Food web patterns and their consequences. *Nature* **350**:669–674.

Pimm, S. L., and P. Raven. 2000. Biodiversity: Extinction by numbers. *Nature* **403**:843–845.

Pimm, S. L., G. J. Russell, J. L. Gittleman, and T. M. Brooks. 1995. The future of biodiversity. *Science* **269**:347–350.

Pinto, S. M., and A. S. MacDougall. 2010. Dispersal limitation and environmental structure interact to restrict the occupation of optimal habitat. *American Naturalist* **175**:675–686.

Pitman, N. C., J. W. Terborgh, M. R. Silman, P. Núñez V, D. A. Neill, C. E. Cerón, W. A. Palacios, and M. Aulestia. 2001. Dominance and distribution of tree species in upper Amazonian terra firme forests. *Ecology* **82**:2101–2117.

Poff, N. L. 1997. Landscape filters and species traits: Towards mechanistic understanding and prediction in stream ecology. *Journal of the North American Benthological Society* **16**:391–409.

Polis, G. A. 1999. Why are parts of the world green? Multiple factors control productivity and the distribution of biomass. *Oikos* **86**:3–15.

Polis, G. A., W. B. Anderson, and R. D. Holt. 1997. Toward an integration of landscape and food web ecology: The dynamics of spatially subsidized food webs. *Annual Review of Ecology and Systematics* **28**:289–316.

Polis, G. A., and S. D. Hurd. 1996. Linking marine and terrestrial food webs: Allochthonous input from the ocean supports high secondary productivity on small islands and coastal land communities. *American Naturalist* **147**:396–423.

Polis, G. A., M. E. Power, and G. R. Huxel. 2004. *Food webs at the landscape level*. University of Chicago Press, Chicago.

Poorter, L., L. Bongers, and F. Bongers. 2006. Architecture of 54 moist-forest tree species: Traits, trade-offs, and functional groups. *Ecology* **87**:1289–1301.

Poorter, L., I. McDonald, A. Alarcon, E. Fichtler, J. C. Licona, M. Pena-Claros, F. Sterck, Z. Villegas, and U. Sass-Klaassen. 2010. The importance of wood traits and hydraulic conductance for the performance and life history strategies of 42 rainforest tree species. *New Phytologist* **185**:481–492.

Poorter, L., S. J. Wright, H. Paz, D. D. Ackerly, R. Condit, G. Ibarra-Manriques, K. E. Harms, et al. 2008. Are functional traits good predictors of demographic rates? Evidence from five Neotropical forests. *Ecology* **89**:1908–1920.

Post, D. M., M. L. Pace, and N. G. Hairston. 2000. Ecosystem size determines food-chain length in lakes. *Nature* **405**:1047–1049.

Poulin, R. 1997. Species richness of parasite assemblages: Evolution and patterns. *Annual Review of Ecology and Systematics* **28**:341–358.

Prasad, S., P. Manasa, S. Buddhi, S. M. Singh, and S. Shivaji. 2011. Antagonistic interaction networks among bacteria from a cold soil environment. *FEMS Microbiology Ecology* **78**:376–385.

Presley, S. J., L. M. Cisneros, B. D. Patterson, and M. R. Willig. 2011a. Vertebrate metacommunity structure along an extensive elevational gradient in the tropics: A comparison of bats, rodents and birds. *Global Ecology and Biogeography* **21**:968–976.

Presley, S. J., C. L. Higgins, C. Lopez-Gonzalez, and R. D. Stevens. 2009. Elements of metacommunity structure of Paraguayan bats: Multiple gradients require analysis of multiple ordination axes. *Oecologia* **160**:781–793.

Presley, S. J., C. L. Higgins, and M. R. Willig. 2010. A comprehensive framework for the evaluation of metacommunity structure. *Oikos* **119**:908–917.

Presley, S. J., M. R. Willig, C. P. Bloch, I. Castro-Arellano, C. L. Higgins, and B. T. Klingbeil. 2011b. A complex metacommunity structure for gastropods along an elevational gradient. *Biotropica* **43**:480–488.

Preston, F. W. 1948. The commonness, and rarity, of species. *Ecology* **29**:254–283.

Preston, F. W. 1960. Time and space and the variation of species. *Ecology* **41**:611–627.

Preston, F. W. 1962. The canonical distribution of commonness and rarity: Part 1. *Ecology* **43**:185–215.

Price, P. W., I. R. Diniz, H. C. Morais, and E.S.A. Marques. 1995. The abundance of insect herbivore species in the tropics: The high local richness of rare species. *Biotropica* **27**:468–478.

Pringle, R. M., D. F. Doak, A. K. Brody, R. Jocque, and T. M. Palmer. 2010. Spatial pattern enhances ecosystem functioning in an African savanna. *PLoS Biology* **8**:e1000377.

Proches, S., J.R.U. Wilson, D. M. Richardson, and M. Rejmanek. 2008. Searching for phylogenetic pattern in biological invasions. *Global Ecology and Biogeography* **17**:5–10.

Proulx, M., and A. Mazumder. 1998. Reversal of grazing impact on plant species richness in nutrient-poor vs. nutrient-rich ecosystems. *Ecology* **79**:2581–2592.

Ptacnik, R., S. D. Moorthi, and H. Hillebrand. 2010. Hutchinson reversed, or why there need to be so many species. *Advances in Ecological Research* **43**:1–43.

Pulliam, H. R. 1988. Sources, sinks, and population regulation. *American Naturalist* **132**:652–661.

Pulliam, H. R. 2000. On the relationship between niche and distribution. *Ecology Letters* **3**:349–361.

Purschke, O., B. C. Schmid, M. T. Sykes, P. Poschlod, S. G. Michalski, W. Durka, I. Kühn, M. Winter, and H. C. Prentice. 2013. Contrasting changes in taxonomic, phylogenetic and functional diversity during a long-term succession: Insights into assembly processes. *Journal of Ecology* **101**:857–866.

Pyron, R. A., and F. T. Burbrink. 2014. Ecological and evolutionary determinants of species richness and phylogenetic diversity for island snakes. *Global Ecology and Biogeography* **23**:848–856.

Qian, H., and R. E. Ricklefs. 2007. A latitudinal gradient in large-scale beta diversity for vascular plants in North America. *Ecology Letters* **10**:737–744.

Rabosky, D. L., J. Reid, M. A. Cowan, and J. Foulkes. 2007. Overdispersion of body size in Australian desert lizard communities at local scales only: No evidence for the Narcissus effect. *Oecologia* **154**:561–570.

Rahbek, C., and G. R. Graves. 2001. Multiscale assessment of patterns of avian species richness. *Proceedings of the National Academy of Sciences USA* **98**:4534–4539.

Redfield, A. C. 1958. The biological control of chemical factors in the environment. *American Scientist* **46**:205–221.

Rees, M. 1995. Community structure in sand dune annuals: Is seed weight a key quantity? *Journal of Ecology* **83**:857–863.

Reich, P. B., D. Tilman, F. Isbell, K. Mueller, S. E. Hobbie, D.F.B. Flynn, and N. Eisenhauer. 2012. Impacts of biodiversity loss escalate through time as redundancy fades. *Science* **336**:589–592.

Reich, P. B., I. J. Wright, J. Cavender-Bares, J. M. Craine, J. Oleksyn, M. Westoby, and M. B. Walters. 2003. The evolution of plant functional variation: Traits, spectra, and strategies. *International Journal of Plant Sciences* **164**:S143–S164.

Reichenbach, T., M. Mobilia, and E. Frey. 2007. Mobility promotes and jeopardizes biodiversity in rock–paper–scissors games. *Nature* **448**:1046–1049.

Resetarits, W. J., and C. A. Binckley. 2013. Patch quality and context, but not patch number, drive multi-scale colonization dynamics in experimental aquatic landscapes. *Oecologia* **173**:933–946.

Resetarits, W. J., and A. Silberbush. 2016. Local contagion and regional compression: Habitat selection drives spatially explicit multiscale dynamics of colonization in experimental metacommunities. *Ecology Letters* **19**:191–200.

Reynolds, C. S. 2002. Resilience in aquatic ecosystems—Hysteresis, homeostasis, and health. *Aquatic Ecosystem Health & Management* **5**:3–17.

Reynolds, H. L., and S. W. Pacala. 1993. An analytical treatment of root-to-shoot ratio and plant competition for soil nutrient and light. *American Naturalist* **141**:51–70.

Reznick, D. N., and C. K. Ghalambor. 2001. The population ecology of contemporary adaptations: What empirical studies reveal about the conditions that promote adaptive evolution. *Genetica* **112**:183–198.

Richardson, J. L., M. C. Urban, D. I. Bolnick, and D. K. Skelly. 2014. Microgeographic adaptation and the spatial scale of evolution. *Trends in Ecology and Evolution* **29**:165–176.

Richgels, K.L.D., J. T. Hoverman, and P.T.J. Johnson. 2013. Evaluating the role of regional and local processes in structuring a larval trematode metacommunity of *Helisoma trivolvis*. *Ecography* **36**:854–863.

Ricklefs, R. E. 1987. Community diversity: Relative roles of local and regional processes. *Science* **235**:167–171.

Ricklefs, R. E. 2004. A comprehensive framework for global patterns in biodiversity. *Ecology Letters* **7**:1–15.

Ricklefs, R. E. 2006. The unified neutral theory of biodiversity: Do the numbers add up? *Ecology* **87**:1424–1431.

Ricklefs, R. E. 2008. Disintegration of the ecological community. *American Naturalist* **172**:741–750.

Ricklefs, R. E. 2011. Applying a regional community concept to forest birds of eastern North America. *Proceedings of the National Academy of Sciences USA* **108**: 2300–2305.

Ricklefs, R. E. 2012. Species richness and morphological diversity of passerine birds. *Proceedings of the National Academy of Sciences USA* **109**:14482–14487.

Ricklefs, R. E. 2013. Habitat-independent spatial structure in populations of some forest birds in eastern North America. *Journal of Animal Ecology* **82**:145–154.

Ricklefs, R. E. 2015. Intrinsic dynamics of the regional community. *Ecology Letters* **18**:497–503.

Ricklefs, R. E., and F. L. He. 2016. Region effects influence local tree species diversity. *Proceedings of the National Academy of Sciences USA* **113**:674–679.

Ricklefs, R. E., R. E. Latham, and H. Qian. 1999. Global patterns of tree species richness in moist forests: Distinguishing ecological influences and historical contingency. *Oikos* **86**:369–373.

Ricklefs, R. E., and I. J. Lovette. 1999. The roles of island area per se and habitat diversity in the species-area relationship of four Lesser Antillean faunal groups. *Journal of Animal Ecology* **68**:1142–1160.

Ricklefs, R. E., and S. S. Renner. 2012. Global correlations in tropical tree species richness and abundance reject neutrality. *Science* **335**:464–467.

Ricklefs, R. E., and D. Schluter, editors. 1993. *Species diversity in ecological communities: Historical and geographical perspectives.* University of Chicago Press, Chicago.

Riibak, K., T. Reitalu, R. Tamme, A. Helm, P. Gerhold, S. Znamenskiy, K. Bengtsson, E. Rosén, H. C. Prentice, and M. Pärtel. 2015. Dark diversity in dry calcareous grasslands is determined by dispersal ability and stress-tolerance. *Ecography* **38**:713–721.

Riley, G. A. 1963. *Marine biology: Proceedings of the first international interdisciplinary conference.* American Institute of Biological Sciences, Washington, DC..

Ripa, J., L. Stolind, P. Lundberg, and J. S. Brown. 2009. Niche co-evolution in consumer-resource communities. *Evolutionary Ecology Research* **11**:305–323.

Ritchie, M. E. 2010. *Scale, heterogeneity, and the structure and diversity of ecological communities.* Princeton University Press, Princeton, NJ.

Robinson, J. V., and J. E. Dickerson. 1987. Does invasion sequence affect community structure? *Ecology* **68**:587–595.

Robinson, N., T. Kadlec, M. D. Bowers, and R. P. Guralnick. 2014. Integrating species traits and habitat characteristics into models of butterfly diversity in a fragmented ecosystem. *Ecological Modelling* **281**:15–25.

Rodriguez, A., G. Jansson, and H. Andren. 2007. Composition of an avian guild in spatially structured habitats supports a competition-colonization trade-off. *Proceedings of the Royal Society. Series B: Biological Sciences* **274**:1403–1411.

Rogers, A. D., P. A. Tyler, D. P. Connelly, J. T. Copley, R. James, R. D. Larter, K. Linse, et al. 2012. The discovery of new deep-sea hydrothermal vent communities in the Southern Ocean and implications for biogeography. *PLoS Biology* **10**:e1001234.

Rojas-Echenique, J., and S. Allesina. 2011. Interaction rules affect species coexistence in intransitive networks. *Ecology* **92**:1174–1180.

Rooney, N., and K. S. McCann. 2012. Integrating food web diversity, structure and stability. *Trends in Ecology and Evolution* **27**:40–46.

Rosenzweig, M. L. 1995. Species diversity in space and time. Cambridge University Press, Cambridge.

Rosenzweig, M. L. 2003. *Win-win ecology: How the Earth's species can survive in the midst of human enterprise.* Oxford University Press, Oxford.

Rosenzweig, M. L., and Z. Abramsky. 1993. How are diversity and productivity related? Pages 52–65 *in* R. E. Ricklefs and D. Schluter, editors, *Species diversity in ecological communities.* University Chicago Press, Chicago.

Rosenzweig, M. L., M. J. Donoghue, Y. M. Li, and C. Yuan. 2011. Estimating species density. Pages 276–288 *in* B. J. McGill and A. Magurran, editors, *Biological diversity: Frontiers in measurement and assessment.* Oxford University Press, Oxford.

Rosindell, J., and S. J. Cornell. 2007. Species–area relationships from a spatially explicit neutral model in an infinite landscape. *Ecology Letters* **10**:586–595.

Rosindell, J., and S. J. Cornell. 2009. Species-area curves, neutral models and long distance dispersal. *Ecology* **90**:1743–1750.

Rosindell, J., S. J. Cornell, S. P. Hubbell, and R. S. Etienne. 2010. Protracted speciation revitalizes the neutral theory of biodiversity. *Ecology Letters* **13**:716–727.

Rosindell, J., L. J. Harmon, and R. S. Etienne. 2015. Unifying ecology and macroevolution with individual-based theory. *Ecology Letters* **18**:472–482.

Rosindell, J., S. P. Hubbell, and R. S. Etienne. 2011. The unified neutral theory of biodiversity and biogeography at age ten. *Trends in Ecology and Evolution* **26**:340–348.

Roslin, T., G. Varkonyi, M. Koponen, V. Vikberg, and M. Nieminen. 2014. Species-area relationships across four trophic levels—Decreasing island size truncates food chains. *Ecography* **37**:443–453.

Rossberg, A. G., R. Ishii, T. Amemiya, and K. Itoh. 2008. The top-down mechanism for body-mass abundance scaling. *Ecology* **89**:567–580.

Roughgarden, J., 2009. Is there a general theory of community ecology? *Biology and Philosophy* **24**:521–529.

Ruger, N., U. Berger, S. P. Hubbell, G. Vieilledent, and R. Condit. 2011. Growth strategies of tropical tree species: Disentangling light and size effects. *PLoS One* **6**:e25330.

Rundle, H. D., and P. Nosil. 2005. Ecological speciation. *Ecology Letters* **8**:336–352.

Ruokolainen, K., and H. Tuomisto. 2002. Beta-diversity in tropical forests. *Science* **297**:1439.

Ruppert, J.L.W., M. J. Travers, L. L. Smith, M. J. Fortin, and M. G. Meekan. 2013. Caught in the middle: Combined impacts of shark removal and coral loss on the fish communities of coral reefs. *PLoS One* **8**:e74648.

Ryabov, A. B., and B. Blasius. 2011. A graphical theory of competition on spatial resource gradients: Competition on spatial resource gradients. *Ecology Letters* **14**:220–228.

Ryall, K. L., and L. Fahrig. 2006. Response of predators to loss and fragmentation of prey habitat: A review of theory. *Ecology* **87**:1086–1093.

Ryberg, W. A., and J. M. Chase. 2007. Predator-dependent species-area relationships. *American Naturalist* **170**:636–642.

Ryberg, W. A., K. G. Smith, and J. M. Chase. 2012. Predators alter the scaling of diversity in prey metacommunities. *Oikos* **121**:1995–2000.

Rybicki, J., and I. Hanski. 2013. Species-area relationships and extinctions caused by habitat loss and fragmentation. *Ecology Letters* **16**:27–38.

Rychtecká, T., V. Lanta, I. Weiterová, and J. Lepš. 2014. Sown species richness and realized diversity can influence functioning of plant communities differently. *Naturwissenschaften* **101**:637–644.

Sabo, J. L., and M. E. Power. 2002. River-watershed exchange: Effects of riverine subsidies on riparian lizards and their terrestrial prey. *Ecology* **83**:1860–1869.

Saito, V. S., J. Soininen, A. A. Fonseca-Gessner, and T. Siqueira. 2015. Dispersal traits drive the phylogenetic distance decay of similarity in Neotropical stream metacommunities. *Journal of Biogeography* **42**:2101–2111.

Sanderson, J. G., J. Diamond, and S. L. Pimm. 2011. Response to Collins et al. (2011). *Journal of Biogeography* **38**:2397–2397.

Sanderson, J. G., J. M. Diamond, and S. L. Pimm. 2009. Pairwise co-existence of Bismarck and Solomon landbird species. *Evolutionary Ecology Research* **11**:771–786.

Sarnelle, O., and R. A. Knapp. 2004. Zooplankton recovery after fish removal: Limitations of the egg bank. *Limnology and Oceanography* **49**:1382–1392.

Savage, V. M., C. T. Webb, and J. Norberg. 2007. A general multi-trait-based framework for studying the effects of biodiversity on ecosystem functioning. *Journal of Theoretical Biology* **247**:213–229.

Sax, D. F., R. Early, and J. Bellemare. 2013. Niche syndromes, species extinction risks, and management under climate change. *Trends in Ecology and Evolution* **28**: 517–523.

Sax, D. F., and S. D. Gaines. 2003. Species diversity: from global decreases to local increases. *Trends in Ecology and Evolution* **18**:561–566.

Sax, D. F., B. P. Kinlan, and K. F. Smith. 2005. A conceptual framework for comparing species assemblages in native and exotic habitats. *Oikos* **108**:457–464.

Schade, J. D., J. F. Espeleta, C. A. Klausmeier, M. E. McGroddy, S. A. Thomas, and L. X. Zhang. 2005. A conceptual framework for ecosystem stoichiometry: Balancing resource supply and demand. *Oikos* **109**:40–51.

Schaffer, W. M. 1981. Ecological abstraction: The consequences of reduced dimensionality in ecological models. *Ecological Monographs* **51**:383–401.

Scheffer, M. 1990. Multiplicity of stable states in fresh-water systems. *Hydrobiologia* **200**:475–486.

Scheffer, M., S. Carpenter, J. A. Foley, C. Folke, and B. Walker. 2001. Catastrophic shifts in ecosystems. *Nature* **413**:591–596.

Scheffer, M., and E. H. van Nes. 2006. Self-organized similarity, the evolutionary emergence of groups of similar species. *Proceedings of the National Academy of Sciences USA* **103**:6230–6235.

Scheffer, M., R. Vergnon, E. H. van Nes, J.G.M. Cuppen, E.T.H.M. Peeters, R. Leijs, and A. N. Nilsson. 2015. The evolution of functionally redundant species; evidence from beetles. *PLoS One* **10**:e0137974.

Scheiner, S. M. 2003. Six types of species-area curves. *Global Ecology and Biogeography* **12**:441–447.

Scheiner, S. M. 2004. A melange of curves—Further dialogue about species-area relationships. *Global Ecology and Biogeography* **13**:479–484.

Scheiner, S. M., A. Chiarucci, G. A. Fox, M. R. Helmus, D. J. McGlinn, and M. R. Willig. 2011. The underpinnings of the relationship of species richness with space and time. *Ecological Monographs* **81**:195–213.

Scheiner, S. M., and M. R. Willig. 2011. *The theory of ecology*. University of Chicago Press, Chicago.

Schimel, D. S., G. P. Asner, and P. Moorcroft. 2013. Observing changing ecological diversity in the Anthropocene. *Frontiers in Ecology and the Environment* **11**:129–137.

Schimper, A.F.W. 1898. *Pflanzen-geographie auf physiologischer grundlage*. G. Fisher, Jena, Germany.

Schluter, D. 2000. *The ecology of adaptive radiation*. Oxford University Press, Oxford.

Schluter, D., and P. R. Grant. 1984. Determinants of morphological patterns in communities of Darwin's finches. *American Naturalist* **123**:175–196.

Schluter, D., and J. D. McPhail. 1992. Ecological character displacement and speciation in sticklebacks. *American Naturalist* **140**:85–108.

Schnitzer, S. A., and W. P. Carson. 2001. Treefall gaps and the maintenance of species diversity in a tropical forest. *Ecology* **82**:913–919.

Schoener, T. W. 1974. Resource partitioning in ecological communities. *Science* **185**:27–39.

Schoener, T. W. 1983. Field experiments on interspecific competition. *American Naturalist* **122**:240–285.

Schoener, T. W. 1986. Mechanistic approaches to community ecology: A new reductionism? *American Zoologist* **26**:81–106.

Schoener, T. W. 1989a. The ecological niche. Pages 79–113 *in* J. M. Cherrett, editor, *Ecological concepts*. Blackwell Scientific, Oxford.

Schoener, T. W. 1989b. Food webs from the small to the large. *Ecology* **70**:1559–1589.

Schoener, T. W. 2011. The newest synthesis: Understanding the interplay of evolutionary and ecological dynamics. *Science* **331**:426–429.

Schoener, T. W., and D. A. Spiller. 2006. Nonsynchronous recovery of community characteristics in island spiders after a catastrophic hurricane. *Proceedings of the National Academy of Sciences USA* **103**:2220–2225.

Schoener, T. W., D. A. Spiller, and J. B. Losos. 2001a. Natural restoration of the species-area relation for a lizard after a hurricane. *Science* **294**:1525–1528.

Schoener, T. W., D. A. Spiller, and J. B. Losos. 2001b. Predators increase the risk of catastrophic extinction of prey populations. *Nature* **412**:183–186.

Schramski, J. R., A. I. Dell, J. M. Grady, R. M. Sibly, and J. H. Brown. 2015. Metabolic theory predicts whole-ecosystem properties. *Proceedings of the National Academy of Sciences USA* **112**:2617–2622.

Schreiber, S. J., and T. P. Killingback. 2013. Spatial heterogeneity promotes coexistence of rock–paper–scissors metacommunities. *Theoretical Population Biology* **86**:1–11.

Schroder, A., L. Persson, and A. M. De Roos. 2005. Direct experimental evidence for alternative stable states: A review. *Oikos* **110**:3–19.

Schuler, M. S., J. M. Chase, and T. M. Knight. 2015. More individuals drive the species energy-area relationship in an experimental zooplankton community. *Oikos* **124**: 1065–1070.

Schuler, M. A., J. M. Chase, and T. M. Knight. 2017. Habitat size modulates the influence of heterogeneity on species richness patterns in a model zooplankton community. *Ecology* **98**:1651–1659. doi:10.1002/ecy.1833.

Schulze, E. D., and H. A. Mooney. 1994. *Biodiversity and ecosystem function*. Springer-Verlag, Berlin.

Schweitzer, J. A., J. K. Bailey, B. J. Rehill, G. D. Martinsen, S. C. Hart, R. L. Lindroth, P. Keim, and T. G. Whitham. 2004. Genetically based trait in a dominant tree affects ecosystem processes. *Ecology Letters* **7**:127–134.

Sedio, B. E., and A. M. Ostling. 2013. How specialised must natural enemies be to facilitate coexistence among plants? *Ecology Letters* **16**:995–1003.

Seidler, T. G., and J. B. Plotkin. 2006. Seed dispersal and spatial pattern in tropical trees. *PLoS Biology* **4**:2132–2137.

Sepkoski, J. J., Jr. 1996. Competition in macroevolution: The double wedge revisited. Pages 211–255 *in* J. W. Valentine, D. Jablonski, D. H. Erwin, and J. H. Lipps, editors, *Evolutionary paleobiology*. University of Chicago Press, Chicago.

Shade, A., J. G. Caporaso, J. Handelsman, R. Knight, and N. Fierer. 2013. A meta-analysis of changes in bacterial and archaeal communities with time. *ISME Journal* **7**:1493–1506.

Shen, G. C., M. J. Yu, X. S. Hu, X. C. Mi, H. B. Ren, I. F. Sun, and K. P. Ma. 2009. Species-area relationships explained by the joint effects of dispersal limitation and habitat heterogeneity. *Ecology* **90**:3033–3041.

Shevtsov, J., K. Wickings, and B. C. Patten. 2013. Evaluating the role of biotic interactions in structuring communities using a gradient analysis of multiple interacting guilds. *Oikos* **122**:1594–1605.

Shinen, J. L., and S. A. Navarrete. 2014. Lottery coexistence on rocky shores: Weak niche differentiation or equal competitors engaged in neutral dynamics? *American Naturalist* **183**:342–362.

Shipley, B. 2010. *From plant traits to vegetation structure: Chance and selection in the assembly of ecological communities*. Cambridge University Press, Cambridge.

Shipley, B., C.E.T. Paine, and C. Baraloto. 2012. Quantifying the importance of local niche-based and stochastic processes to tropical tree community assembly. *Ecology* **93**:760–769.

Shmida, A., and S. Ellner. 1984. Coexistence of plant species with similar niches. *Vegetatio* **58**:29–55.

Shmida, A., and M. V. Wilson. 1985. Biological determinants of species diversity. *Journal of Biogeography* **12**:1–20.

Shoemaker, L. G., and B. A. Melbourne. 2016. Linking metacommunity paradigms to spatial coexistence mechanisms. *Ecology* **97**:2436–2446.

Shorrocks, B. 1991. Competition on a divided and ephemeral resource: A cage experiment. *Biological Journal of the Linnean Society* **43**:211–220.

Shulman, R. S., and J. M. Chase. 2007. Increasing isolation reduces predator:prey species richness ratios in aquatic food webs. *Oikos* **116**:1581–1587.

Shurin, J. B. 2000. Dispersal limitation, invasion resistance, and the structure of pond zooplankton communities. *Ecology* **81**:3074–3086.

Shurin, J. B. 2001. Interactive effects of predation and dispersal on zooplankton communities. *Ecology* **82**:3404–3416.

Shurin, J. B., and E. G. Allen. 2001. Effects of competition, predation, and dispersal on species richness at local and regional scales. *American Naturalist* **158**:624–637.

Shurin, J. B., P. Amarasekare, J. M. Chase, R. D. Holt, M. F. Hoopes, and M. A. Leibold. 2004. Alternative stable states and regional community structure. *Journal of Theoretical Biology* **227**:359–368.

Shurin, J. B., S. E. Arnott, H. Hillebrand, A. Longmuir., B. Pinel-Alloul, M. Winder, and N. D. Yan. 2007. Diversity-stability relationship varies with latitude in zooplankton. *Ecology Letters* **10**:127–134.

Siefert, A., C. Ravenscroft, M. D. Weiser, and N. G. Swenson. 2013. Functional beta-diversity patterns reveal deterministic community assembly processes in eastern North American trees. *Global Ecology and Biogeography* **22**:682–691.

Siepielski, A. M., K.-L. Hung, E. E. Bein, and M. A. McPeek. 2010. Experimental evidence for neutral community dynamics governing an insect assemblage. *Ecology* **91**: 847–857.

Siepielski, A. M., and M. A. McPeek. 2010. On the evidence for species coexistence: a critique of the coexistence program. *Ecology* **91**:3153–3164.

Siepielski, A. M., A. N. Mertens, B. L. Wilkinson, and M. A. McPeek. 2011. Signature of ecological partitioning in the maintenance of damselfly diversity. *Journal of Animal Ecology* **80**:1163–1173.

Silvertown, J., Y. Araya, and D. Gowing. 2015. Hydrological niches in terrestrial plant communities: A review. *Journal of Ecology* **103**:93–108.

Silvertown, J., M. E. Dodd, D.J.G. Gowing, and J. O. Mountford. 1999. Hydrologically defined niches reveal a basis for species richness in plant communities. *Nature* **400**:61–63.

Silvertown, J., K. McConway, D. Gowing, M. Dodd, M. F. Fay, J. A. Joseph, and K. Dolphin. 2006. Absence of phylogenetic signal in the niche structure of meadow plant communities. *Proceedings of the Royal Society. Series B: Biological Sciences* **273**:39–44.

Simberloff, D. 1970. Taxonomic diversity of island biotas. *Evolution* **24**:23–47.

Simberloff, D. 1976. Species turnover and equilibrium island biogeography. *Science* **194**:572–578.

Simberloff, D. 1978. Using island biogeographic distributions to determine if colonization is stochastic. *American Naturalist* **112**:713–726.

Simberloff, D. 1980. Succession of paradigms in ecology: Essentialism to materialism and probabilism. *Synthese* **43**:3–39.

Simberloff, D. 1988. The contribution of population and community biology to conservation science. *Annual Review of Ecology and Systematics* **19**:473–511.

Simberloff, D. 2004. Community ecology: Is it time to move on? *American Naturalist* **163**:787–799.

Simberloff, D., and L. G. Abele. 1982. Refuge design and island biogeographic theory: Effects of fragmentation. *American Naturalist* **120**:41–50.

Simberloff, D., and L. G. Abele. 1976. Island biogeography and conservation: Strategy and limitations. *Science* **193**:1032–1032.

Simberloff, D., and L. G. Abele. 1984. Conservation and obfuscation: Subdivision of reserves. *Oikos* **42**:399–401.

Simberloff, D., and W. Boecklen. 1981. Santa Rosalia reconsidered: Size ratios and competition. *Evolution* **35**:1206–1228.

Simberloff, D., and E. O. Wilson. 1969. Experimental zoogeography of islands: Colonization of empty islands. *Ecology* **50**:278–296.

Simberloff, D., and E. O. Wilson. 1970. Experimental zoogeography of islands: A two-year record of colonization. *Ecology* **51**:934–937.

Simova, I., D. Storch, P. Keil, B. Boyle, O. L. Phillips, and B. J. Enquist. 2011. Global species-energy relationship in forest plots: Role of abundance, temperature and species climatic tolerances. *Global Ecology and Biogeography* **20**:842–856.

Simpson, G. G. 1944. *Tempo and mode in evolution*. Columbia University Press, New York.

Siqueira, T., L. M. Bini, F. O. Roque, and K. Cottenie. 2012a. A metacommunity framework for enhancing the effectiveness of biological monitoring strategies. *PLoS One* 7:e43626.

Siqueira, T., L. M. Bini, F. O. Roque, S. R. Marques Couceiro, S. Trivinho-Strixino, and K. Cottenie. 2012b. Common and rare species respond to similar niche processes in macroinvertebrate metacommunities. *Ecography* 35:183–192.

Sirot, C., S. Villeger, D. Mouillot, A. M. Darnaude, J. Ramos-Miranda, D. Flores-Hernandez, and J. Panfili. 2015. Combinations of biological attributes predict temporal dynamics of fish species in response to environmental changes. *Ecological Indicators* 48:147–156.

Šizling, A. L., D. Storch, E. Šizlingová, J. Reif, and K. J. Gaston. 2009. Species abundance distribution results from a spatial analogy of central limit theorem. *Proceedings of the National Academy of Sciences USA* 106:6691–6695.

Sjogren, P. 1991. Extinction and isolation gradients in metapopulaitons: The case of the pool frog (*Rana lessonae*). *Biological journal of the Linnean Society* 42:135–147.

Skellam, J. 1951. Random dispersal in theoretical populations. *Biometrika* 38:196–218.

Skelly, D. K., E. E. Werner, and S. A. Cortwright. 1999. Long-term distributional dynamics of a Michigan amphibian assemblage. *Ecology* 80:2326–2337.

Slatkin, M. 1974. Competition and regional coexistence. *Ecology* 55:128–134.

Smith, A. B., B. Sandel, N.J.B. Kraft, and S. Carey. 2013. Characterizing scale-dependent community assembly using the functional-diversity-area relationship. *Ecology* 94:2392–2402.

Smith, F. A., S. K. Lyons, S. K. Morgan Ernest, and J. H. Brown. 2008. Macroecology: More than the division of food and space among species on continents. *Progress in Physical Geography* 32:115–138.

Smith, G. K. 2013. The coexistence of ecologically similar species. Ph.D. diss., University of Texas at Austin.

Smith, K. G., K. R. Lips, and J. M. Chase. 2009. Selecting for extinction: Nonrandom disease-associated extinction homogenizes amphibian biotas. *Ecology Letters* 12:1069–1078.

Smith, T. W., and J. T. Lundholm. 2010. Variation partitioning as a tool to distinguish between niche and neutral processes. *Ecography* 33:648–655.

Snyder, R. E., E. T. Borer, and P. Chesson. 2005. Examining the relative importance of spatial and nonspatial coexistence mechanisms. *American Naturalist* 166:E75–E94.

Snyder, R. E., and P. Chesson. 2003. Local dispersal can facilitate coexistence in the presence of permanent spatial heterogeneity. *Ecology Letters* 6:301–309.

Sobral, F. L., and M. V. Cianciaruso. 2015. Functional and phylogenetic structure of forest and savanna bird assemblages across spatial scales. *Ecography* 39:533–541.

Soininen, J. 2010. Species turnover along abiotic and biotic gradients: patterns in space equal patterns in time? *BioScience* 60:433–439.

Soininen, J. 2014. A quantitative analysis of species sorting across organisms and ecosystems. *Ecology* 95:3284–3292.

Soininen, J. 2016. Spatial structure in ecological communities—A quantitative analysis. *Oikos* 125:160–166.

Soininen, J., R. McDonald, and H. Hillebrand. 2007. The distance decay of similarity in ecological communities. *Ecography* 30:3–12.

Sokol, E. R., E. F. Benfield, L. K. Belden, and H. M. Valett. 2011. The assembly of ecological communities inferred from taxonomic and functional composition. *American Naturalist* **177**:630–644.

Soliveres, S., F. T. Maestre, W. Urlich, P. Manning, S. Boch, M. A. Bowker, D. Prati, et al. 2015. Intransitive competition is widespread in plant communities and maintains their species richness. *Ecology Letters* **18**:790–798.

Soranno, P. A., K. S. Cheruvelil, K. E. Webster, M. T. Bremigan, T. Wagner, and C. A. Stow. 2010. Using landscape limnology to classify freshwater ecosystems for multi-ecosystem management and conservation. *BioScience* **60**:440–454.

Spasojevic, M. J., S. Copeland, and K. N. Suding. 2014. Using functional diversity patterns to explore metacommunity dynamics: A framework for understanding local and regional influences on community structure. *Ecography* **37**:939–949.

Spasojevic, M. J., and K. N. Suding. 2012. Inferring community assembly mechanisms from functional diversity patterns: The importance of multiple assembly process. *Journal of Ecology* **100**:652–661.

Spencer, M., A. R. Blaustein, M. W. Schwartz, and D. M. Cohen. 1999. Species richness and the proportion of predatory animal species in temporary freshwater pools: relationships with habitat size and permanence. *Ecology Letters* **2**: 157–166.

Spiller, D. A., J. B. Losos, and T. W. Schoener. 1998. Impact of a catastrophic hurricane on island populations. *Science* **281**:695–697.

Spiller, D. A., and T. W. Schoener. 1998. Lizards reduce spider species richness by excluding rare species. *Ecology* **79**:503–516.

Srivastava, D. S., and J. H. Lawton. 1998. Why more productive sites have more species: An experimental test of theory using tree-hole communities. *American Naturalist* **152**:510–529.

Srivastava, D. S., and M. Vellend. 2005. Biodiversity-ecosystem function research: Is it relevant to conservation? *Annual Review of Ecology, Evolution, and Systematics* **36**:267–294.

Stanton, M. L., T. M. Palmer, and T. P. Young. 2002. Competition-colonization trade-offs in a guild of African acacia-ants. *Ecological Monographs* **72**:347–363.

Steffen, W., J. Grinevald, P. Crutzen, and J. McNeill. 2011. The Anthropocene: Conceptual and historical perspectives. *Philosophical Transactions of the Royal Society. Series A: Mathematical, Physical and Engineering Sciences* **369**:842–867.

Stegen, J. C., B. J. Enquist, and R. Ferriere. 2009. Advancing the metabolic theory of biodiversity. *Ecology Letters* **12**:1001–1015.

Stegen, J. C., A. L. Freestone, T. O. Crist, M. J. Anderson, J. M. Chase, L. S. Comita, H. V. Cornell, K. F. Davies, S. P. Harrison, and A. H. Hurlbert. 2013. Stochastic and deterministic drivers of spatial and temporal turnover in breeding bird communities. *Global Ecology and Biogeography* **22**:202–212.

Stegen, J. C., and A. H. Hurlbert. 2011. Inferring ecological processes from taxonomic, phylogenetic and functional trait beta-diversity. *PLoS One* **6**:e20906.

Stein, A., K. Gerstner, and H. Kreft. 2014. Environmental heterogeneity as a universal driver of species richness across taxa, biomes and spatial scales. *Ecology Letters* **17**:866–880.

Stein, C., H. Auge, M. Fischer, W. W. Weisser, and D. Prati. 2008. Dispersal and seed limitation affect diversity and productivity of montaine grasslands. *Oikos* **117**:1469–1478.

Steinbauer, M. J., K. Dolos, B. Reinekin, C. Beierkuhnlein. 2012. Current measures for distance decay in similarity of species composition are influenced by study extent and grain size. *Global Ecology and Biogeography* **21**:1203–1212.

Steiner, C. F. 2014. Stochastic sequential dispersal and nutrient enrichment drive beta diversity in space and time. *Ecology* **95**:2603–2612.

Steiner, C. F., and M. A. Leibold. 2004. Cyclic assembly trajectories and scale-dependent productivity-diversity relationships. *Ecology* **85**:107–113.

Stemberger, R. S. 1995. Pleistocene refuge areas and postglacial dispersal of copepods of the northeastern United States. *Canadian Journal of Fisheries and Aquatic Sciences* **52**:2197–2210.

Sterck, F., L. Markesteijn, F. Schieving, and L. Poorter. 2011. Functional traits determine trade-offs and niches in a tropical forest community. *Proceedings of the National Academy of Sciences USA* **108**:20627–20632.

Sterner, R. W., and J. J. Elser. 2002. *Ecological stoichiometry: The biology of elements from molecules to the biosphere.* Princeton University Press, Princeton, NJ.

Stier, A. C., S. W. Geange, K. M. Hanson, and B. M. Bolker. 2013. Predator density and timing of arrival affect reef fish community assembly. *Ecology* **94**:1057–1068.

Stier, A. C., K. M. Hanson, S. J. Holbrook, R. J. Schmitt, and A. J. Brooks. 2014a. Predation and landscape characteristics independently affect reef fish community organization. *Ecology* **95**:1294–1307.

Stier, A. C., A. M. Hein, V. Parravicini, and M. Kulbicki. 2014b. Larval dispersal drives trophic structure across Pacific coral reefs. *Nature Communications* **5**:6575. doi: 10.1038/ncomms6575.

Stoks, R., and M. A. McPeek. 2006. A tale of two diversifications: Reciprocal habitat shifts to fill ecological space along the pond permanence gradient. *American Naturalist* **168**:S50–S72.

Stomp, M., J. Huisman, F. de Jongh, A. J. Veraart, D. Gerla, M. Rijkeboer, B. W. Ibelings, U.I.A. Wollenzien, and L. J. Stal. 2004. Adaptive divergence in pigment composition promotes phytoplankton biodiversity. *Nature* **432**:104–107.

Stone, L., T. Dayan, and D. Simberloff. 1996. Community-wide assembly patterns unmasked: The importance of species' differing geographical ranges. *American Naturalist* **148**:997–1015.

Stoner, K. E., K. Vulinec, S. J. Wright, and C. A. Peres. 2007. Hunting and plant community dynamics in tropical forests: A synthesis and future directions. *Biotropica* **39**:385–392.

Storch, D., K. L. Evans, and K. J. Gaston. 2005. The species-area-energy relationship. *Ecology Letters* **8**:487–492.

Storch, D., P. Keil, and W. Jetz. 2012. Universal species—area and endemics—area relationships at continental scales. *Nature* **488**:78–81.

Strong, D. R., D. Simberloff, L. G. Abele, and A. B. Thistle, editors. 1984. *Ecological communities: Conceptual issues and the evidence.* Princeton University Press, Princeton, NJ.

Strong, D. R., L. A. Szyska, and D. S. Simberloff. 1979. Test of community-wide character displacement against null hypotheses. *Evolution* **33**:897–913.

Stuart, Y. E., and J. B. Losos. 2013. Ecological character displacement: Glass half full or half empty? *Trends in Ecology and Evolution* **28**:402–408.

Stump, S. M., and P. Chesson. 2015. Distance-responsive predation is not necessary for the Janzen-Connell hypothesis. *Theoretical Population Biology* **106**:60–70.

Sugihara, G. 1980. Minimal community structure: an explanation of species abundance patterns. *American Naturalist* **116**:770–787.

Supp, S. R., and S.K.M. Ernest. 2014. Species-level and community-level responses to disturbance: A cross-community analysis. *Ecology* **95**:1717–1723.

Supp, S. R., X. Xiao, S.K.M. Ernest, and E. P. White. 2012. An experimental test of the response of macroecological patterns to altered species interactions. *Ecology* **93**: 2505–2511.

Suzuki, M., T. Miyashita, H. Kabaya, K. Ochiai, M. Asada, and Z. Kikvidze. 2013. Deer herbivory as an important driver of divergence of ground vegetation communities in temperate forests. *Oikos* **122**:104–110.

Svenning, J. C., S. Normand, and F. Skov. 2008. Postglacial dispersal limitation of widespread forest plant species in nemoral Europe. *Ecography* **31**:316–326.

Svenning, J. C., and S. J. Wright. 2005. Seed limitation in a Panamanian forest. *Journal of Ecology* **93**:853–862.

Svensson, J. R., M. Lindegarth, P. R. Jonsson, and H. Pavia. 2012. Disturbance-diversity models: What do they really predict and how are they tested? *Proceedings of the Royal Society. Series B: Biological Sciences* **279**:2163–2170.

Svirezhev, I.U.M., and D. O. Logofet. 1983. *Stability of biological communities*. Mir Publishers, Moscow.

Swenson, N. G., P. Anglada-Cordero, and J. A. Barone. 2011. Deterministic tropical tree community turnover: Evidence from patterns of functional beta diversity along an elevational gradient. *Proceedings of the Royal Society. Series B: Biological Sciences* **278**:877–884.

Swenson, N. G., and B. J. Enquist. 2009. Opposing assembly mechanisms in a Neotropical dry forest: Implications for phylogenetic and functional community ecology. *Ecology* **90**:2161–2170.

Swenson, N. G., B. J. Enquist, J. Pither, J. Thompson, and J. K. Zimmerman. 2006. The problem and promise of scale dependency in community phylogenetics. *Ecology* **87**:2418–2424.

Swenson, N. G., B. J. Enquist, J. Thompson, and J. K. Zimmerman. 2007. The influence of spatial and size scale on phylogenetic relatedness in tropical forest communities. *Ecology* **88**:1770–1780.

Swenson, N. G., D. L. Erickson, X. C. Mi, N. A. Bourg, J. Forero-Montana, X. J. Ge, R. Howe, J et al. 2012. Phylogenetic and functional alpha and beta diversity in temperate and tropical tree communities. *Ecology* **93**:S112–S125.

Swenson, N. G., M. D. Weiser, L. F. Mao, S. Normand, M. A. Rodriguez, L. X. Lin, M. Cao, and J. C. Svenning. 2016. Constancy in functional space across a species richness anomaly. *American Naturalist* **187**:E83–E92.

Symons, C., and S. Arnott. 2013. Regional zooplankton dispersal provides spatial insurance for ecosystem function. *Global Change Biology* **19**:1610–1619.

Takimoto, G., and D. M. Post. 2013. Environmental determinants of food-chain length: A meta-analysis. *Ecological Research* **28**:675–681.

Takimoto, G., D. A. Spiller, and D. M. Post. 2008. Ecosystem size, but not disturbance, determines food-chain length on islands of the Bahamas. *Ecology* **89**:3001–3007.

Tamme, R., I. Hiiesalu, L. Laanisto, R. Szava-Kovats, and M. Partel. 2010. Environmental heterogeneity, species diversity and co-existence at different spatial scales. *Journal of Vegetation Science* **21**:796–801.

Tansley, A. 1917. On competition between *Galium saxatile* L. (*G. hercynicum* Weig.) and *Galium sylvestre* Poll. (*G. asperum* Schreb.) on different types of soil. *Journal of Ecology* **5**:173–179.

Tansley, A. G. 1935. The use and abuse of vegetational concepts and terms. *Ecology* **16**:284–307.

Taub, F. B. 1974. Closed ecological systems. *Annual Review of Ecology and Systematics* **5**:139–160.

Taylor, W. D., L. R. Williams, S. C. Hern, and V. W. Lambou. 1979. *Phytoplankton water quality relationships in U.S. lakes. Part 1: Methods, rationale, and data limitations.* U.S. Environmental Protection Agency, Washington, DC.

Ter Steege, H., N. Pitman, D. Sabatier, H. Castellanos, P. Van der Hout, D. C. Daly, M. Silveira, et al. 2003. A spatial model of tree alpha-diversity and tree density for the Amazon. *Biodiversity and Conservation* **12**:2255–2277.

Terborgh, J. W. 2015. Toward a trophic theory of species diversity. *Proceedings of the National Academy of Sciences USA* **112**:11415–11422.

Terborgh, J. W., and J. Faaborg. 1980. Saturation of bird communities in the West Indies. *American Naturalist* **116**:178–195.

Terborgh, J., K. Feeley, M. Silman, P. Nunez, and B. Balukjian. 2006. Vegetation dynamics of predator-free land-bridge islands. *Journal of Ecology* **94**:253–263.

Terborgh, J., L. Lopez, P. Nunez, M. Rao, G. Shahabuddin, G. Orihuela, M. Riveros, et al. 2001. Ecological meltdown in predator-free forest fragments. *Science* **294**: 1923–1926.

terHorst, C. P., T. E. Miller, and E. Powell. 2010. When can competition for resources lead to ecological equivalence? *Evolutionary Ecology Research* **12**:843–854.

Tessier, A. J., and R. J. Horwitz. 1990. Influence of water chemistry on size structure of zooplankton assemblages. *Canadian Journal of Fisheries and Aquatic Sciences* **47**:1937–1943.

Tessier, A. J., M. A. Leibold, and J. Tsao. 2000. A fundamental trade-off in resource exploitation by Daphnia and consequences to plankton communities. *Ecology* **81**:826–841.

Tessier, A. J., A. Young, and M. Leibold. 1992. Population-dynamics and body-size selection in *Daphnia. Limnology and Oceanography* **37**:1–13.

Tews, J., U. Brose, V. Grimm, K. Tielborger, M. C. Wichmann, M. Schwager, and F. Jeltsch. 2004. Animal species diversity driven by habitat heterogeneity/diversity: The importance of keystone structures. *Journal of Biogeography* **31**:79–92.

Thebault, E., and C. Fontaine. 2010. Stability of ecological communities and the architecture of mutualistic and trophic networks. *Science* **329**:853–856.

Thibault, K. M., S.K.M. Ernest, and J. H. Brown. 2010. Redundant or complementary? Impact of a colonizing species on community structure and function. *Oikos* **119**:1719–1726.

Thompson, K., J. G. Hodgson, J. P. Grime, and M. J. Burke. 2001. Plant traits and temporal scale: Evidence from a 5-year invasion experiment using native species. *Journal of Ecology* **89**:1054–1060.

Thompson, J. N. 1998. Rapid evolution as an ecological process. *Trends in Ecology and Evolution* **13**:329–332.

Thompson, J. N. 2005. *The geographic mosaic of coevolution.* University of Chicago Press, Chicago.

Thompson, J. N. 2014. *Interaction and coevolution.* University of Chicago Press, Chicago.

Thum, R. A. 2007. Reproductive interference, priority effects and the maintenance of parapatry in *Skistodiaptomus* copepods. *Oikos* **116**:759–768.

Tilman, D. 1976. Ecological competition between algae: Experimental confirmation of resource-based competition theory. *Science* **192**:463–465.

Tilman, D. 1977. Resource competition between planktonic algae: An experimental and theoretical approach. *Ecology* **58**:338–348.

Tilman, D. 1982. *Resource competition and community structure*. Princeton University Press, Princeton, NJ.

Tilman, D. 1988. *Plant strategies and the dynamics and structure of plant communities*. Princeton University Press, Princeton, NJ.

Tilman, D. 1994. Competition and biodiversity in spatially structured habitats. *Ecology* **75**:2–16.

Tilman, D. 1997. Community invasibility, recruitment limitation, and grassland biodiversity. *Ecology* **78**:81–92.

Tilman, D., and J. A. Downing. 1994. Biodiversity and stability in grasslands. *Nature* **367**: 363–365.

Tilman, D., F. Isbell, and J. M. Cowles. 2014. Biodiversity and ecosystem functioning. *Annual Review of Ecology, Evolution, and Systematics* **45**:471–493.

Tilman, D., and P. M. Kareiva. 1997. *Spatial ecology: The role of space in population dynamics and interspecific interactions*. Princeton University Press, Princeton, NJ.

Tilman, D., C. L. Lehman, and K. T. Thomson. 1997. Plant diversity and ecosystem productivity: Theoretical considerations. *Proceedings of the National Academy of Sciences USA* **94**:1857–1861.

Tilman, D., and S. Pacala. 1993. The maintenance of species richness in plant communities. Pages 13–25 *in* R. E. Ricklefs and D. Schluter, editors, *Species diversity in ecological communities*. University Chicago Press, Chicago.

Tilman, D., P. B. Reich, and F. Isbell. 2012. Biodiversity impacts ecosystem productivity as much as resources, disturbance, or herbivory. *Proceedings of the National Academy of Sciences USA* **109**:10394–10397.

Tilman, D., P. B. Reich, J. Knops, D. Wedin, T. Mielke, and C. Lehman. 2001. Diversity and productivity in a long-term grassland experiment. *Science* **294**:843–845.

Tilman, D., D. Wedin, and J. Knops. 1996. Productivity and sustainability influenced by biodiversity in grassland ecosystems. *Nature* **379**:718–720.

Tittensor, D. P., C. Mora, W. Jetz, H. K. Lotze, D. Ricard, E. Vanden Berghe, and B. Worm. 2010. Global patterns and predictors of marine biodiversity across taxa. *Nature* **466**:1098–1107.

Tokeshi, M. 1996. Power fraction: A new explanation of relative abundance patterns in species-rich assemblages. *Oikos* **75**:543–550.

Tokolyi, J., J. Schmidt, and Z. Barta. 2014. Climate and mammalian life histories. *Biological Journal of the Linnean Society* **111**:719–736.

Triantis, K. A., F. Guilhaumon, and R. H. Whittaker. 2012. The island species-area relationship: Biology and statistics. *Journal of Biogeography* **39**:215–231.

Triantis, K. A., M. Mylonas, K. Lika, and K. Vardinoyannis. 2003. A model for the species-area-habitat relationship. *Journal of Biogeography* **30**:19–27.

Trisos, C. H., O. L. Petchey, and J. A. Tobias. 2014. Unraveling the interplay of community assembly processes acting on multiple niche axes across spatial scales. *American Naturalist* **184**:593–608.

Troia, M. J., and K. B. Gido. 2015. Functional strategies drive community assembly of stream fishes along environmental gradients and across spatial scales. *Oecologia* **177**:545–559.

Troll, C. 1939. Luftbildplan und ökologische Bodenforschung [Aerial photography and ecological studies of the Earth]. *Berlin Zeitschrift der Gesellschaft für Erdkunde* **8**:241–298.

Tscharntke, T., I. Steffan-Dewenter, A. Kruess, and C. Thies. 2002. Characteristics of insect populations on habitat fragments: A mini review. *Ecological Research* **17**:229–239.

Tscharntke, T., J. M. Tylianakis, T. A. Rand, R. K. Didham, L. Fahrig, P. Batary, J. Bengtsson, et al. 2012. Landscape moderation of biodiversity patterns and processes—Eight hypotheses. *Biological Reviews* **87**:661–685.

Tucker, C. M., M. W. Cadotte, S. B. Carvalho, T. J. Davies, S. Ferrier, S. A. Fritz, R. Grenyer, et al. 2016. A guide to phylogenetic metrics for conservation community ecology and macroecology. *Biological Reviews* **92**:698–715. doi:10.1111/brv.12252.

Tucker, C. M., L. G. Shoemaker, K. F. Davies, D. R. Nemergut, and B. A. Melbourne. 2016. Differentiating between niche and neutral assembly in metacommunities using null models of β-diversity. *Oikos* **125**:778–789.

Tuomisto, H. 2010a. A diversity of beta diversities: Straightening up a concept gone awry. Part 1. Defining beta diversity as a function of alpha and gamma diversity. *Ecography* **33**:2–22.

Tuomisto, H., 2010b. A diversity of beta diversities: Straightening up a concept gone awry. Part 2. Quantifying beta diversity and related phenomena. *Ecography* **33**:23–45.

Tuomisto, H., L. Ruokolainen, and K. Ruokolainen. 2012. Modelling niche and neutral dynamics: On the ecological interpretation of variation partitioning results. *Ecography* **35**:961–971.

Tuomisto, H., K. Ruokolainen, and M. Yii-Halla. 2003. Dispersal, environment, and floristic variation of western Amazonian forests. *Science* **299**:241–244.

Turgeon, J., R. Stoks, R. A. Thum, J. M. Brown, and M. A. McPeek. 2005. Simultaneous Quaternary radiations of three damselfly clades across the Holarctic. *American Naturalist* **165**:E78–E107.

Turnbull, L. A., M. J. Crawley, and M. Rees. 2000. Are plant populations seed-limited? A review of seed sowing experiments. *Oikos* **88**:225–238.

Turnbull, L. A., M. Rees, and M. J. Crawley. 1999. Seed mass and the competition/colonization trade-off: A sowing experiment. *Journal of Ecology* **87**:899–912.

Turner, M. G., R. H. Gardner, and R. V. O'Neill. 2001. *Landscape ecology in theory and practice*. Springer, New York.

Tylianakis, J. M., T. Tscharntke, and O. T. Lewis. 2007. Habitat modification alters the structure of tropical host-parasitoid food webs. *Nature* **445**:202–205.

Ulrich, W. 2004. Species co-occurrences and neutral models: Reassessing J. M. Diamond's assembly rules. *Oikos* **107**:603–609.

Ulrich, W., A. Baselga, B. Kusumoto, T. Shiono, H. Tuomisto, and Y. Kubota, Y. 2017. The tangled link between β-and γ-diversity: A Narcissus effect weakens statistical inferences in null model analyses of diversity patterns. *Global Ecology and Biogeography* **26**:1–5.

Ulrich, W., and N. J. Gotelli. 2013. Pattern detection in null model analysis. *Oikos* **122**:2–18.

Urban, D. L., R. V. Oneill, and H. H. Shugart. 1987. Landscape ecology. *BioScience* **37**:119–127.

Urban, M. C., and L. De Meester. 2009. Community monopolization: local adaptation enhances priority effects in an evolving metacommunity. Proceedings of the Royal Society. Series B: Biological Sciences **276**:4129–4138.

Urban, M. C., L. De Meester, M. Vellend, R. Stoks, and J. Vanoverbeke. 2012. A crucial step toward realism: Responses to climate change from an evolving metacommunity perspective. *Evolutionary Applications* **5**:154–167.

Urban, M., M. Leibold, P. Amarasekare, L. Demeester, R. Gomulkiewicz, M. Hochberg, C. Klausmeier, N. Loeuille, C. Demazancourt, and J. Norberg. 2008. The evolutionary ecology of metacommunities. *Trends in Ecology and Evolution* **23**: 311–317.

Urban, M. C., and D. K. Skelly. 2006. Evolving metacommunities: Toward an evolutionary perspective on metacommunities. *Ecology* **87**:1616–1626.

Uriarte, M., R. Condit, C. D. Canham, and S. P. Hubbell. 2004. A spatially explicit model of sapling growth in a tropical forest: Does the identity of neighbours matter? *Journal of Ecology* **92**:348–360.

Valencia, R., R. B. Foster, G. Villa, R. Condit, J. C. Svenning, C. Hernandez, K. Romoleroux, E. Losos, E. Magard, and H. Balslev. 2004. Tree species distributions and local habitat variation in the Amazon: Large forest plot in eastern Ecuador. *Journal of Ecology* **92**:214–229.

Valladares, G., L. Cagnolo, and A. Salvo. 2012. Forest fragmentation leads to food web contraction. *Oikos* **121**:299–305.

van de Leemput, I. A., E. H. van Nes, and M. Scheffer. 2015. Resilience of alternative states in spatially extended ecosystems. *PLoS One* **10**:e0116859.

van der Gast, C. J. 2013. Microbial biogeography and what Baas Becking should have said. *Microbiology Today* **40**:108–111.

van der Gast, C. J. 2015. Microbial biogeography: The end of the ubiquitous dispersal hypothesis. *Environmental Microbiology* **17**:544–546.

van der Gast, C. J., D. Ager, and A. K. Lilley. 2008. Temporal scaling of bacterial taxa is influenced by both stochastic and deterministic ecological factors. *Environmental Microbiology* **10**:1411–1418.

van der Plas, F., T. Janzen, A. Ordonez, W. Fokkema, J. Reinders, R. S. Etienne, and H. Olff. 2015. A new modeling approach estimates the relative importance of different community assembly processes. *Ecology* **96**:1502–1515.

van der Putten, W. H., R. D. Bardgett, J. D. Bever, T. M. Bezemer, B. B. Casper, T. Fukami, P. Kardol, et al. 2013. Plant-soil feedbacks: The past, the present and future challenges. *Journal of Ecology* **101**:265–276.

Van der Veken, S., J. Bellemare, K. Verheyen, and M. Hermy. 2007. Life-history traits are correlated with geographical distribution patterns of western European forest herb species. *Journal of Biogeography* **34**:1723–1735.

van Kleunen, M., W. Dawson, F. Essl, J. Pergl, M. Winter, E. Weber, H. Kreft, et al. 2015. Global exchange and accumulation of non-native plants. *Nature* **525**:100–103.

van Nes, E. H., and M. Scheffer. 2003. Alternative attractors may boost uncertainty and sensitivity in ecological models. *Ecological Modelling* **159**:117–124.

van Noordwijk, C.G.E., W.C.E.P. Verberk, H. Turin, T. Heijerman, K. Alders, W. Dekoninck, K. Hannig, et al. 2015. Species–area relationships are modulated by trophic rank, habitat affinity, and dispersal ability. *Ecology* **96**:518–531.

van Nouhuys, S., and E. Punju. 2010. Coexistence of competing parasitoids: Which is the fugitive and where does it hide? *Oikos* **119**:61–70.

Van Valen, L. M. 1973. A new evolutionary law. *Evolutionary Theory* **1**:1–30.

Vandermeer, J. 1973. Regional stabilization of locally unstable predator-prey relationships. *Journal of Theoretical Biology* **41**:161–170.

Vandermeer, J. 1975. Interspecific competition: A new approach to the classical theory. *Science* **188**:253–255.

Vandermeer, J., and S. Yitbarek. 2012. Self-organized spatial pattern determines biodiversity in spatial competition. *Journal of Theoretical Biology* **300**:48–56.

Vannote, R. L., G. W. Minshall, K. W. Cummins, J. R. Sedell, and C. E. Cushing. 1980. River continuum concept. *Canadian Journal of Fisheries and Aquatic Sciences* **37**:130–137.

Vannote, R. L., and B. W. Sweeney. 1980. Geographic analysis of thermal equilibria: A conceptual model for evaluating the effect of natural and modified thermal regimes on aquatic insect communities. *American Naturalist* **115**:667–695.

Vanoverbeke, J., M. C. Urban, and L. De Meester. 2015. Community assembly is a race between immigration and adaptation: Eco-evolutionary interactions across spatial scales. *Ecography* **38**:1–13.

Vasseur, D. A., and J. W. Fox. 2011. Adaptive dynamics of competition for nutritionally complementary resources: Character convergence, displacement, and parallelism. *American Naturalist* **178**:501–514.

Veech, J. A., and T. O. Crist. 2010. Toward a unified view of diversity partitioning. *Ecology* **91**:1988–1992.

Vellend, M. 2010. Conceptual synthesis in community ecology. *Quarterly Review of Biology* **85**:183–206.

Vellend, M. 2016. *The theory of ecological communities*. Princeton University Press, Princeton, NJ.

Vellend, M., L. Baeten, A. Becker-Scarpitta, V. Boucher-Lalonde, J. L. McCune, J. Messier, I. H. Myers-Smith, D. F. Sax. 2017. Plant biodiversity change across scales during the Anthropocene. *Annual Review of Plant Biology* **68**:563–586.

Vellend, M., L. Baeten, I. H. Myers-Smith, S. C. Elmendorf, R. Beausejour, C. D. Brown, P. De Frenne, K. Verheyen, and S. Wipf. 2013. Global meta-analysis reveals no net change in local-scale plant biodiversity over time. *Proceedings of the National Academy of Sciences USA* **110**:19456–19459.

Vellend, M., E.B.M. Drummond, and J. L. Muir. 2009. Ecological differentiation among genotypes of dandelions (*Taraxacum officinale*). *Weed Science* **57**:410–416.

Vellend, M., D. S. Srivastava, K. M. Anderson, C. D. Brown, J. E. Jankowski, E. J. Kleynhans, N.J.B. Kraft, et al. 2014. Assessing the relative importance of neutral stochasticity in ecological communities. *Oikos* **123**:1420–1430.

Venail, P. A., R. C. MacLean, T. Bouvier, M. A. Brockhurst, M. E. Hochberg, and N. Mouquet. 2008. Diversity and productivity peak at intermediate dispersal rate in evolving metacommunities. *Nature* **452**:210–214.

Venail, P. A., A. Narwani, K. Fritschie, M. A. Alexandrou, T. H. Oakley, and B. J. Cardinale. 2014. The influence of phylogenetic relatedness on species interactions among freshwater green algae in a mesocosm experiment. *Journal of Ecology* **102**:1288–1299.

Verreydt, D., L. De Meester, E. Decaestecker, M. J. Villena, K. Van Der Gucht, P. Vannormelingen, W. Vyverman, and S. A. Declerck. 2012. Dispersal-mediated trophic interactions can generate apparent patterns of dispersal limitation in aquatic metacommunities. *Ecology Letters* **15**:218–226.

Vincent, T. L., and J. S. Brown. 2005. *Evolutionary game theory, natural selection and Darwinian dynamics*. Cambridge University Press, Cambridge.

Violle, C., A. Bonis, M. Plantegenest, C. Cudennec, C. Damgaard, B. Marion, D. Le Coeur, and J.-B. Bouzille. 2011. Plant functional traits capture species richness variations along a flooding gradient. *Oikos* **120**:389–398.

Volkov, I., J. R. Banavar, F. He, S. P. Hubbell, and A. Maritan. 2005. Density dependence explains tree species abundance and diversity in tropical forests. *Nature* **438**:658–661.

Volkov, I., J. R. Banavar, S. P. Hubbell, and A. Maritan. 2003. Neutral theory and relative species abundance in ecology. *Nature* **424**:1035–1037.

Von Humboldt, A. 1807. *Essai sur la géographie des plantes: accompagné d'un tableau physique des régions équinoxiales*. Chez Levrault, Schoell et compagnie, Paris.

von Wehrden, H., D. J. Abson, M. Beckmann, A. F. Cord, S. Klotz, and R. Seppelt. 2014. Realigning the land-sharing/land-sparing debate to match conservation needs: Considering diversity scales and land-use history. *Landscape Ecology* **29**:941–948.

Walker, B. H., and D. Salt. 2012. *Resilience practice: Building capacity to absorb disturbance and maintain function*. Island Press, Washington, DC.

Wallace, A. R. 1876. *The geographical distribution of animals: With a study of the relations of living and extinct faunas as elucidating the past changes of the Earth's surface*. Macmillan and Co., London.

Walsh, J. J., and K. A. Steidinger. 2001. Saharan dust and Florida red tides: The cyanophyte connection. *Journal of Geophysical Research-Oceans* **106**:11597–11612.

Wang, X., T. Wiegand, N. J. Kraft, N. G. Swenson, S. J. Davies, Z. Hao, R. Howe, et al. 2016. Stochastic dilution effects weaken deterministic effects of niche-based processes in species rich forests. *Ecology* **97**:347–360.

Wang, X., T. Wiegand, A. Wolf, R. Howe, S. J. Davies, and Z. Hao. 2011. Spatial patterns of tree species richness in two temperate forests. *Journal of Ecology* **99**:1382–1393.

Wang, Z., W. Ye, H. Cao, Z. Huang, J. Lian, L. Li, S. Wei, and I. F. Sun. 2009a. Species-topography association in a species-rich subtropical forest of China. *Basic and Applied Ecology* **10**:648–655.

Wang, Z. H., J. H. Brown, Z. Y. Tang, and J. Y. Fang. 2009b. Temperature dependence, spatial scale, and tree species diversity in eastern Asia and North America. *Proceedings of the National Academy of Sciences USA* **106**:13388–13392.

Wardle, D. A. 1999. Is "sampling effect" a problem for experiments investigating biodiversity-ecosystem function relationships? *Oikos* **87**:403–407.

Wardle, D. A. 2016. Do experiments exploring plant diversity-ecosystem functioning relationships inform how biodiversity loss impacts natural ecosystems? *Journal of Vegetation Science* **27**:646–653.

Wardle, D. A., K. I. Bonner, and G. M. Barker. 2000. Stability of ecosystem properties in response to above-ground functional group richness and composition. *Oikos* **89**:11–23.

Wardle, D. A., O. Zackrisson, G. Hörnberg, and C. Gallet, C. 1997. The influence of island area on ecosystem properties. *Science* **277**:1296–1299.

Warren, B. H., D. Simberloff, R. E. Ricklefs, R. Aguilee, F. L. Condamine, D. Gravel, H. Morlon, et al. 2015. Islands as model systems in ecology and evolution: Prospects fifty years after MacArthur-Wilson. *Ecology Letters* **18**:200–217.

Warren, P. H. 1996. The effects of between-habitat dispersal rate on protist communities and metacommunities in microcosms at two spatial scales. *Oecologia* **105**:132–140.

Warren, P. H., and K. J. Gaston. 1992. Predator-prey ratios: A special case of a general pattern? *Philosophical Transactions of the Royal Society. Series B: Biological Sciences* **338**:113–130.

Warton, D. I., F. G. Blanchet, R. B. O'Hara, O. Ovaskainen, S. Taskinen, S. C. Walker, and F.K.C. Hui. 2015. So many variables: Joint modeling in community ecology. *Trends in Ecology and Evolution* **30**:766–779.

Wasserberg, G., Z. Abramsky, N. Valdivia, and B. Kotler. 2005. The role of vegetation characteristics and foraging substrate in organizing a centrifugal gerbil community. *Journal of Mammalogy* **86**:1009–1014.

Watt, A. S. 1947. Pattern and process in the plant community. *Journal of Ecology* **35**:1–22.

Webb, C. O., D. D. Ackerly, M. A. McPeek, and M. J. Donoghue. 2002. Phylogenies and community ecology. *Annual Review of Ecology and Systematics* **33**:475–505.

Webb, C. T., J. A. Hoeting, G. M. Ames, M. I. Pyne, and N. LeRoy Poff. 2010. A structured and dynamic framework to advance traits-based theory and prediction in ecology. *Ecology Letters* **13**:267–283.

Weiher, E. 1999. The combined effects of scale and productivity on species richness. *Journal of Ecology* **87**:1005–1011.

Weiher, E., D. Freund, T. Bunton, A. Stefanski, T. Lee, and S. Bentivenga. 2011. Advances, challenges and a developing synthesis of ecological community assembly theory. *Philosophical Transactions of the Royal Society. Series B: Biological Sciences* **366**:2403–2413.

Weiher, E., and P. A. Keddy. 1995. Assembly rules, null models, and trait dispersion: New questions from old patterns. *Oikos* **74**:159–164.

Weinstein, B. G., B. Tinoco, J. L. Parra, L. M. Brown, J. A. McGuire, F. G. Stiles, and C. H. Graham. 2014. Taxonomic, phylogenetic, and trait beta diversity in South American hummingbirds. *American Naturalist* **184**:211–224.

Wellborn, G. A. 1994. Size-biased predation and prey life-histories: A comparative study of freshwater amphipod populations. *Ecology* **75**:2104–2117.

Wellborn, G. A. 2002. Trade-off between competitive ability and antipredator adaptation in a freshwater amphipod species complex. *Ecology* **83**:129–136.

Wellborn, G. A., and R. D. Cothran. 2004. Phenotypic similarity and differentiation among sympatric cryptic species in a freshwater amphipod species complex. *Freshwater Biology* **49**:1–13.

Wellborn, G. A., and R. D. Cothran. 2007. Niche diversity in crustacean cryptic species: Complementarity in spatial distribution and predation risk. *Oecologia* **154**:175–183.

Wellborn, G. A., D. K. Skelly, and E. E. Werner. 1996. Mechanisms creating community structure across a freshwater habitat gradient. *Annual Review of Ecology and Systematics* **27**:337–363.

Werner, E. E. 1977. Species packing and niche complementarity in three sunfishes. *American Naturalist* **111**:553–578.

Werner, E. E. 1998. Ecological experiments and a research program in community ecology. Pages 3–26 *in* W. J. Restetarits and J. Bernardo, editors, *Experimental ecology: Issues and perspectives.* Oxford University Press, New York.

Werner, E. E., and D. J. Hall. 1977. Competition and habitat shift in two sunfishes (Centrarchidae). *Ecology* **58**:869–876.

Werner, E. E., and D. J. Hall. 1979. Foraging efficiency and habitat switching in competing sunfishes. *Ecology* **60**:256–264.

Werner, E. E., and M. A. McPeek. 1994. Direct and indirect effects of predators on two anuran species along an environmental gradient. *Ecology* **75**:1368–1382.

Werner, E. E., D. K. Skelly, R. A. Relyea, and K. L. Yurewicz. 2007. Amphibian species richness across environmental gradients. *Oikos* **116**:1697–1712.

West, G. B., J. H. Brown, and B. J. Enquist. 1997. A general model for the origin of allometric scaling laws in biology. *Science* **276**:122–126.

Westoby, M., and I. J. Wright. 2006. Land-plant ecology on the basis of functional traits. *Trends in Ecology and Evolution* **21**:261–268.

White, E. P., P. B. Adler, W. K. Lauenroth, R. A. Gill, D. Greenberg, D. M. Kaufman, A. Rassweiler, J. A. Rusak, M. D. Smith, and J. R. Steinbeck. 2006. A comparison of the species–time relationship across ecosystems and taxonomic groups. *Oikos* **112**: 185–195.

White, E. P., S. K. Ernest, P. B. Adler, A. H. Hurlbert, and S. K. Lyons. 2010. Integrating spatial and temporal approaches to understanding species richness. *Philosophical Transactions of the Royal Society. Series B: Biological Sciences* **365**:3633–3643.

White, E. P., S. K. Ernest, A. J. Kerkhoff, and B. J. Enquist. 2007. Relationships between body size and abundance in ecology. *Trends in Ecology and Evolution* **22**:323–330.

White, E. P., and A. H. Hurlbert. 2010. The combined influence of the local environment and regional enrichment on bird species richness. *American Naturalist* **175**:E35–E43.

Whitham, T. G., J. K. Bailey, J. A. Schweitzer, S. M. Shuster, R. K. Bangert, C. J. LeRoy, E. V. Lonsdorf, et al. 2006. A framework for community and ecosystem genetics: from genes to ecosystems. *Nature Reviews Genetics* **7**:510–523.

Whitham, T. G., G. D. Martinsen, K. D. Floate, H. S. Dungey, B. M. Potts, and P. Keim. 1999. Plant hybrid zones affect biodiversity: Tools for a genetic-based understanding of community structure. *Ecology* **80**:416–428.

Whittaker, R. H. 1960. Vegetation of the Siskiyou Mountains, Oregon and California. *Ecological Monographs* **30**:279–338.

Whittaker, R. H. 1967. Gradient analysis of vegetation. *Biological Reviews* **42**:207–264.

Whittaker, R. H. 1975. *Communities and ecosystems.* MacMillan Publishing, New York.

Whittaker, R. J. 2010. Meta-analyses and mega-mistakes: Calling time on meta-analysis of the species richness-productivity relationship. *Ecology* **91**:2522–2533.

Whittaker, R. J., R. J. Ladle, M. B. Araujo, J. M. Fernandez-Palacios, J. D. Delgado, and J. R. Arevalo. 2007. The island immaturity–speciation pulse model of island evolution: An alternative to the "diversity begets diversity" model. *Ecography* **30**:321–327.

Wiegand, T., A. Huth, S. Getzin, X. G. Wang, Z. Q. Hao, C.V.S. Gunatilleke, and I. Gunatilleke. 2012. Testing the independent species' arrangement assertion made by theories of stochastic geometry of biodiversity. *Proceedings of the Royal Society. Series B: Biological Sciences* **279**:3312–3320.

Wiegand, T., and K. A. Moloney. 2014. *A handbook of spatial point pattern analysis.* CRC Press, Boca Raton, FL.

Wiens, J. A. 1989. Spatial scaling in ecology. *Functional Ecology* **3**:385–397.

Wiens, J. J., D. D. Ackerly, A. P. Allen, B. L. Anacker, L. B. Buckley, H. V. Cornell, E. I. Damschen, et al. 2010. Niche conservatism as an emerging principle in ecology and conservation biology. *Ecology Letters* **13**:1310–1324.

Wiens, J. J., and M. J. Donoghue. 2004. Historical biogeography, ecology and species richness. *Trends in Ecology and Evolution* **19**:639–644.

Wiersma, P., A. Munoz-Garcia, A. Walker, and J. B. Williams. 2007. Tropical birds have a slow pace of life. *Proceedings of the National Academy of Sciences USA* **104**:9340–9345.

Wilbur, H. M. 1972. Competition, predation, and structure of *Abystoma-Rana sylvatica* community. *Ecology* **53**:3–21.

Wilbur, H. M. 1997. Experimental ecology of food webs: Complex systems in temporary ponds. *Ecology* **78**:2279–2302.

Wilcove, D. S., D. Rothstein, J. Dubow, A. Phillips, and E. Losos. 1998. Quantifying threats to imperiled species in the United States. *BioScience* **48**:607–615.

Williams, C. B. 1964. *Patterns in the balance of nature and related problems in quantitative ecology.* Academic Press, New York.

Williams, N. M., E. E. Crone, T. H. Roulston, R. L. Minckley, L. Packer, and S. G. Potts. 2010. Ecological and life-history traits predict bee species responses to environmental disturbances. *Biological Conservation* **143**:2280–2291.

Willig, M. R., D. M. Kaufman, and R. D. Stevens. 2003. Latitudinal gradients of biodiversity: Pattern, process, scale, and synthesis. *Annual Review of Ecology, Evolution and Systematics* **34**:273–309.

Willig, M. R., S. J. Presley, C. P. Bloch, I. Castro-Arellano, L. M. Cisneros, C. L. Higgins, and B. T. Klingbeil. 2011. Tropical metacommunities along elevational gradients: Effects of forest type and other environmental factors. *Oikos* **120**:1497–1508.

Wilsey, B. J., and H. W. Polley. 2003. Effects of seed additions and grazing history on diversity and productivity of subhumid grasslands. *Ecology* **84**:920–931.

Wilson, D. S. 1992. Complex interactions in metacommunities, with implications for biodiversity and higher levels of selection. *Ecology* **73**:1984–2000.

Wilson, J. B. 1995. Null models for assembly rules: The Jack-Horner effect is more insidious than the Narcissus effect. *Oikos* **72**:139–144.

Wilson, W., P. Lundberg, D. Vázquez, J. Shurin, M. Smith, W. Langford, K. Gross, and G. Mittelbach. 2003. Biodiversity and species interactions: Extending Lotka–Volterra community theory. *Ecology Letters* **6**:944–952.

Wilson, W., C. Osenberg, R. Schmitt, and R. Nisbet. 1999. Complementary foraging behaviors allow coexistence of two consumers. *Ecology* **80**:2358–2372.

Wilson, W. G., and P. A. Abrams. 2005. Coexistence of cycling and dispersing consumer species: Armstrong and McGehee in space. *American Naturalist* **165**:193–205.

Wootton, J. T. 1994. The nature and consequences of indirect effects in ecological communities. *Annual Review of Ecology and Systematics* **25**:443–466.

Wootton, J. T. 1998. Effects of disturbance on species diversity: A multitrophic perspective. *American Naturalist* **152**:803–825.

Wootton, J. T. 2005. Field parameterization and experimental test of the neutral theory of biodiversity. *Nature* **433**:309–312.

Wright, D. H. 1983. Species-energy theory: An extension of species-area theory. *Oikos* **41**:496–506.

Wright, J. P., and C. G. Jones. 2006. The concept of organisms as ecosystem engineers ten years on: Progress, limitations, and challenges. *BioScience* **56**:203–209.

Xiao, X., D. J. McGlinn, and E. P. White. 2015. A strong test of the maximum entropy theory of ecology. *American Naturalist* **185**:E70–E80.

Yaacobi, G., Y. Ziv, and M. L. Rosenzweig. 2007. Habitat fragmentation may not matter to species diversity. *Proceedings of the Royal Society. Series B: Biological Sciences* **274**:2409–2412.

Yang, J., N. G. Swenson, G. Zhang, X. Ci, M. Cao, L. Sha, J. Li, J. W. Ferry Slik, and L. Lin. 2015. Local-scale partitioning of functional and phylogenetic beta-diversity in a tropical tree assemblage. *Scientific Reports* **5**:12731. doi:10.1038/srep12731.

Yang, J., G. Zhang, X. Ci, N. G. Swenson, M. Cao, L. Sha, J. Li, C. C. Baskin, J.W.F. Slik, and L. Lin. 2014. Functional and phylogenetic assembly in a Chinese tropical tree community across size classes, spatial scales and habitats. *Functional Ecology* **28**:520–529.

Yee, D. A., and S. A. Juliano. 2007. Abundance matters: A field experiment testing the more individuals hypothesis for richness-productivity relationships. *Oecologia* **153**: 153–162.

Yeh, Y.-C., P. R. Peres-Neto, S.-W. Huang, Y.-C. Lai, C.-Y. Tu, F.-K. Shiah, G.-C. Gong, and C.-h. Hsieh. 2015. Determinism of bacterial metacommunity dynamics in the southern East China Sea varies depending on hydrography. *Ecography* **38**:198–212.

Yoshida, T., L. E. Jones, S. P. Ellner, G. F. Fussmann, and N. G. Hairston. 2003. Rapid evolution drives ecological dynamics in a predator-prey system. *Nature* **424**:303–306.

Young, H. S., D. J. McCauley, R. B. Dunbar, M. S. Hutson, M. Ter-Kuile, and R. Dirzo. 2013. The roles of productivity and ecosystem size in determining food chain length in tropical terrestrial ecosystems. *Ecology* **94**:692–701.

Young, T. P., J. M. Chase, and R. T. Huddleston. 2001. Community succession and assembly comparing contrasting and combining paradigms in the context of ecological restoration. *Ecological Restoration* **19**:5–18.

Yu, D. W., H. B. Wilson, and N. E. Pierce. 2001. An empirical model of species coexistence in a spatially structured environment. *Ecology* **82**:1761–1771.

Zeigler, B. P. 1977. Persistence and patchiness of predator-prey systems induced by discrete event population exchange mechanisms. *Journal of Theoretical Biology* **67**:687–713.

Zeiter, M., A. Stampfli, and D. M. Newbery. 2006. Recruitment limitation constrains local species richness and productivity in dry grassland. *Ecology* **87**:942–951.

Zhou, S. R., and D. Y. Zhang. 2008. A nearly neutral model of biodiversity. *Ecology* **89**:248–258.

Ziv, Y., B. P. Kotler, Z. Abramsky, and M. L. Rosenzweig. 1995. Foraging efficiencies of competing rodents: Why do gerbils exhibit shared preference habitat selection. *Oikos* **73**:260–268.

Zobel, M. 1997. The relative role of species pools in determining plant species richness: An alternative explanation of species coexistence? *Trends in Ecology and Evolution* **12**:266–269.

Zobel, M., M. Öpik, M. Moora, and M. Pärtel. 2006. Biodiversity and ecosystem functioning: It is time for dispersal experiments. *Journal of Vegetation Science* **17**: 543–547.

Zokan, M., and J. M. Drake. 2015. The effect of hydroperiod and predation on the diversity of temporary pond zooplankton communities. *Ecology and Evolution* **5**:3066–3074.

Index

α-diversity, 262–63; changes in, 184; metacommunity assembly processes influencing, 276–77

Abele, L. G., 16, 283

Abrams, P. A., 147, 204, 219, 305, 363

Abramsky, Z., 79, 261

abundance, 26, 267, 272–73. *See also* species-abundance distribution (SAD): anthropogenic activities and, 302; biodiversity and, 272–73; body and range size and, 301, 339–40; in different metacommunity models, 42–43; diversity of, 26; functional traits and, 189–90; in heterogeneous metacommunities, 269–70; interspecific interaction influencing, 21; on island habitats, 279; in novel patch, 210; predators and, 333; of predators in isolation, 317, 332–33; predicting from traits, 197–200; scale influence on, 5; traits related to, 189

Ackerly, D. D., 153, 168–69, 170, 172

Adams, H. E., 350

adaptive evolution: ecological processes and, 202–38; in ecological speciation, 239; in metapopulations and metacommunities, 205–7

adaptive radiation, 216, 241–42, 251–52

adaptive trait change, rates of, 209

Adler, F. R., 324

Adler, P. B., 25, 38–39, 52, 75, 112, 132, 135–36, 158, 161–62, 166, 172, 261, 296

Ai, D., 388

Alexander, H. M., 127, 352

Alexandrou, M. A., 159–60

Algar, A. C., 170

Allan, E., 159

Allee effects, 141–42, 149, 245, 249

Allen, A. P., 7, 9, 262, 305, 379

Allesina, S., 144, 319

allopatric speciation, 98, 171, 174–75, 241

Allouche, O. M., 270, 279, 282, 295n3

Alonso, D., 19, 61, 112, 274

Amarasekare, P., 18, 25, 32–33, 141–42, 174, 305–6, 323, 388–89

Amezcua, A. B., 305

analysis bias, 105, 192

Anderson, M. G., 332–33

Anderson, M. J., 179n.1, 187, 277

Anderson, R. S., 364

Andrewartha, H. G., 391

Angert, A. L., 53, 161, 163

Anthropocene era: biodiversity in, 283–94; habitat and biodiversity loss in, 286–91; habitat loss, fragmentation, and conservation in, 283–86; invasions, extinctions, and diversity change in, 292

anthropogenic activities. *See also* Anthropocene era: effects of, 336; in environmental change, 292; processes influenced by, 294; regional species dispersal and, 293, 294; species abundances and, 302; in species dispersal, 293; in taxonomic β-diversity, 184

aquatic systems. *See also* freshwater ponds/lakes; lake ecosystems; ponds: microcosmic, 70–71, 371–72; nutrient inputs and biostability in, 366

Araújo, M. B., 63

Araya, Y. N., 79

archetypes, 23–24, 47–48, 380, 386. *See also specific archetypes*: comparing and contrasting, 34–37; core assumptions of, 23, 50, 51; inadequacy of, 150; interpatch habitat heterogeneity and connectivity in, 37; key propositions and predictions for, 35; metacommunity ecology at interface of, 37–42; in occupancy model, 44–46; overview and analysis of, 27–34; synthesis of, 25–26, 42–46; temporal fluctuations in, 132–34; variation partitioning testing of, 109–12

archipelagos: adaptive radiation on, 247–48, 251; biogeographic distribution on, 254; species diversity on, 278

area per se: energy and, 297–99; in habitat loss and fragmentation, 285–86; heterogeneity and, 279–81, 286, 297; shaping nested SAR, 277, 278

Armstrong, R. A., 47, 147
Arnan, X., 161, 183
Arnott, S. E., 358
Arrhenius, O., 256
arrivals: extinctions and, 36; local adaptation and, 208; rates of in different metacommunity models, 36; timing of, 26, 208, 210, 218
Askins, R. A., 283, 288
assembly rules, 16, 391; Diamond's, 90, 95–96; trait- and phylogeny-based, 154–58
Atmar, W., 99
Azaele, S. A., 267, 277

β-diversity, 262; changes in, 184; distance-delay relationship of, 255; high, 178; in high-productivity habitats, 327–31; metacommunity assembly processes influencing, 276–77; reduction in, 292; scale dependent, 188–89; spatially clumped species in, 267; taxonomic patterns in, 152; taxonomic versus functional, 181–82; temporal changes in, 177
Baas Becking hypothesis, 73–74
bacterial communities, variations in, 122
Badgley, C., 190
Baho, D. L., 357
Bailey, J. K., 204
Baiser, B., 184, 185, 187
Baker, J. P., 315
Baldeck, C. A., 114, 118, 183
Baldi, A., 285
Barabás, G., 224
Barbour, M. A., 205
Barfield, M., 305–6
Barone, J. A., 102, 105
Barro Colorado Island, Panama, 80–84; data from, 93; disturbances and diversity in, 271–72; environmental and spatial effects in, 117; habitat heterogeneity in, 118, 120; macroecological patterns in, 275; NT archetype in, 113–16; species pool in, 126
Baselga, A., 277
Bazzaz, F. A., 172
Beck, J., 271, 274
BEF. See biodiversity-ecosystem functioning (BEF) experiments; biodiversity-ecosystem functioning (BEF) relationships
Begley-Miller, D. R., 307
Beisner, B. E., 59
Bell, G., 16, 18, 24, 27–28, 165, 206
Belmaker, J., 5, 171–73, 316

Bengtsson, J., 60, 390
Benitez, M. S., 325
Bennett, J. A., 159–60
Bennett, J. R., 109, 124, 127
Berga, M., 309, 358
Best, R. J., 160
Bever, J., 9, 141, 148–49, 273, 324–25
biased dispersal, 387
Binckley, C. A., 351
biodiversity. See also biodiversity macroecology; habitat heterogeneity: along ecological gradients, 294–300; anthropogenic effects on, 283–94, 336; conservation of in island-like systems, 283–86; destruction and conservation of, 1–2; dispersal rate and, 71–72; ecological drift and dispersal limitations in, 5; ecosystem functioning and, 336–37; in food webs, 318; functional, at taxonomic level, 183–84; habitat loss and loss of, 286–91; integrating evolutionary and ecological processes affecting, 250–54; local versus regional controls on, 87–88; in neutral versus SS environments, 182; niche-based vs. regional/dispersal-based factors in, 64–65; patterns of, 3; productivity and, 337–38; scale-dependent changes in, 292–94; spatial processes in, 13–15; speciation "priming the pump" of, 50; species-abundance distribution patterns and metrics for, 92–94; and species–area relationship, 276–77; species pool size and, 124–26; taxonomic measures of, 178; through time, 240
biodiversity-ecosystem functioning (BEF) experiments, 336–37; pattern from, 337–38; species pool diversity in, 354–55
biodiversity-ecosystem functioning (BEF) relationships: changing, 354–56; ME dynamics in, 351; negative, 347–48, 355; spatial processes mediating, 339–42
biodiversity macroecology, 259, 260, 265–66; clumping of species influencing, 273; component interactions in, 273–74; components of, 267; dissecting, 267–74; metacommunity assembly processes in, 274–76; relative abundance of individuals influencing, 272–73; synthesizing processes and patterns of, 262–67; total number of individuals influencing, 270–72; total number of species influencing, 268–70
biodiversity trait distributions, 392

biogeographic scales, 245–50
biogeography: as community structure driver, 245; contemporary, 259–61, 294–300; versus environmental factors, 249–50; historical, 239, 260, 261; interacting with phylogeny, 239; macroecology and, 384; metacommunity and, 239, 245–50; in species interactions, 16
biotic homogenization, 71, 184, 292
biotic interactions, 61, 64, 67, 90, 166,178
Birch, L. C., 391
bird species: distributions of, 95; temporal turnover of, 137; trait-based perspective of, 155; traits of correlated to environmental conditions, 190
birth rates: death rates and, 8, 54–55; immigration-emigration rates and, 349–50
Bismarck Archipelago habitat: bird distributions in, 95; species matrices on, 98; trait-based perspective of, 155
bistability, macrophyte-phytoplankton, 366
Blackburn, T. M., 190, 256, 258
Blanchet, F. G., 109
Blasius, B., 348, 350
body size: convergence of, 224, 225; correlated to environmental drivers, 190; distributions of, 256, 261–62, 270, 300–301; feeding performance and, 172; partitioning of, 14; in response to environmental and spatial factors, 121–22; in species abundance, 301; species coexistence and, 155
body size ratios, 14, 224
body size relations, 2, 8, 14, 391
Boecklen, W. J., 16, 155, 278
Böhnke, M., 192, 193
Bohonak, A. J., 115
Bolnick, D. I., 204
Bond, E. M., 341, 348
Bonsall, M. B., 153, 160
Borcard, D., 108, 191
Borda-de-Agua, L., 274
Bormann, F. H., 370
Borregaard, M. K., 246
boundary clumping, 99–104
Bourret, A., 244–45
Boyce, M. S., 55
Brind'Amour, A., 153, 192
Britton, N., 149
Brose, U., 332
Brown, A. M., 194, 197, 262, 295

Brown, B. L., 10, 25, 92, 119, 127, 381
Brown, C. D., 115, 288
Brown, J. H., 2, 18, 137–38, 156, 181, 206, 256–58, 262, 301, 354, 365, 379, 392
Brown, J. M., 245
Brown, J. S., 363
Brown, W. L., 155, 204, 219
Brownlee, J., 359
Buenau, K. E., 367
Burbrink, F. T., 72
Burke, I. C., 371
Burke, M. J., 64, 66
Burns, J. H., 159
Burns, K. C., 61
Buss, L. W., 144
butterflies: traits and species distributions of, 199–200; traits-phylogeny relationships in, 160

Cadenasso, M. L., 373
Cadotte, M. W., 7, 70–71, 75, 160, 189, 272, 306
CAFÉ dynamics. *See* community assembly and the functioning of ecosystems (CAFÉ) dynamics
Cahill, J. F., Jr., 159–60
Calcagno, V., 9, 61, 126, 143, 332
Capinha, C., 292, 293
Capitán, J. A., 327, 328, 330
Cardinale, B. J., 299, 337, 339, 342, 344
Cardoso, P., 187
Carlson, A., 60
Carson, W. P., 271
Case, T. J., 28, 155, 319, 391
Casini, M., 333
Cassey, P., 293–94
Caswell, H., 16, 27–28, 148
Cavender-Bares, J., 153, 155, 156, 160, 167–70, 173, 261–62
Ceballos, G., 292
Cedar Creek BEF experiment, 351
Center for Tropical Forest Science–Forest Global Earth Observatory network, 80–81, 82
Chalcraft, D. R., 142, 296, 299, 327
Chalmandrier, L., 160
Chang, L. W., 110, 118
Channel Island archipelago, bird species turnover in, 137
Chao, A., 179n1, 277
character convergence, 219; in neutral dynamics, 225

character displacement, 204, 219; community-wide, 155; evolution of, 220–22; with interspecific competition, 222–24; in SS models, 225

Charles-Dominque, T., 192–94

Chase, J. M., 2, 5–8, 16, 18, 19, 24, 25, 29–30, 35, 38, 52, 55, 59, 73, 75, 88, 92–93, 107, 112–15, 118, 120, 124, 132, 133, 141–43, 148, 162, 163, 169, 172, 178, 180, 181, 186–87, 259, 263, 267, 273, 277, 282, 296, 299, 304–9, 311–13, 316, 317, 327–30, 332–33, 341, 343, 348, 377, 385, 392

Chave, J., 5, 7, 18–19, 32, 33, 93, 154, 259, 263

checkerboard distributions, 90, 95–99, 100–101, 105, 155, 259, 260

Chesson, P. L., 6, 28, 30, 33–34, 38, 43, 47, 52, 55, 132, 157, 161, 162, 166, 169, 272, 304–5, 324, 350

Chiarucci, A., 299

Chinese subtropical forest, fourth-corner analysis of, 192–94

Chisholm, R. A., 93, 257, 259

Chiu, C. H., 187

churning food chains, 325–30

Chuyong, G. B., 82–83

Cianciaruso, M. V., 171

Cisneros, L. M., 157

clades: adaptive radiations of, 251–52; competition in, 251; diversification of, 242–44

Clark, C. M., 64–66, 128

Clark, J. S., 55

Clauss, M., 190

Clavel, J., 196

Clements, F. E., 17, 19, 28, 99, 189, 381, 384–85

Clementsian gradients, 99–105

Cleveland, C. C., 358

climate: biome distribution and, 77, 80; dispersal limitation and, 356–59; diversity and, 261, 263; productivity and, 338

climate change: conservation actions and, 389–91; dispersal limitation and, 356–59

cluster Poisson theory, 264

Cody, M. L., 15, 155

coevolution: diffuse, 222; enemy-victim, 325; frequency-dependent, 226–28; geographical mosaic theory of, 204, 205, 227–28; host-pathogen, 254, 325

coexistence. *See also* local coexistence: of amphipod species, 57, 233–36; competition and, 13–15; dispersal and habitat

heterogeneity in, 305; at equilibrium, 16; fragile, 38–39; frequency-dependent in spatially continuous metacommunities, 147–49; of inferior and superior competitors, 32–34; interactions of scales in, 6; neutral, 2, 56–58, 208, 219, 222, 229, 235, 242; niche-based perspectives of, 28–29; patterns of, 3; resource competition and, 31; spatial distributions of resources and, 29–30; spatial processes and trophic interactions in, 304–6; species combinations in, 389; species traits and, 16; stabilizing and equalizing factors in, 38–39, 219; trait differences in, 161–65; traits and phylogenies in, 151–76

coexistence theory: modern vs. classic, 169; predicting trait and phylogenetic over- or underdispersion, 162–67

Cohen, J. E., 24, 31

coherence, 100–101

Collinge, S. K., 74

Collins, M. D., 97–98, 105–6

colonization: early-stage stochasticity in, 59; species abundance and timing of, 210

colonization-competition trade-offs, 8, 14, 26, 30–31, 33–35, 39–40, 74–75, 174; in habitat heterogeneity, 270; stabilizing coexistence strength in, 43

colonization-extinction dynamics, 15–16, 18, 23, 24, 55; differential, 32; environmental and resource constraints in, 75–77; ETIB in, 313–14; frequent, 133; in island biogeography, 134; multiple stable equilibria and, 142–43; stochastic, 60–62; in temporal turnover, 139; turnover pattern and, 137–38

colonization–extinction equilibrium, 59–62

colonization–extinction rates, 36–37, 40–41, 56; predators altering, 311

colonization rates, 10, 36, 43, 56; in demographic stochasticity and ecological drift, 59–62; environmental filtering and, 40, 41; in food webs, 318; habitat isolation and, 70–74; stochastic drift and, 36; stochastic extinction and, 59–62

Colwell, R. K., 271, 274, 277

Comita, L. S., 54, 324

community. *See also* metacommunities: biomass of, 354–56; churning, 325–28, 330; definition of, 3; influence of scale in, 4–5; local adaptation and disturbances in, 202; niche-based vs. regional/dispersal-based factors in diversity of, 64–65; open, 354–55; predictability versus randomness of, 384–85; spatial processes in, 13–15; subjectivity of, 4

community assembly: complexity of, 388; deterministic, time influence on, 140; dispersal effects on, 384; ecosystem functioning and, 335–68; humpty-dumpty effect in, 330–31; key components of, 214; metacommunity ecology in, 240; phylogenetic and trait-based information to understand, 158–61; variation in, 368

community assembly and the functioning of ecosystems (CAFÉ) perspective, 340, 342; mass effects altering, 350; in metacommunity context, 342–59

community composition: dispersal- vs. niche-based theories of, 106–8; local fluctuations in, 140; variation partition of, 110

community ecology: focus of, 1–2; key concepts of, 2; versus metacommunity ecology, 24

community genetics, 204–5

community-level patterns, 3; spatiotemporal design of, 3–4

community monopolization effect, 210–13; empirical evidence for, 213–19; followed by niche differentiation in patches, 241–42; interaction of with neutral evolution, 228–36

community monopolization hypothesis, 202, 207–19

community monopolization theory, three-patch model, 208–10

community phylogeny approach, 249

community selection, 205, 227

competition, 21. See also colonization-competition trade-offs: among closely related species, 158–60; coexistence and, 13–15; diffuse, 204, 220, 224, 226; functional traits and, 160–61; habitat type and, 233–35; interspecific, 146, 151, 222–24; phylogenetic information as predictor of, 159–60; species dispersal and, 165–66; trait overdispersion and, 167

competition–fecundity trade-offs, 39–40

competitive effect, evolution of, 222, 225

competitive exclusion, 162, 344; principles of, 13, 17; misconception about, 30; species traits and, 156–57

competitive interactions: between different species, 57–59; trait overdispersion and, 169–72

competitive response, evolution of, 222, 225

competitors: convergence of, 222–23; dominant, 340, 342; good, 32; poor, 340, 343, 349; subdominant, 340, 353–54

complementarity, 347–48, 382

complex adaptive systems (CASs), 335; ecosystems within metacommunities as, 359–63; features of, 339

complexity-stability approach, 303

Condit, R., 5, 7, 28, 106–8, 114, 263, 273

congeneric species pairs, 14, 95–97

connectance: dispersal rates and, 122–24; in food webs, 318

connectivity, 383; manipulation of, 18; in type of metacommunity model, 36, 37

Connell, J. H., 5, 9, 13, 32, 52, 148, 227, 261, 308, 324

Connor, E. F., 5, 16, 60, 91–92, 95–96, 98, 99, 277, 278

conservation: actions for, 390–91; of matter, 3; SLOSS habitats in, 283–84

consumer-resource models, 24, 29, 38, 162, 344

contemporary biogeography, 259–61; biodiversity along ecological gradients in, 294–300

contingency, 21

continuous habitats, 12

continuum theory, 263–64

co-occurrence: null models and, 94–99; scale dependence in trait dispersion in, 167–68, 170–71; studies of, 92

Cooper, J. K., 305–6

Cooper, N., 156, 175

coral reefs: goby species in, 56; predatory fish in, 316

core assumptions, 23, 49. See also archetypes; specific archetypes

core mechanisms, 49. See also specific mechanisms

core-satellite dynamics, 39–40

core-satellite hypothesis, 32

core species, 91

Corixa species, coexistence and competition among, 13–15

Corlett, R. T., 389

Cornell, H. V., 5, 94

Cornwell, W. K., 153, 159, 160, 172

Cothran, R. D., 57, 233

Cottenie, K., 85, 92, 109, 110, 111–12, 115–17, 122–23, 127, 140, 178, 218–19, 273, 343

Craine, J. M., 153

Crespi, B. J., 249
Crist, T. O., 267, 276, 277
Crooks, K. R., 332
Crutsinger, G. M., 205, 216
Culver, D., 24, 31, 40, 388
Currie, D. J., 258, 261, 263, 271, 295, 296
Curtis, J. T., 286–87, 288, 290
Cusens, J., 295, 296

Daily, G. C., 288
Dallas, T., 104
Dalling, J. W., 115
D'Amen, M., 257
Damschen, E. I., 74
damselfly species: neutrally coexisting, 56–57, 58; speciation in, 242–43
Damuth, J., 301
Dangremond, E. M., 68
Daphnia species: biogeography and environmental effects on, 250; population differentiation in, 214–16, 217
Darwin, C., *xiii–xiv*, 155, 158, 189, 203, 256
daSilva, P. G., 183
Daufresne, T., 344
Davidson, D. W., 75
Davies, K. F., 161
Davies, S. J., 83
Davies, T. J., 154, 160
Dayan, T., 204
de Bello, F., 311
De Bie, T., 122
De Caceres, M., 116, 118, 124, 183, 277
de Knegt, H. J., 121
de la Sancha, N. U., 104
de Mazancourt, C., 319
De Meester, L., 115, 207–10, 213, 214, 216, 217, 219, 229, 230, 236, 241
De Schrijver, A., 272
DeAngelis, K. M., 358
death. *See also* birth, death rates and: commonality of (Naeem Triptych), *xiii–xiv*; rates of, 54–55
Debinski, D. M., 285
Declerck, S. A., 358
demographic properties, 54–55
demographic stochasticity, 28, 49, 50, 51; in community turnover, 139; versus environmental stochasticity, 55–56; evidence for, 54–62; in multiple stable equilibria, 141–42; stochastic extinction and colonization rates in, 59–62

Denslow, J. S., 270
desert rodent assemblage, 354
destabilizing effects, 151, 169, 175, 305, 375, 379; in food webs, 305–6; frequency-dependent, 169–72, 202
destabilizing niche relations, 151, 162–63
destabilizing trait differences, 162
determinism: of extinction rates, 60–61; of interspecific competition, 95, 128; lack of, 56, 178; neutral theory as null model against, 28; of population properties, 55; in response to environmental change, 61–62; in species interactions, 59; versus stochasticity, 15–18, 55, 138
deterministic community assembly, 59, 130, 131–34, 140
detritus diffusion, 374–76
Devictor, V., 187
Diamond, J. M., 15, 16, 91–92, 94–96, 98, 99, 137, 154–55, 181, 283, 391
Diaz, S., 190
Dickerson, J. E., 59
Didham, R. K., 285
diet breadth, 317–18
differentiation rates over time, 242–44
Diniz, J.A.F., 116
Dionne, K., 57, 235
Dirzo, R., 283
dispersal, 2–3, 26, 51, 382, 383; anthropogenic activities and, 293, 294; in BEF relationships, 339–40; biased, 387; biodiversity patterns and, 5; in coexistence, 305; in colonization and extinction events, 16; effects of on community assembly and niche-trait evolution, 384; factors important to, 50; in food-web versus competition models, 322–23; habitat heterogeneity and, 40, 84–87; human-mediated, 390–91; influences of, 24–25; interspecific effects and, 74–77; predation and, 306–7; regional, 293, 294; resources levels and, 348–50; stability and, 322, 323; strong versus weak, 126–27
dispersal limitation, 7, 49, 123, 335; in community patterns, 28; environmental change and, 356–59; evidence for, 62–74; factors in, 69; in habitat heterogeneity, 270; habitat isolation and dispersal rates in, 70–74; importance of, 385; in local ecosystems, 361–62; multiple stable equilibria and, 142; positive effects of, 65–67; species interactions in, 68–70; in

species pool unpredictability, 181; species sort and, 358; species variation in, 67–68; in taxonomic β-diversity, 184; at taxonomic level, 182
dispersal-limited metacommunities, 352–56
dispersal rates, 50; along habitat corridors, 74; in dispersal limitation, 70–74; in local community diversity, 40–42; metacommunity connectance and, 122–24; patch heterogeneity and, 386; in response to environmental and spatial factors, 121–22; scale dependence of, 7–8; in type of metacommunity model, 36
dispersal sufficiency, 7–8, 335
dispersal surplus, 8, 335; mass effect under, 348–52; in open ocean systems, 362
dispersal traits, 151, 154, 387
dispersers, keystone, 323, 388–89
distance–decay relationship (DDR), 266–67, 268; human activities and, 292, 293; predicting from neutral model, 275–76
distribution, 90; archetype interaction in, 91; biogeographic, 91–92; competition and, 165–66; in dispersal limitation, 63; habitat heterogeneity and, 77–84; interspecific interaction influencing, 21; scale influence on, 5; traits related to, 189
Diver, C., 17
diversity. *See* α-diversity; β-diversity; biodiversity; γ-diversity
diversity-biomass relations, 347–48, 348
diversity partitioning, 276–77
diversity–stability theory, 319–23
diving beetles, clumping traits in, 224, 226
Doledec, S., 192
Donald, D. B., 364
Donohue, I., 261, 299
Donoso, D. A., 168, 171
Dornelas, M., 28, 132, 133, 135, 138–39, 150, 184, 292, 338
Douglas, W. Y., 9
Downing, J. A., 336, 337, 365
Drakare, S., 278, 296, 298
Drake, J. A., 59, 309
Dray, S., 109, 194, 197
drift, 51, 383. *See also* ecological drift: evidence for, 54–62; factors important to, 50; predation and, 308–9; with scale, 385; in temporal turnover, 139
Duarte, L.D.S., 105–6, 246
Duivenvoorden, J. F., 107

Durrett, R., 144
Dybzinski, R., 362

EACs. *See* endpoint assembly cycles (EACs)
EAR. *See* endemics-area relationship (EAR)
early-divergence radiation, 252
eco-evolution: building, 204–5; concepts of, 203–4; dynamics of, 2–3, 202–38; processes of, 20–21, 27, 205, 210, 219, 235
eco-evolutionary feedbacks, 202, 203, 237–38, 384
ecological drift, 383; biodiversity patterns and, 5; demographic stochasticity and, 55–56; in different species, 57–59; in habitat heterogeneity, 270; multiple stable equilibria and, 142; predation and, 308–9; stochastic extinction and colonization rates in, 59–62; stochasticity leading to, 56–59; in taxonomic β-diversity, 181, 184; at taxonomic level, 182
ecological performance, functional traits and, 160–61
ecological processes: integration evolutionary processes with, 250–54; in speciation, 240
ecological speciation: in metacommunity processes, 241–42; in SS dynamics, 239
ecological subsidies, 370
ecological traits: evolution of, 222–24; trade-offs of, 49
ecology: artistic representation of, *xiii–xiv*; synthetic, 20, 47, 117, 177, 200, 258–59, 359, 370, 379, 380, 382, 385, 392
Economo, E. P., 386–87
Ecosphere, 371n1
ecosystem consequences, 22
ecosystem functioning, 392; community assembly and, 335–68; in dispersal-limited metacommunities, 352–56; diversity and, 351; environmental change and dispersal limitation in, 356–59; local species richness and, 341–42; in metacommunity context, 342–59; in SS models, 343–48
ecosystems. *See also* local ecosystems: adaptive response of, 361–62; anthropogenic effects on, 336; artificial, 371–72; as CASs, 359–63; community assembly and, 368; emergent patterns in, 364–65; material and energy inputs into, 372; regular features of, 363–67; spatial dynamics in, 371–73; tipping points in, 335, 339, 363–67, 368; varied component movement rates in, 369
Edenhamn, P., 60

edge effects, 61, 280, 284
Edwards, K. F., 190, 198–99
Egler, F. E., 101–5
Ehrenfeld, J. G., 141
Ehrlén, J., 65
Eisenhauer, N., 338, 339
Ejrnaes, R., 142
Elahi, R., 292
Ellis, A. M., 133
Ellner, S. P., 85, 203, 237–38
Elmendorf, S. C., 67, 68, 352
Elser, J. J., 2, 365, 379
Elton, C. S., 2, 17, 28, 52, 155, 158–59, 189, 319, 332, 363, 391
embedded absences, 100
emigration, 321, 372, 373; dispersal and, 349–50; preventing extinction, 206; rates of, 55, 349–50
endemics-area relationship (EAR), 266, 268; predicting extinction rates, 284
endpoint assembly cycles (EACs), 131–32, 144–46, 150, 251; food-web-induced, 327–31; regional context of, 145–46; slow assembly process leading to, 325–27
endpoint communities, 352–54
enemies, differential shared, 47
energy: inputs of, 372, 373; as limiting resource, 294
energy flow constraints, 3
energy richness, 294–95; species richness and, 297–99; strongly positive, 296–97
Enquist, B. J., 155, 172, 175, 189
environment. *See also* ecosystems; habitat: assembly history and productivity of, 145–46; in functional and phylogenetic patterns, 182–84; mosaic nature of, 14–15; nonlinear patterns of, 109n2; number of measured variables of, 117–18, 119; organic material movement and homogeneity of, 377; species distribution in, 197–98
environmental change: dispersal limitation and, 356–59; evolutionary response to, 206–7; local richness and, 358–59; resource supply and, 357; in temporal turnover, 139–40
environmental conditions: community composition and, 143–46; dispersal rates and, 122–23; functional traits and, 189–92; influence of, 77–84; in metacommunity structuring, 113–17; priority effects in, 142–43; variation partitioning in diagnosing, 106–12

environmental differences, 28, 95; mass effects and, 85; response to, 362; spatial distance and, 106–9; trait differences and, 169, 180–81
environmental distance, 50, 106–8, 123–24, 182
environmental disturbances, 196; in extinction, 218; in relative abundance of individuals, 272; species diversity and, 271–72
environmental filtering, 40, 41, 159, 162, 172; ecological speciation and, 242; functional traits and phylogenies in, 151; importance of, 385; trait underdispersion and, 169
environmental gradients, biodiversity along, 294–300
environmental heterogeneity, 50; organic material movement in reduction of, 377; spatiotemporal, 51; species size and ability to partition, 112–13
environmental stochasticity, 139; versus demographic stochasticity, 55–56
environmental tolerances, 172; in taxonomic β-diversity, 184
environmental variation: community composition and, 228–30; functional and taxonomic pattern response to, 200; plant strategies across, 189–90; temporal turnover and, 133–34, 150
equalizing differences/factors: in coexistence, 38–39, 219; fitness, 158, 162–63, 304; stabilizing differences and, 56, 165–67
equalizing factors, 38, 162, 165, 159, 219
equilibrium, single stable, 325
equilibrium coexistence, 16
equilibrium theory of island biogeography (ETIB), 256, 310–11; in colonization-extinction dynamics, 313–14; predator and prey diversity in, 316; trophic, 317–18
Ernest, S. M., 184, 354, 358–59
Eros, T., 104
Essl, F., 293
ETIB. *See* equilibrium theory of island biogeography (ETIB)
Etienne, R. S., 154, 173, 174–75, 274
Evans, K. L., 261, 271, 295, 296
evenly spaced gradients, 99, 100
evolution. *See also* co-evolution: adaptive, 202–39; artistic representation of, *xiii–xiv*; frequency-dependent, 207, 226–28; neutral, interacting with community monopolization, 228–36; of neutrality, 202,

207; toward neutrality, 219–26; of traits, 203–4, 209–10, 216, 223, 235, 384
evolutionary divergence, types of, 229
evolutionary-ecological process integration, 250–54
evolutionary ecology. *See* eco-evolution
evolutionary priority effects, 209, 216, 244
evolutionary processes, 39, 153, 224, 383.
See also eco-evolution, processes of; macroevolutionary processes: coexistence and, 219; creating genetic variation, 204–5; 349–50, ecological processes and, 237–38; large-scale, 240; local, 362; long-term, 203; spatial, 244, 392; stabilizing effect of, 39
evolutionary relationships. *See* phylogenetic relationships
evolutionary rescue, 206
evolved neutrality, 230
evolving metacommunity approach, 205, 207, 237, 386
Ewers, R. M., 285
exotic species, dispersal of, 390–94
extinction. *See also* colonization-extinction dynamics: anthropogenic activity and, 292, 294; dispersal limitations in, 16; with environmental disturbances, 218; island biogeography versus metacommunity-like structures predicting, 291; periodic, 229; recolonization after, 60–61; regional or local, 344–45, 352; stochastic, 28, 30–31, 49, 56, 59–62
extinction rates. *See also* colonization-extinction rates: colonization rates and, 41; with habitat loss, 284; stochastic drift and, 36

Faaborg, J., 64
facilitation, 21, 304, 368
Fagan, W. F., 306, 332
Fahimipour, A. K., 332–33
Fahrig, L., 3, 285, 286
Farjalla, V. F., 122
Farneda, F. Z., 161
feedbacks, 21, 27, 35; between biogeographic/evolutionary and ecological drivers, 350, 354; eco-evolutionary, 202, 203–6, 227–28, 236–37, 246; in food webs, 303, 304, 324–27; intra- vs. interspecific, 38; negative, 141–43; plant-soil, 141, 144, 148; positive, 141, 151; between trophic interactions and spatial processes, 304, 324–27
Feeley, K., 156

Feener, D. H., Jr., 75
Fellers, J. H., 75
Ferdy, J.-B., 141–42
Fernandes, I. M., 104
Fine, P.V.A., 5, 79, 178–79, 261
Fischer, J. D., 291
Fišer, Ž., 235
fish communities: traits correlated to environmental conditions in, 190; variation partitioning and residuals in, 129
Fisher, R. A., 135, 136, 256
fitness, habitat gradient and, 78–81
fitness–density covariance, 43
fitness differences, 52; in near-neutrality, 230–32; species richness levels and, 351
Fitzpatrick, C. R., 205
Fletcher, R. J., 36, 39, 57, 308
flight capacity, 183
Flinn, K. M., 126–27
Fontaine, C., 376
food discovery–interspecific dominance trade-off, 75
food-web metaecology, 373–74
food webs, 2, 21–22; community assembly dynamics in, 144–45; diversity and stability in, 319–23; interactions in, 303, 304; in metacommunities, 303–34; metacommunity assembly in, 327–31; metacommunity processes in, 332–33; structure of, 332–33, 392; trophic interactions in, 324–27
Forbes, A. E., 59, 115
Forbes, S. A., 363
forest habitats. *See also* tropical rainforests: pairwise species associations in, 166; plot variation and species pool effects in, 124–25; species associated with variables of, 82–83
Foster, B. L., 27, 88, 113, 165, 339–40, 347
Fournier, B., 44–45, 127, 386–87, 388, 390
fourth-corner analysis, 191–98, 247
Fox, B. J., 156, 181
Fox, D. L., 190
Fox, J. W., 57, 219, 226–27
Fraser, L. H., 261, 296
frequency dependence, 26, 128, 129; of competition, 56–57; in community assembly, 140; in habitat heterogeneity, 270; negative, 148–49; positive, 149; spatial variation and, 309; in spatially continuous metacommunities, 132, 147–49; in taxonomic β-diversity, 184; between trophic interactions and spatial processes, 324–27; weak negative, 148–49

frequency-dependent evolution, 207, 226–28
frequency-dependent interactions, 9, 140, 141, 147, 149, 169–72, 299, 324–25
frequency-dependent selection, 202, 226–27
freshwater algae fitness differences, 160, 164
freshwater ponds/lakes, 113; bacterioplankton community in, 357–58; coexistence in, 13–14, 233; dispersal limitation in, 64–65, 67; metacommunity structures in, 102–3; in niche theory development, 113; performance trade-offs in, 79
Fritschie, K. J., 159–60, 163
Fukami, T., 55, 59, 142, 178, 179, 180, 216, 218, 241, 327
Fukumori, K., 87, 310
functional β-diversity, 185–86, 187
functional-diversity-area relationships (FAR), 173
functional information: in spatiotemporal change analysis, 184–89; in taxonomic pattern analysis, 179–84
functional macroecology, 260, 261–62
functional redundancy, 178, 185, 187, 200, 181–82
functional species groups, 156
functional-trait approach, 153; with taxonomic approach, 177
functional trait dispersions: scaling dependence of, 167–73; in species sort metacommunity, 167–73
functional traits: not phylogenetically conserved, 159–60; in old growth vs. disturbed tropical forests, 179–80; performance and, 160–61; phylogenetically conserved, 158–61; phylogenies and, 151; in species abundance and distribution, 189–97; species variations with, 154; taxonomic patterns and, 177–201
fundamental niche, 63

γ-diversity, 262; β-diversity and, 292
Gaines, S. D., 292, 293
Galapagos Islands: finches of, *xiii–xiv*, 251; land snail distribution in, 247–48, 252; in Naeem Triptych, *xiii–xiv*
Gardezi, T., 296, 297, 298, 327, 330, 385
Garnier, E., 159, 197
Garzon-Lopez, C. X., 83, 120
Gaston, K. J., 256, 258, 295, 364
Gauch, H. G., 263–64
Gause, G. F., 13, 14, 17, 52

Gavin, M. C., 72–73
Gavish, Y., 286
Gavrilets, S., 242, 244–45
generalized fractal theory, 263, 264
geographic variation, 26
geographical mosaic theory of coevolution, 204, 227–28
geometric constraints, 166–67, 174
geometric scaling processes, 2–3
Gerhold, P., 160
Gerla, D. J., 141–42
Ghalambor, C. K., 203
Ghazoul, J., 288, 291
Ghilarov, A. M., 219
Gibson, L., 61
Gido, K. B., 168, 171
Giladi. I., 286
Gilbert, B., 70–71, 73, 109, 124, 127, 140, 273
Gilbert, F., 74, 314–15
Gilbert-Norton, L., 74
Gillespie, R. G., 241–42, 251–53
Gillman, L. N., 297
Gillooly, J. F., 262, 379
Gilpin, M. E., 24, 91–92, 95–96, 283
Gleason, H. A., 17, 19, 28, 99, 179, 189, 256, 264, 381, 384–85, 391
Gleasonian gradients, 99, 100, 101, 102–4, 105
global monopolization, 212–13
Godoy, O., 159–60
Goldberg, D. E., 161
Goldenfeld, N., 274
Gomulkiewicz, R., 206
Goncalves-Souza, T., 170
Gonzalez, A., 18, 61, 74, 206, 296–98, 327, 330, 332, 338, 385
Gonzalez-Caro, S., 171
Goodnight, C. J., 205
Gotelli, N. J., 95, 96, 105, 174, 191, 257, 271, 274
Götzenberger, L., 101
Gouhier, T. C., 147, 305–6
Gounand, I., 305–6, 375
Grace, J. B., 28, 339
Graham, C. H., 175, 178–79
Grainger, T. N., 70–71, 73, 273
Grant, B. R., 251
Grant, P. R., 155, 204, 251
grasslands: regional controls in, 88; taxonomic and functional dissimilarities in, 179–80
Gravel, D., 19, 25, 38, 112, 316–23, 372, 374–76
Graves, G. R., 5

Gray, J. S., 277
Green, J. L., 258, 274
Grime, J. P., 64, 66, 189, 196
Grinnell, J., 52, 155
Groening. J., 244
Gross, K., 40, 342, 344
Grover, J. P., 30
growth rates: of different populations, 215;
    greater than zero, 63; low-density, 43–44;
    niche differences in, 57–58; population,
    43–44, 63, 174, 206; rapid dispersal and,
    174; species trade-offs in, 227
Gueze, M., 116
Gunatilleke, C.V.S., 82, 83
Gurevitch, J., 52
Gyllenberg, M., 18, 32, 91, 174, 259, 263

habitat: competition and type of, 233–35;
    connectivity of, 18; filtrating of in
    taxonomic β-diversity, 185–86; food chain
    length and productivity of, 332; isolation of
    in dispersal limitation, 70–74; matrix, 10,
    12, 284; partitioning of, 13, 57, 235; traits
    related to, 151
habitat amount hypothesis, 285
habitat corridors, 74, 88, 288, 290, 390
habitat-fitness trade-offs, 78–81
habitat fragmentation, 118–21, 255; ISAR in,
    283–86; isolated islands of, 290; metacom-
    munity perspective on, 291; species richness
    in, 289; in trophic cascades, 332–33
habitat heterogeneity, 269–70, 383; different
    species' response to, 120–21; dispersal and,
    84–87; dispersal heterogeneity and, 40;
    distribution of, 169; effects of, 77–84,
    118–22; energy and, 297–99; in food-web
    structure, 333; island size and, 282; on
    islands, 279; in mass effects models, 32–33;
    predation and, 307–8; principle role of, 121;
    reductions in, 272; in SAR shaping, 277–78;
    species sorting due to, 18
habitat loss: biodiversity loss and, 286–91;
    ISAR in, 283–86
habitat tolerances, 172
Haddad, N. M., 61, 272, 284, 285, 286, 391
Haegeman, B., 305–6, 348–49, 351
Hagen, M., 332
Hairston, N. G., 203
Hajek, M., 122, 125
Hall, D. J., 113
Halley, J. M., 266, 283, 288

Hamond, H., 226
Hanski, I., 18, 24, 31–32, 60, 75, 91, 174, 206,
    259, 263, 266, 284, 285–87
Hanson, C. A., 73
Hardin, G., 13, 30
Hardy, O. J., 173, 174, 175
harlequin-type landscape, 388
Harmon, L. J., 261
Harms, K. E., 7, 64, 65, 82, 83, 116
Harpole, W. S., 53, 153, 272, 299
Harrison, R. D., 69–71
Harrison, S., 28, 59–60, 86, 261
Harte, J., 94, 266, 274, 275, 277, 284, 392
Harvey, P. H., 156
Hastings, A., 7, 9, 24, 31, 32, 47, 305
Hatosy, S. M., 139
Hatton, I. A., 332
Hawkins, B. A., 190, 295, 296
He, F. L., 263, 267, 274, 284, 288
Heath, S. K., 390–91
Hector, A., 337–38
Hedin, L. O., 344
Heinlein, J. M., 307
Heino, J., 104, 122, 178, 390
Helmus, M. R., 173
Henderson, P. A., 365
Hendry, A. P., 2–3, 203
Henriques-Silva, R., 100, 103, 105–6, 257
herbivores, 214, 316, 364; environmental
    disturbances and, 120; in food web
    structure, 272; habitat associations and
    performance trade-offs of, 79; insect,
    314–16; movement of in metaecosystems,
    374–76; in plant diversity, 306–7; in plant
    extinction, 311, 332
Herkert, J. R., 286, 289
Hernández, M.I.M., 183
heterogeneity, 382. See also habitat heteroge-
    neity: coexistence and, 40, 305; extrinsically
    generated, 8–9; in habitat loss and
    fragmentation, 285–86; influences of,
    24–25; intrinsically generated, 9; patch
    dynamics and, 87; in type of metacommu-
    nity model, 36, 37
heterogeneity–diversity relationship, 77–84
heterogeneous metacommunity, 13, 35–36,167;
    species abundance in, 269–70; trait
    distributions in, 169–70
heteromyopia, 148
Hibbing, M. E., 144
Higgs, A. J., 283

Hill, J. L., 286, 289
Hillebrand, H., 261, 272, 295
HilleRisLambers, J., 6, 52, 54, 156, 158, 162, 163, 166, 169, 172
historical biogeographic hypothesis, 249
historical biogeography, 260, 261; versus contemporary biogeography, 294–300; interacting with metacommunity processes, 239; in metacommunity structure, 251
historical effects, 245–50
Hochkirch, A., 244
Holling, C. S., 363
Holt, B. G., 245
Holt, R. D., 8, 30, 40, 47, 147, 206, 285, 305–6, 311, 313–14, 316–17, 319, 325–26, 328, 330, 332, 373, 375, 376
Holyoak, M., 7, 8, 15, 18, 25, 305, 311, 324, 380–83, 385, 390–91
homogenization–synchronizing effect, 322
homogenizing gene frequencies, 206
Hooper, D. U., 337
Hoopes, M. F., 332
Horn, H. S., 24, 40, 388
Hortal, J., 8–9, 77, 272, 278, 282
Horwitz, R. J., 315
Hoshino, J., 310
host–pathogen relations, 148, 227–28, 254, 304, 325, 334
Hoverman, J. T., 102, 113, 115
Howeth, J. G., 85–86, 115, 350
Hoyle, M., 74, 314–15
Hubbell, S. P., 2, 7, 17–18, 24, 27–28, 30, 39, 52, 92–93, 113–14, 128, 139, 165, 168, 171, 174, 182, 204, 219, 220–22, 224, 229, 259, 263, 266, 270, 271, 274, 284, 288, 392
Hubert, N., 171
Huffaker, C., 7, 8, 9, 18, 32, 305, 311, 324
Hughes, A. R., 204
Hugueny, B., 60
Huisman, J., 47, 147
human activities. See Anthropocene era; anthropogenic activities
humpty-dumpty effect, 330–31
Huntly, N., 28
Hurd, S. D., 333
Hurlbert, A. H., 5, 109, 127, 271
Hurtt, G. C., 165–66, 174
Huston, M. A., 337
Hutchinson, G. E., xiii–xiv, 13–15, 17–18, 30–31, 52, 63, 113, 155, 189, 203, 224, 256
Hyallela species, 233

Hyatt, L. A., 324
hysteresis patterns, 366, 367

Ibanez, C., 190
Ibarra, J. T., 187
Iljon, T., 141–42
immigrant sink populations, 351
immigrants, 320, 372
immigration rates, 54–55, 312, 314, 350, 372
indirect effects, 21–22, 333, 372, 375–78
individual-based rarefaction curve, 274
individuals: relative abundance of in metacommunity, 267, 272–73; spatial aggregation of, 267, 273; total number in metacommunity, 267, 270–72
inferior competitors, 14–15; displacement of, 32–33; survival of, 32–33, 41, 75
Ingram, T., 153, 156, 172
Ings, T. C., 376
inorganic nutrient movement, 369, 373–75, 377
interaction network connectance, 317–18
intermediate disturbance hypothesis, 32
interspecific competition: functional traits and phylogenies in, 151; through time, 222–24; unlikely combination of, 146
interspecific dominance–food discovery trade-off, 75
interspecific effects, 162; versus intraspecific effects, 141; stochasticity and dispersal interactions with, 74–77
interspecific interactions, 128; in abundance and distribution, 21; multiple stable equilibria and, 142
intraspecific aggregation of species, 273
intraspecific effects, 162; versus interspecific effects, 141
invasions, 64n1, 145, 208, 292; diversity and, 67; by exotic species, 293–94; resistance to, 214–16, 218, 230–32, 362; vulnerability to, 326, 328
Isbell, F., 358, 391
Ishii, Y., 148
island biogeography, 16; colonization-extinction dynamics in, 134; equilibrium theory of, 27–28, 56, 256; trait-based perspective of, 155
island biogeography theory, 279; distance from mainland and biodiversity of, 70–74; niche theory and, 282; predicting extinctions, 291; stochastic extinctions in, 60
island species–area curves, 282

island species–area relationship (ISAR), 265–66, 311–13; habitat loss and fragmentation in, 283–86; in lakes, 315; metacommunity assembly and, 279–82; versus nested SAR, 279; predators altering, 311–13

islands: co-occurrence patterns on, 95–99; extinction-recolonization patterns in, 60–62; size and heterogeneity of, 282, 311; size of, 279; true versus habitat, 279–80

isolated ecosystems, tipping points in, 335

Ives, A. R., 173

Jabot, F., 154
Jaccard's index, 268
Jackrel, S. L., 205
Jackson, J.B.C., 144
Jamil, T., 190, 198, 199
Janzen, D. H., 148, 227, 324
Janzen-Connell effects, 129, 148–49, 169, 270
Janzen-Connell hypothesis, 254
Janzen-Connell mechanism of coexistence, 227, 324–25
Jean, K., 72–73
Jeffries, M. J., 364
Jena experiment, 347–48, 348, 354–455
Jenkins, D. G., 115, 245–46
Jeppesen, E., 367
Jetz, W., 5, 171, 172–73, 190, 261, 271
Jiang, L., 59, 142
John, R., 54, 116
Johnson, P.T.J., 60, 103
Jones, C. G., 9
Jones, M. M., 5
Jost, L., 277
Juliano, S. A., 271

Kadmon, R., 86, 270, 282, 295n3
Kallimanis, A. S., 278
Kalyuzhny, M., 139
Kardol, P., 141
Kareiva, P. M., 18
Karlson, R. H., 261
Kaspari, M., 8
Katano, I., 307
Keddy, P. A., 156, 157, 167–69, 181
Keil, P., 5, 266, 284
Keith, S. A., 103, 188
Keitt, T. H., 141–42, 386–87
Kelly, C. K., 39
Kembel, S. W., 154, 160, 168, 171, 173, 175
Kerr, B., 144, 146

keystone dispersers, 323, 388–89
Khalig, I., 159
Killingback, T. P., 144, 146
Kimura, M., 27–28
Kinnison, M. T., 203
Kinzig, A. P., 266, 284
Klausmeier, C. A., 18n.1, 190
Kleyer, M., 191
Klironomos, J. N., 141
Knapp, R. A., 249
Kneitel, J. M., 73, 306, 307
Knight, T. M., 267, 273, 333
Kodric-Brown, A., 206
Koh, L. P., 288, 291
Kohn, D. D., 282
Kolasa, J., 133
Korhonen, J. J., 132, 135
Kraft, N. J. B., 124, 157, 161–64, 166, 168–70, 172, 182, 263, 277
Kreft, H., 70, 72, 258
Kremen, C., 291
Kruess, A., 18, 314, 316
Kuang, J. J., 305
Kulmatiski, A., 324–25
Kunin, W. E., 85
Kunstler, G., 161

Lack, D., 17
Lai, J., 83
lake ecosystems. See also freshwater ponds/lakes: dispersal limitation and environmental change in, 361–62; nutrient inputs and turbidity in, 366–67; zooplankton diversity in, 331
Lakshadweep Islands extinction-colonization model, 61–62
Lamouroux, N., 190
Lan, Z., 299, 300
land-sharing scenario, 291
land-sparing scenario, 291
Landa, K., 161
landscape ecosystem ecology, 373, 374
Lankau, R. A., 144, 227
Lauenroth, W. K., 135, 136, 371
Lavorel, S., 190, 197
Law, R., 141, 144–48, 325, 327, 330
Lawler, S. P., 7, 18, 305, 311, 324
Lawlor, L. R., 376
Lawton, J. H., 1–2, 4, 13, 17, 271, 277, 391
Lechowicz, M. J., 109, 140
LeCraw, R. M., 332

Legendre, L. F., 109
Legendre, P., 109, 110, 118, 140, 191, 192, 194, 267, 277
legumes, nitrogen fixation in, 159
Lei, G. C., 75
Leibold, M. A., 2, 6–8, 15, 16, 18, 19, 24–27, 29–30, 34–40, 47, 52, 85–87, 92–93, 99, 101, 105–6, 112–15, 120, 128, 132, 141, 144–47, 157, 160, 162, 163, 169, 172, 181, 182, 186, 191, 210–13, 214, 228–30, 236, 246, 247, 249, 250, 254, 296, 299, 304–5, 307, 327, 328, 329, 330, 340–46, 350, 351, 353, 356, 358, 360–62, 364, 377, 380–83, 385–87, 388, 390, 392
Lepidoptera species, temporal relationships of, 135–36
Leroux, S. J., 333
Leslie, P. H., 57
Lessard, J. P., 79, 173
Letten, A. D., 38, 39, 52, 157, 160, 163, 169, 201
Levi, T., 370
Levin, S. A., 24, 25, 32, 47, 52, 144, 359, 360, 363, 385
Levine, J. M., 54, 63–64, 144, 156, 158, 161, 162, 166, 169
Levins, R., 16, 24, 29, 31, 40, 47, 52, 56, 155, 283, 286–87, 388
Ley, R. E., 316
Li, W.K.W., 365
Lieberman, A. S., 370
life-history differences, 32–33
Likens, G. E., 370
Lindenmayer, D., 389–90
Lindström, E. S., 350
Litchman, E., 190
Liu, X., 183
Livingston, G., 75–76, 310
local adaptations: versus assembly, 217–19; evolution in, 236; by homogenizing gene frequencies, 206; and metacommunity disturbances, 202
local coexistence, 5, 8, 38, 124; adaptive evolution and, 202; dispersal rates and, 341; mechanisms of, 46–47; in metacommunities, 220–21; principles necessary for, 304–5; scale and, 166–67; from stabilizing processes, 52–54; temporal variation and, 147
local community composition, control of, 184, 350, 361
local competition–environmental filtering tension, 157–58

local ecosystems: as complex adaptive systems, 360–61; determining diversity, 87–88; features of, 360; turnover patterns and diversity of, 137–39
local effects, 50
local feeding performance traits, 172
local monopolization, 212–13, 217–18
local-scale diversity, 19, 46–47, 291, 335, 351
local scales, 1, 6, 9–10, 49, 380–81; bridge to regional/global scales from, 257, 298; competition at, 343; defined, 257, 262; ecosystem function at, 335, 351, 354, 360; equivalent species at, 232, 235; food webs at, 322, 330; niche differentiation at, 228–29; processes of, 9–10, 28, 49, 50–52, 88, 237–38, 251, 318
local selection, interaction-dependent, 360
local variation, 360
localities: contributing to metacommunity pattern, 126–27; within metacommunity, 5–6
localized interactions, 360, 361
localness, emphasis on, 13
Lockwood, J. L., 184, 185, 187, 292
Loeuille, N., 40, 87, 210–13, 228–30, 236, 325, 386–87, 388, 390
Logofet, D. O., 320
Logue, J. B., 25, 34–35, 50, 70–71, 92, 343, 350
Lomolino, M. V., 261, 265–66, 277
López-González, C., 103
Loreau, M., 3, 5, 8, 15, 25, 32, 33, 40–42, 71, 85, 174, 259, 270, 273, 292, 305–6, 319, 322–23, 333, 336, 337–38, 341–43, 346, 348–51, 362, 368, 370–73, 375
Losos, J., 159, 204, 251, 261, 310–11
Lotka, A. J., 2, 52, 256
Lotka-Volterra competition equations, 38, 240
Lotka-Volterra competition models, 141, 391
Lovette, I. J., 8, 60, 261, 279
Luckinbill, L. S., 305
Lukaszewski, Y., 67
Lundholm, J. T., 109, 127
Lutscher, F., 141–42
Luza, A. L., 170
Lynch, J. F., 60

MacArthur, J. W., 77
MacArthur, R. H., 7, 13–18, 24, 27, 29, 30, 40, 47, 52, 56, 60–62, 70, 77, 93, 132, 134, 137, 155, 158–59, 184, 256, 259, 261, 279, 283, 286–87, 310–12, 319, 388

MacArthur's paradox, 15–18
MacDougall, A. S., 83, 88, 306
Mackey, R. L., 261
macroecological patterns, 300–302
macroecology: biodiversity, 259, 260, 262–67; biogeography and, 384; definition of, 257–62; focus of, 256–57; four conceptual realms of, 259–62; functional, 260–62; metabolic and functional, 260, 261–62; of metacommunities, 255–302; origin of, 255–56; unified theories of, 263–64
macroevolution, 239–54
macrophyte-phytoplankton interactions, 366
Maguire, B., 70
Magurran, A. E., 184, 365
Malcom, J. W., 203, 214
mammals, traits of correlated to environmental conditions, 190
Mangan, S. A., 54
Margalef, R., 189, 363
Markl, J. S., 69
Marleau, J. N., 372, 375–76
Maron, J. L., 306, 339–40, 347, 358
Martin, H. G., 187, 274
Martinson, H. M., 306, 332
Martiny, J.B.H., 73
mass balance, 372
mass effects (ME), 23; in CAFÉ dynamics, 350; dispersal surplus under, 348–52; evidence for, 85–86; habitat heterogeneity-dispersal interactions in, 84–87; in metacommunities under dispersal surplus, 348–52; processes of, 26; in synthesized occupancy model, 44–46
mass effects (ME) models, 24–25, 32–34, 47; assembly processes in, 201; critical assumptions of, 51; environmental and trait variation in, 154; habitat heterogeneity in, 277–78; key propositions and predictions for, 35; local species richness and productivity in, 341–42; local versus regional diversity controls in, 87–88; spatial and environmental effects in, 110, 112; temporal turnover in, 133; trait and phylogenetic dispersions in, 174
Massol, F., 370, 371, 373, 374, 377, 378, 387
mate recognition, speciation reinforcement through, 244–45
material fluxes, 370–71
material inputs, 372–73; homogenizing effect of, 377

Matias, M. G., 266, 282, 285
mating interference, 244–45
matrix habitats, 10, 12; adjacent, 284
Matthews, T. J., 279
Maurer, B. A., 17, 256, 257–58
maximum entropy (MaxEnt) theory, 264, 275–76, 392
May, F., 258, 265, 273, 275, 276
May, R. M., 267, 283, 303, 319–23
Mayfield, M. M., 156, 158, 161, 162, 166, 169
Mayr, E., 96
May's paradox of diversity, 319, 322
Mazel, F., 188–89
Mazumder, A., 272, 307
McBride, P. D., 296, 327
McCabe, D. J., 95, 96, 191
McCann, K. S., 319, 320
McCauley, S. J., 102, 113, 115
McCoy, M. W., 60, 277, 278, 333
McGehee, R., 47, 147
McGill, B. J., 2, 19, 26, 92–94, 153, 155, 158, 166, 174, 175, 189, 220, 256, 259, 261–65, 267, 274, 301, 392
McGlinn, D. J., 275
McIntosh, R. P., 17, 101
McKinney, M. L., 184, 292
McLachlan, J. S., 390–91
McPeek, M. A., 15, 19, 25, 28, 38, 52, 56, 57, 79, 112, 113, 115, 181, 182, 242–45, 305
McPhail, J. D., 204
McQueen, D. J., 364
ME. *See* mass effects (ME); mass effects (ME) models
Melbourne, B. A., 43–44, 77
Menezes, S., 161
Menken, S. B., 40
mesocosms: colonization histories of, 142, 143; habitat type and competition among, 233–35
meta-May hypothesis, 320
meta-May matrix, 320–21
metabolic theory of macroecology, 260, 261–62, 379, 392
metabolism, 2
metacommunities: adaptive evolution in, 205–7; discrete versus continuous, 9–13; generalized view of, 6; multiple processes at multiple scales in, 5–9; sensu lato, 132, 390; sensu stricto, 9–10, 147, 387; ubermodel of, 42–46; unified framework for, 25–26

metacommunity archetype—metacommunity ecology interface, 37–42

metacommunity archetypes. *See* archetypes

metacommunity assembly: in biodiversity macroecology patterns, 274–76; island SAR and, 279–82; mechanisms of, 385; nested SAR and, 277–78; trait-based theories of, 189–97

metacommunity dynamics: functional traits and phylogenies in, 151–76; functional traits and taxonomic patterns in, 177–201; with and without evolution, 231

metacommunity ecology: accomplishments of, 382–84; from basic to applied, 389–92; versus community ecology, 24; focus of, 256–57; history of, 13–15; as hub for synthetic ecology, 20; macroecology and, 255; mechanisms of, 383; prospectus for, 19–20; synthesis through, 384–85; transition in, 380–92; unifying community and ecosystems ecology, 2–3

metacommunity evolution, Hubbell's model of, 220–22

metacommunity–historical biogeography links, 251

metacommunity patterns: different species and localities contributing to, 126–27; factors influencing variation in, 117–26; number of measured variables in, 117–18; spatial factors in, 129–30

metacommunity perspective: current status and limitations of, 385–87; future of, 387–89; on habitat fragmentation and loss, 291; reemergence of (1990s–2000s), 18–19; theory frameworks of, 23

metacommunity processes: in food-web structure, 332–33; phylogenetic and trait-based information correlation in, 158–61; in phylogeny and radiations, 241–45; trophic interactions influencing, 306–10

metacommunity structure: elements of, 99–106; explaining, 127–29; multidimensional variation patterns in, 388; multiscalar, 388; null models and co-occurrence in, 94–99; spatially explicit, 387; trophical, 310–18; variation in, 112–17

metaecosystems, 336, 369–79; burgeoning theory on, 369; elements of, 373–77; principles of, 377–78; schematic summary of, 378

metapopulation, 287; adaptive evolution in, 205–7; models of, 56, 59–60, 263–64, 284; theories of, 10, 24, 31, 264, 380, 390

metapopulation dynamics hypothesis, 32

metapopulation–metacommunity perspective, 372

Methot, G., 330

Meynard, C. N., 103, 105–6, 182, 257

Michels, E., 85–86

Michigan ponds: environmental and spatial effects in, 113, 114–15, 116, 117; environmental variables in, 118, 119; habitat heterogeneity in, 119–20, 121; species pool in, 126; taxa- and trait-level variation partition in, 180–81

microbes: dispersal limited, 73–74; temporal turnover of, 136

microcosms: aquatic, 70–71, 371–72; Gausse's experimental, 14; Park's experimental, 57–59

microevolutionary processes, 178

micropatches, 40–42

migration rates, 54–55, 312, 314, 350. *See also* emigration rates; immigration rates: species richness and, 85, 86

migrations, assisted, 390–91

Mikkelson, G. M., 99, 101, 191

Millennium Ecosystem Assessment, 287, 288

Miller, T. E., 73, 85, 86, 222, 225, 306, 341

Miller, T. J., 60

Mims, M. C., 190

Mittelbach, G. G., 3, 17, 21, 239–42, 250, 261, 295, 296, 327

Moles, A. T., 65

Molofsky, J., 9, 141–42, 148–49, 273

Moloney, K. A., 388

Mooij, W. M., 141–42

Mooney, H. A., 336

Moore, K. A., 67, 68, 352

Mordecai, E. A., 87

more individuals hypothesis (MIH), 271

Morin, P. J., 3, 17, 21

Morisita's index, 101

Morlon, H., 188–89, 266

Morrison, L. W., 61

Morton, R. D., 141, 144–45, 325, 327, 330

mosaic habitat, 290–91

Mouillot, D., 182, 189

Moulton, M. P., 167

Mouquet, N., 8, 15, 25, 32, 33, 40–42, 71, 85, 93, 153, 155, 160, 174, 259, 261–62, 270, 273, 292, 322–23, 341–43, 348–52, 389

movement. *See* spatial dynamics
movement ecology, 379
Muller-Landau, H. C., 324
multidimensional approach, 90
multiple stable equilibria (MSEs), 131–32,
     141–44, 162, 251, 325; food-web-induced,
     327–31; in temporal turnover, 150
multiscale perspective, 390
Munday, P. L., 56
Münkemüller, T., 105–6, 153, 160, 171, 178
Murakami, M., 370
Murrell, D. J., 63–64, 148
Muschick, M., 225
mutualisms, 21, 48, 363, 368
mutualistic interactions, 50, 68
mutualists, differential use of, 9, 47–48, 240,
     316; soil, 325
Myers, J. A., 7, 19, 25, 28, 38, 64, 65, 88, 110,
     124–25, 263

Naeem, S., *xiii–xiv*, 155, 336–37, 368, 379
Naeem Triptych, *xiii–xiv*, 336–37
Naeslund, B., 358
Nagelkerke, C. J., 40
Naiman, R. J., 370
Nakano, S., 333, 370
Narcissus effect, 156
Narwani, A., 159–60, 163, 164
Nathan, R., 379
National Ecology Analysis and Synthesis
     convention, 25
natural selection, 2, 28, 57; in niche conver-
     gence, 219–26
Naughton, H. R., 159–60
Navarrete, S. A., 56
Naveh, Z., 370
NCEAS Metacommunity Working Group,
     381–82
near equivalence, 242–45
near-neutrality, 230
nearly neutral coexistence/co-occurrence, 57,
     233
nearly neutral models, 39, 51, 190, 230, 233
Nekola, J. C., 256, 263, 265, 266, 274, 301
nested species-area relationship (SAR), 255,
     265–66; versus island SAR, 279; metacom-
     munity assembly and, 277–78
nested subsets, 99, 105
nestedness, 100, 102, 104, 105, 259
Neufeld, C. J., 61
Neutel, A.-M., 319

neutral assembly, 15
neutral coexistence, 57–58, 208: community
     monopolization and, 229; convergence
     and, 219, 222; diversity and, 242; in
     natural metacommunity, 235; unified
     theories of, 2
neutral evolution–community monopolization
     interaction, 228–36
neutral redundancy, 190
neutral theory (NT), 18–19, 23, 24, 27–28, 47,
     264, 392; assembly processes in, 200, 201;
     versus niche theory, 19, 82–83; in predicting
     species-abundance distributions, 90;
     speciation parameter of, 28; versus species
     sort perspective, 384–85
neutral theory (NT) models, 25; biogeographic
     effects in, 249; critical assumptions of, 51;
     diversity in, 182; eco-evolutionary dynamics
     leading to, 219–26; environmental effects in,
     106–7; key propositions and predictions for,
     35; local versus regional diversity controls
     in, 87–88; versus niche models, 19, 82–83;
     nonecological speciation in, 239; predicting
     multiple macroecology biodiversity patterns,
     275–76; predictions from, 29; processes of,
     26; SAD shape in, 153; spatial patterning in,
     110, 112; species–abundance distribution in,
     93–94; species–area relations in, 33; species
     variations with functional traits in, 154; in
     synthesized occupancy model, 44–46;
     temporal turnover in, 133; trait and
     phylogenetic dispersions in, 174–75
neutrality, 56–57; evolution toward, 207,
     219–26; maintenance of in metacommunity,
     242–44; versus selection, 19
Newbold, T., 161
Newman, E. A., 274, 275
Newmark, W. D., 61
Newton, A. C., 103
niche, 2; diversification of, 216; limitation of ,
     64–65; plasticity of, 122
niche assembly, 15
niche-based archetypes, 25, 28–29, 113
niche-based traits, 154
niche conservatism, 159
niche construction paradigm, 205, 363
niche constructors, 9
niche convergence, evolution toward, 219–26
niche differences, 57; destabilizing, 162;
     fitness, 164; in local coexistence, 52;
     stabilizing, 162–63, 164

niche differentiation, 49; after community monopolization in isolated communities, 241–42; energy providing, 294; evolution of, 220–22

niche evolution dynamics, 388

niche monopolization, 236

niche monopolization hypothesis, 207–8

niche-overlap, 16

niche partitioning, 14–15, 155, 222, 235–36, 240; coexistence and, 58; diversity and, 82–83; evolving, 224; over-dispersion, 157; in tropical forests, 82–83

niche selection, 2–3, 26; dispersal in, 40–42; at SS-PD interface, 39–40; in type of metacommunity model, 36

niche theory: island biogeography theory and, 282; misconception about, 30; versus neutral theory, 19, 82–83, 182

niche-trait approach, 153

niche traits: clumping, 224, 226; evolution of, 384

Nilsson, I. N., 60

Nilsson, S. G., 60

Niphargus amphipods, niche partitioning among, 235–36

Nisbet, R. M., 18, 25, 32–33, 174

nitrogen cycle, 371

nitrogen-phosphorus relationships, 365

nondivergence radiation, 252

nonlinear competitive variances, 43

nonspatial mechanism, 43, 44

nontransitive competition scenario, 144–45

nontransitive interactions, 146

Noonburg, E. G., 305

Norberg, J., 206, 356, 358, 360–62

North American Breeding Bird Survey, 139

Nosil, P., 241

Novotny, V., 316

NT. *See* neutral theory (NT); neutral theory (NT) models

null expectation, 91, 99, 101, 124, 157, 167, 173, 258

null hypothesis, dynamic, 175

null model wars, 16–18, 19

null models, 27–28, 90, 91; co-occurrence and, 94–99; problems with, 156

nutrient inputs, 369, 373–75, 377: in lake ecosystem, 377; in metaecosystems, 374–76, 378

occupancy models, 24, 42–43; using four archetypes, 44–46

ocean systems, dispersal limitation and environmental change in, 361–62

O'Connor, M., 294

Odling-Smee, J., 9, 205

Odum, E. P., 363, 373

Ohashi, H., 310

Oksanen, J., 270

Oksanen, T., 333

Olden, J. D., 184, 190, 292

Olff, H., 311

Olson, V. A., 190

organic material movement, homogenizing effect of, 377

organisms, movement of, 372

Orrock, J. L., 36, 39, 55, 57, 68, 142, 308, 385

Ossola, A., 192

Ostling, A. M., 324

Östman, Ö., 74, 305–6, 350

Ouin, A., 286, 289

Ozinga, W. A., 352

Pacala, S. W., 30, 33, 93, 141, 165–66, 174, 257, 259

Pacific islands, bird species richness on, 72

Paine, R. T., 24, 32, 47, 307, 308

pairwise species associations, 166

Palmgren, A., 17

Pandit, S. N., 133

Panhuis, T. M., 242

Pantel, J. H., 214–16, 217

paradox of the plankton, 30

Paramecium, competitive interactions among, 13, 14

parasite-host interactions, 316

Parent, C. E., 246–47, 248, 249

Park, T., 57

Park Grass experimental plots, 85

Parmentier, I., 168–69, 170

Parravicini, V., 261

Pärtel, M., 124

Passy, S. I., 257, 270n.2, 301

Pastor, J., 299

patch colonization theory, 30–31

patch dynamics (PD) models, 18, 23, 24–25, 30–32, 47, 235–37; biogeographic effects in, 249; critical assumptions of, 51; extinction in, 218; heterogeneity and, 87; interspecific trait variation, stochasticity, and dispersal in, 74–77; key propositions and predictions for, 35; local versus regional diversity controls in, 87–88; processes of, 26; spatial patterning in, 110, 112; in synthesized occupancy model, 44–46; temporal turnover in, 133; traits influence in, 154; trait and phylogenetic dispersions in, 174

patch heterogeneity: dispersal rate and, 386; in type of metacommunity ecology archetype, 37

patches: differences in, 126; disturbances in, 32; stochasticity of, 56

Patel, S. N., 59, 142

pathogens: host relations with, 148, 227–28, 254, 304, 325, 334; resistance to, 22

pattern-based studies, 90

pattern-first approach, 258

Patterson, B. D., 99

Pavoine, S., 153, 160, 179n.1

PD models. *See* patch dynamics (PD) models

Pearce-Duvet, J. C., 75

Peay, K. G., 72, 159

Pelletier, F., 203

Pereira, H. M., 266, 283, 284, 287, 288

Peres-Neto, P. R., 109, 110, 116, 128, 160, 191, 246–47, 392

permanence, 114–15, 120, 325, 328

permanent generalization, 228

permanent specialization, 228

Perry, G. L. W., 165, 166

Petermann, J. S., 324, 332, 354–55, 356

Peterson, A. T., 63

Pfennig, D. W., 204

Pfennig, K. S., 204

phylogenetic dispersions: in non-SS metacommunities, 173–85; scaling dependence of, 167–71; in SS metacommunity, 167–73

phylogenetic divergence time, 160

phylogenetic-diversity-area relationship (PDAR), 173

phylogenetic relationships, 152; in NT model, 154

phylogenies, 152; as community structure driver, 245; correlated with traits, 158–61; functional traits and, 151; in land snail distribution, 247–48; over- and underdispersion of, 156–57, 162–67

phylogeny-based assembly rules, 154–58

phylogeny–biogeography interaction, 239

phytoplankton: environmental variation and traits of, 189–90; performance traits of, 198–99

Pickett, S.T.A., 172, 373

Pigot, A. L., 154, 173, 174–75, 182

Pilière, A.F.H., 161

Pillai, P., 317, 318, 319, 332

Pillar, V. D., 105–6, 246

Pimm, S. L., 167, 266, 283, 287, 288, 292, 363

Pinel-Alloul, B., 330

Pinto, S. M., 83, 88

pioneer species, 330–31

Pitman, N. C., 5

plant biomass, 352, 364, 376

plant material export, 373–74

plants: herbivores and extinction of, 311; immigration of, 373–74; movement of in metaecosystems, 378; nutrient-dependent growth of, 373–74; seed-dispersal limited, 65–67; strongly versus weakly dispersing, 126–27

plant–soil feedbacks, 148, 324–25

plant–soil microbe interactions, 141

Plotkin, J. B., 115, 273, 274

Poff, N. L., 196

Polis, G. A., 319, 333, 370

Polley, H. W., 340

ponds. *See also* freshwater ponds/lakes; Michigan ponds: environmental and spatial effects in, 114–16, 117; environmental variables in, 118, 119; habitat heterogeneity in, 119–20, 121; low-productivity, 187; permanence of, 114–15, 120; predator versus prey SARs in, 314–15, 317; predatory fish reducing diversity in, 308–10; productivity-diversity relationships in, 327–29; species pool in, 126; SS metacommunity archetype in, 113; taxa- and trait-level variation partition in, 180–81; zooplankton of connected by streamflows, 85–86

poorly competitive species, 343

Poorter, L., 161

populations: demographic properties of, 54–55; differentiation of, 204, 214–16, 240–41; interactions of, 372–73; mass effects on growth of, 86

Post, D. M., 332

Poulin, R., 316

Power, M. E., 333

prairie succession patterns, 75–76

Prasad, S., 144

predation: modifying dispersal effects, 306–7; modifying drift effects, 308–9; modifying habitat heterogeneity, 307–8

predator–prey diversity ratios, 364

predator–prey interactions, 8, 21, 32, 226, 316; stabilizing, 305

predator–prey pairs, 326

predator–prey richness ratios, 316

predator–prey theory, 332

predators: abundance of in isolation, 317; altering island species-area relationship of prey, 311–13; dispersal limited, 7; in food webs, 304–5; habitat heterogeneity and, 307–8; metacommunity processes and, 306–7, 310; versus no predators, 186; prevalence of, 272–73; prey extinction and, 311; reduced abundance of, 332–33; removal of top, 307–8; resistance to, 181; seed, 68, 148, 272; space and energy needs of, 332; strong, 73; top, 306–10

Presley, S. J., 99–100, 102, 104, 105–6, 191

Preston, F. W., 136, 256, 267

prey. *See also* predator-prey interactions: extinction of, 311; homogenization of communities of, 307–8; richness of in isolation, 317

Price, P. W., 316

Pringle, R. M., 370

priority effects, 9, 32, 59, 218, 299, 388; community monopolization and, 208; diversity and, 185; evolutionary, 209–10, 216, 244–45; local clumping and, 273; in metacommunities, 141–44, 245; MSEs and, 141–44, 162; species coexistence and, 149; stochasticity and, 59; temporal, 131, 140

producers: nutrient levels depleted by, 377; trait-by-environment correlations in, 195–96; in two-patch metaecosystem, 374, 376

productive space hypothesis, 332

productivity: diversity and, 337–38; local species richness and, 341; metaecosystem dynamics and, 378

productivity–diversity relationship, 295. *See also* species-energy relationship: scale-dependent, 327–31

Proulx, M., 272, 307

Pseudomonas fluorescens assembly studies, 216, 218

Ptacnik, M., 343, 352

publication bias, 105

Pulliam, H. R., 8, 32, 85

Punju, E., 75

Purschke, O., 178

Pyron, R. A., 72

Qian, H., 263

quasi-Clementsian gradients, 101–5

quasi-nested pattern, 103

R* principle, 29–30, 75, 141

Rabosky, D. L., 156

radiation, 251–52; adaptive, 216, 241–42, 251–52; early-divergence, 252; more-complex, 251–52; no, 251–52; nondivergence, 252

Rahbek, C., 5

"random" communities, 174

random effects, 127–28. *See also* residual variation: variation partitioning in diagnosing, 106–12

randomness, 99, 105, 385

range size, 258, 259, 301

rarity, 69, 90, 93, 256, 259, 267

Raven, P. H., 283, 287

realized niche, 63

recolonization, 60–61, 284

"red queen" dynamics, 236

Redfield, A. C., 363, 365

Rees, M., 75

regional effects: dispersal rates and, 350; in metacommunity assembly, 50

regional environment heterogeneity, 50, 87–88, 189

regional extinction: by competitive exclusion, 344; dispersal limitation and, 352; expanded habitat distribution and, 344–45

regional pool size, 124–26

regional scales, 10, 29, 40, 42, 257; coexistence at, 74, 173; convergent evolution at, 225; defines, 257, 262; diversity in, 292, 330–31, 343; food webs at, 318, 327; processes acting on, 50–51

regional species dispersal, 293, 294

Reich, P. B., 190, 352

Reichenbach, T., 146

Renner, S. S., 19, 28, 154

rescue effects, 74, 206

Resetarits, W. J., 351, 387

residual effects, 210, 230, 309

residual variation, 110, 112, 114, 116, 124, 140, 199, 210, 368; MSEs and, 143; spatial variation and, 227, 232, 386; unexplained/ unstructured, 127–29

resource competition: endpoint community in, 352–53; in metacommunity, 342–59; species convergence with, 222–23; for two resources, 344

resource-competition models, 335, 352–53

resource generalists, 222

resource-ratio hypothesis, 299

resource ratios, varying, 344–45

resource specialists, 222

resource-use traits, 151

resource utilization and response, 50, 51; evidence for, 51–54; evolutionary convergence in, 219

resources: coexistence and competition for, 31; differential use of, 52–54; dispersal effects on, 348–50; environmental change and, 357; partitioning, 47, 57; speciation and abundance of, 30

Reynolds, C. S., 190

Reynolds, H. L., 141

Reznick, D. N., 203

Richardson, J. L., 203

Richgels, K.L.D., 103

Ricklefs, R. E., 1–3, 8, 13, 17–19, 28, 60, 64, 87, 88, 154, 227, 240, 245–46, 249, 250, 254, 261, 263, 279, 310–11, 325, 339, 391

Riibak, K., 67–68

Riley, G. A., 17–18, 219

Ripa, J., 363

Ritchie, M. E., 8, 30, 311

river continuum concept, 370

RLQ analysis, 191–97

Robinson, J. V., 59

Robinson, N., 199

rock-scissors-paper scenarios, 140, 144–45, 146

Rodriguez, A., 75

Rogers, A. D., 72–73

Rojas-Echenique, J., 144

Rooney, N., 184, 292, 319

Rosalia, S., *xiii–xiv*

Rosenzweig, M. L., 261, 276, 290–91, 295

Rosindell, J., 27, 28, 29, 52, 55, 92, 94, 182, 240

Roslin, T., 315

Rossberg, A. G., 365

Rothamsted experimental site, 136

Roughgarden, J., 33, 258

Ruger, N., 54

Rundle, H. D., 241

Ruokolainen, K., 107

Ruppert, J. L. W., 310

Ryabov, A. B., 348, 350

Ryall, K. L., 316

Ryberg, W. A., 114, 115, 186, 282, 308, 309, 311–13, 317

Rybicki, J., 266, 284

Rychtecká, T., 347, 348, 354

Sabo, J. L., 333

SAD. *See* species-abundance distribution (SAD)

Saito, V. S., 183

salinity perturbations, 357–58

Salt, D., 389–90

sample-based rarefaction curve, 274

sampling curves, 266n1

sampling effects: in habitat loss and fragmentation, 285–86; in island SAR, 279–81; versus metacommunity-level effects, 286; shaping nested SAR, 277

Sanderson, J. G., 96, 97–98

SAR. *See* species-area relationship (SAR)

Sarnelle, O., 249

Sarracenia purpurea studies, 222–24

satellite species, 91

Savage, V. M., 262

savannah, fourth-corner analysis of, 193–94

Sax, D. F., 63, 292, 293, 356, 390

scale dependence, 151; of productivity-diversity relationship, 327–31

scale-dependent biodiversity patterns, 292–300

scales, 380, 383. *See also* local scales; regional scales; spatial scales: dispersal rates and, 7–8; influences of, 3–5, 24–25; multiple, 5–9; spatial heterogeneity and, 8–9

scaling approach, 255

Schade, J. D., 377

Schaffer, W. M., 363, 376

Scheffer, M., 219, 224, 226, 363, 366, 367

Scheiner, S. M., 21, 265–66, 277, 279

Schemske, D. W., 239–42, 250

Schiesari, 390

Schimel, D. S., 389

Schimper, A.F.W., 189

Schluter, D., 18, 155, 204, 240, 241, 261

Schnitzer, S. A., 271

Schoener, T. W., 5, 15, 47, 52, 61, 155, 203, 306, 307, 311, 332

Schramski, J. R., 262

Schreiber, S. J., 144, 146

Schroder, A. L., 141

Schuler, M., 271, 282

Schulze, E. D., 336

Schweitzer, J. A., 205

secondarily adapted species, 358

Sedio, B. E., 324

seed-dispersal limitation, 65; animal dispersers and, 68–70; animal predators in, 68; positive effects of, 65–67; species variations in, 67–68

seed dispersers, 68–70

seed predators, 68

Seidler, T. G., 115, 273

selection, 383. *See also* natural selection: complementarity and, 347–48; destabilizing frequency-dependent, 202; factors important to, 50; interaction-dependent local, 360; versus neutrality, 19

Sepkoski, J. J., Jr., 240

Shade, A., 135, 136

Shen, G. C., 278

Shevtsov, J., 104

Shimada, M., 148

Shinen, J. L., 56

Shipley, B., 94, 182, 392

Shmida, A., 8, 32, 85

Shoemaker, L. G., 43–44, 77

Shorrocks, B., 75

Shulman, R. S., 115, 316, 317

Shurin, J. B., 7, 9, 18, 40, 64–65, 67, 115, 135, 142–43, 153, 156, 172, 305, 306, 331

Sibanda, N., 72–73

Siefert, A., 178–79, 183

Siepielski, A. M., 39, 52, 56, 57, 58

Silberbush, A., 387

Silvertown, J., 79, 153, 172

Simberloff, D., 4, 5, 13, 16, 60, 91–92, 95–96, 155, 159, 204, 283

Simova, I., 263, 271

Simpson, G. G., 240

Siqueira, T., 390

Sirot, C., 161

size–abundance scaling, 365

Šizling, A. L., 274

Sjogren, P., 60

Skellam, J., 30–31

Skelly, D. K., 113, 205

sky islands, alpine plant communities of, 182

Slatkin, M., 24, 31

SLOSS habitats, 283–86

Smith, A. B., 188–89

Smith, D. C., 57

Smith, F. A., 274

Smith, G. K., 57, 173, 233, 234, 235

Smith, K. G., 161, 188

Smith, T. W., 109, 127

snails, ecological processes in distribution of, 247–48

Snyder, R. E., 33–34, 43

Sobral, 171

Sobral, F. L., 171

soil feedbacks, 141

Soininen, J., 9, 92, 109, 110, 116, 118, 122, 125, 127, 132, 133, 140, 178, 218–19, 265, 267, 273, 343

Sokol, E. R., 178, 182

Soliveres, S., 146

Solomon Islands, species matrices on, 98

Soranno, P. A., 377

Soule, M. E., 332

source-sink relations, 8, 23, 24–25, 35, 43, 49, 89, 133, 212, 273, 377. *See also* mass effects (ME)

space. *See also under spatial:* in functional and phylogenetic patterns, 182–84; in metacommunity patterns, 90–130; structure of, 386–87; temporal patterns and, 131–50; variation in, 31, 43, 114, 116, 147, 153, 210, 227, 232, 310, 327, 369, 379

Spasojevic, M. J., 153, 172, 178, 182

spatial clines, 244

spatial clustering, 69–70, 71

spatial distance, 50, 106–8, 123–24, 182

spatial distribution, 26; in competition and coexistence, 13–15

spatial dynamics: importance of in ecosystems, 371–73; in metaecosystems, 378, 379

spatial effects: dispersal rates and, 122–23; functional information in analysis of, 184–89; in metacommunity structuring, 113–17; in species distribution, 248–49; variation partitioning in diagnosing, 106–12

spatial–environmental relationship, 50, 106–12, 123–24,182

spatial heterogeneity: in continuous metacommunities, 147–48; species scale and properties in, 8–9

spatial insurance effect, 346–47

spatial movement, 370, 372, 374; small-scale, 370

spatial processes, 381–82; in competition and coexistence, 13–15; diversity–stability theory and, 319–23; hysteresis patterns and,

367; mediating BEF relationships, 339–42; in metaecosystems, 369; reemergence of, 18–19; trophic interactions and, 304–6, 324–27

spatial scales: development of, 256; in EMS analysis, 105; for energy and diversity measures, 296; importance of, 392; macroecology focus on, 255; metacommunity processes and, 245–47

spatial storage effects, 43, 44, 350

spatial turnover: deterministic features in, 131–32; temporal turnover and, 140

spatially clumped species, 267

spatially continuous metacommunities, 132, 147–49

spatially mediated frequency dependence, 270

spatiotemporal fluctuations, 15

spatiotemporal heterogeneities, 1, 206–7

spatiotemporal perspective, 15, 18–19

specialist enemies, 9, 148, 324

specialization, 44; continuous escalation of, 228–29; habitat, 49, 69, 82, 88, 210; niche, 222; permanent, 228; of predators, 308–9, 324

speciation, 2–3, 26, 383; allopatric, 98, 171, 174–75, 241; in biodiversity, 50; counteracting stochastic extinctions, 28; ecological, 239–42; ecological processes in, 240; nonecological, 242–45

species: arrivals of, 26, 36, 208, 210, 218; coexistent combinations of, 389; contributing to metacommunity pattern, 126–27; differences of, 126–27, 386–87; frequencies of, 227–28; geographic variation of, 26; rarity of, 69, 90, 93, 256, 259, 267; spatially clumped, 26, 100, 149; spatiotemporal variations of, 140; total number in metacommunity, 267, 268–70; tracking, 212

species abundance. *See* abundance

species-abundance distribution (SAD), 92, 255, 256, 268; biodiversity components shaping, 273–74; diversity metrics and, 92–94; energy shaping, 297; in grassland communities, 153; "hollow-shaped," 264–65; number of individuals shaping, 271; predicting from neutral model, 275–76; predicting rates of, 90; relative abundance of individuals shaping, 272–73; studies of, 92

species–area relationship (SAR), 92, 135, 149–50, 173, 255, 265–66, 268, 287. *See also* island species-area relationship (ISAR); nested species-area relationship (SAR):

biodiversity components affecting, 267, 269, 273–74; diversity partitioning and, 276–77; energy shaping, 297; in estimating biodiversity loss, 288; island, 265–66, 279–82; nested, 265–66, 277–78; for predator and prey, 316; predicting from neutral model, 275–76; relative abundance of individuals shaping, 272; species clumping shaping, 273

species clumping in metacommunity, 267, 273

species dispersal. *See* dispersal

species distributions: along spatial and environmental gradients, 92; biogeographic in, 247–49; checkerboard-like patterns of, 95–99; predicting from traits, 197–200

species–energy relationship, 294–300

species equivalence, 230–32, 242–45

species interactions, 3, 5–6; biogeographic patterns in, 16; in dispersal limitation, 68–70; in heterogeneity, 9; mechanisms of, 384; niche-based perspectives of, 28–29; scenarios for, 24; in taxonomic β-diversity, 184, 185–86; traits related to, 151

species pool: functional redundancy in, 187; historic effects on, 245–50; local abundance and size of, 339–40; Narcissus effect in, 156

species pool effects, 124–26

species-pool size, 348; regional, 124–26; species richness in plant biomass and, 352

species richness: with area, 256; changes in, 184; components of, 267–73; energy richness and, 297–99; local, 165, 337–40; pairwise associations and, 165; productivity and, 341

species size, 112–13. *See also* body size: in response to environmental and spatial factors, 121–22

species sorting (SS) models, 16, 23, 24, 28–30, 47; adaptive evolution in, 239; assembly processes in, 200–201; biogeographic effects in, 249; critical assumptions of, 51; dispersal limitation and, 352–56, 358; dispersal limitation in, 63; due to habitat heterogeneity, 18; ecosystem function in, 343–48; evaluating importance of, 90; habitat heterogeneity in, 77–84, 277–78; high dispersal in, 213; key propositions and predictions for, 35; local diversity controls in, 87–88; versus neutral theory, 384–85; phylogenetic and functional-trait dispersions in, 167–73; processes of, 26; SAD shape in, 153; spatial and environmental effects in,

species sorting (SS) models (cont.)
110, 112; spatial effects in, 106; species-abundance distribution in, 92–94; species composition and environmental variation in, 218–19; strong trait-by-environment relationships and, 197; in synthesized occupancy model, 44–46; temporal turnover in, 133–34; trait and relatedness similarities in, 151

species sorting (SS)–mass effect (ME) interface, 40–42

species sorting (SS)–neutral theory (NT) interface, 38–39

species sorting (SS)–patch dynamic (PD) interface, 39–40

species–time–area relationship (STAR), 135, 150

species–time relationship (STR), 135, 136, 137, 149–50

species traits. *See* traits

Spencer, M., 314

Spiller, D. A., 61, 307

spillover: predation, 333; species, 25, 84–85, 154, 174

Srivastava, D. S., 271, 338

SS models. *See* species sorting (SS) models

stability: coexistence and, 38–39; diversity and, 319–23

stabilizing coexistence strength, 43, 44

stabilizing differences: equalizing differences and, 166–67; niche, 164, 165; trait, 163–65

stabilizing–equalizing balance, 38–39, 166–67; in local coexistence, 52; niche differentiation and, 49

stabilizing processes: in coexistence in food webs, 305–6; in local coexistence, 52–54

Stanton, M. L., 75

steady-state behavior, 325

Steffen, W., 283

Stegen, J. C., 109, 127, 132, 133, 262

Steidinger, K. A., 371

Stein, A. K., 8–9, 77, 78, 261, 269, 272, 278

Stein, C. H., 339–40

Steinbauer, M. J., 119

Steiner, C. F., 144–45, 146, 299, 327, 329, 330

Stemberger, R. S., 249

Sterck, F., 161

Sterner, R. W., 2, 365, 379

Stier, A. C., 308, 316

stochastic dilution effect, 165–67, 174, 220

stochastic ecological drift, 2–3, 26, 383; in community patterns, 28; in smaller communities, 36

stochastic extinction, 49, 56; in demographic stochasticity and ecological drift, 59–62; in patch colonization theory, 30–31; speciation and, 28

stochastic processes, 26, 55, 381–82

stochasticity. *See also* demographic stochasticity; environmental stochasticity: in coexistence, 43; in colonization and extinction, 16; versus determinism, 15–18, 17–18; existence of, 55; interactions of with dispersal and interspecific effects, 74–77; leading to ecological drift, 56–59

stoichiometric balance/imbalance, 299, 335

stoichiometry, 2, 203, 219, 363, 379

Stoks, R., 56

Stomp, M., 53

Stone, L., 156

Stoner, K. E., 69

storage effects, 43, 44, 350

Storch, D., 296, 299, 327

Strauss, S. Y., 144, 159, 227

stream insect diversity, 183

Strong, D. R., 15, 16, 155, 391

Stuart, Y. E., 204

Stump, S. M., 324

subdominant species, 42, 340, 343, 353–54

subtropical forest, fourth-corner analysis of, 192–94

Suding, K. N., 172

Sugihara, G., 93

Supp, S. R., 184, 258, 272, 311–12, 358–59

Suzuki, M., 307

Svenning, J. C., 67–68, 115–16

Svensson, J. R., 261, 271–72

Svirezhev, I.U.M., 320

Sweeney, B. W., 189

Swenson, N. G., 167–73, 178–79, 183, 261

symmetric fitness, 224

Symons, C., 358

synthetic hub, 2

Takimoto, G., 332

Tamme, R., 282

Tang, S., 319

Tansley, A., 13, 28, 101–5, 189

Taub, F. B., 371

taxomomic scale, 173

taxon pulses, 228

taxonomic approaches, 152, 153; functional-trait approach and, 177

taxonomic β-diversity: different scenarios of, 186–88; environmental factors in, 182; mechanisms underlying, 184–96

taxonomic diversity, temporospatial, 184–89

taxonomic patterns: functional information in analysis of, 179–84; functional-trait patterns and, 177–201; in metacommunity assembly, 189

taxonomic scale, 105

taxonomic turnover, 179–80

Taylor, W. D., 315

temporal fluctuations: extrinsic, 47; in four archetypes, 132–34; functional information in analysis of, 184–89; intrinsic, 47

temporal processes, 140

temporal turnover, 131, 233; environmental variation and, 150; factors in, 131; interpretation of, 131; multiple mechanisms in, 150; patterns of, 134–40; predictions of, 133–34; spatial turnover and, 140; variations in, 131

temporospatial interactions, 131–50

ter Steege, H., 263

Terborgh, J. W., 64, 307, 332

terHorst, C. P., 219, 222–23, 244

Tessier, A. J., 79, 113, 163, 181, 203, 214, 315, 351

testable predictions, 40, 382

Tews, J., 269

Thebault, E., 376

Thibault, K. M., 354, 355

Thompson, J. N., 66, 203, 204, 227, 240

Thum, R. A., 244, 249

Tielborger, K., 86

Tilman, D., 7, 9, 18, 24, 29–30, 32, 47, 52, 64, 66–67, 75, 76, 99, 113, 141, 153, 174, 223, 272, 299, 336, 337, 342–46, 351, 377

Tilman, D., 53

time. See temporal fluctuations; temporal turnover

tipping points, 335, 339, 363–67, 368

Tittensor, D. P., 296

Tokeshi, M., 93

Tokolyi, J., 190

trade-offs, 49. See also colonization–competition trade-offs: ecological trait, 49; food discovery-interspecific dominance, 75; habitat-fitness, 78–81; in local coexistence, 52–54

trait-based assembly rules, 154–58

trait-based theory, 151–53: correlated with phylogeny, 158–61; of metacommunity assembly, 189–97

trait-by-environment correlations, 239

trait-by-environment matching, 383

trait-by-environment relationships: in aquatic ecosystems, 195; with disturbed environmental conditions, 196; with large vs. small environmental gradients, 197; in predicting species abundance and distribution, 197–98; in producers versus animals, 195–96; sampling effects in, 194–95; statistical approaches to, 191–94

trait dispersions: in non-SS metacommunities, 173–85; over- versus under-, 173; in species sort metacommunity, 167–73

trait matching, 230

traits: clumpy distributions of, 224, 226; competition and overdispersion of, 167.169–72; with competitive exclusion, 156–57; continuous divergence of, 229; differences of, 152, 161–65; environmental filtering and underdispersion of, 169, 172; equalizing, 38–39, 56, 158, 162–63, 165–67, 219, 304; evolution of, 203–4, 209–10, 216, 223, 235, 384; niche-based, 154, 384; optimization of, 362–63; over- or underdispersion of, 162–67; phylogenetically conserved, 159; predicting species abundance and distribution from, 197–200; in species coexistence and competition, 16, 161–65

tree clusters, dispersal limitation and, 69–70, 71

Triantis, 259, 279

Tribolium beetles, 57

Trisos, C. H., 170, 172

Troia, M. J., 168, 171

Troll, C., 370

trophic cascade, habitat-loss-induced, 310, 332–33

trophic ecosystem structure, 335, 364

trophic interactions, 303, 334; in metacommunity processes, 306–10; in MSEs and EACs, 328; spatial processes and, 304–6, 324–27

trophic pyramids, 2, 363, 391

trophically structured metacommunities theory, 310–18

tropical forests: niche partitioning in, 80–81, 82–83; species pool in, 124–26; taxonomic turnover in, 179–80 tropical rainforests: environmental and spatial effects in, 116; tree metacommunities of, 113–14

Tscharntke, T., 18, 314, 316, 332

Tucker, C. M., 153, 179n1

Tuomisto, H., 107–8, 109, 119, 127, 140, 187, 277

Turgeon, J., 245

Turnbull, L. A., 64, 65, 75

Turner, M. G., 370

turnover, 60, 99–104, 315; dispersal and, 42; local species, 144–45, 242, 262; metacommunity, 150; patch, 202, 232–33, 237; of population, 351; spatial, 140, 143–45, 331; taxonomic, 179; temporal, 131–40, 142–45, 149–50, 233, 299, 329, 330–31

two-patch scenarios, 32–34, 376

Tylianakis, J. M., 332

ubermodel, 23, 42–46

Ulrich, W., 95, 105, 174

unified ecological theories, 2, 263–64, 391–92

unpredictability, 16, 166, 190, 391

Urban, M. C., 203, 205, 206, 208–10, 213, 216, 219, 229, 230, 236, 241, 370

Uriate, M., 54

Usher, M. B., 283

Valencia, R., 83

Valladares, G., 332

van de Leemput, I. A., 367

van der Gast, C. J., 73–74, 133

van der Plas, F., 178

van der Putten, W. H., 324–25

Van der Veken, S., 67–68

van Kleunen, M., 292

van Nes, E. H., 219, 224, 367

van Noordwijk, C.G.E., 315

van Nouhuys, S., 75

Van Valen, L. M., 236

Vandermeer, J., 32, 38, 144

Vannote, R. L., 189, 370

Vanoverbeke, J., 229, 230, 232, 236, 241–42

variation partitioning, 90–91, 232–33, 239, 381–82, 385–86; caveats in, 129–30; comparing taxonomic and functional diversity, 180–81; in diagnosing spatial, environmental, and random effects, 106–12; limitation of, 388; from ponds with and without top predatory fish, 308–10; with and without evolution, 232

variation patterns: in metacommunities, 112–26; multidimensional, 388

Vasseur, D. A., 219, 226–27

Veech, J. A., 267, 276, 277

Vellend, M., 2–3, 19, 21, 25–26, 50, 55, 57, 59, 128, 184, 203, 205–6, 213, 216, 258, 292–94, 338, 370, 383–84

Venail, P. A., 159–60, 350

Verreydt, D., 358

Vincent, T. L., 363

Violle, C., 159

Volkov, I., 28, 93, 94, 258, 259

Von Humboldt, A., 189

von Wehrden, H., 291

Walker, B. H., 389–90

Wallace, A. R., xiii–xiv, 155, 189, 245, 256

Wallace Line, 245

Walsh, J. J., 282, 371

Wang, X., 166–67, 174, 183

Wang, Z., 83, 296, 298

Wardle, D. A., 337, 338, 339

Warner, R. R., 47, 55

Warren, B. H., 310–11

Warren, P. H., 18, 364

Warton, D. I., 389

Wasserberg, G., 79

Watling, J. I., 36, 39, 55, 57, 142, 308, 385

Watts, C., 226

Webb, C. T., 153–55, 157–59, 175, 189, 261–62

Weiher, E., 153, 155–57, 167–69, 172, 181, 189, 299

Weinstein, B. G., 182

Weiser, M. D., 261

Weissing, F. J., 47, 147

Wellborn, G. A., 57, 79, 113, 115, 181, 187, 196, 233, 306

Werner, E. E., 18, 102, 113, 115, 153, 181

West, G. B., 262, 392

Westoby, M., 65, 189, 190

wetlands, strongly- versus weakly-dispersing plants in, 126–27

White, E. P., 5, 132, 133, 135, 136, 137, 265, 266, 301

Whitham, T. G., 204, 205

Whittaker, R. H., 77, 80, 91, 101, 263–64, 295

Whittaker, R. J., 261

Wiegand, T., 166–67, 174, 388

Wiens, J. A., 158, 261, 385
Wiersma, P., 190
Wilbur, H. M., 18, 113
Wilcove, D. S., 283
Williams, C. B., 155
Williams, N. M., 161
Willig, M. R., 21, 103, 105, 261, 295
Wilsey, B. J., 340
Wilson, D. S., 26
Wilson, E. O., 7, 10, 16, 27, 56, 60–62, 70, 72, 132, 134, 137, 155, 158–59, 184, 204, 219, 256, 259, 261, 279, 287, 306, 310–12
Wilson, H. B., 9
Wilson, J. B., 156, 205
Wilson, M. V., 8, 32
Wilson, W., 16, 18–19, 93, 147, 148, 172, 259, 263, 265
Wootton, J. T., 30, 53, 376
Wright, D. H., 261, 271, 297
Wright, I. J., 189, 190
Wright, J. P., 9
Wright, S. J., 115–16

Xiao, X., 275

Yaacobi, G., 286
Yang, J., 171, 183
Yee, D. A., 271
Yeh, Y.-C., 104
Yitbarek, S., 144
Yoshida, T., 203
Young, H. S., 332
Young, T. P., 331
Yu, D. W., 9, 55–56

Zeigler, B. P., 32
Zeiter, M., 339–40
zero-net-growth isoclines (ZNGIs), 344, 345, 349, 353
zero-sum constraints, 270
Zhang, G., 39
Zhou, S. R., 39
Ziv, Y., 79, 286
Zobel, M., 124, 339
Zokan, M., 309

MONOGRAPHS IN POPULATION BIOLOGY

1. *The Theory of Island Biogeography*, by Robert H. MacArthur and Edward O. Wilson
2. *Evolution in Changing Environments: Some Theoretical Explorations*, by Richard Levins
3. *Adaptive Geometry of Trees*, by Henry S. Horn
4. *Theoretical Aspects of Population Genetics*, by Motoo Kimura and Tomoko Ohta
5. *Populations in a Seasonal Environment*, by Steven D. Fretwell
6. *Stability and Complexity in Model Ecosystems*, by Robert M. May
7. *Competition and the Structure of Bird Communities*, by Martin L. Cody
8. *Sex and Evolution*, by George C. Williams
9. *Group Selection in Predator-Prey Communities*, by Michael E. Gilpin
10. *Geographic Variation, Speciation, and Clines*, by John A. Endler
11. *Food Webs and Niche Space*, by Joel E. Cohen
12. *Caste and Ecology in the Social Insects*, by George F. Oster and Edward O. Wilson
13. *The Dynamics of Arthropod Predator-Prey Systems*, by Michael P. Hassel
14. *Some Adaptations of Marsh-Nesting Blackbirds*, by Gordon H. Orians
15. *Evolutionary Biology of Parasites*, by Peter W. Price
16. *Cultural Transmission and Evolution: A Quantitative Approach*, by L. L. Cavalli-Sforza and M. W. Feldman
17. *Resource Competition and Community Structure*, by David Tilman
18. *The Theory of Sex Allocation*, by Eric L. Charnov
19. *Mate Choice in Plants: Tactics, Mechanisms, and Consequences*, by Nancy Burley and Mary F. Wilson
20. *The Florida Scrub Jay: Demography of a Cooperative-Breeding Bird*, by Glen E. Woolfenden and John W. Fitzpatrick
21. *Natural Selection in the Wild*, by John A. Endler
22. *Theoretical Studies on Sex Ratio Evolution*, by Samuel Karlin and Sabin Lessard
23. *A Hierarchical Concept of Ecosystems*, by R. V. O'Neill, D. L. DeAngelis, J. B. Waide, and T.F.H. Allen
24. *Population Ecology of the Cooperatively Breeding Acorn Woodpecker*, by Walter D. Koenig and Ronald L. Mumme
25. *Population Ecology of Individuals*, by Adam Lomnicki
26. *Plant Strategies and the Dynamics and Structure of Plant Communities*, by David Tilman

27. *Population Harvesting: Demographic Models of Fish, Forest, and Animal Resources*, by Wayne M. Getz and Robert G. Haight

28. *The Ecological Detective: Confronting Models with Data*, by Ray Hilborn and Marc Mangel

29. *Evolutionary Ecology across Three Trophic Levels: Goldenrods, Gallmakers, and Natural Enemies*, by Warren G. Abrahamson and Arthur E. Weis

30. *Spatial Ecology: The Role of Space in Population Dynamics and Interspecific Interactions*, edited by David Tilman and Peter Kareiva

31. *Stability in Model Populations*, by Laurence D. Mueller and Amitabh Joshi

32. *The Unified Neutral Theory of Biodiversity and Biogeography*, by Stephen P. Hubbell

33. *The Functional Consequences of Biodiversity: Empirical Progress and Theoretical Extensions*, edited by Ann P. Kinzig, Stephen J. Pacala, and David Tilman

34. *Communities and Ecosystems: Linking the Aboveground and Belowground Components*, by David Wardle

35. *Complex Population Dynamics: A Theoretical/Empirical Synthesis*, by Peter Turchin

36. *Consumer-Resource Dynamics*, by William W. Murdoch, Cheryl J. Briggs, and Roger M. Nisbet

37. *Niche Construction: The Neglected Process in Evolution*, by F. John Odling-Smee, Kevin N. Laland, and Marcus W. Feldman

38. *Geographical Genetics*, by Bryan K. Epperson

39. *Consanguinity, Inbreeding, and Genetic Drift in Italy*, by Luigi Luca Cavalli-Sforza, Antonio Moroni, and Gianna Zei

40. *Genetic Structure and Selection in Subdivided Populations*, by François Rousset

41. *Fitness Landscapes and the Origin of Species*, by Sergey Gavrilets

42. *Self-Organization in Complex Ecosystems*, by Ricard V. Solé and Jordi Bascompte

43. *Mechanistic Home Range Analysis* by Paul R. Moorcroft and Mark A. Lewis

44. *Sex Allocation*, by Stuart West

45. *Scale, Heterogeneity, and the Structure of Diversity of Ecological Communities*, by Mark E. Ritchie

46. *From Populations to Ecosystems: Theoretical Foundations for a New Ecological Synthesis*, by Michel Loreau

47. *Resolving Ecosystem Complexity*, by Oswald J. Schmitz

48. *Adaptive Diversification*, by Michael Doebeli

49. *Ecological Niches and Geographic Distributions*, by A. Townsend Peterson, Jorge Soberón, Richard G. Pearson, Robert P. Anderson, Enrique Martínez-Meyer, Miguel Nakamura, and Miguel Bastos Araíjo.

50. *Food Webs*, by Kevin S. McCann

51. *Population and Community Ecology of Ontogenetic Development*, by André M. de Roos and Lennart Persson

52. *Ecology of Climate Change: The Importance of Biotic Interactions*, by Eric Post

53. *Mutualistic Networks*, by Jordi Bascompte and Pedro Jordano

54. *The Population Biology of Tuberculosis*, by Christopher Dye

55. *Quantitative Viral Ecology: Dynamics of Viruses and Their Microbial Hosts*, by Joshua Weitz

56. *The Phytochemical Landscape: Linking Trophic Interactions and Nutrient Dynamics*, by Mark D. Hunter

57. *The Theory of Ecological Communities*, by Mark Vellend

58. *Evolutionary Community Ecology: The Dynamics of Natural Selection and Community Structure*, by Mark A. McPeek

59. *Metacommunity Ecology*, by Mathew A. Leibold and Jonathan M. Chase